S0-BVP-598

The Swift Fox

The Swift Fox

Ecology and Conservation of Swift Foxes in a Changing World

Edited by: Marsha A. Sovada and Ludwig Carbyn

Published by:
Canadian Plains Research Center
2003

Canadian Plains Research Center
University of Regina
Regina, Saskatchewan S4S 0A2
Canada
Tel: (306) 585-4758
Fax: (306) 585-4699
e-mail: canadian.plains@uregina.ca
http://www.cprc.uregina.ca

National Library of Canada Cataloguing in Publication Data

The swift fox: ecology and conservation of swift foxes in a changing world /Marsha A. Sovada, Ludwig Carbyn, editors

(Canadian plains proceedings 0317-6401 34)
Proceedings of a conference held in Regina, Feb. 18-19, 1998
Includes bibliographical references and index.
ISBN 0-88977-154-5

1. Kit fox—Congresses. I. Carbyn, Ludwig N. II. Sovada, Marsha A., 1954- III. University of Regina. Canadian Plains Research Center. IV. Series.
QL737.C22S94 2003 599.775 C2003-910455-9.

Cover photo by Robert L. Harrison
Cover design by Donna Achtzehner, Canadian Plains Research Center
The image of a swift fox used on the page starting each new section of this book was drawn by Paul Geraghty
Index by Patricia Furdek, Indexing and Abstracting Services, Coquitlam, British Columbia
Printed and bound in Canada by Houghton Boston, Saskatoon
Printed on acid-free, recycled paper.

Contents

Preface

The Canadian Wildlife Service, Edmonton, Alberta, and the U.S. Geological Survey's Northern Prairie Wildlife Research Center, Jamestown, North Dakota, convened an International Symposium on Swift Foxes in Saskatoon, Saskatchewan, in 1998. The symposium was aimed at fostering information exchange and identifying the "state-of-the-science" of swift fox ecology and status in North America. Biologists and endangered species experts from fifteen states, three provinces, and seven countries participated in the symposium to discuss current distribution, population dynamics, characteristics of dispersal, habitat selection, disease, taxonomy, legal status, and conservation of this species, as well as information about the closely related kit fox. A specific goal of the meeting was to formulate a strategy that best addressed ways in which management could be applied to increase the distribution of the species and the viability of existing populations.

At the time of the meeting, swift foxes were a candidate species for listing under the Endangered Species Act in the United States but were precluded by higher priorities. In Canada, the successful reintroduction program, which began in 1983, was being transformed into a monitoring program. Research and monitoring was underway in both countries, and much has been learned about the swift fox since 1998. In 2002, the U.S. Fish and Wildlife Service removed the swift fox from candidate status because of the indicated viability and vitality of populations, which sufficiently reduced the immediacy of threats to the species. In 1998, in Canada, the status of swift foxes was changed from "extirpated" to "endangered," reflecting the continually improving and expanding population that followed the initial reintroduction of foxes into an area where they had been extirpated. Continued vigilance and commitment of scientists and resource managers with special interest in the conservation of swift foxes will be needed to monitor the status of the species in both the U.S. and Canada to ensure the viability of swift fox populations into the future.

This volume brings together the papers presented at the symposium and others written after the symposium. Each chapter was reviewed by the symposium editors along with two peer reviewers and revised in consideration of advice received. As editors, we hope we have developed a volume that is useful and informative to scientists, resource managers, and students.

The topics in this book range from a general consideration of conservation issues (Setting the Stage—Part I) to what is known about the current distribution of the species on the North American continent (Part II). Part III is devoted to evaluations of scent stations for surveying fox presence and abundance. That section in particular emphasizes the need for more research to provide better tools to efficiently and effectively determine the numerical abundance and distribution of swift foxes in native prairies. Also, included in Part III is a review on mitigating capture-related injuries due to trapping and handling of foxes in field studies.

A major portion of the book provides information on the population ecology of swift foxes. It is a subject that attempts to better understand the dynamics of changes in distribution and abundance of swift foxes on the continent, particularly as related to loss of suitable habitats, ecosystem modification brought on by human activities, and climate change. The plight of many endemic prairie species, like the swift fox, suggests that there is an urgent need for all governments, and society as a whole, to redress the loss of native prairies on the continent. Actively conserving swift foxes and other native endemic species will contain, and in some cases reverse, the trends of prairie degradation that marked much of human activities in the twentieth century. As such these native species, particularly those with a high profile, can be used as "flagship species" to promote conservation. The final section of the book is devoted to our current knowledge of swift fox taxonomy, physiology, and disease-related issues. All are subjects that need to be more fully explored in future research.

Publication of this volume would not have been possible without the generous contributions of many individuals and financial assistance from several organizations. We extend our thanks to the many peer reviewers who greatly improved the content of these papers:

Steven H. Allen, North Dakota Game and Fish Department

David E. Andersen, USGS, University of Minnesota Cooperative Research Unit

Cheryl S. Asa, St. Louis Zoo

Jane E. Austin, USGS, Northern Prairie Wildlife Research Center

William E. Berg, Minnesota Department of Natural Resources

Bonnie S. Bowen, Iowa State University

Steven Brechtel, Alberta Department of Environmental Protection

Molly J. Burns, USGS, Northern Prairie Wildlife Research Center

Brian L. Cypher, San Joaquin Valley Endangered Species Recovery Program

Lloyd B. Fox, Kansas Department of Wildlife and Parks

Eric M. Gese, USDA, National Wildlife Research Center

Brian Giddings, Montana Department of Fish, Wildlife and Parks

Robert L. Harrison, University of New Mexico

Stephen Herrero, University of Calgary

Laurence D. Igl, USGS, Northern Prairie Wildlife Research Center

Patrick A. Kelly, San Joaquin Valley Endangered Species Recovery Program

Craig J. Knowles, Fauna West Wildlife Consultants (Montana)

Terry J. Kreeger, University of Wyoming

Kyran Kunkle, Turner Endangered Species Foundation (Montana)

Frederick G. Lindzey, USGS, University of Wyoming Cooperative Research Unit

Mark Lomolino, Oklahoma Biological Survey, University of Oklahoma

David W. Macdonald, Oxford University

Gary Matson, Matson's Laboratory (Montana)

L. David Mech, USGS, Northern Prairie Wildlife Research Center

Axel Moehrenschlager, Calgary Zoological Park

Michael K. Phillips, Turner Endangered Species Foundation (Montana)

Michael L. Phillips, USGS, Northern Prairie Wildlife Research Center

Katherine Ralls, Smithsonian Institution

Christiane C. Slivinski, Kansas Department of Wildlife and Parks

Glen A. Sargeant, USGS, Northern Prairie Wildlife Research Center

Eileen Dowd Stukel, South Dakota Department of Game, Fish and Parks

Rebecca L. Telesco, University of Tennessee

Patrick J. White, NPS, Yellowstone National Park

Amy L. Zimmerman, USGS, Northern Prairie Wildlife Research Center

We appreciate the assistance of Peter Jonkers and Robert O. Woodward for their efforts in organizing the symposium, and Brian K. Scheick, Amy L. Zimmerman, and Brian Mlazgar for assisting in the production of this book. We thank Ronald E. Kirby, Janet Keough, Glenn R. Guntenspergen, and Gerald McKeating for encouragement and support.

The U.S. Geological Survey (Northern Prairie Wildlife Research Center), the Canadian Wildlife Service (Environment Canada), The Wildlife Society, the Fort Worth Zoo, and the Swift Fox Conservation Society provided funds for the symposium and the production of this book.

Marsha A. Sovada and Ludwig Carbyn
June 2003

Part I – Setting the Stage

A Review of the Ecology, Distribution, and Status of Swift Foxes in the United States

■ David Allardyce and Marsha A. Sovada

Abstract: Swift foxes are native to the shortgrass and mixed-grass prairies of the Great Plains of North America. They are particularly suited to the dynamic conditions of the prairies because they: 1) are opportunistic and generalist foragers, 2) use dens for shelter from extreme climatic conditions (heat and cold) and for protection from predators, and 3) apparently require little or no open water for survival. There is controversy surrounding the taxonomy of swift foxes, i.e., whether there are 2 subspecies (northern and southern range) based on morphological and perhaps behavioral differences. Knowledge of swift fox biology is important for developing strategies to conserve the swift fox and their critical habitats. We provide a review of swift fox ecology, historic distribution, and populations status in the United States and identify potential threats to populations. We include a brief summary of actions taken to conserve the species in the Great Plains Region of the United States.

Taxonomic Status

The swift fox was first described by Thomas Say in James (1823). Say named the fox *Canis velox*, but it was reassigned to the genus *Vulpes* by Audubon and Bachman (1851). Soon afterwards, Merriam (1888) described and named the kit fox (*Vulpes macrotis*). Hall and Kelson (1959) suggested that swift foxes and kit foxes were conspecifics, being subspecifically distinct. Similar conclusions were reported by Dragoo et al. (1990) using morphometric analysis coupled with electrophoretic protein analysis. However, other studies using multivariate morphologic analysis have distinguished swift and kit foxes (Rohwer and Kilgore 1973, Thornton and Creel 1975, Egoscue 1979, McGrew 1979, Stromberg and Boyce 1986). A recent evaluation of taxonomic status, using mitochondrial DNA restriction-site and sequence analyses, concluded that the kit fox and the swift fox should be recognized as separate species (Mercure et al. 1993). They suggested the Rocky Mountains may have been a barrier to gene flow between kit and swift foxes because populations on either side belong to 2 divergent mitochondrial DNA clades. Dragoo and Wayne (2003) suggest that swift fox and kit fox should be considered as one species based on recent microsatellite and mtDNA analysis. For the purposes of this paper, we consider the swift fox as a separate species from the kit fox.

Merriam (1902) classified 2 subspecies of the swift fox as the northern swift fox (*V. v. hebes*) and the southern swift fox (*V. v. velox*). Merriam's classification was still in use when the northern subspecies was listed as endangered in 1970 (Federal Register 1970). The endangered listing was removed in the United States in 1980 (Federal Register 1980); however, this designation remains in Canada (50 CFR 17.11). These subspecific designations have been a point of controversy among taxonomists (see Dragoo and Wayne 2003). Stromberg and Boyce (1986) investigated the question of the existence of 2 subspecies of swift fox (*V. v. velox* and *V. v. hebes*). After examining 250 specimens, they concluded that subspecific classification of the southern swift fox and northern swift fox was probably not justified, although significant geographic variation was noted. These authors suggested that such variation may reflect genetic variation and warrants conservation efforts to preserve genetic diversity, and urged caution in reintroduction programs because of risk of potential loss of genetic adaptation to the severities of northern climates. Dragoo and Wayne (2003) advocate many fewer subspecies classifications of swift and kit foxes, yet agree with the recommendation to limited gene flow between populations to maintain potentially adaptive differences among populations.

Life History and Ecology

Swift foxes seem to be monogamous and likely pair for life (Kilgore 1969). They live in pairs or occasionally as trios or groups of 2 males and 2–3 females, with 1 breeding female and nonbreeding "helpers" (Kilgore 1969, Covell 1992). Swift foxes are monestrous (Asa and Valdespino 1998), with breeding beginning in late December or early January in the southern portion of their range to March in the northern portion of their range (Kilgore 1969, Hines 1980, Carbyn et al. 1994). Gestation is estimated to be 51 days (Hayssen et al. 1993). Average litter sizes of 2.4–5.7 have been reported based on counts of pups at natal dens (Kilgore 1969, Hillman and Sharps 1978, Covell 1992, Carbyn et al. 1994, Roell 1999, Schauster et al. 2000, Andersen et al. 2003). Both members of the pair provide for the young, and young foxes remain with the adults for 4–6 months (Rongstad et al. 1989, Covell 1992), which is longer than other North American canids. Covell (1992) reported that male pups had a higher frequency of dispersal from September–

January (23% versus 10%) and dispersed significantly farther (9.4 km versus 2.1 km) than female pups. In western Kansas, 3 pups radiomarked at dens in August 1996 had dispersed approximately 25 km, 22 km, and 18 km (male, female, female, respectively) by January of 1997 when monitoring ceased (M. Sovada, unpublished data). Four other pups (2 males, 2 females) monitored during the same period remained within 4 km of their natal denning area.

Reported annual mortality rates range from 0.47 to 0.57 (Covell 1992, Sovada et al. 1998, Andersen et al. 2003), which is similar to rates reported for other North American foxes (Lord 1961, Storm et al. 1976, Cypher and Scrivner 1992, Disney and Spiegel 1992, Ralls and White 1995). Coyotes (*Canis latrans*) have been identified as the principal cause of swift fox mortality (Laurion 1988, Covell 1992, Carbyn et al. 1994, Sovada et al. 1998, Kitchen 1999, Andersen et al. 2003). Interference competition rather than a food resource is the likely causal factor based on the infrequent consumption of swift foxes killed by coyotes (Sovada et al. 1998, Kitchen 1999). Both Kitchen (1999) and Andersen et al. (2003) speculated that coyote-caused mortality may suppress fox populations, especially in times of low prey availability. Other predators of swift foxes that have been identified are golden eagles (*Aquila chrysaetos*) and American badgers (*Taxidea taxus*; Carbyn et al. 1994, Andersen et al. 2003). Potential predators include red foxes (*Vulpes vulpes*), bobcats (*Lynx rufus*), large hawks, and great horned owls (*Bubo virginianus*). Mortality factors associated with human activities include vehicle collision, secondary poisoning, shooting, and trapping (Kilgore 1969, Rongstad et al. 1989, Carbyn et al. 1994, Sovada et al. 1998). Collisions with automobiles are a significant mortality factor for young animals in some landscapes (Sovada et al. 1998).

Swift foxes host a variety of internal and external parasites (Kilgore 1969, Pybus and Williams 2003). Fleas (*Pulex spp., Opisocrostos hirsutus*) are the most common and abundant ectoparasite. Kilgore (1969) suggested that the large numbers of fleas found in swift fox dens may contribute to the frequent changes in dens used by foxes. Other parasites include hookworms (*Ancylostoma caninum, Uncinaria* sp.) and whipworms (*Trichuris vulpis*), as well as miscellaneous protozoans and ectoparasite species (Pybus and Williams 2003). There is little known about the occurrence of diseases in swift foxes, however it is likely that swift foxes are susceptible to most diseases that occur in other canids (see Pybus and Williams 2003).

Swift foxes are among the most fossorial members of the North American canid family, and unlike most other canids, dens are used throughout the year (Cutter 1958a, Kilgore 1969, Egoscue 1979, Hines 1980). Swift foxes either excavate their own dens or enlarge the burrows of other animals (Rongstad et al. 1989, Hines and Case 1991). Swift foxes typically use dens daily, have multiple dens, and members of a family group may be found together in a den (Hines and Case 1991, Andersen et al. 2003). Dens are largely located in shortgrass and mixed-grass prairie, although dens also have been found in cultivated fields and other human-made habitats (Cutter 1958a, Kilgore 1969, Uresk and Sharps 1986, Jackson and Choate 2000). Dens serve several functions, such as providing escape cover from predators, protection from extreme climate conditions in both summer and winter, and shelter for raising young (Rongstad et al. 1989, Fox 1991). Notably, the distribution and density of dens which are used for protection from predators are considered important components of swift fox habitat requirements, and may influence predation rates and growth rates of swift fox populations (Herrero et al. 1991, Swift Fox Conservation Team 1997a, Kitchen et al. 1999, Pruss 1999).

Swift foxes are primarily nocturnal, although limited daytime activity may occur near den sites (Kilgore 1969, Laurion 1988, Kitchen et al. 1999, Andersen et al. 2003). Andersen et al. (2003) observed that much of the night-time activity also occurred near dens. Members of family groups forage separately (Hines and Case 1991), but home ranges of family members are similar (Andersen et al. 2003). Swift foxes tend to travel greater distances in the breeding season (Hines and Case 1991, Kitchen 1999), which Kitchen (1999) speculated would increase potential encounters with coyotes and thus vulnerability to coyote predation. Her observation of higher coyote-caused mortality during the breeding season supports this premise.

There are few quantitative analyses of habitat use by swift fox, although studies have described suitable habitats as shortgrass and mixed-grass prairies in gentle rolling to level terrain (Cutter 1958a, Kilgore 1969, Hillman and Sharps 1978, Hines 1980). Kitchen (1999) found swift foxes only in open plains habitat, and did not capture any foxes or observe their activity in nearby piñón pine-juniper habitat. Additional favorable conditions likely include low potential for contact with human activities and low densities of predators such as coyotes (Hillman and Sharps 1978).

Published estimates of swift fox home ranges are quite variable and difficult to compare because different techniques and criteria were used to estimate home-range size. Hines and Case (1991) reported an average home-range size of 32.3 km^2 (range 7.7–79.3 km^2) for 7 swift foxes in Nebraska using minimum convex polygon method (MCP; Mohr 1947), but 4 animals were followed ≤5 nights in winter or very early spring. Andersen et al. (2003) reported a similar average MCP home-range size of 29.0 km^2 (range 12.8–34.3 km^2) on the Piñón Canyon Maneuver Site in southeastern Colorado (1986–87) for 5 swift foxes with ≥34 locations ($\bar{x}$ = 188 locations) over minimum period of 7 months. A smaller estimate (MCP) of average home range, 25.1 km^2 (range 8.7–43.0 km^2), was determined for 22 swift foxes in western Kansas (Sovada, unpublished data) using only foxes with >60 locations. Zimmerman et al. (2003) estimated average MCP home-range size of 10.4 km^2 (range 7.3–16.9 km^2) for 5 swift foxes in Montana. Using the 95% adaptive kernel method (Worton 1989), Kitchen et al. (1999) reported average home-range size of 7.6 km^2 for foxes (with >60 locations per season) on the Piñón Canyon Maneuver Site during 1997–98. Early studies suggested that swift foxes were not territorial (Hines 1980, Cameron 1984), although more recent data have provided evidence of territoriality. Andersen et al. (2003) reported nearly total exclusion of an individual swift fox's core activity area to other same-sex individuals.

Swift foxes, like other canids, are opportunistic foragers, feeding on a wide variety of mammals, birds, arthropods, plants and carrion (Cutter 1958b, Kilgore 1969, Hillman and Sharps 1978, Zumbaugh et al. 1985, Uresk and Sharps 1986, Hines and Case 1991, Kitchen et al. 1999, Sovada et al. 2001). Mammals usually dominate swift fox diets; especially important are lagomorphs (Cutter 1958b, Kilgore 1969, Cameron 1984, Hines and Case 1991) and rodents (Uresk and Sharps 1986, Kitchen et al. 1999). Insects and birds become important food items in late summer and early fall (Uresk and Sharps 1986, Kitchen et al. 1999); several studies have reported use of carrion throughout the year (Kilgore 1969, Zumbaugh et al. 1985, Uresk and Sharps 1986, Hines and Case 1991). The generalist foraging behavior of swift foxes makes food an unlikely limiting factor, yet there is no evidence to refute or support food availability as a factor limiting populations. Evidence from kit foxes suggests that food scarcity may result in temporary or local declines (White and Ralls 1993). Kitchen (1999) speculated that increased distances traveled and time spent foraging during periods of limited food availability may increase vulnerability of swift foxes to coyote-caused mortality and may contribute to population decline.

Historic Distribution

The swift fox is native to the shortgrass and mixed-grass prairies of the Great Plains Region of North America (Egoscue 1979), and historically occurred in all or portions of North Dakota, South Dakota, Montana, Nebraska, Wyoming, Colorado, Kansas, Oklahoma, New Mexico, and Texas and the southern prairie region of Alberta, Manitoba, and Saskatchewan (Hall and Kelson 1959, Banfield 1974, Egoscue 1979, Sovada and Scheick 1999).

The historic distribution of the swift fox in the Great Plains has been based on limited information found in museum and fur-trade records and by accounts of early naturalists and explorers. This information about historic distribution is especially fragmentary and not all observations can be verified. Many contemporary biologists (Swift Fox Conservation Team 1997a) believe that the historic swift fox distribution was influenced by and likely restricted to the expanse of shortgrass and mixed-grass prairie (see Risser et al. 1981). This assessment is largely based on observations of swift fox behaviors, habitat use, and locations of confirmed records. The eastern boundary of the historic distribution of swift foxes remains particularly unclear, which may be attributed to the naturally shifting geographic boundary between the mixed-grass and tall grass prairies; this shifting largely occurred as a result of climatic variations (Küchler 1972, 1985; Weaver et al. 1996). Thus, swift fox distribution in the east may have shifted with the changing boundary between the mixed-grass and tall grass prairies. In the southern portion of the historical range, in extreme western Texas and New Mexico, the swift fox range overlapped with that of the kit fox, and hybridization between the 2 species apparently occurred along the Pecos River (Hall and Kelson 1959, Rohwer and Kilgore 1973, Mercure et al. 1993). It is important to note, as Hubbard (1994) reported, that the contact zone between swift and kit foxes "is as enduring as it is broad," existing for several thousand years, yet abrupt morphological differences between swift and kit foxes exist in this zone.

Some historical range descriptions include swift foxes in Minnesota and Iowa, however, there are no verifiable records of swift fox occurrence in either state (Dinsmore 1994; E. Birney, Bell Museum of Natural History, University of Minnesota, personal communication). Swanson et al. (1945) suggested swift foxes may have occasionally ventured into Minnesota based on their occurrence only 56 km west of Minnesota, as reported in Alexander Henry's journals (Reid and Gannon 1928). Similarly in Iowa, there are only second-hand reports of swift fox observations (Dinsmore 1994). Furthermore, no records of swift foxes have been confirmed for counties in South Dakota or Nebraska adjacent to Iowa. Minnesota and Iowa may have been on the fringe of swift fox range, with foxes venturing into these states intermittently and at low densities.

The first published record of the swift fox was in 1801 from Alexander Henry's fur shipment records, from Pembina Post of the Northwest Company's Red River District (Reid and Gannon 1928) in North Dakota. The main post was located at the junction of the Pembina and Red rivers, with branch posts to the west in the Hair Hills (Pembina Hills) and the mouth of the Reed River in Canada. Alexander Henry's journals do not identify exact trapping locations, but rather likely reflect the locations where fox pelts were traded. Specifically, there is no mention of swift or kit foxes in Henry's journals, only references to red foxes. "Kit" (swift) foxes were only noted in tabulation of yearly acquisitions (Reid and Gannon 1928). The small number taken in the 8 years of trapping records suggests swift foxes were not common in the vicinity of northeastern North Dakota during Henry's operation at the Pembina Post. Moreover, Reid and Gannon (1928) suggested that swift foxes likely were not common in northeastern North Dakota, stating "...being a plains animal it is quite probable they were more common farther west." However, Bailey (1926) quotes Charles Cavileer from "A Story of '53" describing the fur-trade in Walhalla, as obtaining 400–600 "kit foxes" per year from the Pembina mountain region for a period before the American bison (*Bison bison*) disappeared.

John James Audubon reported swift foxes near Fort Clark in North Dakota in 1833 (*in* Bailey 1926) and again in 1843 (Audubon and Bachman 1854). In 1850, Thaddeus A. Culbertson collected a specimen near Fort Union (Bailey 1926). F.V. Hayden (1862 *in* Bailey 1926) reported capturing 50–100 swift foxes each winter near each of the trading posts along the Missouri River. Elliot Coues reported swift foxes as "common" along the Souris (Mouse) River (*in* Bailey 1926) and collected 5 specimens now located in the National Museum (Washington, D.C.). In a report from an 1873 military expedition from Fort Rice (on the Missouri River near the geographic center of North Dakota) due west to and along the Musselshell River to the Yellowstone River, Allen (1874) described swift foxes as "quite frequent." George B. Grinnell (1914, *in* Knowles et al. 2003) found swift foxes abundant along the Little Missouri in 1874. Ludlow (1875) traveled from Fort Lincoln (near present-day Mandan, North Dakota) to the Black Hills of South Dakota, and described swift foxes as "abundant everywhere on the plains," though not often seen, because of its small size and furtive disposition, in preference to running.

All of South Dakota was historically considered as part of the range of the swift fox (Over and Churchill 1941, Hall and Kelson 1959, Egoscue 1979, Hall 1981); however, E. Birney, (Bell Museum of Natural History, University of Minnesota, Minneapolis, personal communication) and J. Knox Jones (Texas Tech University, Lubbock) could find no record of swift foxes in the easternmost counties of South Dakota while researching for their 1988 book, *Mammals of the North-central States*. Archeological remains were found at the Mobridge site (ca. 1650–1700) and the Walth Bay site (ca. 1550–1600), both in Walworth County (K. Lippincott, South Dakota Archeological Consultant, Pierre, personal communication). Perhaps the first published account of swift foxes in South Dakota was from Pierre-Antoine Tabeau, a member of a fur-trading expedition from St. Louis to the upper Missouri River from 1803–1805, who observed both red and "a kind of little gray fox" (presumably swift fox) as very common in the upper Missouri (*in* Abel 1939). Large numbers of swift fox pelts were traded at the American Fur Company's Upper Missouri Outfit (near the confluence of the Big Sioux and Missouri rivers) from 1835–1838 (Johnson 1969). Ludlow (1875) reported that swift foxes were "abundant" on the plains while he was traveling from Fort Lincoln, North Dakota, through Corson and Harding counties, South Dakota, to the Black Hills. Visher (1914) described swift fox as present but very rare in Harding County, but noted that furbearer trapping reports were dependent on bounty prices. The Smithsonian Institution records show a specimen collected northwest of Pierre, Hughes County, in 1917 (USNM 300300). This record is the easternmost historic record that we could find for the state of South Dakota.

The swift fox was thought to occur in much of Nebraska prior to European settlement (Jones 1964, Hines and Case 1980, Hall 1981). However, much of the far eastern portion of Nebraska was considered tall grass prairie (Risser et al. 1981), which is inconsistent with habitat typically occupied by swift foxes. There are few historic records of swift foxes in Nebraska (Sovada and Scheick 1999). Cary (1902) reported that swift foxes "occur sparingly" in the Hat Creek Basin in northwestern Nebraska.

Zumbaugh and Choate (1985) reviewed historical accounts of the swift fox in Kansas. The swift fox occurred in at least 36 counties and perhaps 44 counties (including verified and unverified records) throughout most of the western half of the state in shortgrass and mixed-grass prairie habitat (Carter 1939 *in* Zumbaugh and Choate 1985, Hall and Kelson 1959, Bee et al. 1981, Zumbaugh and Choate 1985, Fox and Roy 1996).

In Oklahoma, the swift fox was considered to occur historically throughout the panhandle region (Cimarron, Texas, and Beaver counties) and western portions of three adjacent counties (Harper, Woodward, and Ellis counties; Duck and Fletcher 1945, Hall 1981, Caire et al. 1989).

Reports from 2 early expeditions to parts of Oklahoma, excluding the panhandle, did not include presence of swift foxes in their observations of mammals, suggesting that swift foxes did not historically occur in other parts of Oklahoma. In the first expedition in 1835, Washington Irving joined a military expedition from Fort Gibson (northeast Oklahoma) to the center of the state. In his book *A Tour of the Prairies*, Irving (1835) did not note the presence of swift foxes in his accounting of mammals. In 1852, Captain Randolph B. Marcy explored the Red River, which defines the present southern border of Oklahoma. Marcy's list of mammals encountered did not include swift foxes (Marcy 1854).

The first record of swift fox occurrence in Oklahoma was in 1888 located in the "Neutral Strip" (panhandle; Caire et al. 1989). Several agencies conducted biological explorations of the lands that were opened to settlers during the land runs of 1889 and 1893. Charles P. Rowley led a party from the American Museum of Natural History that spent several weeks in October and November 1889 in the western part of the panhandle collecting American bison. The party also collected a number of small mammals near Corrumpa and Seneca creeks in the southwestern part of present-day Cimarron County, but they did not note swift foxes. Similarly, from the same period, collections in Wood County for the Field Museum of Natural History in

Chicago, and in Woodward County for the U.S. Bureau of the Biological Survey made no mention of swift foxes.

Bailey (1905) provided the first published report of swift foxes in Texas, reporting swift foxes in Stanton, Martin, Midland, Oldham, and Armstrong counties. Egoscue (1979), Hall (1981), and Jones et al. (1987) defined the historic range of the swift fox in Texas to include the panhandle region south into the west-central portion of the state; approximately 78 counties were included. Jones et al. (1987) indicated only 28 counties in Texas had reliable records of swift fox based on the literature, trapping records, or museum specimens, and they estimated that half of the historic range of the swift fox in Texas (high plains below the 34th parallel) was no longer suitable for swift foxes due to intensive agriculture. Our accounting of historical and current records include swift foxes in 26 counties with another observation from the convergence of Crane, Pecos and Upton counties (see Sovada and Scheick 1999). Certainly foxes occurred in other counties lacking recorded observations or specimens, but the striking absence of records from the grassland type of Southern Mixed-Grass Prairie with Shrubs (Risser et al. 1981; *see* Sovada and Scheick 1999) suggests this shrubby grassland habitat may be less suitable for swift foxes. Similar to the shifting nature of the boundary between mixed-grass and tall grass prairies described above, encroachment of shrubs in the southern mixed-grass prairie (Archer 1994) may have influenced swift fox distribution.

Historically, the swift fox was considered common in the shortgrass to mixed-grass prairies east of the Rocky Mountains of Montana (Swift Fox Conservation Team 1997a, Knowles et al. 2003). Meriwether Lewis and William Clark observed swift foxes along the Marias and Missouri rivers in 1805 and 1806 (Burroughs 1961). In the late 1800s, Coues (1878) reported that swift foxes were common between the Milk River and the Canadian border. There are many other reports of swift foxes in Montana in the late 1800s and early 1900s (e.g., Audubon and Bachman 1854, Allen 1874, Custer 1875, McChesney 1879; see Knowles et al. 2003). The last historical record of swift foxes in Montana was in 1918, when Bailey and Bailey (1918) noted the swift fox commonly occurring on the plains along the eastern edge of Glacier National Park. Hall and Kelson (1959) indicated the range of the species crossed the mountains of western Montana and extended into British Columbia. This determination was contradicted by Soper (1964) and Hoffmann et al. (1969), who did not list the species as part of British Columbia's fauna. Banfield (1974) describes Canadian distribution at the time of settlement as the southern prairies of Canada from the Pembina Hills of Manitoba to the foothills of the Rocky Mountains.

In Wyoming and Colorado, swift foxes originally occupied the short and mixed-grass prairie regions in the eastern halves of each state (Cary 1911, Long 1965, Armstrong 1972, Hall 1981, Lindberg 1986). Distribution in Wyoming included all or portions of 17 counties; historic records defined the western edge of distribution along the eastern portions of Big Horn, Washakie, Freemont, and Sweetwater counties (Long 1965, Hall 1981, Lindberg 1986). Cary (1911) described accounts of swift foxes as far west as Boulder County in central Colorado.

In New Mexico, swift foxes likely occurred in 12 counties including Colfax, Union, Mora, Harding, San Miguel, Guadalupe, Quay, De Baca, Curry, Roosevelt, Chaves, and Lea (Swift Fox Conservation Team 1997a); vegetative classification in this region was Plains-Mesa Grasslands (Dick-Peddie 1993). Bailey (1931), Egoscue (1979), and Hall (1981) described the species as occurring east of the Pecos River drainage in the extreme eastern portion of New Mexico. There are no records from 1894 to 1952, except for a report from Santa Rosa labeled *V. macrotis*, which Bailey (1931) believed was *V. velox*; and a museum record (Museum of Southwest Biology [MSB], University of New Mexico, #BRD101289) from a fox collected 13 km southwest of Albuquerque, which is substantially outside of the accepted historic range of the swift fox. Further examination of this specimen has identified it as *V. macrotis* (Robert Harrison, University of New Mexico, Albuquerque, personal communication).

Status of Populations

The swift fox was considered common or abundant in much of its original range until the late 1800s to the early 1900s. Records of the American Fur Company's Upper Missouri Outfit (near the confluence of the Big Sioux and Missouri Rivers) from 1835 to 1838 included 10,427 swift fox pelts compared to 1,051 red fox pelts and 13 gray fox (*Urocyon cinereoargenteus*) pelts received during the same period (Johnson 1969). Alexander Henry's journals noted the take of 117 "kit" foxes from 1800 to 1806 in northeastern North Dakota with an additional 120 "kit" foxes received from the Hudson's Bay Company at Pembina in 1905–1906 (Reid and Gannon 1928).

Swift fox numbers and distribution declined dramatically by the late 1800s and early 1900s. Zumbaugh and Choate (1985) provided evidence that, in Kansas, swift foxes were extremely abundant in the mid-1800s, but became less abundant by the turn of the twentieth century. The decline continued and there were only sporadic reports in the 1930s (Carter 1939, Tihen and Sprague 1939). The species was believed to have been extirpated from Kansas by the 1940s (Black 1937, Cockrum 1952, Hall 1955, Sovada and Scheick 1999). Jones (1964) suggested that the swift fox may have been extirpated from Nebraska by the early 1900s because there were no records from 1901 (Merritt Cary, Manuscript on file, U.S. Fish and Wildlife Service, Washington, D.C., *in* Jones

1964) to 1953–54, when a female and 2 pups were taken in Morril County near Bridgeport. Blair (1939) did not list any known records of swift fox in Oklahoma. In the panhandle of Oklahoma, Duck and Fletcher (1945) reported that "in earlier days it [swift fox] was frequently seen throughout this area," but noted that they were "very rare" in 1945. Cockrum (1952) considered them extinct in Oklahoma. Bailey (1905) noted that Texas ranchers indicated that swift foxes were scarce compared to their numbers in previous years. A specimen from Armstrong County in 1905 (Bailey 1905) was the last recorded swift fox in Texas until 1948, when one was collected in Swisher County (Glass 1956). By 1900, swift fox range had contracted significantly in Colorado (Cary 1911, Lechleitner 1969, Armstrong 1972, Hall 1981), yet Cary (1911) reported that swift foxes were common in many Colorado counties. In 1915–1916, swift foxes were collected in Adams and Las Animas counties, but there was no subsequent documentation in the state until 1941 in Crowley County (*see* Sovada and Scheick 1999). In Montana, the lack of confirmed records since the report of Bailey and Bailey (1918) and no records of swift fox occurrence in 16 years of fur harvest data prompted Hoffmann et al. (1969) to conclude the species was probably extinct in Montana. In Wyoming, swift fox numbers began a marked decline by the late 1800s, and there are no reports of swift foxes from 1898 until a report from Laramie County in 1958 (Long 1965). There were no reports of swift foxes in South Dakota between 1914 and 1966 (Hillman and Sharps 1978), in North Dakota between 1915 and 1970 (Pfeifer and Hibbard 1970), and in Nebraska between 1901 and 1953 or 1954 (Jones 1964). Swift foxes were last seen in Canada in 1938 (Soper 1964); the last confirmed specimen came from near Govenlock, Saskatchewan in 1928 (Carbyn 1998).

Many factors have been attributed to the population decline of the swift fox that followed the arrival of European settlers to the prairie regions (Zumbaugh and Choate 1985, Scott-Brown et al. 1987). Perhaps the most important direct cause of swift fox population decline was inadvertent poisoning from strychnine-laced baits, which were widely used to control gray wolves (*Canis lupus*) in western North America (Seton 1909, Young 1944). Swift foxes readily accepted poisoned baits and thus died by the thousands (Bailey 1926, Young 1944). Grinnell (1914) commented on taking 3000 "wolves" (1/4 gray wolves and 3/4 coyotes), as well as "several bales" of swift fox skins, during a poisoning campaign in Rush County, Kansas, during 1860–1861. Intense trapping was another factor contributing to declines in swift fox populations. Hudson's Bay Company records showed that 117,025 swift fox pelts were sold in London between 1853 and 1877 (Rand 1948). In Canada, commercial trapping of swift foxes continued into the early 1900s, but with appreciably lower success; only 508 pelts were sold from all of the Canadian prairie provinces in 1925.

Modification, degradation, loss, and fragmentation of native grasslands and the associated declines in prey species also have been implicated in the decline of swift foxes (Egoscue 1979). The settlers converted large expanses of prairies to cropland (Samson and Knopf 1994, Samson et al. 1998), first in much of the tall grass prairie and later in portions of the mixed-grass prairie. The drier areas of shortgrass prairie were less suitable for grain farming but were suitable for cattle production. However, despite less conversion of native prairies to cropland in the shortgrass prairies (Samson et al. 1998), native grazers such as American bison and prairie dogs (*Cynomys* spp.) were largely replaced by domestic cattle, which have very different grazing behaviors (Schwartz and Ellis 1981). It is unknown how this change affected swift foxes and their habitat requisites. Furthermore, northern populations of swift foxes may have relied more heavily on carrion, such as bison killed by gray wolves or dying from natural causes, to survive severe winter months. This substantial food source was no longer available once wolves and bison were eradicated from the region (Carbyn 1986, Klausz et al. 1996). While suffering declines in northern and eastern parts of their range, swift fox populations were able to persist in the southwestern portion of their historical range.

Swift fox numbers appeared to be recovering over portions of their former range beginning in the 1950s (Martin and Sternberg 1955, Glass 1956, Anderson and Nelson 1958, Andersen and Fleharty 1964, Long 1965, Kilgore 1969, Sharps 1977, Egoscue 1979, Hines 1980). Reported observations were increasing in the 1950s and 1960s, a trend that continued into the late 1970s and 1980s (Long 1965, Blus et al. 1967, Pfeifer and Hibbard 1970, Van Ballenberghe 1975, Moore and Martin 1980, Floyd and Stromberg 1981, Zumbaugh and Choate 1985, Lindberg 1986, Schmitt 1995, *also see* Sovada and Scheick 1999). The return of swift foxes in areas of their historical distribution was attributed to cancellation of all registration of predicides (U.S. Fish and Wildlife Service 1978), declines in fur value resulting in less trapping, and restrictions on methods of fur harvest (Kilgore 1969). Floyd and Stromberg (1981) suggested an additional factor could be the significant decline in the number of farms and ranches and thus, reduction in human activities detrimental to foxes.

Distribution data, largely gathered by the Swift Fox Conservation Team (SFCT; comprised of biologists representing state and federal wildlife conservation agencies, *see* "State Agencies Response to Federal Actions" below) in the mid to late 1990s, indicated a more extensive distribution than estimated by the U.S. Fish and Wildlife Service (USFWS, Federal Register 1995), particularly in the core of the species, historical range (i.e., Colorado, Kansas, and southeastern Wyoming) and other parts of the historical distribution (see SFCT Annual Reports 1996-99).

Results of swift fox surveys by the SFCT, on a county-by-county basis, confirmed that swift fox populations in this core area, which includes a significant portion of the current range, are likely not isolated, although densities vary from region to region (Sovada and Scheick 1999). However, some populations likely are isolated, especially in Montana and parts of Wyoming (Merrill et al. 1996, Redmond et al. 1998).

The updated information indicated that swift fox populations currently occupy a substantial portion of their historic range, and possibly have reoccupied some of their former habitat throughout nearly all of the historic range in Wyoming, eastern Colorado, and several counties in the western one-third of Kansas. Evaluations in New Mexico and Oklahoma indicated that swift foxes are present throughout most of their former range. Swift fox populations appear to be reoccupying and expanding in at least 2–3 counties in north-central Montana with increasing confirmed and unconfirmed reports in 2–3 southeastern counties. Texas has been able to confirm swift fox presence in at least 3 counties in the northern panhandle. It is believed that swift fox populations continue to be absent from North Dakota, are declining in South Dakota, and are present in minimal numbers in at least 2–3 counties in western Nebraska. Some populations where swift fox are reoccupying viable habitats or have been recently reintroduced (e.g., Blackfeet Indian Reservation in Montana, Giddings 1998) may still be isolated.

Potential Threats to Populations

Present or Threatened Destruction or Modification of Habitat

The USFWS (Federal Register 1995) estimated that 45% of the swift fox's habitat throughout its historic range has been lost as a result of prairie conversion based on Soil Conservation Service (1989) data. Where native prairies remain, they often are fragmented into smaller and isolated areas, reducing available habitat and prey while increasing predation and competition for swift foxes. These remnant prairies are not the same ecosystems that were present prior to European settlement of the region. Replacement of bison with domestic livestock and the suppression of fire have resulted in changes in plant community composition and landscape patterns (Bragg and Steuter 1996, Weaver et al. 1996). Agriculture, residential development, and other commercial development have been pervasive in the shortgrass and mixed-grass prairies and are considered a cause of population declines, yet swift fox populations are relatively widespread. However, the further expansion of cultivated acreage beyond the sustainable land base onto lands that are marginal for agricultural purposes, poses a severe threat to remaining swift fox habitat. For example, the changeover from dryland agriculture systems to irrigated crops that is occurring in western Kansas could potentially have a substantial and negative impact on swift fox populations although the extent of the impact is yet to be examined. We make this speculation based on the use of fallow fields by swift foxes as sites for dens and foraging and on their limited use of irrigated croplands, such as corn (M. Sovada, unpublished data). With the continuing development of irrigated crop systems, the presence of fallow fields will be lost, which potentially could impact swift fox populations. Further investigation is needed to verify this speculation. Moreover, the planting of tall dense vegetation in fields enrolled in the Conservation Reserve Program, which is common in western Kansas, also may have negative impacts overall on swift fox populations because swift foxes avoid this tall dense perennial cover (M. Sovada, unpublished data).

Distribution and occurrence has been confirmed in a large number of counties throughout the species' historic range since 1995 (Sovada and Scheick 1999), particularly in the states of Colorado, Kansas, New Mexico, Oklahoma, and Wyoming. These states are considered to be the species' current core area and may be contributing swift foxes to adjacent states. The species' distribution is relatively widespread; however, distributions and associated densities appear to be highly variable among the 9 occupied states (Swift Fox Conservation Team 1997a). Although it is generally agreed that the swift fox evolved in the shortgrass and mixed-grass prairie ecosystem and numerous studies have provided qualitative ecological data, definition of range-wide habitat requirements in an altered landscape has not been adequately addressed.

Habitats within the shortgrass and mixed-grass prairie ecosystems are recognized as being able to provide the "essentials" of a diverse prey base, level to slightly undulating topography which affords long viewing distances to detect predators, and firm, friable soils that are suitable for the excavation and maintenance of multiple den sites for year-round use. Habitats in today's altered landscapes can vary significantly, and their suitability for swift foxes are difficult to define. The swift fox has adapted to a variety of habitat types, including the establishment of populations in a mixture of agriculture and rangeland. Kilgore (1969) and Hines (1980) reported swift fox populations inhabiting habitats of mixed agricultural use in Oklahoma and Nebraska, respectively. In western Kansas, swift foxes are commonly found in cropland-dominated landscapes, which included fragments of shortgrass prairie, but are largely comprised of fallow cropland, wheat, sunflower, and irrigated crop fields (Fox and Roy 1996, Sovada et al. 1998, Jackson and Choate 2000). However, not all habitats are used by swift foxes in proportion to their availability (M. Sovada, unpublished data). Jones et al. (1987) and Jackson and Choate (2000) found swift foxes denning in fallow cropland fields. In Wyoming, Woolley et al. (1995) found swift foxes in shortgrass, mixed-grass, sagebrush-grassland, and sagebrush-greasewood habitat types with topography ranging from flat to badlands-like terrain. In Montana, no den sites were observed in any habitat other than native rangeland, but 1 capture out of 16 occurred in mixed habitat during the 1996 trapping efforts (Giddings and Zimmerman 1996). Paradoxically, a recent study in New Mexico did not document the use of mixed agricultural/rangeland habitats by swift foxes (Harrison and Schmitt 1997).

The conversion of native grassland prairies has been implicated as one of the most important factors for the contraction of the swift fox range (Hillman and Sharps 1978). We believe that alteration of the landscape likely influences local and seasonal prey availability, increases risk of predation on swift foxes, and leads to interspecific competition with other predators such as the coyote and red fox. However, in Kansas and perhaps portions of Oklahoma and Colorado, a mixed agricultural/rangeland landscape does not appear to necessarily diminish the habitat value of associated grasslands from a forage availability standpoint. In fact, Kansas biologists believe that agricultural systems on privately owned lands are crucial to swift fox conservation because publicly owned lands in Kansas are either too small or inadequate to support swift fox populations (C. Roy, Kansas Department of Wildlife and Parks, Emporia, personal communication). The SFCT has suggested that "it is not solely the conversion of prairie to cropland that hinders current swift fox restoration efforts but also juxtaposition of the remaining prairies, management of rangelands, cropping patterns of farmlands, and changes in canid communities that occur in response to the conversion of prairie habitat to cropland" (Swift Fox Conservation Team 1997a).

Just as farm policies of the past have encouraged conversion of native prairies to cropland (Baydack et al. 1996), farm policies of the future could impart restoration of habitat suitable for swift foxes. The Conservation Reserve Program (CRP) established under the 1985 Farm Bill and renewed under the 1996 extension has stimulated conversion of millions of croplands to perennial grassland cover (Young and Osborn 1990). However, in many areas of the shortgrass prairie, CRP fields were planted to tallgrass prairie species or non-native grasses (Sovada et al. 2003). When these fields are not grazed, mowed, or burned, they develop dense tall stands of grasses that are not suitable for use by swift foxes (Swift Fox Conservation Team 1997a). Current regulations for CRP lands do not provide adequate habitat guidelines that would benefit swift foxes. New CRP guidelines could provide incentives for participants in the CRP to plant native shortgrass species that are better suited for use by swift foxes.

Recreational/Commercial Harvest

Determining the magnitude and significance of harvest on swift fox populations as a result of trapping, hunting, predator control, and other activities has been difficult because of limited data available across the species' range. Although we acknowledge that private predator control activities result in swift fox mortalities, it is unknown if these activities are a major source of mortality that directly impacts local populations. Predator control activities conducted by the U.S. Department of Agriculture (Wildlife Services) targeting coyotes are responsible for a relatively small percentage of swift fox incidental take and may in fact be benefitting some local populations. Many wildlife biologists concur that annual mortality resulting from these activities is a very small percentage of total swift fox mortality (Swift Fox Conservation Team 1997a). In some parts of the swift fox range, where mortality may be exceptionally high as a result of predation or other factors, incidental harvest could become a concern when considered from a cumulative aspect. However, studies conducted in different parts of swift fox range have confirmed that predation by coyotes is the most significant mortality factor (Laurion 1988, Carbyn et al. 1994, Sovada et al. 1998, Kitchen et al. 1999). Additional interspecific competition with red foxes may exacerbate this problem (Ralls and White 1995; M. Sovada, unpublished data). Currently, 2 studies are being conducted to examine the impact of coyote control on swift fox populations (Mote et al. 1998, Seidel 1998).

Swift foxes are legally protected under State laws in all 10 states which encompass the species' historical range and are currently protected from harvest through laws or regulations in 7 of these states (Swift Fox Conservation Team 1997a). States that do provide harvest opportunities regulate harvest by season length and monitor harvest numbers annually. Kansas, New Mexico, and Texas provide a regulated harvest season and estimate annual harvest figures. Colorado, Montana, North Dakota, and Oklahoma list swift fox as furbearers but the harvest season is closed all year. Nebraska lists swift fox as "endangered," and in South Dakota they are "threatened." Wyoming lists swift fox in their nongame regulations, and only incidental harvest is allowed to provide additional distribution data. Trapper education programs are becoming more available to fur harvesters, and fur harvester education courses are currently required in several states, which SFCT believes may help to reduce incidental harvest across the swift fox range.

There is insufficient information to weigh the impact of harvest on swift fox distribution or population densities; therefore, the importance of harvest in limiting or regulating swift fox populations is unknown. The evidence available suggests that regulated harvest has had no impact in limiting swift fox populations. For example, swift fox populations in Colorado have remained widespread despite 55 years of regulated harvest. No noticeable reduction in distribution has occurred in Kansas since the opening of a trapping season on swift fox in 1982. In comparison, swift fox have been protected from harvest in South Dakota, Nebraska, and Oklahoma, with no apparent increase in distribution or population densities during the same period. Further, pelt prices during the last 10 years have remained low, varying from $3 to $10 (U.S. currency, Swift Fox Conservation Team 1997a). Thus, there is little interest or incentive to actively harvest swift fox. Most swift fox are taken incidental to coyote trapping activities and are not considered a target species.

Disease and Predation

Parasite and disease agents in wild swift fox populations have not been extensively studied; however, there is no indication that parasites or diseases are significant factors in the population dynamics of wild foxes. It is believed that swift foxes share a community of parasites and diseases with sympatric canids and have not developed a specialized suite of agents. Various disease agents have been documented serologically (e.g., sylvatic plague and canine distemper) (Pybus and Williams 2003); however, there are few cases of confirmed overt diseases in wild swift foxes. One report of canine distemper was reported in Wyoming in 1999 (T. Olson, Wyoming Cooperative Fish and Wildlife Research Unit, University of Wyoming, Laramie, personal communication).

Although predation by mammal and avian predators such as American badgers and golden eagles has been documented (Carbyn et al. 1994, Andersen et al. 2003), predation by coyotes is the most important natural mortality factor for swift fox populations in the United States and the reintroduced populations in Canada (Laurion 1988, Covell 1992, Carbyn et al. 1994, Sovada et al. 1998). The reported annual mortality rates (range from 0.47 to 0.57; Covell 1992, Sovada et al. 1998, Andersen et al. 2003) may seem high, yet they are similar to rates reported for other North American foxes (Lord 1961, Storm et al. 1976, Cypher and Scrivner 1992, Disney and Spiegel 1992, Ralls and White 1995). It has been suggested that potentially high reproductive rates may compensate for high mortality rates (Sovada et al. 1998). In many situations, control of coyotes may enhance distribution and abundance of swift fox populations. However, managers should proceed cautiously. Cypher and Scrivner (1992) evaluated reduction of coyote numbers to increase kit fox survival but were unsuccessful in reducing coyote numbers sufficiently to affect kit fox populations.

Other Natural or Man-made Factors

Predation by and interspecific competition with coyotes and expansion of red fox populations may be the 2 most serious limiting factors to swift fox recolonization of suitable habitat identified within the species' historic range. Sovada (1995), in a discussion of future research needs, stated that "significant changes in landscape (increased agriculture, lack of corridors) may result in increased risk of predation. Competition with coyote and red fox confer a likely ecological barrier for settling into new areas." Coyote killing of swift foxes significantly affected efforts to reintroduce swift foxes in Canada (Scott-Brown et al. 1986, Carbyn et al. 1994). The relationship between red foxes and swift foxes is unknown, although preliminary data analysis of an experimental study examining this relationship suggests that red foxes are a barrier to swift fox populations expanding into unoccupied, but suitable, areas (M.A. Sovada, unpublished data). Ralls and White (1995) reported that although coyote predation on kit foxes can be severe, red foxes may pose an even greater threat to kit fox populations because where red foxes rapidly moved into areas occupied by kit foxes, the red foxes appeared to displace the kit foxes. This observation of kit fox–red fox relations may serve as a model for the relations between the swift fox and red fox. Based on known interspecific relationships between other canids, the red fox may be a substantial barrier to swift fox range expansion and may be more detrimental to swift foxes than coyotes (Ralls and White 1995). Without understanding interspecific relations between the swift fox and red fox, we risk ineffective or possible failure of management programs and loss of time and funds, which may compromise the conservation of the swift fox.

Swift foxes are frequently observed along roadways. Several studies have indicated that swift foxes frequently use roadways as travel lanes and for foraging activities, and they may build dens nearby (Hines and Case 1991, Pruss 1999). These roadway associations can be major sources of vehicle-related mortality for juvenile foxes (Sovada et al. 1998). The significance of this mortality factor to the overall question of maintaining population viability has not been studied in any detail. Vehicle-caused mortality does not appear to be a significant adverse problem from a range-wide perspective. In Kansas, where road densities are fairly high, state biologists believe that factors such as road densities, distance traveled, and driver speed may increase the rate of swift fox mortalities. Kansas has for several years utilized vehicle-caused mortalities per unit time as a means to calculate population trend index. However, annual road mortalities in Kansas do not appear to be affecting distribution and status of this species (C. Roy, personal communication).

It is generally accepted that the widespread use of strychnine intended to kill wolves and coyotes was the major cause of decline in swift fox populations in the 1800s (Scott-Brown et al. 1987). Significant changes have occurred in predator-control programs since the days of indiscriminate use of strychnine and compound 1080. Federal and State predator-control methods during the 1950s replaced non-selective poisons like strychnine with more selective toxicants, contributing to a rebound of swift fox populations in the 1960s through the early 1980s. Also, the 1972 Presidential ban (Executive Order 11643) on predator toxicant use (strychnine, compound 1080, etc.) on Federal lands is considered to have been a positive factor to swift fox conservation. Further, current Federal and State predator-control programs have developed selective techniques, such as pan-tension devices, snare stops, and bait placement techniques, in an attempt to exclude swift fox from traps. The U.S. Department of Agriculture maintains records of the number of swift fox killed during Federal predator control activities, and swift foxes comprise a very small percentage of total mortalities (U.S. Department of Agriculture, Wildlife Services; unpublished data).

Legal predator control activities conducted by private individuals and landowners using leg-hold traps, snares, and shooting also occur, although the magnitude and range-wide effect of these types of measures is unknown. However, it also must be acknowledged that illegal control of predators through the use of carcasses of livestock laced with insecticides, toxicants, or banned chemicals still occurs throughout the western Great Plains. These methods are nonselective in nature and take a wide variety of wildlife species such as coyote, red fox, bald eagle (*Haliaéetus leucocéphalus*), and golden eagle, and it is believed that swift fox mortalities do occur as a result of primary or secondary contact (*see* Sovada et al. 1998). Unfortunately, the extent or magnitude of this problem relative to overall swift fox population viability is unknown but probably varies in any given year depending on the perceived level of the depredation problems.

Rodent control activities, specifically prairie dog control actions which have eliminated 98% of the historic population levels (Marsh 1984), may have had adverse impacts on swift fox populations in the northern portion of their distribution. This does not appear to be the case in Wyoming, Colorado, and in the majority of states in the southern portions of the species' range. Studies indicate that the swift fox, as an opportunistic forager, successfully adapts its diet to availability of food items, and research has documented diets high in insects, vegetation, certain agricultural crops, and a variety of small birds and mammals (Cutter 1958a, Kilgore 1969, Zumbaugh et al. 1985, Uresk and Sharps 1986, Hines and Case 1991, Kitchen et al. 1999, Sovada et al. 2001). In South Dakota, where some of the most viable black-tail prairie dog (*Cynomys ludovicianus*) populations in the country remain, a close association between the swift fox and the prairie dog has been reported (Hillman and Sharps 1978, Uresk and Sharps 1986); here the prairie dog ecosystem may have provided a stable, year-round source of food in northern portions of the species' historic distribution. In the north, where climatic conditions and the duration of temperature extremes are generally regarded as being more harsh and pervasive, the prairie dog ecosystem may have had greater importance to the long-term stability and viability of swift fox populations. However, it is apparent from the studies done by SFCT and the individual States during the past 3 to 5 years that swift fox populations in today's altered landscape are not necessarily dependent upon the availability of prairie dog towns and complexes (Cutter 1958a, Kilgore 1969, Zumbaugh et al. 1985, Hines and Case 1991, Kitchen et al. 1999). Prairie dogs were largely absent or at best rare in areas of Canada that were historically occupied by swift foxes (Banfield 1974).

Summary of Federal and State Governments' Actions in the United States

Federal Action

On 3 March 1992, the USFWS received a petition requesting that the swift fox be listed as an endangered species in the northern portion of its range, if not its entire range, in the United States. The petitioner asserted that swift foxes once occurred in abundant numbers throughout the Great Plains; however, concurrent with settlement of the region, populations declined and the species is now considered rare in the northern portion of its range. The petitioner indicated that the swift fox was extremely vulnerable to human activities such as trapping, hunting, automobiles, agricultural conversion of suitable habitat, and prey reduction from rodent control programs. The petitioner requested that, at a minimum, the swift fox be listed as an endangered species in Montana, North Dakota, South Dakota, and Nebraska. Justification for such action as cited by the petitioner includes the present status of the species and its habitat in the petitioned area, the strong link to the prairie dog ecosystem, the large distance from the kit–swift fox hybrid zone, and the potential for these populations to contain the northern subspecies of swift fox (*Vulpes velox hebes*).

The general deficiency of data on the current status and distribution within the species' historic range, limited biological information on the species, and insufficient genetic data to confirm the occurrence of both a northern and southern subspecies of the swift fox resulted in an unusually long initial review process. A 90-day finding was published (Federal Register 1994) on 1 June 1994. On the basis of the best scientific information available at the time, the USFWS found that listing the swift fox was warranted throughout its entire range, but listing was precluded by work on other species that had higher priority for listing. The notice indicated that the USFWS would conduct a status review of the swift fox and requested that any additional information on the status, population, trend, distribution, and habitat use of this species be submitted to the Field Supervisor (USFWS, Pierre, South Dakota).

A notice of a 12-month petition finding was published (Federal Register 1995) on 16 June 1995, announcing that the USFWS had determined that listing of the swift fox was warranted but precluded by higher priority actions. The species' status was elevated from a Category 2 to a Category 1 species, and a listing priority of 8 was assigned based on a perceived low- to medium-magnitude of threats and the immediacy of actual threats to the species. The USFWS estimated that the swift fox had been extirpated from 80% of its historical range, and that remaining populations existed in scattered, isolated pockets of remnant shortgrass to mid-grass prairie habitats. Seventy to 75% of remaining swift fox populations were believed to reside on private lands, with the remaining populations on Federal lands belonging to the U.S. Forest Service, the National Park Service, the Bureau of Land Management, and the Department of the Army (Federal Register 1995). The status of the swift fox was changed by the USFWS to a listing priority of 9 in September 1997 (Federal Register 1997). Additionally, processing of a final determination on the proposed rule to list the swift fox was delayed by the moratorium on final listings imposed on 10 April 1995 (Public Law 104-6), and the development of the USFWS's listing priority guidance (Federal Register 1996) to clarify the order in which they process rule makings.

State Agencies Response to Federal Actions

In response to the 90-day Administrative Petition Finding, on 1 August 1994, the Directors of the 10 state wildlife agencies (Colorado, Kansas, Montana, Nebraska, New Mexico, North Dakota, Oklahoma, South Dakota, Texas, and Wyoming) affected by the finding sent a letter to the USFWS indicating their belief that an error had been made in the 90-day finding, both in omission of key information and in a lack of rigorous analysis of the existing data. The letter identified 2 key points that were omitted from the 90-day finding: (1) both the range and the distribution of the swift fox have increased over the past 30 years, although the distribution is less than the presumed distribution prior to settlement; and (2) more than 75% of swift fox habitat is on private property, and although recognizing that land status is not a consideration for listing, a practical assessment of this situation suggests that landowner practices could have significant effects on swift fox populations. In their letter, the Directors described their commitment to developing a Conservation Strategy Plan as a cooperative venture that would pool resources and seek to develop a regional management plan.

In October 1994, the 10 affected state wildlife management agencies and interested cooperators (several Federal agencies, including the USFWS) formed the Swift Fox Conservation Team. The goal of the SFCT was to develop a Conservation Assessment and Conservation Strategy (CACS) document for the swift fox. The CACS was intended to provide the framework to direct swift fox conservation as an alternative to a federally mandated recovery effort.

On 29 December 1994, the USFWS received a draft CACS from the SFCT, which outlined short- and long-range goals, objectives, and strategies for management of the swift fox throughout its range. The CACS, similar to a long-term recovery plan with a number of prioritized objectives and strategies, is considered a working document which will be modified annually based on SFCT accomplishments and data needs. That is, objectives and strategies are prioritized and accomplishment dates are set based, in part, on the USFWS's recommendations and suggestions to the SFCT; accomplishments are reviewed and reported on an annual basis. According to the CACS, attainment of conservation strategy objectives is intended to be accomplished by 2015 if adequate funding and resources are available.

The primary objective of the SFCT during the late 1990s was to compile and collate current distribution, occurrence, and status data while attempting to develop standardized survey methodology that may be applied across the range of the swift fox (see Swift Fox Conservation Team Annual reports 1995, 1996, 1997b, 1998, 1999, and papers in 2002). In some states such as Colorado, Kansas, Montana, New Mexico, Texas, and Wyoming, biological data have been collected about natal dens, territories and habitat selection, food habits, dispersal, litter sizes, recruitment, mortality, and survey techniques.

Although only limited data are available to project short- or long-term population viability of swift foxes on a range-wide basis, recent observations indicate the presence of numerous connected populations across a large portion of the historic range. Evaluations conducted by the SFCT have demonstrated nearly continuous distribution of swift fox populations from Wyoming south throughout eastern Colorado, western Kansas, the Oklahoma panhandle, eastern New Mexico, and in 3 counties in the extreme northern panhandle of Texas. Scattered populations of swift fox can be found in Montana, South Dakota, and Nebraska. In Montana, 2 of the 3 areas where swift foxes occur are the result of either reintroduction efforts or recolonization and emigration from reintroduced populations in southern Saskatchewan and Alberta, Canada. In August 1998, The Defenders of Wildlife, the Blackfeet Indian Reservation, and the Cochrane Wildlife Reserve in Alberta, Canada, initiated a swift fox reintroduction effort within the Blackfeet Indian Reservation in Glacier County.

In 2001, the USFWS reviewed the status of swift fox populations and determined that the magnitude and immediacy of threats to the species were not such that warranted listing under the Endangered Species Act (Federal Register 2001). The ruling submitted that the continuity of populations indicates an apparent viability and vitality which demonstrates that the magnitude and immediacy of threats to the species may have been sufficiently reduced to a level that precludes the necessity of listing. However, vigilance in monitoring populations is recommended to ensure conservation of swift foxes. Recognizing the need for vigilance, state and federal agencies have reaffirmed their commitment to accomplishing the goals established in the CACS, and their support of necessary conservation actions that ensure healthy populations of the swift fox.

Literature Cited

Abel, A.H. 1939. Tabeau's narrative of Loisel's expedition to the upper Missouri. University of Oklahoma Press, Norman, Oklahoma.

Allen, J.A. 1874. Notes on the mammals of portions of Kansas, Colorado, Wyoming and Utah. Part 1. On the mammals of middle and western Kansas. Bulletin of Essex Institute 6:45–52.

Andersen, D.E., T.R. Laurion, J.R. Cary, R.S. Sikes, M.A. McLeod, and E.M. Gese. 2003. Aspects of swift fox ecology in southeastern Colorado. Pp. 139–148 *in* M.A. Sovada and L.N. Carbyn, editors. The Swift Fox: ecology and conservation of swift foxes in a changing world. Canadian Plains Research Center, University of Regina, Saskatchewan.

Andersen, K.W., and E.D. Fleharty. 1964. Additional fox records for Kansas. Transactions of the Kansas Academy of Science 67:193–194.

Anderson, S., and B.C. Nelson. 1958. Additional records of mammals of Kansas. Transactions of the Kansas Academy of Science 61:302–312.

Archer, S. 1994. Woody plant encroachment into southwestern grasslands and savannas: rates, patterns, and proximate causes. Pp. 13–68 *in* M. Vavra, W.A. Laycock and R.D. Pieper, editors. Ecological implications of livestock herbivory in the west. Society for Range Management, Denver, Colorado.

Armstrong, D. M. 1972. Distribution of mammals in Colorado. University of Kansas, Museum of Natural History, Monograph No. 3.

Asa, C.S., and C. Valdespino. 1998. Canid reproductive biology: an integration of proximate mechanisms and ultimate causes. American Zoology 38:251–259.

Audubon, J.J., and J. Bachman. 1851. The viviparous quadrupeds of North America. Volume 2. V.G. Audubon, New York, New York.

——. 1854. The quadrupeds of North America. Volume 3. V.G. Audubon, New York, New York.

Bailey, V. 1905. Biological survey of Texas. North American Fauna 25:1–222.

——. 1926. A biological survey of North Dakota. North American Fauna 49:1–226.

——. 1931. Mammals of New Mexico. North American Fauna 53:1–412.

Bailey, V., and F.M. Bailey. 1918. Wild animals of Glacier National Park. U.S. Government Printing Office, Washington, DC.

Banfield, A.W.F. 1974. The mammals of Canada. University of Toronto Press, Toronto, Ontario.

Baydack, R.K., J.H. Patterson, C.D. Rubec, A.J. Tyrchniewicz, and T.W. Weins. 1996. Management challenges for prairie grasslands in the twenty-first century. Pp. 249–259 *in* F. B. Samson and F.L. Knopf, editors. Prairie conservation: preserving North America's most endangered ecosystem. Island Press, Washington, D.C.

Bee, J.W., G.E. Glass, R.S. Hoffmann, and R.P. Patterson. 1981. Mammals in Kansas. University of Kansas, Museum of Natural History, Public Education Series Number 7.

Black, J.D. 1937. Mammals of Kansas. Kansas State Board of Agriculture, Biennial Report 35:116–217.

Blair, W.F. 1939. Faunal relationships and geographic distribution of mammals in Oklahoma. American Midland Naturalist 22:85–133.

Blus, L.J., G.R. Sherman, and J.D. Henderson. 1967. A noteworthy record of the swift fox in McPherson County, Nebraska. Journal of Mammalogy 48:471–472.

Bragg, T.B., and A.A. Steuter. 1996. Prairie ecology–the mixed prairie. Pp. 53–65 *in* F.B. Samson and F.L. Knopf, editors. Prairie conservation: preserving North America's most endangered ecosystem. Island Press, Washington, D.C.

Burroughs, R.D. 1961. The natural history of the Lewis and Clark expedition. Michigan State University, East Lansing, Michigan.

Caire, W., J.D. Tyler, B.P. Glass, and M.A. Mares. 1989. Pp. 288–291 *in* Mammals of Oklahoma. University of Oklahoma Press, Norman, Oklahoma.

Cameron, M.W. 1984. The swift fox (*Vulpes velox*) on the Pawnee National Grasslands: its food habits, population dynamics, and ecology. Thesis. University of Northern Colorado, Greeley, Colorado.

Carbyn, L.N. 1986. Some observations on the behavior of swift foxes in reintroduction programs within the Canadian prairies. Alberta Naturalist 16:37–41.

——. 1998. Updated COSEWIC status report: swift fox, *Vulpes velox*. Committee on the Status of Endangered Wildlife in Canada. Ottawa, Ontario.

Carbyn, L.N., H.J. Armbruster, and C. Mammo. 1994. Swift fox reintroduction program in Canada from 1983 to 1992. Pp. 247–271 *in* M.L. Bowles and C.J. Whelan, editors. Restoration of endangered species: conceptual issues, planning and implementation. Cambridge University Press, Cambridge, UK.

Carter, F.L. 1939. A history of the changes in population of certain mammals in western Kansas. Thesis. Fort Hayes State College, Hayes, Kansas.

Cary, M. 1902. Some general remarks upon the distribution of life in northwest Nebraska. Nebraska Ornithologists' Union Proceedings 3:63–75.

——. 1911. A biological survey of Colorado. North American Fauna 33:1–256.

Cockrum, E.L. 1952. Mammals of Kansas. University of Kansas, Museum of Natural History.

Coues, E. 1878. Field notes on birds observed in Dakota and Montana along the forty-ninth parallel during the seasons of 1873 and 1874. Article XXV. Pp. 545–661 *in* Bulletin of the U.S. Geological and Geographical Survey Volume IV. Government Printing Office, Washington D.C.

Covell, D.F. 1992. Ecology of the swift fox (*Vulpes velox*) in southeastern Colorado. Thesis. University of Wisconsin, Madison, Wisconsin.

Custer, G.A. 1875. Report of the Chief of Army Engineers. *In* Annual report of the Secretary of War. 43rd Congress, 2nd session. H.R. Executive Document, Part II (1874–1875), Appendix KK: preliminary report of reconnaissance to the Black Hills, St. Paul, Minnesota, 7 September 1984. (McFarling, L. 1955. Exploring the northern plains, 1804–1876).

Cutter, W.L. 1958a. Denning of the swift fox in northern Texas. Journal of Mammalogy 39:70–74.

——. 1958b. Food habits of the swift fox in northern Texas. Journal of Mammalogy 39:527–532.

Cypher, B.L., and J.H. Scrivner. 1992. Coyote control to protect endangered San Joaquin kit foxes at the Naval Petroleum Reserves, California. Proceedings of the Vertebrate Pest Conference 15:42–47.

Dick-Peddie, W.A. 1993. New Mexico vegetation, past, present and future. University of New Mexico, Albuquerque, New

Mexico.
Dinsmore, J.J. 1994. A country so full of game: the story of wildlife in Iowa. University of Iowa Press, Iowa City, Iowa.
Disney, M., and L.K. Spiegel. 1992. Sources and rates of San Joaquin kit fox mortality in western Kern County, California. Transactions of the Western Section of the Wildlife Society 28:73–82.
Dragoo, J.W., J.R. Choate, T.L. Yates, and T.P. O'Farrell. 1990. Evolutionary and taxonomic relationships among North American arid-land foxes. Journal of Mammalogy 71:318–332.
Dragoo, J.W., and R. Wayne. 2003. Systematics and population genetics of swift and kit foxes. Pp. 207–221 *in* M.A. Sovada and L.N. Carbyn, editors. The Swift Fox: ecology and conservation of swift foxes in a changing world. Canadian Plains Research Center, University of Regina, Saskatchewan.
Duck, L.G., and J.B. Fletcher. 1945. A survey of the game and furbearing animals of Oklahoma. Oklahoma Game and Fish Commission, Pittmann-Robertson Series No. 2, State Bulletin No. 3. Oklahoma City, Oklahoma.
Egoscue, H.J. 1979. *Vulpes velox*. Mammalian Species 122:1–5.
Federal Register. 1970. Conservation of endangered species and other fish or wildlife. Federal Register 35:8491–8498.
——. 1980. Endangered and threatened wildlife and plants, proposed endangered status for U.S. populations of five species. Federal Register 45:49844–49847.
——. 1994. Endangered and threatened wildlife and plants: 90-day finding for a petition to list the swift fox as endangered. Federal Register 59:28328–28330.
——. 1995. Endangered and threatened wildlife and plants: 12-month finding for a petition to list the swift fox as endangered. Federal Register 60:31663–31666.
——. 1996. Endangered and threatened wildlife and plants: restarting the listing program and final listing priority guidance. Federal Register 61:24722–24728.
——. 1997. Endangered and threatened wildlife and plants: review of plant and animal taxa, proposed rule. Federal Register 62:49397–49411.
——. 2001. Endangered and threatened wildlife and plants: annual notice of findings on recycled petitions. Federal Register 66:1295–1300.
Floyd, B.L., and M.R. Stromberg. 1981. New records of the swift fox (*Vulpes velox*) in Wyoming. Journal of Mammalogy 62:250–251.
Fox, L.B. 1991. Return of the swiftest fox. Kansas Department of Wildlife and Parks Publication. January–February:4–9.
Fox, L.B., and C.C. Roy. 1996. Swift fox (*Vulpes velox*) management and research in Kansas: 1995 annual report. Pp. 39–51 *in* S. Allen, J. Whitaker Hougland, and E. Dowd Stukel, editors. Annual report of the Swift Fox Conservation Team 1995. North Dakota Game and Fish Department, Bismarck, North Dakota.
Giddings, B. 1998. Swift fox management activities in Montana. Pp. 18–22 *in* C. Roy, editor. Annual report of the Swift fox Conservation Team 1998. Kansas Department of Wildlife and Parks, Emporia, Kansas.
Giddings, B., and A. Zimmerman. 1996. Distribution and investigations of swift fox in Montana. Pp. 25–29 *in* R. Luce, and F. Lindzey, editors. Annual report of the Swift Fox Conservation Team 1996. Wyoming Game and Fish Department, Lander, Wyoming.
Glass, B.P. 1956. Status of the kit fox, *Vulpes velox*, in the High Plains. Proceedings of Oklahoma Academy of Science 37:162–163.
Grinnell, G.B. 1914. The wolf hunter. Grosset and Dunlap Publishers, New York, New York.
Hall, E.R. 1955. Handbook of mammals of Kansas. University of Kansas Museum of Natural History, Lawrence, Kansas.
——. 1981. The mammals of North America. Volume 2. John Wiley and Sons, New York, New York.
Hall, E.R., and K.R. Kelson. 1959. The mammals of North America. John Wiley and Sons, New York, New York.
Harrison, R.L., and C.G. Schmitt. 1997. Swift fox investigations in New Mexico, 1997. Pp. 97–106 *in* B. Giddings, editor. Annual report of the Swift Fox Conservation Team 1997. Montana Department of Fish, Wildlife, and Parks, Helena, Montana.
Hayssen, V., A. van Tienhoven, and A. van Tienhoven. 1993. Asdell's patterns of mammalian reproduction. Cornell University Press, Ithaca, New York.
Herrero, S., C. Mamo, L.N. Carbyn, and J.M. Scott-Brown. 1991. Swift fox reintroduction into Canada. Pp. 246–252 *in* G.L. Holroyd, G. Burns, and H.C. Smith, editors. Proceedings of the second endangered species and prairie conservation workshop. Provincial Museum of Alberta, Natural History Section, Occasional Paper No. 15, Edmonton, Alberta.
Hillman, C.N., and J.C. Sharps. 1978. Return of swift fox to northern Great Plains. Proceedings of South Dakota Academy of Science 57:154–162.
Hines, T.D. 1980. An ecological study of *Vulpes velox* in Nebraska. Thesis. University of Nebraska, Lincoln, Nebraska.
Hines, T.D., and R.M. Case. 1991. Diet, home range, movements, and activity periods of swift fox in Nebraska. Prairie Naturalist 23:131–138.
Hoffmann, R.S., P.L. Wright, and F.W. Newby. 1969. The distribution of some mammals in Montana. I. Mammals other than bats. Journal of Mammalogy 50:597–604.
Hubbard, J.P. 1994. The status of the swift fox in New Mexico. Unpublished report, New Mexico Department of Game and Fish, Albuquerque, New Mexico.
Jackson, V.L., and J.R. Choate. 2000. Dens and den sites of the swift fox, Vulpes velox. Southwestern Naturalist 45:212–220.
James, E. 1823. Account of an expedition from Pittsburgh to the Rocky Mountains: performed in the years 1819 and '20, by order of Honorable J.C. Calhoun, Secretary of War; under the command of Major Stephen H. Long. Volume 1, pp 484–487. H.C. Cary and I. Lea, Philadelphia, Pennsylvania.
Johnson, D.R. 1969. Returns of the American Fur Company, 1835–1839. Journal of Mammalogy 50:836–839.
Jones, J.K., Jr. 1964. Distribution and taxonomy of mammals of Nebraska. University of Kansas Museum of Natural History, Lawrence, Kansas.
Jones, J.K., Jr., C. Jones, R.R. Hollander, and R.W. Manning. 1987. The swift fox in Texas. Unpublished report. Texas Tech University, Lubbock, Texas.
Kilgore, D.L., Jr. 1969. An ecological study of swift fox (*Vulpes velox*) in the Oklahoma Panhandle. American Midland Naturalist 81:512–534.
Kitchen, A.M. 1999. Resource partitioning between coyotes and swift foxes: space, time, and diet. Thesis. Utah State University, Logan, Utah.
Kitchen, A.M., E.M. Gese, and E.R. Schauster. 1999. Resource partitioning between coyotes and swift foxes: space, time, and diet. Canadian Journal of Zoology 77:1645–1656.
Klausz, E.E., R.W. Wein, and L.N. Carbyn. 1996. Interaction of vegetation structure and snow conditions on prey availability for swift fox in the northern mixed-grass prairies: some

hypotheses. Pp. 281–286 *in* W.D. Willms and J.F. Dormaar, editors. Proceedings of the fourth prairie conservation and endangered species workshop. Provincial Museum of Alberta, Natural History Occasional Paper 23.

Knowles, C.J., P.R. Knowles, B. Giddings, and A.R. Dood. 2003. The historic and recent status of the swift fox in Montana. Pp. 41–47 *in* M.A. Sovada and L.N. Carbyn, editors. The Swift Fox: ecology and conservation of swift foxes in a changing world. Canadian Plains Research Center, University of Regina, Saskatchewan.

Küchler, A.W. 1972. The oscillations of the mixed prairie in Kansas. Erdkunde 26:120–129.

——. 1985. Potential national vegetation. National Atlas of the United States of America. Map. U.S. Department of Interior, U.S. Geological Survey, Reston, Virginia.

Laurion, T R. 1988. Underdog. Natural History 97:66–70.

Lechleitner, R.R. 1969. Wild mammals of Colorado. Pruett Publishers, Bolder, Colorado.

Lindberg, M.S. 1986. Swift fox distribution in Wyoming: a biogeographical study. Thesis. University of Wyoming, Laramie, Wyoming.

Long, C.A. 1965. The mammals of Wyoming. University of Kansas, Museum of Natural History 14:493–758.

Lord, R.D., Jr. 1961. A population study of the gray fox. American Midland Naturalist 66:87–109.

Ludlow, W.M. 1875. Report of a reconnaissance of the Black Hills of Dakota made in the summer of 1874. Government Printing Office, Washington, D.C.

Marcy, R.B. 1854. Exploration of the Red River of Louisiana in the year 1852 (Appendix F, Zoology, pp. 200–201, Mammals). B. Tucker, Senate Printer, Washington D.C.

Marsh, R.E. 1984. Ground squirrels, prairie dogs and marmots as pests on rangeland. Pp. 195–208 *in* Proceedings of the conference for organization and practice of vertebrate pest control, Hampshire, England, Plant Protection Division, Fernherst, UK.

Martin, E.P., and G.F. Sternberg. 1955. A swift fox, *Vulpes velox velox* (Say), from western Kansas. Transactions of the Kansas Academy of Science 58:345–346.

McChesney, C.E. 1879. Report on the mammals and birds of the general region of the Bighorn River and mountains of the Montana Territory. Report of the Chief of Engineers, U.S. Army. Appendix SS3.

McGrew, J.C. 1979. *Vulpes macrotis*. Mammalian Species 123:1–6.

Mercure, A., K. Ralls, K.P. Koepfli, and R.K. Wayne. 1993. Genetic subdivisions among small canids: mitochondrial DNA differentiation of swift, kit, and arctic foxes. Evolution 47:1313–1328.

Merriam, C.H. 1888. Description of a new fox from southern California. Proceedings of the Biological Society of Washington 15:73–74.

——. 1902. Three new foxes of the kit fox and desert fox groups. Proceedings of the Biological Society of Washington 15:73–74.

Merrill, E.H., T.W. Kohley, M.E. Herdendorf, W.A. Reiners, K.L. Driese, R.W. Marrs, and S.H. Anderson. 1996. Wyoming Gap Analysis: a geographic analysis of biodiversity. Final report, Wyoming Cooperative Fish Wildlife Unit, University of Wyoming, Laramie, Wyoming.

Mohr, C.O. 1947. Table of equivalent populations of North American small mammals. American Midland Naturalist 37:223–249.

Moore, R.E., and N.S. Martin. 1980. A recent record of the swift fox (*Vulpes velox*) in Montana. Journal of Mammalogy 61:161.

Mote, K., J. Kamler, W. Ballard, and R. Gilliland. 1998. Swift fox investigation in Texas, 1998. Pp. 46–50 *in* C. Roy, editor. Annual report of the Swift Fox Conservation Team 1998. Kansas Department of Wildlife and Parks, Emporia, Kansas.

Over, W.H., and E.P. Churchill. 1941. Mammals of South Dakota. University of South Dakota, Museum and Department of Zoology. Vermillion, South Dakota.

Pfeifer, W.K., and E.A. Hibbard. 1970. A recent record of the swift fox (*Vulpes velox*) in North Dakota. Journal of Mammalogy 51:835.

Pruss, S.D. 1999. Selection of natal dens by the swift fox (*Vulpes velox*) on the Canadian prairies. Canadian Journal of Zoology 77:646–652.

Pybus, M.J., and E.S. Williams. 2003. Parasites and diseases of wild swift foxes. Pp. 231–236 *in* M.A. Sovada and L.N. Carbyn, editors. The Swift Fox: ecology and conservation of swift foxes in a changing world. Canadian Plains Research Center, University of Regina, Saskatchewan.

Ralls, K., and P.J. White. 1995. Predation on endangered San Joaquin kit foxes by larger canids. Journal of Mammalogy 76:723–729.

Rand, A.L. 1948. Mammals of the eastern Rockies and western Plains of Canada. Natural Museum of Canada Bulletin 108.

Redmond, R.L., M.M. Hart, J.C. Winne, W.A. Williams, P.C. Thornton, Z. Ma, C.M. Tobalske, M.M. Thornton, K.P. McLaughlin, T.P. Tady, F.B. Fisher, and S.W. Running. 1998. The Montana Gap Analysis Project: final report. Unpublished report. Montana Cooperative Wildlife Research Unit, University of Montana, Missoula.

Reid, R., and C.G. Gannon. 1928. Natural history notes on the journals of Alexander Henry. North Dakota Historical Quarterly 2:168–201.

Risser, P.G., E.C. Birney, H.D. Blocker, S.W. May, W.J. Parton, and J.A. Wiens. 1981. The true prairie ecosystem. Hutchinson Ross Publishing, Stroudsburg, Pennsylvania.

Roell, B.J. 1999. Demography and spatial use of swift fox (*Vulpes velox*) in northeastern Colorado. Thesis. University of Northern Colorado, Greeley, Colorado.

Rohwer, S.A., and D.L. Kilgore. 1973. Interbreeding in arid-land foxes, *Vulpes velox* and *V. macrotis*. Systematic Zoology 22:157–166.

Rongstad, O.J., T.R. Laurion, and D.E. Andersen. 1989. Ecology of swift fox on the Piñon Canyon Maneuver site, Colorado. Final report to the U.S. Army, Directorate of Engineering and Housing, Fort Carson, Colorado.

Samson, F.B., and F.L. Knopf. 1994. Prairie conservation in North America. BioScience 44:418–421.

Samson, F.B., F.L. Knopf, and W.R. Ostlie. 1998. Grasslands. Pp. 437–472 *in* M. Mac, P. Opler, C. Puckett Haecker, and P. Doran, editors. Status and trends of the nation's biological resources. Volume 2. U.S. Department of the Interior, U.S. Geological Survey, Reston, Virginia.

Schauster, E.R., E.M. Gese, and A.M. Kitchen. 2000. Population ecology of swift foxes in southeastern Colorado. P. 181 *in* Abstracts of The Wildlife Society seventh annual conference. Nashville, Tennessee. (*Abstract only*)

Schmitt, C.G. 1995. Swift fox investigation in New Mexico. Pp. 27–31 *in* S.H. Allen, J.W. Harland, and E.D. Stukel, editors. Annual report of the Swift Fox Conservation Team 1995. North Dakota Department of Game and Fish, Bismarck, North Dakota.

Schwartz, C.C., and J.E. Ellis. 1981. Feeding ecology and niche separation in some native and domestic ungulates on the shortgrass prairie. Journal of Applied Ecology 18:343–353.

Scott-Brown, J.M., S. Herrero, and C. Mamo. 1986. Monitoring of released swift foxes in Alberta and Saskatchewan. Final report 1986. Canadian Wildlife Service Report, Edmonton, Alberta.

Scott-Brown, J.M., S. Herrero, and J. Reynolds. 1987. Swift fox. Pp. 433–441 *in* M. Novak, J.A. Baker, M.E. Obbard, and B. Malloch, editors. Wild furbearer management and conservation in North America. Ontario Trappers Association. North Bay, Ontario.

Seidel, J. 1998. Swift fox investigations in Colorado. Pp. 1–3 *in* C. Roy, editor. Annual Report of the Swift Fox Conservation Team 1998. Kansas Department of Game and Parks, Emporia, Kansas.

Seton, E.T. 1909. Life-histories of northern animals: an account of the mammals of Manitoba. Volume 2.–Flesh-eaters. Charles Scribner's Sons, New York, New York.

Sharps, J.C. 1977. The northern swift fox in South Dakota. Unpublished paper presented at Furbearer Management Workshop, Rapid City, South Dakota. South Dakota Department of Game, Fish, and Parks, Pierre, South Dakota.

Soil Conservation Service. 1989. Summary report. 1987 National Resources Inventory. U.S. Department of Agriculture, Soil Conservation Service, Statistical Bulletin No. 790.

Soper, J.D. 1964. The mammals of Alberta. Hamly Press, Edmonton, Alberta.

Sovada, M.A. 1995. National Biological Survey's 1995 annual report for the Swift Fox Conservation Team. Pp. 133–137 *in* S.H. Allen, J.W. Hoagland, and E.D. Stukel, editors. Report of the Swift Fox Conservation Team — 1995. North Dakota Game and Fish Department, Bismarck, North Dakota.

Sovada, M.A., C.C. Roy, J.B. Bright, and J.R. Gillis. 1998. Causes and rates of mortality of swift foxes in western Kansas. Journal of Wildlife Management 62:1300–1306.

Sovada, M.A., C.C. Roy, and D.J. Telesco. 2001. Seasonal food habits of swift foxes in cropland and rangeland habitats in western Kansas. American Midland Naturalist 145:101–111.

Sovada, M.A., and B.K. Scheick. 1999. Preliminary report to the swift fox conservation team: historic and recent distribution of swift foxes in North America. Pp. 80–147 + appendix *in* C.G. Schmitt, editor. 1999 Annual report of the Swift Fox Conservation Team. New Mexico Department of Game and Fish, Albuquerque, New Mexico.

Sovada, M.A., C.C. Slivinski, R.O. Woodward, and M.L. Phillips. 2003. Home range, habitat use, litter size, and pup dispersal of swift fox in two distinct landscapes of western Kansas. Pp. 149–159 *in* M.A. Sovada and L.N. Carbyn, editors. The Swift Fox: ecology and conservation of swift foxes in a changing world. Canadian Plains Research Center, University of Regina, Saskatchewan.

Storm, G.L., R.D. Andrews, R.L. Phillips, R.A. Bishop, D.B. Siniff, and J.R. Tester. 1976. Morphology, reproduction, dispersal, and mortality of midwestern red fox populations. Wildlife Monographs 49.

Stromberg, M.R., and M.S. Boyce. 1986. Systematics and conservation of the swift fox *Vulpes velox* in North America. Biological Conservation 35:97–110.

Swanson, G.A., T. Surber, and T.S. Roberts. 1945. The mammals of Minnesota. Minnesota Department of Conservation, Division of Game and Fish Technical Bulletin 2:1–108.

Swift Fox Conservation Team. 1995. Annual report of the Swift Fox Conservation Team 1995. S.H. Allen, J.W. Harland, and E.D. Stukel, editors. North Dakota Game and Fish Department, Bismarck, North Dakota.

——. 1996. Annual report of the Swift Fox Conservation Team 1996. R. Luce and F. Lindzey, editors. Wyoming Game and Fish Department, Lander, Wyoming.

——. 1997a. Conservation assessment and conservation strategy for swift fox in the United States. R. Kahn, L. Fox, P. Horner, B. Giddings, and C. Roy, technical editors. Montana Fish, Wildlife, and Parks, Helena, Montana.

——. 1997b. Annual report of the Swift Fox Conservation Team 1997. B. Giddings, editor. Montana Department of Fish, Wildlife, and Parks, Helena, Montana.

——. 1998. Annual report of the Swift Fox Conservation Team 1998. C. Roy, editor. Kansas Department of Game and Parks, Emporia, Kansas.

——. 1999. Annual report of the Swift Fox Conservation Team 1999. C.G. Schmitt, editor. New Mexico Department of Game and Fish, Albuquerque, New Mexico.

Thornton, W.A., and G.C. Creel. 1975. The taxonomic status of kit foxes. Texas Journal of Science 26:127–136.

Tihen, J.A., and J.M. Sprague. 1939. Amphibians, reptiles, and mammals of Meade County State Park. Transactions of the Kansas Academy of Science 42:499–512.

U.S. Fish and Wildlife Service. 1978. Predator damage in the west: a study of coyote management alternatives. U.S. Fish and Wildlife Service Report (unnumbered), December, 1978.

Uresk, D.W., and J.C. Sharps. 1986. Denning habitat and diet of swift fox in western South Dakota. Great Basin Naturalist 46:249–253.

Van Ballenberghe, V. 1975. Recent records of the swift fox (*Vulpes velox*) in South Dakota. Journal of Mammalogy 56:525.

Visher, S.S. 1914. The biology of Harding County, northwestern South Dakota. South Dakota Geographic Survey 6:126.

Weaver, T., E.M. Payson, and D.L. Gustafson. 1996. Prairie ecology–the shortgrass prairie. Pp. 67–75 *in* F.B. Samson and F.L. Knopf, editors. Prairie conservation: preserving North America's most endangered ecosystem. Island Press, Washington, D.C.

White, P.J., and K. Ralls. 1993. Reproduction and spacing patterns of kit foxes relative to changing prey availability. Journal of Wildlife Management 57:861–867.

Woolley, T.P., F.G. Lindzey, and R. Rothwell. 1995. Swift fox surveys in Wyoming. Pp. 61–79 *in* S.H. Allen, J.W. Harland, and E.D. Stukel, editors. Annual report of the Swift Fox Conservation Team 1995. North Dakota Game and Fish Department, Bismarck, North Dakota.

Worton, G.J. 1989. Kernel methods for estimating the utilization distribution in home-range studies. Ecology 70:164–168.

Young, C.E., and C.T. Osborn. 1990. Costs and benefits of the Conservation Reserve Program. Journal of Soil and Water Conservation 45:370–373.

Young, S.P. 1944. The wolves of North America. Part 1. Their history, life habits, economic status, and control. Pp. 1–385 *in* S.P. Young and E.A. Goldman, editors, The wolves of North America, Part 1. Dover, New York, New York, American Wildlife Institute, Washington, D.C.

Zimmerman, A.L., L. Irby, and B. Giddings. 2003. The status and ecology of the swift fox in northcentral Montana. Pp. 49–59 in M.A. Sovada and L.N. Carbyn, editors. The Swift Fox: ecology and conservation of swift foxes in a changing world. Canadian Plains Research Center, University of Regina, Saskatchewan.

Zumbaugh, D.M., and J.R. Choate. 1985. Historical biogeography of foxes in Kansas. Transactions of the Kansas Academy of Science 88:1–13.

Zumbaugh, D.M., J.R. Choate, and L.B. Fox. 1985. Winter food habits of swift fox on the Central High Plains. Prairie Naturalist 17:41–47.

The Socio-Economic Context for Swift Fox Conservation in the Prairies of Canada

■ **David A. Gauthier and Daniel S. Licht**

Abstract: Successful wildlife conservation depends in part on an understanding and awareness of past, present, and future land uses and socio-economic conditions. We provide an overview of those factors in the shortgrass and mixed-grass prairies of Canada and the United States, an area generally sympatric with the historic distribution of the swift fox (Vulpes velox). *The region has experienced decades of socio-economic stagnation and decline as evidenced by a decreasing and aging population, low per capita income, idled land, and increased government succor. The result is a singular opportunity for the conservation and restoration of wildlife. Conservation options include partnerships with existing owners and land users, financial incentives, modifications to existing government programs, and establishment of large tracts of public land dedicated to wildlife conservation. Wildlife conservationists are urged to consider socio-economic conditions and to adopt transdisciplinary approaches in the development of conservation programs.*

Introduction

Land use decisions and activities in the Great Plains over the last 2 centuries have largely neglected the conservation of wildlife in both the United States (Samson and Knopf 1994, Joern and Keeler 1995, Samson and Knopf 1996, Licht 1997) and Canada (Gauthier and Henry 1989, Gauthier and Wiken 1998, 2001a). The imperiled and depauperate status of grassland ecosystems and their species is a legacy of this neglect and the dramatic anthropogenic changes on the landscape. Wildlife managers must conserve and restore impoverished flora and fauna within the current and future Great Plains socio-economic framework. Conservation programs that integrate multiple sources of information within a transdisciplinary framework have the greatest chance of success. The humanistic and socio-economic conditions and trends in the Great Plains are singular, and provide unique challenges and opportunities at both the micro and macro scales. Although the primary species of focus of this article is the swift fox (*Vulpes velox*), the issues discussed here are relevant to the conservation of all grassland flora and fauna.

Post-Columbian Settlement

The North American Great Plains extend over the widest latitudinal range of any terrestrial ecological region on the continent (CEC 1997). Licht (1997) used 2 sources of information to delineate a Great Plains of 2.08 million km^2. The shortgrass and mixed-grass region (Fig. 1), an area generally sympatric with the historic range of the swift fox, is 1.76 million km^2, 84% of which is in the United States. Prior to European settlement, the Great Plains may have been North America's richest ecosystem in terms of mammalian biomass (Licht 1997).

Friesen (1984) recognized 3 general eras of European incursion onto the Canadian prairies: exploration and fur trade, ranching, and agricultural settlement. A similar pattern occurred in the United States. The earliest European impact on the prairie ecosystem was the harvest of animals for fur. Fur company records and the journals of early trappers suggest that direct impacts to the swift fox were minor, as trappers focused on beaver (*Castor canadensis*) and other more economically valuable pelts (Rich 1967). However, indirect impacts may have occurred to the swift fox when trappers and hunters began taking large numbers of bison (*Bison bison*) and wolves (*Canis lupus*), disrupting the Great Plains ecosystem. The demise of the bison, a keystone herbivore and source of carrion, occurred around the 1880s (Geist 1996). Wolves persisted on the Great Plains for several more decades; however, they too were

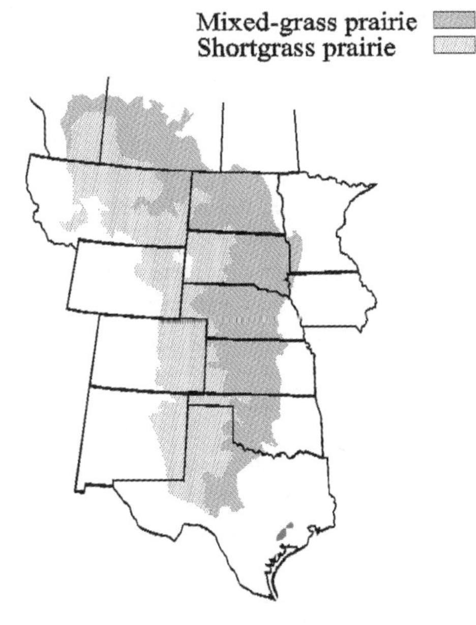

Figure 1. Shortgrass and mixed-grass prairie ecozones. Adapted from Omernik (1987) and World Wildlife Fund Canada (n.d.).

eventually eliminated, victims of an extensive campaign to benefit the ranching industry.

Ranchers followed the trappers and explorers in the commercial exploitation of the shortgrass and mixed-grass regions in both Canada (Breen 1983) and the United States (Hurt 1994). The Canadian government began leasing large tracts of land in the 1880s to individuals and companies wishing to raise cattle. The era lasted for about 25 years until an influx of farmers moved into the region between 1900 and 1921. Similarly, ranchers in the United States had access to vast tracts of "free" range for a few decades, but soon were crowded in by farmers throughout large portions of the region.

Many of the early explorers to the Great Plains had declared the land unfit for cultivated agriculture. For example, John Palliser and Henry Youle Hind both concluded that the southern prairies of Canada were unfit for settlement (Spry 1963). In the United States many explorers concurred with Edwin James' assessment that the region was "almost wholly unfit for cultivation" (Thwaites 1905, 17:147; see also Licht 1997, 10-11). However, some later expeditions into the western plains, such as one by John Macoun, countered those early assessments and claimed that the region was a fertile grassland belt with great potential for agriculture, opening the way for the settlers, railroads, and development (McDougall 1968). Settlement and cultivated agriculture was further encouraged and subsidized in the United States by government programs such as the Homestead Act of 1862 (Licht 1997). In both the United States and Canada, the introduction of widespread cultivated agriculture soon transformed much of the prairie biome into a mosaic of cropland, pasture, and farmsteads. By the 1920s, the extensive, rapid, and often unwise cultivation of the Great Plains had created conditions ripe for an ecological disaster.

Drought plays an important role in the development and maintenance of the prairie ecosystems, and the early explorers to the plains were well aware that drought impacted the region on a regular basis. Yet, the droughts of the 1910s and early 1920s left few homesteaders prepared for the dust bowl of the 1930s (Jones 1987). That period marked one of the greatest human emigrations the world has ever witnessed. In Canada over half a million people moved out of the region. In the United States, perhaps another 2 million emigrated from the plains.

The mass emigrations, dire economic and social conditions, and blowing topsoil all spurred government intervention in both countries. In Canada the Prairie Farm Rehabilitation Administration (PFRA) spearheaded a monumental effort to improve farming operations in the region (Friesen 1984). In the United States the federal government acquired 11.3 million acres of degraded Great Plains farmland between 1933 and 1946 (Wooten 1965). Some of the land was transferred to states or private interests, but much of it was retained, reseeded, and subsequently administered by the U.S. Forest Service (i.e., the National Grasslands) or Bureau of Land Management. Wooten (1965) observed that these federal land utilization programs helped reverse decades of policies encouraging settlement and cultivation on the Great Plains.

Current Conditions and Trends

Biological Diversity

Native biological diversity has been greatly reduced in the Great Plains over the past two centuries. Native grasslands have been degraded, fragmented, or eliminated, with corresponding decreases in species abundance and richness (Biodiversity Science Assessment Team 1994, Mineau et al. 1994). In the United States and Canada several species that historically existed in the Great Plains have been extirpated, including the wolf and grizzly bear (*Ursus arctos*). A much larger number of species indigenous to the region, including the bison, ferruginous hawk (*Buteo regalis*), mountain plover (*Charadrius montanus*), and elk (*Cervus elaphus*), exist in small numbers and/or only on small tracts of habitat. In the U.S. even the black-tailed prairie dog (*Cynomys ludovicianus*), which may have historically occurred on as much as 20% of the shortgrass and mixed-grass prairies, is a candidate for protection under the Endangered Species Act of 1973.

Canada has been impacted more severely than the United States in part because it had less prairie to begin with. Today, the prairie ecozone in Canada is home to comparatively high numbers of threatened and endangered wildlife. The Committee on the Status of Endangered Wildlife in Canada (COSEWIC) identified 14% of the 353 species at risk as occurring in the prairie ecozone (COSEWIC 2000).

In Canada the swift fox was listed in 1978 by COSEWIC as being extirpated. Their status was changed to endangered in 1998 following their reintroduction into an area where they were extirpated (RENEW 1999). Cultivation and industrial development of key prairie habitats are considered major impediments to the continued viability of swift fox populations in Canada (Carbyn 1998; RENEW 1999). In the U.S., although there have been several attempts to list the swift fox as endangered or threatened, it was recently removed as a candidate species (Allardyce and Sovada 2001). Despite its virtual extirpation from the northern part of the range, the animal does not have federal protection.

Swift foxes are species of shortgrass and mixed-grass prairies. Unfortunately, the amount of prairie has declined precipitously, although estimates vary. By some accounts only 25% to 30% of the prairie ecozone of Canada now consists of rangeland or grazing lands (Table 1), the majority of which occurs in southeastern Alberta and southwestern Saskatchewan. Yet prairie continues to be converted to other land types. For example, between 1971

Table 1. Estimates of rangeland in the prairie ecozone of Canada

Rangeland, grazing lands	SAMA 1987 Saskatchewan	Satellite data 1989 Prairie ecozone	Agriculture Census 1996 Prairie ecozone	Statistics Canada Census 1996 Prairie ecozone
Improved	1.8%	-	5.2%	~6%
Unimproved	24.5%	-	21.1%	~25%
Total	26.3%	24.4%	26.3%	~31%

Sources: Saskatchewan Assessment Management Agency (personal communication 1998); 1989 Satellite Data: Manitoba Center for Remote Sensing (personal communication 1989); 1996 Agriculture C ensus Prairie Ecozone (Statistics Canada 1996); 1996 Statistics Canada Census (Statistics Canada 1997b).

and 1996 the area in cropland in the prairie ecozone increased by 28% (48,720 km^2) and the area of improved pasture by 48% (7,248 km^2; Statistics Canada 1996). From 1991 to 1996 the area of unimproved lands (i.e. rangeland) decreased by 6% (Table 2).

In the United States some mixed-grass prairie states have lost >70% of their native grasslands (Samson and Knopf 1994). Shortgrass regions have been generally less affected by the plow; still, some states, such as Texas, have lost >80% of their shortgrass prairie (Samson and Knopf 1994). Cropland (harvested and unharvested) alone accounts for 51% of the mixed-grass region and 26% of the shortgrass region in 1997.

Other impacts have also diminished the biological diversity of the Great Plains. Riemer et al. (1997) found that many remnant prairies in Canada were threatened by exotic plant and brush invasion, heavy grazing, and other factors. Exotic plants such as leafy spurge (*Euphorbia pseudovirgata*), Canada thistle (*Cirsium arvense*), spotted knapweed (*Centourea cyanus*), and sweetclover (*Melilotus* sp.) threaten many remaining prairies. These species may impact swift foxes by changing the composition and abundance of prey, or by making the habitat structure (i.e., height of vegetation) less desirable. Chronic overgrazing has also degraded some native grasslands; however, the impacts on swift foxes from overgrazing may vary from harmful to beneficial (Carbyn 1989). Decades of intensive predator control programs have also impacted the region's wildlife, although many of the programs and methods have become more selective in recent years. For example, during fiscal years 1996–99 the Wildlife Services program of the U.S. Department of Agriculture reported taking an average of 20 swift foxes annually (U.S. Department of Agriculture 2001, http://www.aphis.usda. gov/ws/). Pesticides also impact grassland biodiversity, either directly or indirectly. Insecticides are most relevant to swift foxes because of their diet. In 1997, 17% of the harvested cropland in the mixed-grass and shortgrass region in the United States was sprayed with insecticides (National Agriculture Statistics Service 1999). In addition, non-cropland is also often sprayed with insecticides for purposes of controlling grasshoppers and other insects.

Roads, and more specifically, automobiles, are of special significance to swift foxes since they are a documented mortality factor (Hines and Case 1991, Sovada et al. 1998). In spite of the sparse population, there are few areas in the mixed-grass and shortgrass regions where roads are not currently an issue. In the U.S. Great Plains there may only be seven roadless areas >40,000 ha (Wild Earth 1992). The largest protected roadless area in the region may be the 25,600 ha Wilderness unit in Badlands National Park. In Canada, the Prairies exhibit the highest road density for the country as a result of the rural grid road system.

Human Demographics

In 2000 there were 8.58 million people in the 449 United States counties in the mixed-grass and shortgrass prairies (U.S. Bureau of Census 2001, http://www. census.gov/). In 1996 there were approximately 3.97 million people occupying the seven prairie ecoregions in Canada (Statistics Canada 1997a). However, in both Canada and the United States the vast majority of the prairie population lives in urban areas. For example, in the United States 56% of the population was located in urban counties that comprise only 7% of the landscape (Table 3). In Canada, the proportion of the urban population in the prairie ecozone is 81%, compared to 76% for all of Canada, a remarkable figure given that agricultural activities dominate the landscape and that urban land use occupies only

Table 2. Cropland and pastureland in the prairie ecozone of Canada, 1996

Region	Land in crops (ha)	Improved land for pasture, grazing (ha)	Unimproved land for pasture, grazing, hay (ha)
Alberta	6,894,942	1,268,064	5,183,947
Saskatchewan	14,398,651	1,233,307	5,093,601
Manitoba	4,378,025	316,535	1,454,088
Prairie Ecozone	25,671,618	2,817,906	11,731,636
Change between 1991 and 1996	+1,127,921	+285,874	-563,643

Source: Statistics Canada (1996).

Table 3. Socio-economic status of the Great Plains in the United States

Region	Rural mixed-grass counties [1]	Urban mixed-grass counties [1]	Rural short-grass counties [1]	Urban short-grass counties [1]	United States Total
Total population in 2000	2,418,317	1,379,213	1,366,760	3,418,705	281,421,906
Population per km ^{2} in 2000	3.70	41.77	1.91	54.29	28.98
Rate of population change 1960 -2000	0.90	1.41	0.99	1.98	1.57
Median age in 1990	37.1	32.0	34.4	31.6	32.9
Per capita income in 1987	$8,854	$10,333	$8,710	$11,575	$14,420
Federal payments to individuals per capita in 1989	$1,985	$1,750	$1,699	$1,611	$1,808
Percentage of persons below poverty level in 1989	14.7	12.0	18.6	12.9	13.5

[1] Counties classified as urban are those determined to be metropolitan statistical areas by the U.S. Bureau of Census: for rural mixed-grass $n = 257$; for urban mixed-grass $n = 12$; for rural shortgrass $n = 165$; and for urban shortgrass $n = 15$.
Source: U.S. Bureau of Census (1994), U.S. Bureau of Census (2001, http://www.census.gov/)

0.3% of the region (Government of Canada 1996). The view of a predominantly rural population in the prairies is a myth.

Within the rural Great Plains are some of the least populated areas of the continent. In the shortgrass and mixed-grass regions in the United States there were 229 counties (51%) in 2000 that had <2.34 people per km^2, the threshold that the 1890 census used to define "frontier" (i.e., six people per square mile; U.S. Bureau of Census 2001, http://www.census.gov/). There were 76 counties (17%) which had <0.78 people per km^2, a threshold that the 1890 census used to define "wilderness" (i.e., two people per square mile).

Great Plains rural populations have been declining since the 1930s. Consider that North Dakota was only 21% "frontier" (see definition above) in 1920; in 2000 it was 68% "frontier." Throughout the mixed-grass and shortgrass prairies, 60% of the counties lost population from 1990 to 2000. Of the 100 counties that had the greatest population loss (percentage wise) in the United States from 1990 to 2000, 74 were in the mixed-grass and shortgrass prairies, even though the region only has 14% of all United States counties. Projections for Great Plains states show modest population increases to the year 2025 (Campbell 1997); yet current trends suggest that most of the growth will occur in urban areas with rural areas experiencing population declines. In Prairie Canada, urban growth accounts for 95% of population growth (Statistics Canada 2000a). Decreases in rural populations have been the norm since 1990 and projections are for that to continue with Saskatchewan experiencing the greatest emigration from rural areas (QED Information Systems Inc. 1998). Loss of young people and an increasing proportion of people aged 65 or older are also common trends in rural Western Canada (Roach and Berdahl 2001).

Economics

The demographic decline in the rural Great Plains can be traced directly to the fortunes of the agrarian economy. Farming in the prairie ecozone has been traditionally characterized by a limited variety of crops. For example, in Canada, only 15 field crops (grain, oilseeds, and pulses) and even fewer forage crops occupy more than 95% of the cropped area (Government of Canada 1996). Beef and dairy cattle, swine, horses, chickens, and turkeys are the primary domesticated animals. The agri-food industry contributed $70 billion in goods and services or 8.8% of Canada's gross domestic product (GDP) in 1996. In the United States the mixed-grass and shortgrass region produced $146 billion of agricultural products (market value) in 1997 (National Agricultural Statistics Service 1999). Yet the agrarian economy in both countries has been depressed, and may continue to be for the foreseeable future.

The decline is most simply and cogently explained by the economics of supply and demand. Consider that in 1940 the American farmer produced enough food to feed 10.7 people; in 1980 the same farmer produced enough food to feed 75.7 people, a 707% increase (Cochrane and Runge 1992, Licht 1997). Yet during that same period the country's population increased a comparatively modest 71%. When supply exceeds demand the most marginal producers are the first to suffer: some of the most marginal agricultural producers are found in the shortgrass and mixed-grass plains.

An enormous amount of federal money and resources have gone into supporting the region's economy. From the 1960s through the 1980s, Canadian federal and provincial government payments in direct support of agriculture grew to $4 billion annually, although by the end of the 1990s those payments had been reduced to about $1 billion annually (MacGregor and McRae 2000). In 1997 67% of the farmers in the mixed-grass and shortgrass regions of the U.S. received direct payments from the federal government (1997 Census of Agriculture, http://www.nass.usda.gov/census/). Of those that received payments, they received an average of $10,650. In 1997, 45% of the farms in the shortgrass ecozone and 39% in the mixed-grass zone were—when government payments are excluded—deficit farms in that they actually lost money. When government succor is excluded and fallow land accounted for, grazing profits are similar to wheat (i.e., cultivated farming) production (Heimlich and Kula 1991). The amount of aid increases dramatically during drought periods (Licht 1997). In 1988 the United States spent $3.1

Table 4. Farm holdings in the prairie ecozone of Canada in 1996

Region	Number of farms	Area of farms (ha)	Number of individual or family holdings	Corporation (family or non-family) holdings
Alberta	40,368	15,278,434	24,155	4,867
Saskatchewan	56,995	26,569,061	40,418	4,189
Manitoba	22,249	7,045,573	13,887	1,956
Prairie ecozone	119,612	48,893,068	78,460	11,012
Change between 1991 and 1996	-2,727	-225,678	-5,793	+10,067*

* The substantial increase in corporate holdings is at least partly a function of many family farms incorporating.
Sources: Statistics Canada (1991a, 1991b, 1991c, 1996, 1997b).

billion in drought relief. Agricultural losses in Canada from the same drought amounted to $1.8 billion (Wheaton and Arthur 1989). Since droughts are a natural feature of the Prairie environment, these costs are not likely to lessen. An important consideration is that most historical droughts have been longer lasting (~10 years) and more intense than those of the 1930s. In Prairie Canada extreme droughts (e.g., 1930s and worse) occur every 60–100 years, with a 23–45% probability of occurring by 2030 (Leavitt 2001).

Much of the federal assistance to farmers comes in the form of cropland set-asides, programs aimed primarily at reducing production, but with secondary benefits such as reducing soil erosion and providing wildlife habitat. From 1930–90, set-aside programs in the United States averaged 12 million ha annually, with the amount of area increasing in recent years (Licht 1997). The Conservation Reserve Program alone retired approximately 8.1 million ha in the mixed-grass and shortgrass region in 1992. In many parts of the Great Plains the cost of set-aside contracts exceeds the value of the land and buildings on it (Licht 1997). Currently, a significant limitation of these set-aside programs is that they are typically seeded to monotypic stands of exotic grasses, many of which are taller and/or denser than native species and hence, of limited value to swift foxes and other native wildlife. In prairie Canada, PFRA initiated the Permanent Cover Program (PCP) in 1989 as part of the National Soil Conservation Program, in which land marginal for annual cultivation and cereal production was converted to long-term forages. Approximately 15,000 parcels of land totaling 518,000 ha are involved in the program. Also, PFRA's Community Pasture Program, begun in the 1930s to reclaim badly eroded areas, has returned more than 145,000 ha of poor quality cultivated lands to grass cover since 1937 and currently encompasses in excess of 900,000 ha of rangeland.

These economic factors and government programs have affected the demographics, ownership patterns, and life choices of farmers (Tables 4 and 5). For example, off-farm employment for Canadian farmers in the prairie ecozone increased from 33% in 1991 to 37% in 1996. In addition, a process of farm consolidation on the Canadian prairies is evidenced by a decline in the total number of farms between 1991 and 1996, a slight decrease in the total area of farms, and a substantial decrease in the number of individual or family holdings. Thirty-one percent of the farm operators in the mixed-grass and shortgrass region of the United States did not reside on the farms they operated in 1997, up from 27% in 1982. The principal occupation of 34% of the farmers was something other than farming, up from 26% in 1982.

Canada has experienced a decline in net farm income in all 3 prairie provinces, largely as a result of lower commodity prices for grains and oilseeds in Saskatchewan and Manitoba, and higher livestock expenses in Alberta (Statistics Canada 2000b). Off-farm employment has become an increasingly important form of economic support for farm families (Swidinsky et al. 1998) and has increased the dependency of farm families on larger rural communities. Within this context, agricultural producers have often argued that they are expected by society to privately absorb the cost for the conservation of a public good (i.e. wildlife species and habitat) at the expense of losses to crop depredation or lost opportunity costs. Faced with economic pressures and taxation policies that have traditionally favored land conversion to achieve economic production objectives, there have in the past been few incentives for landowners to maintain wildlife habitats.

In the prairie ecozone of Canada, lands under conventional tillage declined by 22% from 1991 to 1996 while lands under conservation tillage increased by 20% (Statistics Canada 1996). The reduction in area under summerfallow and the expansion of tame or seeded pasture from 1981 to 1996 has increased the availability of habitat for some species (Neave et al. 2000). Changes in policies are occurring that recognize the ecological value of lands and encourage landowners to conserve wildlife habitat, for example, through stewardship agreements, conservation easement legislation, and programs such as the North American Waterfowl Management Plan and PFRA's Permanent Cover Program (Agriculture and Agri-Food Canada 1997, Riemer 1993, Wildlife Habitat Canada 2001b).

The U.S. Farm Bill (1985), with amendments in 1990 and 1996, enhanced wildlife benefits of conservation programs (Heard et al. 2000). Under the new U.S. Farm Bill (2002), conservation programmes have been expanded and funding increased substantially. The Conservation Reserve Program (CRP) is re-authorised until 2007, maximum enrolments are expanded from 14.7 million hectares to 15.9 million hectares, and some enrolment criteria have changed. The Environmental Quality Incentive Program

Table 5. Agriculture in the United States portion of the Great Plains in 1997 and rate of change from 1982

Region	Mixed-grass region	Shortgrass region	United States total
Number of farms	156,386 (0.85)	76,710 (0.98)	1,911,859 (0.92)
Area in farms (millions of ha)	60.25 (0.99)	59.05 (0.90)	372.72 (0.97)
Value of farmland and buildings per ha [1]	$196.6 (0.54)	$126.8 (0.57)	$373.2 (1.49)
Total cropland (millions of ha)	34.24 (1.00)	19.98 (1.05)	172.46 (0.97)
Harvested cropland (millions of ha)	25.06 (0.99)	11.86 (0.95)	123.76 (1.10)
Pastureland (millions of ha)	27.44 (1.04)	45.08 (1.04)	196.42 (0.95)
Number of cattle (millions)	15.54 (1.03)	14.54 (1.25)	98.99 (1.03)
Number of farms with net losses	60,809	34,518	926,106
Federal government payments per farm that received payments	$8,957	$14,668	$7,378
Land under CRP[2] or WRP[2] (millions of ha)	2.83	3.27	11.80

[1] In real dollars (i.e., adjusted for inflation) using the Bureau of Labor Statistics consumer price index - urban (CPI-U).
[2] CRP = Conservation Reserve Program; WRP = Wetland Reserve Program.
Source: National Agriculture Statistics Service (1999).

(EQIP), Wetlands Reserve Program (WRP), Wildlife Habitat Incentive Program (WHIP) and Farmland Protection Program (FPP) are all re-authorised through to 2007. A new Grassland Reserve Program, Conservation Security Program (CSP) and other programs have been included in the FSRI Act. Much of the new spending in this Conservation Title is directed at land currently in production rather than to additional land retirement. The 2002 U.S. Farm Bill contains several programs intended to help farmers curb air, water and soil pollution, protect wildlife habitat and defend farmland from development.

Energy production is a distant second to agriculture in much of the North American Great Plains, but it is important locally. For example, in 1999 there were an estimated 104,000 producing oil and gas wells in the prairie ecozone in Canada contributing just under 5% of Canada's GDP. Although energy activities declined significantly throughout the Great Plains in the past decade, the industry is notoriously cyclical and may experience another boom in the region. Yet future booms may be more localized. As noted in a U.S. Forest Service (1989) report, most of the easily recovered supplies have been depleted in the United States. The Swift Fox National Recovery Plan in Canada identifies habitat protection guidelines for protection from energy extraction activities.

Public Lands and Lands Dedicated to Wildlife Conservation

Lands currently in federal ownership comprise about 8% of the shortgrass prairie and 2% of the mixed-grass prairie in the United States (Licht 1997). The Bureau of Land Management, U.S. Forest Service, U.S. Fish and Wildlife Service, and National Park Service administer 49%, 41%, 8%, and 2% of these lands, respectively.

However, the conservation benefits of these lands are greatly compromised by the fragmented ownership patterns. For example, the 1.4 million ha of National Grasslands have an area to perimeter ratio of 1:2.3, with a median size of 128 ha (Licht 1997). In addition, competing land uses and multiple-use missions often compromise wildlife conservation on federal/state/provincial lands. Wildlife lands of small size also often have limited value to conserving wildlife with large spatial needs, although they may serve an important ecological function as "stepping-stones" that allow organisms to move among patches. For example, the approximately 140 Wildlife Management Areas in Nebraska average only 331 ha in size. In the Canadian prairies, approximately 25,000,000 ha (or 5.3% of the prairie ecozone) are contained in >1,000 properties set aside for conservation purposes, but 37% of those lands are <1,000 ha in size (Gauthier and Wiken 2001b).

Non-government entities are playing a larger role in grassland conservation. The Nature Conservancy in the United States and The Nature Conservancy of Canada are actively conserving large tracts of shortgrass and mixed-grass prairie through acquisitions, easements, and other agreements with private landowners. Media mogul Ted Turner has acquired substantial holdings in the shortgrass and mixed-grass region in part to conduct commercial bison operations, and in part to restore the region's biological diversity (approximately 206,400 hectares in South Dakota, Nebraska, Kansas, Oklahoma, and New Mexico and an additional 208,000 hectares in the desert grasslands of New Mexico.

The Future and Opportunities for Swift Fox Conservation

Surveys in both the United States and Canada demonstrate that society has a substantial interest in nature-related activities and places a high value on wildlife (Filion et al. 1993, U.S. Fish and Wildlife Service 1993, Environment Canada 1999). A myriad of federal, state and provincial laws and policies assert that governments must manage wildlife, including the swift fox, for the benefit of present and future generations. The question is how to best do that while also meeting other public needs that may conflict with wildlife conservation.

Public lands can play a significant role in swift fox conservation. They also provide significant outdoor recreation, education, and wildlife viewing opportunities (Wiedner and Kerlinger 1990, U.S. Fish and Wildlife Service 1993). However, a serious limitation of many public lands in the shortgrass and mixed-grass prairies is that the lands are spatially fragmented. Land fragmentation leads to conflicts with neighbors, high levels of impacts via proximal land uses, and an inability to restore some species that depend on large areas. We recommend public land consolidation wherever possible. Indeed, in the United States many local units of the U.S. Forest Service and Bureau of Land Management are aggressively consolidating their lands through exchanges, acquisitions, and other means. In the long run this should facilitate wildlife conservation.

The need and singular opportunities for establishment of large tracts of public lands, or quasi-public lands, in the shortgrass and mixed-grass prairies of the United States for wildlife conservation has been argued by Wallach (1985), Popper and Popper (1987, 1994a, 1994b), Coffman et al. (1990), Licht (1997), and numerous others. These arguments were made in part on the depressed socio-economic conditions in the region, in part on government fiscal policies and programs, in part on the commercial potential of bison, and in part on conservation biology principles. Similar proposals have been made in Canada. For example, a Prairie Conservation Action Plan called for protecting ≥1 large representative area in each of the 4 major prairie ecoregions and ≥10% of the landscape in each habitat subregion (World Wildlife Fund Canada n.d.). However, conservation programming takes place in an atmosphere of socio-economic decline and uncertainty (Gauthier 1994).

The socio-economic decline of the shortgrass and mixed-grass region seems inevitable; whether that will eventually lead to more public land devoted to wildlife conservation remains to be seen. Either way, in the near future wildlife managers in the Great Plains must face the reality of small and scattered public lands within a landscape of private ownership used for commercial purposes.

On a regional scale the Great Plains are one of the most "owned" landscapes in North America. Therefore, conservation programs in the region typically involve a multiplicity of owners, including private owners and lessees, as well as rural and urban municipalities, provincial and federal governments, and a host of interest groups. Throughout the North American prairies, non-government organizations have successfully engaged private agricultural producers (Gauthier 1995). Examples of coordinated inter-agency approaches include the Prairie Conservation Action Plans of the Prairie Provinces (Manitoba Natural Resources 1998, PCAP Committee 1998, PCAP Partnership 2003, Prairie Conservation Forum 1997), the North American Waterfowl Management Plan (U. S. Department of Interior, Environment Canada, and Desarrolo Social México 1994), and Wildlife Habitat Canada's Countryside Canada program (Wildlife Habitat Canada 2001). The larger the geographic region covered, the greater the opportunity, but also the greater the complexity. Gauthier and Wiken (1998) determined that the Great Plains covered in whole or in part 20 state or provincial jurisdictions: 3 in Canada, 14 in the United States and 3 in Mexico (using a slightly larger delineation than Licht [1997]).

Neave et al. (2000) stress that the availability of wildlife habitat in agricultural landscapes is a function of many factors, including changes in land use resulting from changes in world demand, domestic policies, crop prices, availability of new crop varieties, new technology, and the growing use of conservation farming techniques. Evidence in the southern United States suggests that swift foxes can exist with some level of cultivated agriculture (Kilgore 1969, Hines 1980, Sovada et al. 1998). Wildlife managers must engage private landowners and be cognizant of socio-economic factors.

Due to depressed economic conditions, Great Plains farmers and ranchers may be receptive to government and non-government efforts to conserve wildlife if they come with financial incentives. Government payments already make up a large portion of farm income in the United States and Canada. Wildlife managers and conservationists need to take an active role in legislation and programs affecting land use decisions. Yet paradoxically, the dire economic conditions can foster resistance to wildlife conservation. Many failing farmers and ranchers blame environmental regulations and wildlife conservation for their problems. They may be hostile to efforts to restore wildlife or, more specifically, to more government involvement on their property. Efforts to engage private landowners in swift fox conservation need to consider both the opportunities and the potential pitfalls. Wildlife managers in the Great Plains will also find themselves dealing with an increasingly larger number of absentee landowners (Mortensen et al. 1989).

The human depopulation of large portions of the Great Plains seems imminent and inexorable. Yet that does not assure restoration of the ecosystem, and the processes that affect the swift fox. Coyotes are likely to remain the top carnivore, directly killing swift foxes or otherwise outcompeting them. Bison are returning to the plains, yet they are husbanded as livestock (i.e., they are not allowed to die *in situ*, providing carrion for swift foxes and other scavengers). Prairie dogs may continue to be persecuted, resulting in less food for swift foxes and fewer escape holes. Diseases such as canine distemper, plague, and mange may directly or indirectly affect swift foxes. The future of the plains ecosystem is further confounded by global climate change. Some climate forecasts predict as much as 5–7°C rise in annual surface temperature for the prairie provinces of Canada over the next 50 years, further exacerbating the region's ecological and economic woes (Brklacich et al. 1998). Ultimately, the swift fox may, in some parts of its range, serve as an indicator of ecosystem health.

Addressing wildlife and habitat conservation issues in the mixed-grass and shortgrass prairies requires the combined skills of biologists, ecologists, agronomists, political scientists, economists, climatologists, sociologists, philosophers and other disciplinary experts. They require dialogue with decision-makers in industry, business, government and First Nations and Native American tribes. They require dialogue with landowners and land users. Thus while the study of the biology and ecology of species such as the swift fox belongs in the capable hands of biologists, the future of those species and their habitats is dependent upon a much larger partnership of interests.

Acknowledgements

Thanks to Christian Thompson, Wendee Kubik, Lisa Dale-Burnett and Lorena Patino of the Canadian Plains Research Center for their assistance in helping to draw together much of the data used in this chapter.

Literature Cited

Agriculture and Agri-Food Canada. 1997. Profile of production trends and environmental issues in Canada's agriculture and agri-food sector. Agriculture and Agri-Food Canada, Ottawa, Ontario.

Allardyce, D., and M.A. Sovada. 2003. A review of the ecology, distribution, and status of swift foxes in the United States. Pp. 3–18 *in* M.A. Sovada and L.N. Carbyn, editors. Conservation of swift foxes in a changing world. Canadian Plains Research Center, University of Regina, Saskatchewan.

Biodiversity Science Assessment Team. 1994. Biodiversity in Canada—a science assessment for Environment Canada. Canadian Wildlife Service, Environment Canada, Ottawa.

Breen, D.H. 1983. The Canadian prairie west and the ranching frontier 1874–1924. University of Toronto Press, Toronto.

Brklacich, M., C. Bryant, B. Veenhof, and A. Beauchesne. 1998. Implications of global climatic change for Canadian agriculture: a review and appraisal of research from 1984 to 1997 – Volume 1: Synthesis and research needs. Pp. 219–256 *in* G. Koshida and W. Avis, editors. The Canada country study: climate impacts and adaptation: national sectoral volume. Environment Canada, Ottawa, Ontario.

Campbell, P. 1997. Population projections: states, 1995–2025. Current population reports, P25-1131. U.S. Bureau of Census, Washington D.C.

Carbyn, L.N. 1989. Status of the swift fox in Saskatchewan. Blue Jay 47:41–52.

Carbyn, L.N. 1998. Update COSEWIC status report on swift fox *(Vulpes velox)*. Endangered update 1998, Committee on the Status of Endangered Wildlife in Canada (COSEWIC) Secretariat, Canadian Wildlife Service, Environment Canada, Ottawa, Ontario, Canada.

CEC. 1997. Ecological regions of North America. Commission for Environmental Cooperation (CEC), Montreal, Quebec.

Cochrane, W.W., and C.F. Runge. 1992. Reforming farm policy. Iowa State University Press, Ames, Iowa.

Coffman, D., C. Jonkel, and, D.R. Scott. 1990. The big open: a return to grazers of the past. Western Wildlands, fall, 40–44.

COSEWIC. 2000. Canadian species at risk, May 2000. Committee on the Status of Endangered Wildlife in Canada (COSEWIC) Secretariat, Canadian Wildlife Service, Environment Canada, Ottawa, Ontario.

Environment Canada. 1999. The importance of nature to Canadians: survey highlights. Catalogue No. En47-311/1999E, Ottawa, Ontario.

Filion, F., E. Duwors, P. Boxall, P. Bouchard, R. Reid, P. Gray, A. Bath, A. Jacquemot, and G. Legare. 1993. The importance of wildlife to Canadians: highlights of the 1991 survey. Federal-provincial task force on the importance of wildlife to Canadians. Canadian Wildlife Service, Catalogue No. CW 66-103/1993E, Ottawa, Ontario.

Friesen, G. 1984. The Canadian prairies: a history. University of Toronto Press, Toronto, Ontario.

Gauthier, D.A. 1994. The buffalo commons on Canada's plains. Forum for Applied Research and Public Safety 9:118–120.

——. 1995. The sustainability of wildlife. *In* B. Mitchell, editor. Resource and environmental nanagement in Canada: addressing conflict and uncertainty, Oxford Press, Toronto, Ontario.

Gauthier, D.A., and J.D. Henry. 1989. Misunderstanding the prairies. *In* M. Hummel, editor. Canada's endangered spaces, Key Porter Press, Toronto, Ontario.

Gauthier, D.A., and E.B. Wiken. 1998. The Great Plains of North America. Parks 8:9–20.

——. 2001a. Avoiding the endangerment of species: the importance of habitats and ecosystems. *In* K. Beazley and R. Boardman, editors. Politics of the wild: Canada and endangered species, Oxford Press, Toronto, Ontario.

——. 2001b. Monitoring the conservation of grassland habitat, Prairie Ecozone, Canada. Report to the Ecological and Monitoring Assessment Network (EMAN) National Science Symposium, May 2–5, 2001, Calgary, Alberta.

Geist, V. 1996. Buffalo nation: history and legend of the North American bison. Fifth House, Saskatoon, Saskatchewan.

Government of Canada. 1996. Conserving Canada's natural legacy (the state of Canada's environment). CD-ROM. Ottawa.

Heimlich, R.E., and O.E. Kula. 1991. Economics of livestock and crop production on post-CRP lands. *In* L.A. Joyce, J.E. Mitchel, and M.D. Skold, editors. The Conservation Reserve – yesterday, today, and tomorrow, Symposium Proceedings, 14 January 1991. General Technical Report RM-203. Fort Collins, Colorado.

Hines, T.D. 1980. An ecological study of *Vulpes velox* in Nebraska. Thesis. University of Nebraska, Lincoln, Nebraska.

Hines, T.D., and R.M. Case. 1991. Diet, home range, movements, and activity periods of swift fox in Nebraska. Prairie Naturalist 23:131–38.

Hurt, R.D. 1994. American agriculture: a brief history. Iowa University Press, Ames, Iowa.

Joern, A., and K.H. Keeler (editors). 1995. The changing prairie: North American grasslands. Oxford University Press, New York, New York.

Jones, D.C. 1987. Empire of dust. University of Alberta Press, Edmonton, Alberta.

Kilgore, D.L., Jr. 1969. An ecological study of swift fox (*Vulpes velox*) in the Oklahoma Panhandle. American Midland Naturalist 81:512–534.

Leavitt, P. 2001. Sustainable agriculture in western Canada: planning for droughts using the past. NSERC Final Report, University of Regina, Saskatchewan.

Licht, D.S. 1997. Ecology and economics of the Great Plains. University of Nebraska Press, Lincoln, Nebraska.

MacGregor, R.J., and T. McRae. 2000. Driving forces affecting the environmental sustainability of agriculture. Pp. 21–28 *in* T. McRae, C.A.S. Smith, and L.J. Gregorich, editors. Environmental sustainability of Canadian agriculture: report of the Agri-Environmental Indicator Project. Research Branch, Policy Branch, Prairie Farm Rehabilitation Administration, Agriculture and Agri-Food Canada. Ottawa.

Manitoba Natural Resources. 1998. Manitoba's Prairie Conservation Action Plan, 1996–2001. Manitoba Natural Resources, Winnipeg, Manitoba.

McDougall, J.L. 1968. Canadian Pacific: a brief history. McGill University Press, Montreal, Quebec.

Mineau, P., A. Mclaughlin, C. Boutin, M. Evenden, K. Freemark, P. Kevan, G. Mcleod, and A. Tomlin. 1994. Effects of agriculture on biodiversity in Canada. Pp. 59–113 *in* Biodiversity in Canada: A Science Assessment for Environment Canada, Ottawa, Ontario.

Mortensen, T.L., F.L. Leistritz, J.A. Leitch, R.C. Coon, and B.L. Ekstrom. 1989. Landowner characteristics and the economic impact of the Conservation Reserve Program in North Dakota. Journal of Soil and Water Conservation 44:494–497.

National Agricultural Statistics Service. 1999. 1997 Census of agriculture: geographic area series. CD-ROM. U.S. Department of Agriculture, Washington D.C.

Neave, P., E. Neave, T. Weins and T. Riche. 2000. Availability of wildlife habitat on farmland. Pp. 145–156 *in* T. McRae, C.A.S. Smith and L.J. Gregorich, editors. Environmental sustainability of Canadian agriculture: report of the Agri-Environmental Indicator Project, Agriculture and Agri-Food Canada, Ottawa, Ontario.

PCAP Committee. 1998. Saskatchewan Prairie Conservation Action Plan. Canadian Plains Research Center, University of Regina, Regina, Saskatchewan.

PCAP Partnership. 2003. Saskatchewan Prairie Conservation Action Plan, 2003–2008. Canadian Plains Research Center, University of Regina, Saskatchewan.

Popper, D.E., and F.J. Popper. 1987. The Great Plains: from dust to dust. Planning 53:12–18.

——. 1994a. Great Plains: checkered past, hopeful future. Forum for Applied Research on Public Policy. Winter, 89–100.

——. 1994b. The buffalo commons: a bioregional vision of the Great Plains. Landscape Architecture 84:144.

Prairie Conservation Forum. 1997. Alberta Prairie Conservation Action Plan. Prairie Conservation Forum, Lethbridge, Alberta.

QED Information Systems Inc. 1998. Sask Trends Monitor—February 1998. Regina, Saskatchewan.

RENEW. 1999. RENEW Report Number 9, 1998–1999. Committee on the Recovery of Nationally Endangered Wildlife, Ottawa, Ontario.

Rich, E.E. 1967. The fur trade and the northwest to 1857. McClelland and Stewart, Toronto, Ontario.

Riemer, G. 1993. Agricultural policy impacts on land use decision making and options for reform. Pp. 123–132 *in* P. Rakowski and R. Massey, editors. Proceedings of the First National Wildlife Habitat Workshop.

Riemer, G., T. Harrison, L. Hall, and N. Lynn. 1997. The native prairie stewardship program. Pp. 111–116 *in* P. Jonker, J. Vandall, L. Baschak, and D. Gauthier, editors. Caring for home place—protected areas and landscape ecology. Canadian Plains Research Center, University of Regina, and University Extension Press, University of Saskatchewan, Saskatchewan.

Roach, R., and L. Berdahl. 2001. State of the west: western Canadian demographic and economic trends. Canada West Foundation, Calgary, Alberta.

Samson, F.B., and F.L. Knopf. 1994. Prairie conservation in North America. Bioscience 44:418–21.

Samson, F.B., and F.L. Knopf (editors). 1996. Prairie conservation: preserving North America's most endangered ecosystem. Island Press, Washington, D.C.

Sovada, M.A., C.C. Roy, J.B. Bright, and J.R. Gillis. 1998. Causes and rates of mortality of swift foxes in cropland and rangeland habitats in western Kansas. Journal of Wildlife Management 62:1300–1306.

Spry, I.M. 1963. The Palliser expedition. Macmillan Company of Canada, Toronto, Ontario.

Statistics Canada. 1991a. agriculture profile of Alberta, Parts 1 and 2. Catalogue Numbers 95-382 and 95-383, Statistics Canada, Ottawa, Ontario.

——. 1991b. agriculture profile of Manitoba, Parts 1 and 2. Catalogue Numbers 95-363 and 95-364, Statistics Canada, Ottawa, Ontario.

——. 1991c. agriculture profile of Saskatchewan, Parts 1 and 2. Catalogue Numbers 95-370 and 95-371, Statistics Canada, Ottawa, Ontario.

——. 1996. 1996 Census of Agriculture. Catalogue Number 93F0031XCB, Statistics Canada, Ottawa, Ontario.

——. 1997a. 1996 Census of Agriculture. Ottawa, Ontario.

—. 1997b. A national overview: population and dwelling counts. Catalogue Number 93-357-XPB, Statistics Canada, Ottawa, Ontario.

——. 2000a. Human activity and the environment 2000. Statistics Canada, Catalogue no. 11-509-XPB, Ottawa, Ontario.

——. 2000b. Net farm income and farm cash receipts: 1999 and first quarter 2000. The Daily, Thursday May 25, 2000. http://www.statcan.ca/Daily/

Swidinsky, M., W. Howard, and A. Weersink. 1998. Off-farm work by census-farm operators: an overview of structure and mobility patterns. Working Paper #38, Catalogue n. 21-601-MIE98038, Statistics Canada, Ottawa, Ontario.

Thwaites, R.G. 1905. Early western travels, 1748–1846. Volumes 14–21. Arthur H. Clark, Cleveland, Ohio.

U.S. Bureau of Census. 1994. USA counties. Washington, D.C. CD-ROM

U.S. Department of Interior, Environment Canada, and Desarrolo Social México. 1994. 1994 update to the North American waterfowl management plan: expanding the commitment. Unites states Department of Interior, Washington, D.C.

U.S. Fish and Wildlife Service. 1993. 1991 national survey of fishing, hunting and wildlife associated recreation. U.S. Government Printing Office, Washington D.C.

U.S. Forest Service. 1989. An analysis of the minerals situation in the United States, 1989–2040. General Technical Report RM-179. Fort Collins, Colorado.

Wallach, B. 1985. The return of the prairie. Landscape 28:1–5.

Wheaton, E., and L.M. Arthur. 1989. Environmental and economic impacts of the 1988 drought with emphasis on Saskatchewan and Manitoba. Volume 1 SRC No. E-2330-4-E-89. Saskatchewan Research Council, Saskatoon, Saskatchewan.

Wiedner, D., and P. Kerlinger. 1990. Economics of birding: a national survey of active birders. American Birds 4:209–213.

Wild Earth. 1992. National overview: roadless areas map. Wild Earth, special issue.

Wildlife Habitat Canada. 2001a. The status of agricultural landscapes in Canada. Wildlife Habitat Canada (WHC) Wildlife Habitat Status Series. Ottawa, Ontario. http://www.whc.org

——. 2001b. Countryside Canada Stewardship Recognition Program—2000–2001 Annual Report. Wildife Habitat Canada, Ottawa, Ontario.

Wooten, H.H. 1965. The land utilization program, 1934 to 1964: origin, development and present status. Agricultural Economy Report 85. U.S. Department of Agriculture, Economic Research Service, Washington, D.C.

World Wildlife Fund. N.d. Prairie Conservation Action Plan. World Wildlife Fund Canada, Toronto, Ontario.

A Design for Species Restoration—Development and Implementation of a Conservation Assessment and Conservation Strategy for Swift Fox in the United States

■ Eileen Dowd Stukel, Christiane Slivinski and Brian Giddings

Abstract: In 1995, the U.S. Fish and Wildlife Service concluded that federal listing of the swift fox under the Endangered Species Act was warranted but precluded by higher listing priorities. In response to this conclusion, the 10 state fish and wildlife agencies within the swift fox range and several federal wildlife or land management agencies formed the Swift Fox Conservation Team. A primary team function was to conduct a species and habitat assessment and formulate a conservation strategy document to provide the scope and framework for swift fox conservation and recovery. The Swift Fox Conservation Team's goal, as described in this document, is "to maintain or restore swift fox populations within each state to provide the spatial, genetic and demographic structure of the United States swift fox population, throughout at least 50 percent of the suitable habitat available, to ensure long-term species viability and provide species management flexibility." Since 1997, this working document has served as a guide for participating state agencies, federal cooperators, and private partners, outlining objectives, strategies, and specific activities to achieve this primary conservation goal.

In a February 22, 1992 petition, a private citizen requested that the U.S. Fish and Wildlife Service (USFWS) list the swift fox *(Vulpes velox)* as an endangered species in at least the northern portion, if not its entire range (Jon Sharps, 22 February 1992 letter to Director, U.S. Fish and Wildlife Service, Washington, D.C., RE: Petition to list the swift fox *(Vulpes velox)* as an endangered species in a portion of its range). The northern portion of its range was defined in the petition as including the states of Montana, North Dakota, South Dakota, and Nebraska. The USFWS responded on 31 January 1994 with a 90-day finding that the requested action may be warranted throughout the historical range of the swift fox (Federal Register 1994).

Directors of the 10 state fish and wildlife agencies within the swift fox range (Montana, Wyoming, North Dakota, South Dakota, Nebraska, Kansas, Colorado, New Mexico, Oklahoma, and Texas) responded to the 90-day finding in a 1 August 1994 letter to the USFWS. The state directors expressed concern over inaccurate information contained in the 90-day finding and negative impacts of a federal swift fox listing. The state directors described commitments of their respective agencies to swift fox conservation and recovery. Commitments included the development of a multi-state swift fox conservation strategy document and the formation of a swift fox technical working group to formulate and implement the conservation plan. The state directors stated their belief that this cooperative effort would successfully conserve the swift fox in the United States and that this effort would demonstrate a viable, responsible alternative to federal oversight through an Endangered Species Act (ESA) listing.

Following solicitation of public comments, which included receipt of a draft of the conservation strategy document, the USFWS issued a 12-month finding on 16 June 1995. The USFWS concluded that listing of the swift fox throughout its historical range was warranted but precluded from listing by higher listing priorities (Federal Register 1995). The USFWS assigned the species a listing priority of 8, based on a "moderate" magnitude of threats, although the immediacy of threats was considered "imminent." The assigned listing priority resulted partially from the commitments of the state fish and wildlife agency directors and the development of the draft conservation strategy document, which has since been published (Kahn et al. 1997).

The swift fox technical working group was formalized as the Swift Fox Conservation Team (SFCT). The SFCT includes program-level representatives from the 10 state fish and wildlife conservation agencies with management authority for swift fox management. Participating federal agencies include the USFWS, U.S. Geological Survey/Biological Resources Division (USGS/BRD), U.S. Department of Agriculture/Wildlife Services, and the U.S. Forest Service. As of July 1999, the National Park Service, the Bureau of Indian Affairs, and the Bureau of Land Management had not submitted representatives. The SFCT has met annually since 1994, facilitating information exchange on swift fox population status, field monitoring techniques, and research results in an open forum. SFCT activities include an emphasis on the collection and dissemination of sound scientific information.

The SFCT released the Conservation Assessment and Conservation Strategy for Swift Fox in the United States (CACS) as a template for swift fox conservation activities in the United States (Kahn et al. 1997). The overall goal of the CACS is:

to maintain or restore swift fox populations within each state to provide the spatial, genetic and demographic structure of the United States swift fox population, throughout at least 50 percent of the suitable habitat available, to ensure long-term species viability and to provide species management flexibility.

The document's 2 major components are a conservation assessment section and a conservation strategy section. Supporting information includes current national and state swift fox distribution maps, a discussion of historical range, maps of shortgrass and midgrass prairie habitats that may represent suitable swift fox habitat, and a number of research issues that need further investigation.

The conservation assessment section summarizes the recent state of knowledge on swift fox taxonomy, physical description, distribution, population status, management status, life history, habitat relationships, and an evaluation of risk factors that may threaten long-term viability of swift fox. The conservation strategy section of the document provides an outline of objectives, strategies, and activities that will contribute to meeting the above-stated goal. Eleven objectives are identified with associated strategies designed to maintain, conserve, or restore the United States swift fox population.

Strategies have been assigned to general priority ranking to address basic information needs, species and habitat conservation actions, and additional tasks related to the implementation of a successful conservation plan. Activities associated with high priority strategies have been completed or initiated by 1996. Activities associated with high, medium, or low priorities are intended to be fully or partially accomplished in 3-year (1999), 6-year (2002), and 9-year (2005) time frames, respectively (Appendix 1).

In addition to the CACS, the SFCT has produced four peer-edited annual progress reports of activities (Allen et al. 1995, Luce and Lindzey 1996, Giddings 1997, Roy 1998). Findings acquired by state agencies and other cooperators will continue to be compiled annually to provide ongoing species status information and to describe progress in achieving the objectives set forth in the CACS (Kahn et al. 1997).

The SFCT will continue to evaluate individual state progress in achieving CACS objectives, review timelines to complete strategies, and address other relevant issues. To date, three states (Kansas, Montana, and Texas) and two Canadian provinces (Alberta and Saskatchewan) have established state or provincial swift fox working groups (Objective 1). Although not all states may develop swift fox state working groups, shortgrass ecosystem working groups may be formed to address multiple species concerns, including those of the swift fox. In other instances, existing prairie ecosystem groups will include consideration of swift fox habitat and conservation needs.

Current swift fox distribution (Objective 2) has been determined in Kansas, New Mexico, Wyoming, and Colorado. Additionally, South Dakota and Montana have completed evaluation of 25% of potential swift fox range, and Oklahoma has completed 50% of its potential range evaluation. Nebraska and Montana will be initiating statewide surveys within the next 2 years.

Swift fox habitat delineation has been a discussion topic among SFCT members, since there is evidence that swift fox are not necessarily dependent upon shortgrass prairie habitat for survival. Hoagland (1997) reviewed literature on swift fox habitat use to assist SFCT efforts in identifying suitable habitats. Shortgrass prairie habitat mapping (Objective 5) has been completed in Oklahoma, Montana, Kansas, Wyoming, and on Buffalo Gap National Grassland in South Dakota. Participating agencies in Texas, New Mexico, Oklahoma, Montana, and Kansas, and Buffalo Gap National Grassland have initiated efforts to identify and delineate existing swift fox habitat within their respective jurisdictions.

In addition, SFCT representatives from the 10 participating states, with the exceptions of North Dakota, Montana, and Kansas, are involved with the management of other species considered in need of special management or conservation action (Objective 8). These species include the black-tailed prairie dog (*Cynomys ludovicianus*), burrowing owl (*Speotyto cunicularia*), mountain plover (*Charadrius montanus*), and black-footed ferret (*Mustela nigripes*).

Another area of recent SFCT activity was the formation of a subcommittee to develop guidelines for swift fox reintroductions (Objective 7). Because of uncertainty about swift fox genetic variation and incomplete understanding of limiting factors, the SFCT does not currently support swift fox reintroduction projects. However, SFCT-endorsed guidelines will provide recommendations based on the current level of knowledge.

Several products have been developed by SFCT members to promote public support for swift fox conservation (Objective 9). The first annual SFCT newsletter was published and distributed to landowners in several states. An educational video on the shortgrass prairie ecosystem was developed by Kansas and will be available before the end of 1999 for multi-state distribution. The American Zoological Association Canid Technical Advisory Group, a regular participant in SFCT meetings, has been actively involved in developing a swift fox public education program.

Many states have been involved in research projects since 1993 (Objective 10), including Colorado, Kansas, Montana, Nebraska, New Mexico, Texas, Wyoming, and the USGS/BRD Northern Prairie Wildlife Research Center. Research topics have included home range size, habitat use, mortality causes, den site characteristics and use, testing of survey methodology, swift fox juvenile fall dispersal, swift fox and coyote interactions, and swift fox and red fox interaction. Several states, including Colorado,

Kansas, Nebraska, New Mexico, South Dakota, Texas, and Wyoming, have collected blood samples during research studies for disease analysis and/or to compile DNA databases for future genetics studies.

The SFCT does not operate as a separate independent entity, and several considerations merit mention. Despite the commitments of the state fish and wildlife agencies and participating federal agencies, not all participants will fulfill their responsibilities within the same time frame. Some agencies have long-term, ongoing swift fox research studies, while others are at a beginning stage of describing their jurisdiction's swift fox distribution and/or population status. However, the CACS provides the opportunity for some flexibility in accomplishing objectives at different rates. The CACS draws on the state of scientific knowledge as of 1997, and the document will be revised as necessary to reflect updated species/habitat information and research findings.

Funding is certainly a consideration in transforming a planning document to effective, on-the-ground conservation and management activities. This cooperative effort and stated commitments of the SFCT's member agencies will obviously strengthen the likelihood that necessary funding is secured. A document such as the CACS, which describes specific activities and timetables, allows agency work plans to incorporate swift fox efforts and prepare associated budgets.

The CACS and its related cooperative efforts were presented to the USFWS as an alternative to federal listing in the form of an innovative method for states to maintain the lead management role in swift fox conservation and recovery. As of July 1999, these commitments have not been contained in a formalized, signature document, such as a Conservation Agreement, Memorandum of Agreement, or Memorandum of Understanding, partly because of a concern that such agreements do not necessarily withstand legal challenges as substitutes for federal listing under the ESA (P. Gober, USFWS, personal communication). However, following four years (1995–1998) of intensive monitoring, surveys, inventories, and research studies by cooperating state and federal agencies, the USFWS recently evaluated new evidence and concluded that swift fox populations throughout their range do not warrant listing (SFCT/USFWS conference call, 30 April 1998). The USFWS office in Pierre, South Dakota, which has the lead role in the swift fox listing evaluation, has initiated the process for proposing the removal of the species as a candidate species (D. Allardyce, USFWS, personal communication).

If the swift fox is listed despite demonstrated progress toward swift fox conservation and recovery, this cooperative effort will be reevaluated by participating state fish and wildlife agencies. Assuming this cooperative multi-agency approach is adopted in some formal or informal way as the template for swift fox conservation and recovery in the United States, we will have the ingredients of a new approach to successfully conserve and restore an indigenous species.

Literature Cited

Allen, S.H., J.W. Hoagland, and E.D. Stukel, editors. 1995. Report of the swift fox conservation team 1995. North Dakota Game and Fish Department, Bismarck, North Dakota.

Federal Register. 1994. Endangered and threatened wildlife and plants: 90-day finding for a petition to list the swift fox as endangered. Federal Register 59:28328–28330.

——. 1995. Endangered and threatened wildlife and plants: 12-month finding for a petition to list the swift fox as endangered. Federal Register 60:31663–31666.

Giddings, B., editor. 1997. Swift fox conservation team 1997 annual report. Montana Department of Fish, Wildlife and Parks, Helena, Montana.

Hoagland, J.W. 1997. A review of literature related to swift fox habitat use. Pp. 113–125 *in* B. Giddings, editor. Swift fox conservation team 1997 annual report. Montana Department of Fish, Wildlife and Parks, Helena, Montana.

Kahn, R., L. Fox, P. Horner, B. Giddings, and C. Roy, editors. 1997. Conservation assessment and conservation strategy for swift fox in the United States. Montana Department of Fish, Wildlife and Parks, Helena, Montana.

Luce, B. and F. Lindzey, editors. 1996. Annual report of the swift fox conservation team 1996. Wyoming Game and Fish Department, Lander, Wyoming.

Roy, C., editor. 1998. 1998 Annual report of the swift fox conservation team. Kansas Department of Wildlife and Parks, Emporia, Kansas.

Appendix 1
Conservation Strategy Objectives and Associated Activities

	Strategies		Activities	Priority
3.1	Develop and implement statewide monitoring programs that provide population trend information and detect changes in local distribution.	3.1.1	Develop standardized population monitoring techniques.	High
		3.1.2	Coordinate and implement monitoring programs in each state.	
		3.1.3	Implement pelt tagging program in states with legal harvest.	
5.1	Develop swift fox habitat criteria.	5.1.1	Incorporate research results and literature findings in criteria development and employ habitat criteria in defining suitable habitat and in identifying current habitat availability.	High
5.2	Identify and delineate existing suitable swift fox habitat within each state.	5.2.1	States will conduct habitat inventories in cooperation with public and private landowners.	High
		5.2.2	States will delineate habitat on state cover maps using available geo-referenced data systems.	
8.1	Provide distribution and habitat information to other prairie ecosystem efforts through geo-referenced data systems.	8.1.1	Coordinate information exchange with other prairie species working groups, agencies, universities, and conservation organizations.	High
6.1	Identify and delineate public lands with occupied or suitable swift fox habitat.	6.1.1	Evaluate current protection levels for native grassland in federal or state ownership.	Medium
		6.1.2	Initiate habitat protection agreements with land management agencies.	
		6.1.3	Identify habitat corridors and areas between swift fox habitat blocks.	
6.2	Identify and delineate public lands with occupied or suitable swift fox habitat.	6.2.1	Utilize current or develop new land protection programs, including private land partnerships, conservation easements, land exchanges and acquisitions.	Medium
		6.2.2	Cooperate with private landowners to influence range management practices.	
7.2	Monitor and identify new, continued or diminished factors threatening swift fox population expansion.	7.2.1	Review scientific literature on interspecific competition and applicable control methods.	Medium
		7.2.2	Address the adaptability of swift fox to colonize nonnative habitats.	
		7.2.3	Evaluate new, continuing or diminishing threats.	
9.1	Provide scientific basis for swift fox management and an avenue for exchange of technical information.	9.1.1	Collect and compile technical information for distribution to interested individuals and agencies.	Medium
		9.1.2	Provide swift fox management recommendations to government agencies, land managers and planners.	
		9.1.3	Determine need for joint publication promoting prairie species conservation for distribution to wildlife and land managers.	
4.1	Conduct research to determine minimum viable population estimates, monitor genetic diversity and resolve species taxonomic issues.	4.1.1	Identify studies that investigate minimum population viability.	Low
		4.1.2	Form technical committee to resolve taxonomic issues and investigate genetic integrity of U.S. swift fox population.	
		4.1.3	Conduct testing of genetic variation among state populations.	
7.1	Expand distribution of existing populations and restore swift fox to unoccupied suitable habitats.	7.1.1	Develop state criteria and priority areas for restoration to unoccupied suitable habitat.	Low
		7.1.2	Provide recommendation to state working groups.	
		7.1.3	Form technical committee to investigate and review availability of wild/captive foxes and evaluate potential for successful releases.	
8.2	Revise Conservation Assessment and Conservation Strategy document to incorporate related objectives, strategies, or activities of other prairie species conservation plans.	8.2.1	SFCT will update or revise CACS to incorporate new or changing information accordingly.	Low
9.2	Promote public support for swift fox conservation activities.	9.2.1	Develop informational and educational materials with specific target audiences.	Low
		9.2.2	Develop state specific educational materials.	
		9.2.3	Develop informational package and education initiative for private landowners to address habitat and management needs.	
10.1	Investigate biological and ecological parameters of swift fox.	10.1.1	Form technical committee to develop and guide research priorities.	Low
		10.1.2	Evaluate site specific species and habitat needs.	
		10.1.3	Investigate species susceptibility to common diseases and parasites.	
11.1	Develop criteria for removal of swift fox from candidate species list.	11.1.1	Evaluate current species and habitat information with developed criteria for removal of swift fox from candidate species list.	Low
11.2	Develop state-based long-term management plans for swift fox.	11.2.1	Develop management guidelines with species and habitat conservation measures to assure species persistence.	Low

Canada's Experimental Reintroduction of Swift Foxes into an Altered Ecosystem

■ Stephen Herrero

Abstract: Fur trapping records show that swift foxes were once abundant as part of Canada's prairie landscape. Some of the many interacting circumstances that would have influenced, and to a large extent still do influence swift fox numbers are: winter severity; drought; disease; competition and predation by species such as coyotes, bobcats and eagles; and fluctuations in prey availability. Today, Canada's reintroduced swift fox population exists in an environment dramatically different from the historic one. Most of the primary production of the prairie goes to human beings; foods available for swift foxes have declined dramatically. Competitors, especially coyotes, are relatively more abundant, and escape terrain is compromised by a paucity of holes created by ground dwelling rodents and badgers. Whether or not a niche exists for swift foxes in this altered environment can only be determined by monitoring the fate of reintroduced individuals and populations. Because of the complex set of interacting factors that historically influenced swift fox populations, the answer cannot be strictly predicted. Some of the critical parameters that controlled swift fox numbers in historic environments should influence the design of the reintroduction effort. Factors to consider would be providing escape terrain through locally encouraging ground squirrel and badger populations, suppressing competitors during initial release of swift foxes, anticipating disease outbreaks and their consequences, and providing for linked metapopulation structure. There is little doubt that swift foxes would do best in a prairie environment in which key elements are as similar as possible to historic conditions. This might include bison and elk populations generating carrion, wolves to depress currently high coyote populations, and an abundance of ground nesting birds and insects. Given that many facets of the historic prairie environment will not be recreated, the swift foxes reintroduction will remain an experiment in ecological restoration and a comment on today's prairie as a natural system.

Planning for Canada's swift fox (*Vulpes velox*) reintroduction program began in 1975, and the first releases occurred during 1983 (Herrero 1984, Russell and Scotter 1984, Herrero et al. 1986, Herrero et al. 1991). The project was a conjoint effort of Alberta Fish and Wildlife, Canadian Wildlife Service, Cochrane Wildlife Reserve, and the University of Calgary. While many individuals were involved, Miles and Beryl Smeeton of the Cochrane Wildlife Reserve initiated the effort that led to the idea of reintroduction by importing four captive swift foxes into Canada in 1973. A series of graduate students, research associates, and I began working with the Smeeton's during 1975 to propose and plan reintroduction into the wild (Carlington 1980, Reynolds 1983, Schroeder 1985, Scott-Brown et al. 1987, Pruss 1994). By 1979 the Canadian Wildlife Service and Alberta Fish and Wildlife were part of the project. Since the first releases in 1983, the Canadian wild population has now grown to an estimated 289 foxes (95% CI=179-412; Cotterill 1997).

Historically swift foxes in Canada were distributed throughout the mid-grass prairie ecoregion (Carlington 1980). Here their range extended from the Pembina Hills in Manitoba across southern Saskatchewan to the southern foothills of the Rocky Mountains (Fig. 1). Hudson's Bay Company fur trading records suggest the species' former abundance in Canada. Between 1853 and 1877 an average of 4,876 pelts were sold annually (MacFarlane 1905, as cited in Rand 1948).

Herrero et al. (1991) proposed that the demise of swift foxes in Canada began with the extirpation of free-ranging herds of bison and the associated collapse of the prairie ecosystem as it had evolved over 10,000 years since the retreat of Wisconsin glaciation. Poisoning, trapping, and

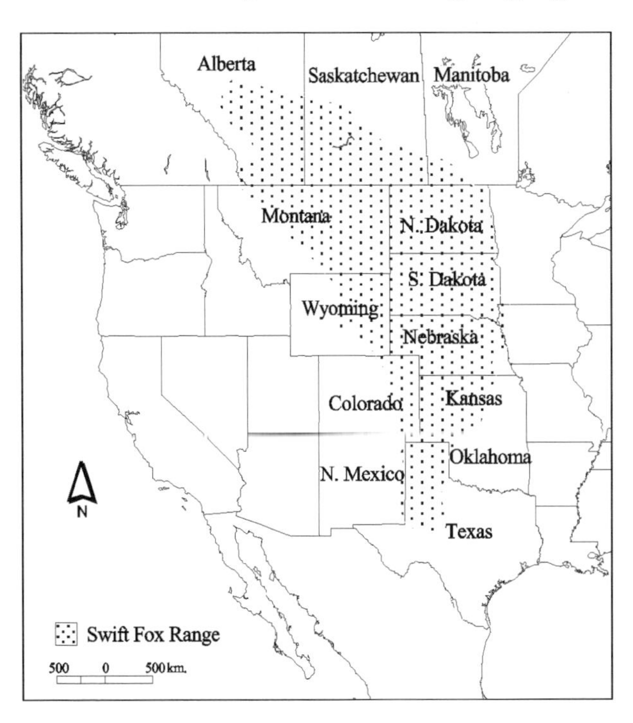

Figure 1. Historic range of the swift fox

shooting of foxes; and massive changes in food availability and competitors, further contributed to the demise of swift foxes (Russell and Scotter 1984). Between 1922 and 1925, an annual average of only 508 swift fox pelts were taken on the Canadian prairie (Statistics Canada Cat.#23207, as cited in Carlington 1980). In Canada, the last recorded swift fox was taken near Govenlock, Saskatchewan, in 1928, and the last generally accepted Canadian sighting was near Manyberries, Alberta, in 1938 (Soper 1964). Despite the extirpation of swift foxes from Canada, the species survived in much of its southern range in the southcentral plains region of the United States.

In this paper I focus on the altered features of the current Canadian prairie environment that affect swift foxes' survivorship, reproduction, and persistence. My proposed conceptual model suggests that historically swift foxes in Canada lived in an environment dramatically different from today. Because of this, and the many interacting factors affecting swift fox population persistence, the outcome of the reintroduction effort cannot be strictly predicted. I propose conditions and management actions that might increase the probability of reintroduced population persistence.

The Prairie Environment: Changes and Constants

The prairie environment of North America has undergone a catastrophic change since the time when the bison was the key grazer. The prairie is the most altered ecosystem in Canada (WWF Canada 1989). Native species have been extirpated from much of their former range and replaced by cultivars (WWF Canada 1989). Most of the primary productivity now goes to support human beings. Landscape design has changed from one in which wide-ranging species could move freely, to one where barriers and filters to movement are extensive. Natural processes such as fire occur less often than during the time of the bison. All of the foregoing and many other changes are interactive, and to reduce their influence on a species such as the swift fox into categories necessarily de-emphasizes the interactions that are a critical part of any system.

What follows is not scientifically rigorous data, but rather a conceptual model evolved from field experience, diverse literature, and data. Each point, except some of the historical information, could be quantified and scientifically tested.

Food

I propose that the food base available to swift foxes has significantly constricted since the elimination of bison and their ecosystem about 120 years ago. When bison were the dominant herbivores on the prairie they would have provided periodic scavenging opportunities for swift foxes, similar to those provided to arctic foxes by polar bear kills. The millions of bison that occupied the Canadian prairie probably were a significant food source for many swift fox populations. The availability of bison carrion would have been aperiodic because of the wide-ranging movements of historic bison herds. When free-ranging herds of bison were extirpated from the Canadian prairie during the 1880s, the number of swift foxes trapped declined precipitously, suggesting a collapse of the swift fox population (Fig. 2).

Other species that were likely killed or scavenged by swift foxes have also dramatically changed in abundance (e.g., elk, deer, ground squirrels, prairie dogs, many species of ground nesting birds, grasshoppers, beetles and other insects, rabbits, and hares). With the preponderance of primary productivity going to human beings, the abundance of forage species and biomass available for swift foxes has been dramatically reduced.

Security and Competition

Swift foxes typically elude potential predators by escaping into the nearest hole that is large enough for them, but not suitable for the pursuing predator. While swift foxes may dig or enlarge dens, they typically do not dig escape holes. The historic prairie would have had many more ground dwelling rodents such as ground squirrels or prairie dogs. One of their specialized predators, the badger, digs holes that swift foxes use to escape potential predators. Both ground-dwelling rodents and badgers are generally undesirable for farming and ranching operations and these animals are suppressed or eliminated (Herrero, unpublished observations). This has probably resulted in major loss of escape terrain for swift foxes which may have significantly amplified the impact of predation (more often interspecific killing) on swift foxes.

Since settlement there has been a dramatic increase in one of the swift foxes' primary predators, the coyote (Dekker 1989, 1997). Prior to settlement, wolves were the most numerous canid predator. Dekker (1997) suggested that before European settlers arrived, the coyote was "largely, perhaps even totally," absent from the Canadian prairie. Dekker cites explorer Anthony Henday, who saw so many gray wolves and bison that he stated it was difficult to tell which were more numerous. He also referred to "little grey foxes" (swift foxes) as far north as central Saskatchewan. Henday did not refer to the coyote. Nor did explorer David Thompson, who spent years exploring Alberta's foothills and mountains (as cited in Dekker 1997). According to Dekker, Thompson accurately described the color phases of other wild canids. Dekker also noted that the earliest fur records for Alberta included pelts of gray wolves, red foxes, and swift foxes, but not coyotes. I conclude that the coyote, if present, was much less numerous than today.

The reason is competitive interactions with wolves (Mech 1966, Allen 1979, Carbyn 1982). Shortly after wolves colonized Isle Royale, the coyote was extirpated, but red foxes survived (Allen 1979). Preliminary data from Yellowstone National Park suggest that within two

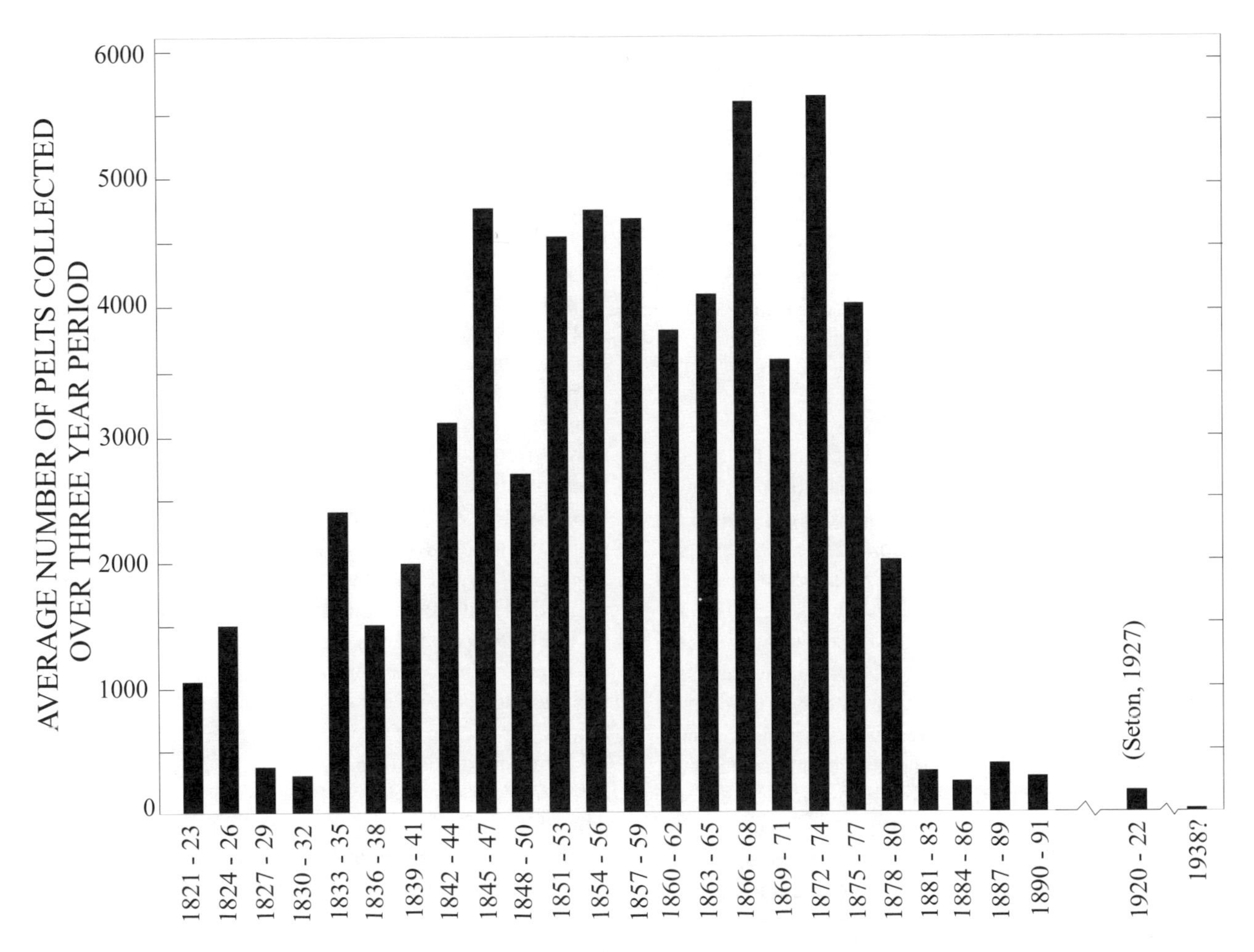

Figure 2. Mean number of swift fox pelts collected per three year intervals, 1821-1938 (Data from Hudson Bay Company Archives, Winnipeg, Manitoba, Canada)

years of the wolf reintroduction the coyote population was reduced to less than half its number pre-reintroduction. Subsequently coyote numbers have recovered somewhat (R. Crabtree, Yellowstone Ecosystem Studies Institute, personal communication). In the absence of wolves, coyote populations expand and thrive in many environments.

Research in Canada and the United States, shows that killing by coyotes is the primary cause of death of radiocollared swift foxes (Herrero et al. 1991, Moehrenschlager et al. 2003, Sovada et al. 2003). Eagles, bobcats, red foxes, and badgers also prey on swift foxes (Herrero et al. 1991, Moehrenschlager et al. 2003).

Coyotes and swift foxes coexist in more southern portions of swift fox range. This suggests possible coexistence in Canada. There are differences however. Winter severity is probably more stressful in Canada. The northern prairie environment is less productive, forcing swift foxes to forage over larger areas and increasing their vulnerability to competitors. In southern swift fox range the species was never extirpated. Thus they occupy a landscape where cultural knowledge of escape holes by swift foxes may have existed for centuries or millennia. At the extreme south of swift fox range, in Mexico, escape holes are significantly more numerous than in Canada (Moehrenschlager et al. 2003).

Climate

Global warming could turn occupied swift fox range in Canada into a more arid, less productive environment, thus further stressing food resources and increasing predation risk for swift foxes. Herrero et al. (1991) reported population declines for swift foxes during a severe, multi-year drought. Cypher et al. (2003) showed that precipitation from three previous years was positively correlated with small mammalian prey availability and kit fox population growth. Up to a point additional precipitation probably increases the number of small mammals available for swift foxes. Larger amounts of precipitation could alter climax vegetation so that it was not favorable to swift fox.

Diseases

Canids are known to be susceptible to a variety of diseases and parasites that may result in significant population declines: canine distemper, bubonic plague, and canine parvo, as well as various parasites, are among the better known (Pybus and Williams 2003). The variety of major diseases that would have affected and still do affect swift foxes is another potential stressor that could interact with other factors and cause loss in several Canadian populations. Transmission of diseases or parasites from domestic dogs could become a concern, as it is for the nearly extirpated population of wild dogs in the Serengeti of eastern Africa (D. Macdonald, Oxford University, personal communication). Caution is needed in inferring the role of various diseases or parasites as Pybus and Williams (2003) found "no evidence that parasites or diseases are significant factors in population dynamics of wild swift foxes."

Landscape/Habitat/Socio-political Environment

Before Europeans brought the concept and creation of private land and agriculture to Canada, the prairie was unfenced, unplowed, and roadless. Most areas that likely once were contiguous swift fox habitat have been converted to cultivated fields or roads. The presence of roads and associated motor vehicles has been shown to be an important source of mortality for swift foxes reintroduced into Canada. This is a difficult source of mortality to manage.

Today's Canadian prairie offers a matrix of agricultural land poorly suited for swift foxes, interspersed with scattered suitable habitat patches (Herrero, unpublished data). The suitable habitat that remains in large patches may provide the habitat base for a metapopulation structure that could protect swift foxes from extirpation due to disease and other interacting stressors. Metapopulation structure requires that if a population goes extinct the area it occupied must be accessible to dispersing individuals from other populations. Distances and conditions must be conducive for recolonization to occur.

Viable Populations

The long-term goal of swift fox reintroduction has been to develop populations as large as possible in areas of assumed suitable habitat (Herrero et al. 1991), the principle being to reach a sufficient number to form a viable population. The more swift foxes, and the more time they spend in an area of potentially suitable habitat, the more likely they are to persist and develop a "culture" that fully utilizes local food, security and other resources.

To date, swift fox population viability models have not been published because required information about population parameters are not available. Swift fox populations have significant growth potential because of the species' relatively high reproductive rate (Scott-Brown et al. 1987). In Canada, growth potential is dampened by a number of factors. I have identified predation, diseases, winter severity, food stress and other environmental stochastic events that can interact to cause major swift fox population declines, probably even leading to local extirpation of populations of several hundred individuals. Because of the difficulty of modeling relevant parameters and their interactions, models of viable swift fox populations would have many assumptions, thus reducing their precision and utility. Swift fox population viability in Canada will probably be determined by long-term monitoring of reintroduced populations.

Conclusions

I have presented data and logic that suggest the current environment of reintroduced swift foxes in Canada is drastically different from historic conditions during the time of free-ranging bison herds and the complement of associated native species of plants and animals. As a result of these changes, a host of important and interacting variables are now different: food, security and competition, climate, landscape and even major ecological processes such as fire. This raises the fundamental question of whether a niche exists for swift foxes in this highly modified environment. A superficial consideration of population viability models and their assumptions suggests that they would be built on so many interrelated and difficult-to-predict assumptions that swift fox population viability in Canada will best be determined by monitoring reintroduced populations.

While prediction of population viability may be difficult, based on my analysis, I propose a number of conditions that could increase the odds of forming long-term viable Canadian populations of swift foxes. Each factor would require substantial research before being considered for implementation.

Restoration of Native Prairie Conditions

With the number of people in the world dependent upon the primary production of the Canadian and United States prairie regions it is almost inconceivable to totally restore North American prairie to pre-European conditions. However, large areas of agriculturally marginally productive prairie exist both in Canada and the United States. Popper and Popper (1994) have proposed that these lands with marginal economic value might best serve society as a "buffalo commons" where native species and processes would be restored. Economically, revenues would be derived from tourism, similar to the Serengeti Plains of Africa. Obviously this would require major changes in current human social and land use tenure systems, something people don't often do willingly. Biologically, the restoration would be equally challenging.

The restoration would likely come in economically marginal lands such as very arid portions of mid- and short-grass prairie, the same habitat preferred by swift foxes. Biologically, this is the Canadian environment in which swift foxes would have the best long-term chances. Quantities of ungulate biomass in the form of bison and elk carrion, suitable for swift fox to scavenge, would be relatively high. Predation pressure from one primary competitor, the coyote, would be low since wolves would keep coyote numbers low. Red fox numbers might rise and this species could, in certain habitats, out-compete swift foxes. Historically, there was a niche for swift foxes on the Canadian prairie. It follows that by recreating these conditions we would recreate suitable habitat for swift foxes.

Immediate Post-release Competitor Suppression to Allow for Better Swift Fox Establishment

The idea behind suppressing numbers of competing and predatory species is to allow a reintroduced species to get established in a new area. This strategy is being used in reintroducing the endangered black-footed ferret in the United States (T. Thorne, Wyoming Fish and Game, personal communication). For a variety of reasons, this strategy has not yet been used in Canada as part of the swift fox reintroduction. It is feasible, but potentially expensive, and possibly socially undesirable to temporarily significantly depress populations of coyotes and possibly red foxes, eagles, and bobcats in swift fox reintroduction sites. For coyotes this happens *de facto* when the price of coyote pelts is high, but the harvest is not consistent over time. Also, coyote suppression alone may lead to expansion of red foxes that are another competitor that may kill or out-compete swift foxes (Cypher et al. 2003). The advantage I see from consistent competitor suppression over perhaps ten years would be better opportunity for reintroduced swift foxes populations to more rapidly expand and then occupy a landscape, thus developing knowledge of food resources and escape terrain. At some point competitor suppression would cease and competitor numbers would quickly increase. Swift fox numbers would decrease, but swift foxes would have had an enhanced opportunity to establish themselves.

Badger and Ground Squirrel Enhancement

This may sound like a nightmare for cattlemen, but increased numbers of badgers and ground squirrels probably would enhance swift fox recovery. While primarily crepuscular and nocturnal (Pruss 1994), swift foxes do include some of the diurnal ground dwelling rodents in their diet. However, the main reason I propose ground squirrel enhancement is to encourage badger numbers and range in a given area. Since the early 1980s one of the important criteria for choosing Canadian swift fox reintroduction sites has been the population density of badgers (Herrero et al. 1991). The Alberta/Saskatchewan border area, where the largest reintroduced Canadian swift fox population is located, was chosen partly with this criterion in mind. Badger holes offer escape for swift foxes when pursued by coyotes and other "predators." Badger enhancement is proposed to increase habitat security for swift foxes. Moehrenschlager et al. (2003) demonstrated the relative paucity of escape holes in re-occupied swift fox habitat in Canada versus occupied swift fox habitat in Mexico.

Linked Metapopulations

The current swift fox recovery plan calls for the establishment of two geographically distinct, but genetically connected core populations (Swift Fox Recovery Team 1996). Given the current population locations and the status of intervening land it is most likely that these populations will remain distinct, and any genetic connection would come from translocation between populations.

Given the probability of the many environmental stochastic events at some point causing at least local population extirpation, consideration could be given to establishing a natural, terrain-linked metapopulation structure. Metapopulation structure occurs when local populations are linked by dispersal corridors. Most of the time there is restricted gene flow along these corridors. However, when one population goes to zero or very low levels, then surrounding populations become sources for recolonization. To implement this type of population structure would require major landscape planning and management. Lacking this, people would have to intervene to create dispersal by translocation.

Acknowledgments

Swift fox reintroduction began because of the encouragement of Miles and Beryl Smeeton of the Cochrane Wildlife Reserve. Five graduate students contributed their understanding, commitment and hard work: Miles Scott-Brown, Bernie Carlington, Shelley Pruss, Jo Anne Reynolds, and Curt Schroeder. The Canadian Wildlife Service was a major partner in the research. My thanks to Lu Carbyn, Dick Russell, and George Scotter. Steve Brechtel, Alberta Environment, helped guide the project for many years. Charles Mamo contributed greatly to the field work. Leonard and Mary-Jane Piotrowski kindly let us use their Lost River Ranch as a base. There were various funders, but the World Wildlife Fund, Canada, was the principal supporter.

Literature Cited

Allen, D.L. 1979. Wolves of Minong—their vital role in a wild community. Houghton Mifflin, Boston.

Carbyn, L.N. 1982. Coyote population fluctuations and spatial distribution in relation to wolf territories in Riding Mountain National Park, Manitoba. Canadian Field-Naturalist 96:176–183.

Carlington, B.G. 1980. The reintroduction of the swift fox, *Vulpes velox*, to the Canadian prairies. Master of Environmental Design Project, University of Calgary, Calgary, Alberta.

Cotterill, S.E. 1997. Population census of the swift fox (*Vulpes velox*) in Canada: winter 1996–1997. Prepared for the Swift Fox National Recovery Team. Alberta Environmental Protection, Natural Resources Service, Wildlife Management Division, Alberta.

Cypher, Brian L., Patrick A. Kelly and Daniel F. Williams. 2003. Factors influencing populations of endangered San Joaquin kit foxes: implications for conservation and recovery. Pp. 125–137 *in* M.A. Sovada and L.N. Carbyn, editors. The Swift Fox: ecology and conservation of swift foxes in a changing world. Canadian Plains Research Center, University of Regina, Saskatchewan.

Dekker, D. 1989. Population fluctuations and spatial relationships among wolves, *Canis lupus*, coyotes, *Canis latrans*, and red foxes, *Vulpes vulpes*, in Jasper National Park, Alberta. Canadian Field-Naturalist 103:261–264.

——. 1997. The canid equation. Pica: The Calgary Field-Naturalists Society 17:9–13.

Herrero, S. 1984. Swift fox once again. Alberta Naturalist 14: 29–32.

Herrero, S., C. Schroeder, and M. Scott-Brown. 1986. Are Canadian foxes swift enough? Biological Conservation 36:159–167.

Herrero, S, C. Mamo, L.N. Carbyn, and M. Scott-Brown. 1991. Swift fox reintroduction into Canada. Pp. 246–252 *in* G.L. Holroyd, G. Bruns, and H.C. Smith, editors. Proceedings of the second endangered species and prairie conservation workshop. The Provincial Museum of Alberta, Edmonton, Alberta.

Mech, L.D. 1966. The wolves of Isle Royale. USDI National Parks Service, Fauna Series 7. 210 pp.

Moehrenschlager, Axel, David W. Macdonald and Cynthia Moehrenschlager. 2003. Reducing capture-related injuries and radio-collaring effects on swift foxes *(Vulpes velox)*. Pp. 107–113 *in* M.A. Sovada and L.N. Carbyn, editors. The Swift Fox: ecology and conservation of swift foxes in a changing world. Canadian Plains Research Center, University of Regina, Saskatchewan.

Popper, D.E. and F.J. Popper. 1994. The buffalo commons: A bioregional vision of the Great Plains. Landscape Architecture 84(4):144.

Pruss, S.D. 1994. An observational study of wild swift fox (*Vulpes velox*) on the Canadian prairie. Master of Environmental Design Project, University of Calgary, Calgary, Alberta.

Pybus, Margo J., and Elizabeth S. Williams. 2003. A review of parasites and diseases of wild swift fox. Pp. 231–236 *in* M.A. Sovada and L.N. Carbyn, editors. The Swift Fox: ecology and conservation of swift foxes in a changing world. Canadian Plains Research Center, University of Regina, Saskatchewan.

Rand, A.L. 1948. Mammals of the eastern Rockies and western plains of Canada. National Museum of Canada Bulletin 108.

Reynolds, J. 1983. A plan for reintroduction of the swift fox to the Canadian prairie. Master of Environmental Design Project, University of Calgary, Calgary, Alberta.

Russell, R.H., and G.W. Scotter. 1984. Return of the native. Nature Canada 13(1):7–13.

Schroeder, C. 1985. A preliminary management plan for securing swift fox for reintroduction into Canada. Master of Environmental Design Project, University of Calgary, Calgary, Alberta.

Scott-Brown, J.M., S. Herrero, and J. Reynolds. 1987. Swift fox. Pp. 433-441 *in* M. Novak, J.A. Baker, M.E. Obbard, and B. Malloch, editors. Wild furbearer management and conservation in North America. Ontario Trappers Association, North Bay, Ontario.

Soper, J.D. 1984. The mammals of Alberta. Hamly Press, Edmonton, Alberta.

Sovada, Marsha A., Christiane C. Slivinski, Robert O. Woodward, and Michael L. Phillips. 2003. Home range, habitat use, litter size, and pup dispersal of swift foxes in two distinct landscapes of western Kansas. Pp. 149–159 *in* M.A. Sovada and L.N. Carbyn, editors. The Swift Fox: ecology and conservation of swift foxes in a changing world. Canadian Plains Research Center, University of Regina, Saskatchewan.

Swift Fox Recovery Team. 1996. National recovery plan for the swift fox. RENEW Report No. 15. Ottawa, Ontario.

World Wildlife Fund (WWF) Canada. 1989. Prairie conservation action plan. World Wildlife Fund Canada, Toronto, Ontario.

Part II – Distribution and Population Shifts

The Historic and Recent Status of the Swift Fox in Montana

■ Craig J. Knowles, Pamela R. Knowles, Brian Giddings and Arnold R. Dood

Abstract: Early records of the swift fox in Montana indicate that it was common and widespread on the prairie grasslands east of the Rocky Mountains. There were no records of swift foxes in Montana from 1918 to 1978, suggesting that the swift fox might have been extirpated from Montana at some point during this period. Since 1978, there have been 53 confirmed and unconfirmed observations of swift foxes in Montana. Recent records and sightings of swift foxes are grouped in southeastern and north-central Montana. Presumably, these observations represent dispersing individuals from populations outside of Montana but recent evidence from north-central Montana suggests that a small resident population now exists in this area. In addition, swift foxes are being reintroduced on the Blackfeet Indian Reservation.

The swift fox (*Vulpes velox*) is believed to originally have been abundant throughout its range on the Great Plains (Johnson 1969). It is presumed to have been extirpated early in this century from the northern portion of its range while remnant populations in the southern portion survived human settlement of the prairies (Warren 1942, Armstrong 1972, Bee et al. 1981). The original cause of this widespread extirpation is attributed to the ready acceptance of poisoned baits and traps placed for coyote (*Canis latrans*) and wolf (*C. lupus*) eradication. Loss of habitat to dryland agriculture, a changing prey base, and increased interspecific competition with coyotes and red foxes (*V. vulpes*) may also have contributed to this decline.

The loss of the swift fox over such a broad area prior to the advent of quantitative ecological studies has resulted in a paucity of ecological information on this species. Until recently, virtually nothing was known about the swift fox in Montana. Its small size and nocturnal habits have also contributed to this lack of even basic biological information. In recent years, swift foxes have increased in numbers in the southern portion of their range, and records of dispersing swift foxes have been reported in formerly unoccupied areas (FaunaWest 1991). In addition, the swift fox has been reintroduced into southern Alberta and Saskatchewan (Carbyn and Killaby 1989, Brechtel et al. 1993) and some of these foxes have moved south into Montana. This paper reviews the historical occurrence of the swift fox in Montana prior to extirpation (1918 last recorded observations) and contemporary Montana swift fox records (1978 first recent record).

Methods

Historical accounts of swift fox observations were obtained by reviewing journal notes recorded by naturalist and explorers who traveled or worked in Montana prior to the extirpation of the swift fox (last observations recorded in 1918). Major museums ($n = 30$) were also asked for any information they had on Montana swift fox specimens. Recent observations (starting from 1978 to the present) were obtained by reviewing the scientific literature and newspaper accounts with reference to swift foxes, consulting with other wildlife professionals, talking to fur dealers, and collecting swift fox specimens taken by the general public incidental to other lawful activities.

Montana Historical Records

There were several general and specific observations of swift foxes in Montana during the 1800s and early 1900s, suggesting that they were widespread and common on the prairies (Table 1). The first swift fox reported in Montana was in 1805 by the Lewis and Clark expedition near the Great Falls of the Missouri River and again at Two Medicine Creek on the Marias River in 1806 (Burroughs 1961). This region is characterized by broad, gently rolling plains dominated by mid- and short-grasses including western wheatgrass (*Agropyron smithii*), needle-and-thread grass (*Stipa comata*), and blue grama (*Bouteloua gracilis*). Generally, big sagebrush (*Artemisia tridentata*) occurs with limited distribution in this area.

Fur returns for the American Fur Company's upper Missouri and Sioux outfits (Montana and the Dakotas) from 1835–1838 show that swift foxes were commonly traded and that they were the most frequently trapped fox in this region (Johnson 1969). Over 10,600 swift fox hides were traded during this period, while only 1,989 red and 108 gray fox (*Urocyon cinereoargenteus*) hides were traded during the same period. An 1875 shipping log from Fort Benton, Montana, recorded the shipment of 350 swift fox hides, 490 red fox hides, and 125 cross fox hides (Williams and Muich 1998). In the same shipment, there were 1,680 wolf hides and 520 coyote hides.

Table 1. List of historic observations of swift foxes in Montana or adjacent areas

Naturalist/Source	Location/Abundance
Lewis & Clark/Burroughs 1961	Great Falls of the Missouri River & Confluence of Two Medicine and Marias Rivers/2 observations
American Fur Co./Johnson 1969	Montana and Dakotas/10,614 skins
Williams and Muich 1998	Fort Benton/350 skins
Audubon & Harris/ Audubon and Bachman 1854, McDermott 1951	Fort Union — northeastern Montana/ 2 observations
Allen/Allen 1874	Yellowstone and Musselshell drainages central and east -central Montana/frequent
Coues/Coues 1878	Montana/Canadian border, Milk River drainage/common
Grinnell/Grinnell 1876	Central Montana and Little Missouri River/common
McChesney/McChesney 1879	Fort Custer (Hardin) — south-central Montana/present in area
National Museum of Natural History/ FaunaWest 1991	Fort Benton area 1880s/one skull Browning area, northwestern plains 1901 –1906/43 specimens
Bailey and Bailey 1918	Eastern footslopes of Glacier National Park/common

In 1843, Audubon (Audubon and Bachman 1854) recorded the swift fox on two occasions near Fort Union at the confluence of the Yellowstone and Missouri Rivers. He also reported finding a swift fox den in this area. Harris, who accompanied Audubon on this expedition (McDermott 1951), gives accurate descriptions of the two swift fox observations.

Coues (1878) reported the swift fox as common during his survey of the area between the Milk River and Canadian border during 1874. This region is rolling, glaciated prairie dominated by mid- and short-grasses similar to that described for the Great Falls area. Allen (1874) described the swift fox as frequent along the plains of the Yellowstone and Musselshell Rivers in 1873. Grinnell (1876), when traveling across central Montana in 1875, considered the swift fox as common on the plains. His route across Montana took him through an area south of the Missouri River dominated by big sagebrush, western wheatgrass, and blue grama, and across foothills and alluvial fans of the Big Snowy and Little Belt Mountain Ranges. These latter areas are dominated by needle-and-thread grass and blue grama. Grinnell also found the swift fox to be abundant along the Little Missouri River when traveling to the Black Hills in 1874 (Custer 1875). McChesney (1879) reported the swift fox to occur around Fort Custer near present day Hardin in southcentral Montana. Collectively, these 4 naturalists traveled much of the prairie region of Montana, and their notes indicate that the swift fox was common on the prairie grasslands. (See Knowles and Knowles [1993] for maps of routes taken by these naturalists.)

One historical specimen (skull only) was taken near Fort Benton in the late 1880s but very little information is known about this specimen. A series of 43 specimens (skins and skulls) were collected between 1901 and 1906 from upper Teton County (now Glacier County) near Blackfoot and Kipp, Montana (vicinity of Browning, Montana). (These specimens are at the U.S. National Museum in Washington, D.C. See FaunaWest [1991] for information on these specimens.) This latter region consists of broad, gently rolling prairies dominated by needle-and-thread grass and blue grama. Later, in this same region, Bailey and Bailey (1918) reported the swift fox as common over the plains along the eastern boundary of Glacier National Park.

The collection of swift foxes from the Browning area and the report by Bailey and Bailey (1918) indicate that the swift fox was still common in Montana early in this century. However, by 1969, Hoffmann et al. (1969) reported the swift fox as extirpated in Montana based on the absence of swift foxes from Montana fur harvest data during a 16-year period. Thus, somewhere between 1918 (last recorded sighting) and 1953 the swift fox went from common to rare in Montana. This would suggest that the decline in swift foxes was rapid and extensive across the state.

The cause of this decline of the swift fox in Montana is unknown and is open to conjecture and speculation. One likely cause is non-target poisoning resulting from coyote and wolf eradication. It would be reasonable to assume that poison baits set for coyotes and wolves were also consumed by swift foxes. Another factor contributing to the decline of swift foxes in Montana would have been rodent (ground squirrel [*Spermophilus* spp.] and prairie dog [*Cynomys* spp.]) control programs. During the 1920s and 1930s, massive rodent eradication programs were conducted in Montana, spreading tons of strychnine-treated grain bait on thousands of acres (BLM 1982, Flath and Clark 1986). Secondary poisoning (*see* Schitoskey 1975), loss of a prey base, and loss of areas with high burrow densities may all have contributed to the decline of the swift fox in Montana.

More than half a century passed after the report by Bailey and Bailey (1918) before another swift fox was recorded in Montana (Moore and Martin 1980; #1, Fig.1). Although the swift fox may never have been completely extirpated from Montana, this hiatus of 60 years between reported observations is sufficient evidence to conclude that the swift fox became very rare as a direct or indirect result of human settlement of Montana's prairie environment.

Recent Montana Records

Since 1978, there have been a series of confirmed and unconfirmed records and sightings of swift foxes in

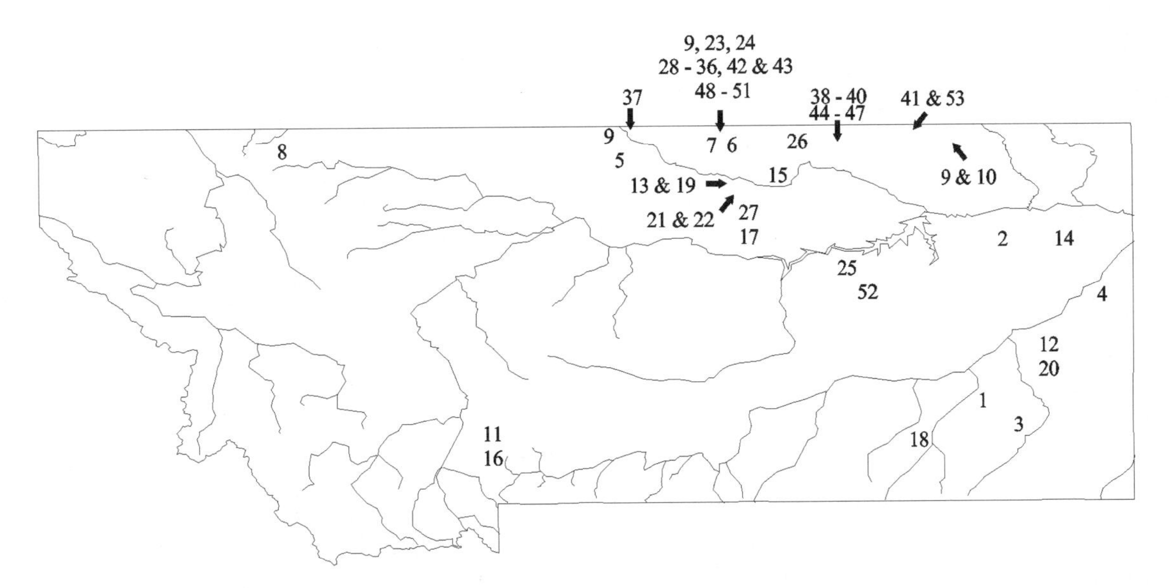

Figure 1. Recent swift fox observations (see Table 2)

Montana. Table 2 summarizes this information and Figure 1 shows the general location of these observations. Although many of the observations were made by people with biological training, all visual observations and uncollected road-killed animals should be considered as unconfirmed sightings. Trapped swift foxes, traded skins, radio-collared swift fox relocations, and photographs of swift foxes are considered as confirmed observations. Until recently (*see* Zimmerman et al. 2003), none of these swift fox observations were followed by systematic surveys to determine if other swift foxes might occur in the immediate area.

In 1978, an adult male swift fox was trapped on the Tongue River bottomlands approximately 58 km south of Miles City (#1, Fig. 1, Table 2; Moore and Martin 1980). This swift fox is now a specimen in the Montana State University vertebrate museum collection (specimen number 6240). It is assumed that this individual dispersed north out of Wyoming. The nearest known swift fox population to this site is in the Gillette-Buffalo area (Rick Pallister, Wyoming Game & Fish Department, personal communication; Lindberg 1986) in the headwaters of the Powder River drainage basin about 161 km to the south. In 1993, another swift fox was observed along the Tongue River in a prairie dog colony on the Northern Cheyenne Reservation about 48 km south of the 1978 specimen.

In 1982, a swift fox was trapped about 32 km north of Circle (#2, Fig. 1, Table 2). This individual was verified by a Montana Department Fish, Wildlife and Parks (MT FW&P) biologist. The animal had mange and was not suitable as a museum skin, but its skull was collected and sent to the MT FW&P laboratory (R. Stoneberg, MT FW&P biologist, personal communication). Apparently, this skull was misplaced as there is no record of the skull entering into the vertebrate museum collection at the MT FW&P laboratory (P. Schladweiler, MT FW&P, personal communication). The closest known swift fox population in 1982 would have been Perkins County, South Dakota (SD), approximately 321 km away.

An additional report was received from this same general area in 1992 (#14, Fig. 1, Table 2). Following an article on swift foxes in the *Billings Gazette* (Henckel 1992), MT FW&P received a letter from a man living near Richey which stated that he had recently observed a swift fox. The letter went on to state that his daughter was a fur dealer and that she had purchased 2 or 3 swift fox skins from trappers in this area. This was followed by an actual specimen that was collected about 20 km northwest of Jordan in January 1995 (#25, Fig.1, Table 2). Another swift fox was sighted in September 1997 approximately 65 km southeast of Jordan (#52, Fig. 1, Table 2; D. Bricco, Bureau of Land Management [BLM] biologist, personal communication). By 1992, swift foxes were released in Saskatchewan (Brechtel et al. 1993) and dispersal distance to this general area could have been as little as 193 km.

Two verified swift fox records were obtained in 1984. In mid-January, an adult female swift fox was trapped north of Broadus and taken to a taxidermist in Forsyth (#3, Fig. 1, Table 2; Henckel 1984). This swift fox was verified by a wildlife biologist (H. Youmans, MT FW&P biologist, personal communication), but the trapper left the taxidermist's office with the fox when informed that the animal was a protected species. As a result, specific details of the capture location for this animal are lacking and the specimen was lost. On 15 November, an adult swift fox was trapped in rolling prairie habitat about 24 km east of Glendive (#4, Fig. 1, Table 2; Vallard 1985). Although this specific area is predominately grasslands, several cultivated sites exist in the general area. This swift fox is now on display in the Miles City MT FW&P office. The closest known population in 1984 was Perkins County, SD—a distance of about 225 km. Despite

Table 2. Recent records or observations of swift foxes in Montana (observation numbers correspond to numbers in Figure 1)

Obs.#	Date	Location	Type of Observation	Source
1	5 March 1978	Tongue River: T2N,R45E,S12	trapped specimen: Adult M	Moore and Martin 1980
2	Winter 1982	N. of Circle: T22N,R48E	trapped specimen: skull collected but subsequently lost	R.Stoneberg pers. comm. (FW&P biologist)
3	January 1984	N. of Broadus, exact location not disclosed	trapped Adult F, specimen not collected	Henckel 1984
4	November 1984	E. of Glendive: T15N,R58E,S2	trapped specimen: Adult	Vallard 1985
5	Winter 1985-86	Gildford, NW of Havre on Sage Creek	trapped skin traded	Carbyn and Killaby 1989
6	Winter 1985-86	Hogeland	visual	Carbyn and Killaby 1989
7	Winter 1985-86	Chinook	trapped skin traded	Carbyn and Killaby 1989
8	11 June to 31 Aug. 1989	Cutbank.Ranger Sta. Glacier NP	visual wearing radio collar	S. Gniadek pers. comm. (NPS biologist)
9	1987-1991	northern Montana	summary report obs. in 7 townships	Brechtel et al.. 1993
10	31 July 1992	S. of Opheim	radio relocation	Henckel 1992
11	24 May 1992	S. of Wilsall: T3N,R9E,S29	visual	D. Quimby pers.comm. (retired mammalogist MSU)
12	9 July 1992	S. of Terry: T11N,R51E,S34	visual, pups at den	B. Heidel pers. comm. (Nat. Heritage Program Botanist)
13	Spring 1992	S. of Fort Belknap Agency along Hwy 66: T32N,R23E,S4	visual, 2 observations	M. Fox pers. comm. (Ft. Belknap Fish & Wildlife)
14	Winter 1992	Richey area	visual & trapped, 2 or 3 skins traded	A Dood pers. comm. (FW&P biologist)
15	1992	near Malta on Hwy 2	road kill, not collected	A Dood pers. comm. (FW&P biologist)
16	April 1993	W of Wilsall: T3N,R7E,S21	visual	A. Harmata pers. comm. (biologist MSU)
17	May 1993	N. of CMR on Hwy 191: T23N,R23E,S25	road kill, not collected	Klien, pers. comm. (U.S. FWS volunteer)
18	2 July 1993	Tongue River S. of Ashland: T3S,R23E,S21	visual	J. Spang pers. comm. (N. Chey. Dept. of Nat. Res.)
19	November 1993	S. of Fort Belknap Agency: T31N,R23E,S17	photo	M. Fox pers. comm. (Ft. Belknap Fish & Game)
20	January 1994	S. of Terry, Hwy 12: T8N,R52E,S28	visual 2 observations	S. Heuther pers. comm.
21	June 1994	SE of Fort Belknap Agency: T31N,R24E,S29	visual 4 observations, includes pups at den	K. Jones, & A. Healey pers. comm. (Ft. Belk . Nat. Res Dept.)
22	September 1994	S. of Fort Belknap Agency: T30N,R23E,S15	visual	M. Fox pers. comm. (Ft. Belknap Fish & Game)
23	November 1994	N. of Chinook: T37N,R18E,S11&12, T36N,R20E,S24	3 trapped: 1 collected, 2 released	J. Peters pers. comm. (BLM biologist)
24	Fall 1994	N. of Chinook: T35N,R18E,S11	2 shot at den: specimens at Chinook Mus.	A. Zimmerman pers. comm. (MSU grad. student)
25	27 January 1995	Northern Garfield Co.: T20N,R36E,S3	1 trapped & collected	B. Giddings pers. comm. (FW&P biologist)
26	24 February 1995	10 km SW Whitewater: T34N,R31E,S21	1 road-killed collected	B. Giddings pers. comm. (FW&P biologist)
27	June 1995	N. of Lodgepole: T27N,R25E,S32	visual	P. Bigby pers. comm. (Fort Belknap Lands Dept.)
28	24 September 1996	N. of Chinook: T35N,R18E,S1	live-trapped: adult M	A. Zimmerman pers. comm. (MSU grad. student)
29	27 September 1996	N. of Chinook: T35N,R18E,S1	live-trapped: juv. M	A. Zimmerman pers. comm. (MSU grad. student)
30	27 September 1996	N. of Chinook: T35N,R18E,S10	live-trapped: adult F	A. Zimmerman pers. comm. (MSU grad. student)
31	3 October 1996	N. of Chinook: T35N,R18E,S10	live-trapped: adult M	A. Zimmerman pers. comm. (MSU grad. student)
32	5 October 1996	N. of Chinook: T35N,R18E,S10	live-trapped: juv. M	A. Zimmerman pers. comm. (MSU grad. student)
33	8 October 1996	N. of Chinook: T35N, R19E,S28	live-trapped: adult F	A. Zimmerman pers. comm. (MSU grad. student)
34	16 October 1996	N. of Chinook: T35N, R19E,S15	live-trapped: juv. M	A. Zimmerman pers. comm. (MSU grad. student)
35	20 October 1996	N. of Chinook: T35N, R19E,S15	live-trapped: adult M	A. Zimmerman pers. comm. (MSU grad. student)
36	5 November 1996	N. of Chinook: T35N, R20E,S21	live-trapped: adult F	A. Zimmerman pers. comm. (MSU grad. student)
37	May 1996	near Wildhorse border crossing	collected, green ear tattoo	B. Giddings pers. comm. (FW&P biologist)
38	May 1996	8 km N. of Whitewater: T35N,R31E,S1	visual, pups at a den	J. Grensten pers. comm. (BLM biologist)
39	5 July 1996	N. of Whitewater: T37N,R30E,S20	radio col. visual	B. Giddings pers. comm. (FW&P biologist)
40	16 December 1996	N. of Whitewater: T37N,R30E,S11	visual collected specimen	B. Giddings pers. comm. (FW&P biologist)
41	December 1996	N. of Hinsdale: T36N,R36E,S21	tracks	B. Giddings pers. comm. (FW&P biologist)
42	10 July 1997	N. of Chinook: T35N,R20,S29	live-trapped: adult M	A. Zimmerman pers. comm. (MSU grad. student)
43	14 July 1997	N. of Chinook: T35N,R20,S29	live-trapped: adult F	A. Zimmerman pers. comm. (MSU grad. student)
44	4 August 1997	N. of Whitewater: T35N,R31,S13	live-trapped: juv. F	A. Zimmerman pers. comm. (MSU grad. student)
45	5 August 1997	N. of Whitewater: T35N,R31,S13	live-trapped: adult M	A. Zimmerman pers. comm. (MSU grad. student)
46	5 August 1997	N. of Whitewater: T35N,R31,S13	live-trapped: adult F	A. Zimmerman pers. comm. (MSU grad. student)
47	7 August 1997	N. of Whitewater: T35N,R31,S13	live-trapped: juv. F	A. Zimmerman pers. comm. (MSU grad. student)
48	13 August 1997	N. of Chinook: T35N,R36,S9	live-trapped: adult M	A. Zimmerman pers. comm. (MSU grad. student)
49	12 September 1997	N. of Chinook: T36N,R19,S16	live-trapped: juv. F	A. Zimmerman pers. comm. (MSU grad. student)
50	10 October 1997	N. of Chinook: T35N,R20,S34	live-trapped: juv. M	A. Zimmerman pers. comm. (MSU grad. student)
51	11 October 1997	N. of Chinook: T36N,R19,S16	live-trapped: juv. F	A. Zimmerman pers. comm. (MSU grad. student)
52	September 1997	SE. of Jordan: T35N,R31,S13	visual	D. Bricco pers. comm. (BLM biologist)
53	25 November 1997	N. of Hinsdale: T36N,R36E,S8	collected specimen	B. Giddings pers. comm. (FW&P biologist)

confirmed observations of swift foxes in these two areas, there have not been any additional swift fox reports from these areas for 15 years.

There were two swift fox reports from an area south of Terry in 1992 and 1994 (#12, 20, Fig. 1, Table 2). The 1992 report was made by a biologist and included 4 young observed at a den site near a ranch headquarters about 14 km south of Terry. The 1994 report was about 24 km southeast of the 1992 sighting and was actually two observations of what could have been the same animal seen on two different days in the same area. Although the animal was observed at the head of a drainage at the edge of the Powder River breaks, the general area consisted of rolling mixed-grass prairie with very little shrub cover. The closest known swift fox populations were approximately 193 km away in Perkins County, South Dakota, and in Campbell County, Wyoming.

Since 1985, there have been numerous reports of swift foxes in north-central Montana, and it is now known that swift foxes are well established in this area (#5, 6–10, 13, 15, 17, 19, 21, 22, 24, 26, 28–36, 38–43, 44–51, 54; Zimmernan 1998, Zimmerman et al. 1998). Reports of swift foxes have come from as far west as the Cut Bank Ranger Station in Glacier National Park and as far east as Opheim. All of these observations (except #54) can probably be attributed to swift foxes released in Alberta and Saskatchewan (Carbyn and Killaby 1989, Henckel 1992, Brechtel et al. 1993) and subsequently dispersing into Montana. Brechtel et al. (1993) show seven townships in northern Montana where radio-marked swift foxes, originally released in Alberta or Saskatchewan, were relocated in Montana. Further evidence of this southward dispersal has come from a radio-marked swift fox observed at the Cut Bank Ranger Station (#8, Fig. 1, Table 2) in Glacier National Park, a radio-marked swift fox collected north of Whitewater (#39, Fig. 1, Table 2), and an ear-tagged swift fox collected near the Wildhorse border crossing (#37, Fig. 1, Table 2). Some of these individuals may have moved up to 80 km from their release sites. For the most part, these observations correspond closely to Canadian swift fox release sites.

Several swift fox sightings have come from the northwestern portion of the Fort Belknap Indian Reservation (#13, 19, 21, 22, Fig. 1, Table 2) from 1992 through 1995. One of these observations included photographic verification which showed that the fox was in glaciated mixed-grass prairie with a moderate density of big sagebrush. Another observation included pups at a den site located along a fence line separating native prairie from dryland agriculture. Although none of these foxes appeared to be marked, if any these foxes had dispersed from Saskatchewan this would represent a dispersal distance of at least 97 km.

During November 1994, three swift foxes were trapped north of Chinook (#23, Fig. 1, Table 2) by Jim Halseth, a Chinook area trapper (J. Peters, BLM biologist, personal communication). One individual, a juvenile male, was found dead in the trap and was given to MT FW&P to be mounted as a display specimen. An adult female was captured and released unharmed in an area where swift fox tracks had been observed the previous winter. The third individual appeared to be an adult but pulled free from the trap when approached. None of these animals wore radio collars or were otherwise marked; indicating that these animals were likely progeny of swift foxes reintroduced into Alberta or Saskatchewan. We later learned that two swift foxes were shot at a den site in this area (#24, Fig. 1, Table 2) during the fall of 1994 and were mounted for display in the Blaine County Museum (A. Zimmerman, Montana State University graduate student, personal communication). Based on tracks and other swift fox signs observed in this area by the trapper, there was reason to believe that swift foxes were established in Montana north of Chinook. This was subsequently verified by Zimmerman et al. (1998). This area is characterized by rolling, glaciated, mixed-grass prairie with limited shrub cover. Approximately half of the land is under Federal land ownership, and cultivation of private lands is minimal.

There are also groupings of swift fox observations in the Whitewater area and north of Hinsdale. One of the swift fox observations near Whitewater included a den with pups (#38, Fig. 1, Table 2). Although our first confirmed record of a swift fox north of Hinsdale came in December 1996 (#41, Fig. 1, Table 2), we received a report from a rancher north of Hinsdale in July 1995 that a small tan fox was remaining in the vicinity of his barn. Although we neglected to follow up on this report, it would indicate that swift foxes had moved into the area north of Hinsdale by 1995 or earlier. Swift foxes were subsequently live-trapped and marked at both these sites (Zimmerman et al. 1998).

Swift fox observations made in 1992 and 1993 near Wilsall (#11, 16, Fig. 1, Table 2) seem unrelated to swift fox observations in north-central and southeastern Montana. Both of these observations were made by biologists well qualified to make an identification of a swift fox. Although the location of these observations is surprising, suitable grassland habitat (needle-and-thread grass/blue grama) does exist near these observations. If these observations represent dispersing individuals, the closest known population is the Alberta/Saskatchewan release site—a distance of over 322 km.

In August 1998, the Blackfeet Tribe selected a grassland area between Two Medicine River and Badger Creek on the eastern portion of their Reservation (#54, Fig. 1, Table 2) to reintroduce swift foxes (Wilkinson 1999). The specific release site was about 25 km west of the swift fox observation recorded by Lewis in 1806 (Burroughs 1961) and 20 km southeast of the collection site for the 43 swift fox specimens collected between 1901 and 1906. Thirty juvenile foxes were released into Richardson's ground squirrel colonies in areas dominated by needle-and-thread grass and blue grama. At least two foxes were subsequently killed by vehicles during the fall of 1998 on Highway 89 southeast of Browning. The foxes used for this reintroduction came from the Cochrane Ecological Institute in Alberta.

Discussion

An examination of historical records indicates that swift foxes were widespread and abundant on the Montana prairies. The fact that several naturalists considered the swift fox common or abundant is remarkable when their nocturnal habits are considered. The swift fox may have been under-represented in historic notes because of their small size and nocturnal habits.

The decline of the swift fox in Montana appears to have been rapid and to have occurred between 1918 and 1953. Information gained on the swift fox in Montana since 1978 would suggest that animals are trapped and skins are sold without the knowledge of MT FW&P. Moreover, the general public, including many trappers, are unknowledgable about the swift fox as a species and would be unlikely to report observations. Similar observations have been reported from Kansas (Zumbaugh 1984) and Texas (Federal Register 1994).

In 1979, the Montana Legislature removed the swift fox from the predator list and placed it on the furbearer list, giving the Fish, Wildlife and Parks Commission management authority to regulate take. The Commission subsequently established a closed season for the swift fox. This management classification has provided some level of protection to the swift fox in Montana. However, trappers may be reluctant to report swift foxes taken incidental to trapping of unprotected species. The swift fox trapped north of Broadus in 1984 is an indication that trappers may be reluctant to report swift fox captures.

The series of swift fox sightings and records since 1978 would suggest that the swift fox occurs in Montana as scattered individuals or as small populations. Estimated minimum dispersal distances of 161–322 km for many of the Montana swift fox records would suggest that there might be populations closer to Montana than generally believed or that these individuals are indicative of resident populations. Recently, Zimmerman et al. (1998) has documented a small swift fox population in north-central Montana.

We estimate over 3,000,000 ha of suitable habitat remains within many of the areas that the swift fox was reported to have occurred in during the 1800s and early 1900s. The two major factors which probably contributed significantly to the original decline of swift foxes have generally been discontinued in Montana. These are predator control with toxicants, and rodent control with toxicants posing secondary toxicity hazards. However, the predator community on the prairie grasslands has been greatly altered, and the niche once occupied by swift foxes may be greatly restricted by increased abundance of coyotes and red foxes.

Although it is likely that many of the recent swift fox observations are dispersing individuals from populations adjacent to Montana, the possibility of additional extant populations in Montana should not be dismissed. It is difficult to imagine that swift fox sightings near Circle (#2, 14, Fig. 1, Table 2), Terry (#12, 20, Fig. 1, Table 2), and Wilsall (#11, 16, Fig. 1, Table 2) were individuals from other populations. The dens with pups near Terry and Fort Belknap Agency are good indications of established swift fox populations breeding in those areas. The recent observations of swift foxes north of Chinook, Whitewater, and Hinsdale (see Fig. 1, Table 2; Zimmermann et al. 1998) would indicate that they are established in this area as a result of the Canadian reintroduction effort. The specific locations for the swift fox observations listed in Table 2 should serve as a starting point for initiating systematic surveys for resident swift fox populations.

Acknowledgments

We want to thank the many people who took the time to report recent swift fox observations to us and the museums that responded to our survey.

Literature Cited

Allen, J.A. 1874. Notes on the natural history of portions of the Dakota and Montana Territories, being the substance of a report to the Secretary of War on the collections made by the North Pacific Railroad Expedition of 1873, Gen. D.S. Stanley, Commander. Birds. Proceedings Boston Society Natural History 27:3–68.

Armstrong, D.M. 1972. Distribution of mammals in Colorado. University of Kansas Museum of Natural History, Monograph No. 3.

Audubon, J.J., and J. Bachman. 1854. The quadrupeds of North America. Vol. 3, Audubon, New York, New York.

Bailey, V., and F.M. Bailey. 1918. Wild animals of Glacier National Park. U.S. Government Printing Office, Washington, D.C.

Bee, J.W., G.E. Glass, R.S. Hoffmann, and R.P. Patterson. 1981. Mammals in Kansas. University of Kansas, Museum of Natural History 7:1–300.

Bureau of Land Management (BLM). 1982. Black-tailed prairie dog control/management in Phillips Resource Area. Programmatic Environmental Assessment. U.S. Department Interior, Bureau Land Management, Lewistown District, Montana.

Brechtel, S.H., L.N. Carbyn, D. Hjertaas, and C. Mamo. 1993. Canadian swift fox reintroduction feasibility study: 1989 to 1992. Scientific and Technical Documents Division, Canadian Wildlife Service, Ottawa, Ontario.

Burroughs, R.D. 1961. The natural history of the Lewis and Clark expedition. Michigan State University Press, East Lansing, Michigan.

Carbyn, L.N., and M. Killaby. 1989. Status of the swift fox in Saskatchewan. Blue Jay 47:41–52.

Coues, E. 1878. Field notes on birds observed in Dakota and Montana along the forty-ninth parallel during the seasons of 1873 and 1874. Article XXV. Pp. 545–661 *in* Bulletin of the U.S. Geological and Geographical Survey Volume IV. Government Printing Office, Washington, D.C.

Custer, G.A. 1875. Report of the Chief of Army Engineers. Annual Report of the Secretary of War. 43rd Congress, 2nd session, H.R. Executive Document I, Part II (1874–1875), Appendix KK: preliminary report of reconnaissance to the Black Hill, St. Paul, Minnesota, 7 September 1874.

FaunaWest. 1991. An ecological and taxonomic review of the swift fox (*Vulpes velox*) with special reference to Montana. Montana Department Fish Wildlife & Parks, Bozeman, Montana.

Federal Register. 1994. Endangered and threatened wildlife and plants: 90-day finding for a petition to list the swift fox as endangered. Federal Register 59:28328–28330.

Flath, D.L., and T.W. Clark. 1986. Historic status of black-footed ferret habitat in Montana. Great Basin Naturalist Memoirs 8:63–71.

Grinnell, G.B. 1876. Zoological report. Pp. 59–92 *in* William Ludlow, editor. Report of a reconnaissance from Carroll, Montana territory, on the upper Missouri, to Yellowstone National Park, and return, made in the summer of 1875. Government Printing Office, Washington, D.C.

Henckel, M. 1984. Swift fox once again roams the prairie. Billings Gazette, 16 February 1984.

Henckel, M. 1992. Swift Foxes, Canadian efforts help to bring them back to Montana. Billings Gazette, 30 April 1992.

Hoffmann, R.S., P.L. Wright, and F.E. Newby. 1969. The distribution of some mammals in Montana. I. Mammals other than bats. Journal of Mammalogy 50:579–604.

Johnson, D.R. 1969. Returns of the American Fur Company, 1835–1839. Journal of Mammalogy 50:836–839.

Knowles, C.J., and P.R. Knowles. 1993. A bibliography of literature and papers pertaining to pre-settlement wildlife and habitat of Montana and adjacent areas. U.S. Forest Service, Missoula, Montana.

Lindberg, M. 1986. Swift fox distribution in Wyoming: a biogeographical study. Thesis, University of Wyoming, Laramie, Wyoming.

McChesney, C.E. 1879. Report on the mammals and birds of the general region of the Bighorn River and Mountains of the Montana Territory. Report of the Chief of Engineers, U.S. Army. Appendix SS3.

McDermott, J.F., editor. 1951. Up the Missouri with Audubon, by Edward Harris. University of Oklahoma Press, Norman, Oklahoma.

Moore, R.E., and N.S. Martin. 1980. A recent record of the swift fox (*Vulpes velox*) in Montana. Journal of Mammalogy 61:161.

National Agricultural Lands Study. 1981. Final report. U.S. Government Printing Office, Washington, D.C.

Schitoskey, F., Jr. 1975. Primary and secondary hazards of three rodenticides to kit foxes. Journal of Wildlife Management 39:416–418.

Vallard, D. 1985. Rare fox trapped near Glendive. Miles City Star, 20 January.

Warren, E.R. 1942. The mammals of Colorado. University of Oklahoma Press, Second edition.

Wilkinson, T. 1999. Rescuing the swift fox. Defenders, Vol 74, No. 1:6–13.

Williams, J., and G. Muich. 1998. In the footsteps of the mountain men. Montana Outdoors, Vol. 29, No. 6:20–23.

Zimmerman, A.L. 1998. Reestablishment of swift fox in north central Montana. Thesis, Montana State University, Bozeman, Montana.

Zimmerman, A.L., L. Irby, and B. Giddings. 1998. The status of the swift fox in north-central Montana. Paper presented at the Swift Fox Symposium. Pp. 49–59 *in* M.A. Sovada and L.N. Carbyn, editors. The Swift Fox: ecology and conservation of swift foxes in a changing world. Canadian Plains Research Center, University of Regina, Saskatchewan.

Zumbaugh, D.M. 1984. Natural history of foxes in Kansas. Thesis, Fort Hays State University, Hays, Kansas.

The Status and Ecology of Swift Foxes in North-central Montana

■ **Amy L. Zimmerman, Lynn Irby and Brian Giddings**

Abstract: Swift foxes were reintroduced in southern Alberta and Saskatchewan from 1983 to 1998 by Canadian wildlife agencies. These reintroductions led to the possibility of individuals dispersing into north-central Montana. We began a study in July 1996 in northern Blaine County, Montana, to determine if swift foxes dispersed from the reintroduced population to establish a resident population in Montana. Secondary sites were surveyed for swift foxes in northern Phillips and Valley counties. We systematically live-trapped swift foxes and radio-collared captured foxes. Sixteen swift foxes were trapped in the fall of 1996 and late summer of 1997. Five captured swift foxes were juveniles and 11 were adults. Trapping success for the two-year study was 0.93 fox/100 trap-nights. In the spring of 1997, 3 litters were produced in northern Blaine County and visually observed throughout the summer. We monitored swift foxes to determine home range and survival rates. In the spring of 1997, we used radio-collared swift foxes to locate rearing den sites. Mean home range sizes were estimated using the 100% minimum convex polygon method (10.4 km², SE = 3.8 km²) and the 95% adaptive kernel method (12.3 km², SE = 4.6 km²). The estimated annual survival rate was 46% (SE = 13%). Scats were collected in the spring, summer, and fall to examine food habits. Swift fox diet was dominated by mammal material, with voles being the most common item in scats. A small population of swift foxes exists in north-central Montana, and evidence suggests that the population is surviving and reproducing. A wider range of habitat in Montana needs to be surveyed to determine swift fox distribution and population density.

Swift foxes (*Vulpes velox*) historically occurred widely on the Great Plains from northeastern New Mexico and northwestern Texas to southern Alberta and Saskatchewan (Johnson 1969, Hall 1981, Allardyce and Sovada 2003). Swift foxes were extirpated or experienced population declines in the northern part of the United States and southern Canada early this century, but persisted in the central and southern portions of the range (Warren 1942, Armstrong 1972, Bee et al. 1981, Allardyce and Sovada 2003). Suggested causes for the decline of swift foxes in their northern range (northern U.S. and southern Canada) included a vulnerability to poisoned baits and traps used for coyote (*Canis latrans*) and gray wolf (*C. lupus*) control, loss of grassland habitat to cultivation, a changing prey base, and increased interspecific competition with coyotes and red foxes (*V. vulpes*, Young 1944:327, Egoscue 1979, Zumbaugh and Choate 1985, Scott-Brown et al. 1987, Allardyce and Sovada 2003).

The species was declared extinct in Montana in 1969 because of a 16-year absence of swift foxes in fur harvest records (Hoffman et al. 1969). No swift foxes were recorded in Montana until 1978, when a male was trapped in Custer County (Moore and Martin 1980). This animal may have been dispersing from northeastern Wyoming or southwestern South Dakota. The swift fox is currently classified in Montana as a state furbearer with a closed season for harvest.

In Canada, the last known swift fox collected prior to reintroductions was in 1928 (Banfield 1974). The swift fox was designated as extirpated by the Committee on the Status of Endangered Wildlife in Canada in 1978 because of the lack of observations since 1928 (Brechtel et al. 1996). A reintroduction program began in southern Alberta and Saskatchewan in 1983–1984 (Brechtel et al. 1996). The Canadian National Recovery Plan for the Swift Fox had three main objectives. The first was to establish two geographically distinct, yet genetically connected core populations with a mean spring density of five adults per township. Other objectives were to ensure the long-term security of key swift fox habitats in two areas of reintroduction and to establish swift foxes in at least 50% of the suitable habitat remaining on the Canadian prairies. The plan's overall goal was to establish a target population of 420 foxes in suitable areas by the year 2000.

Three release areas were established in Alberta and Saskatchewan (Fig. 1; Brechtel et al. 1996). One of the release areas was south of the Cypress Hills along the Alberta-Saskatchewan border, and a second site was on the Wood Mountain Plateau in Grasslands National Park, Saskatchewan. A third release site was located along the Milk River ridge south of Lethbridge, Alberta. This release site was later abandoned (Carbyn et al. 1993). Both Canadian captive-reared foxes and wild foxes obtained from Wyoming were released (Brechtel et al. 1996). Each release consisted of 70 to 100 foxes. In 1993, the wildlife directors of Alberta, Saskatchewan, and the Prairie and Northern Region of the Canadian Wildlife Service decided that the reintroduction effort would continue for another 5 years. A population assessment carried out in the winter of 1996–1997 estimated the Canadian swift fox population at 289 foxes, with a 95% confidence interval of 179–412 foxes (Cotterill 1997).

Observations of swift foxes have been recorded recently in north-central and southeastern Montana (Fig. 2).

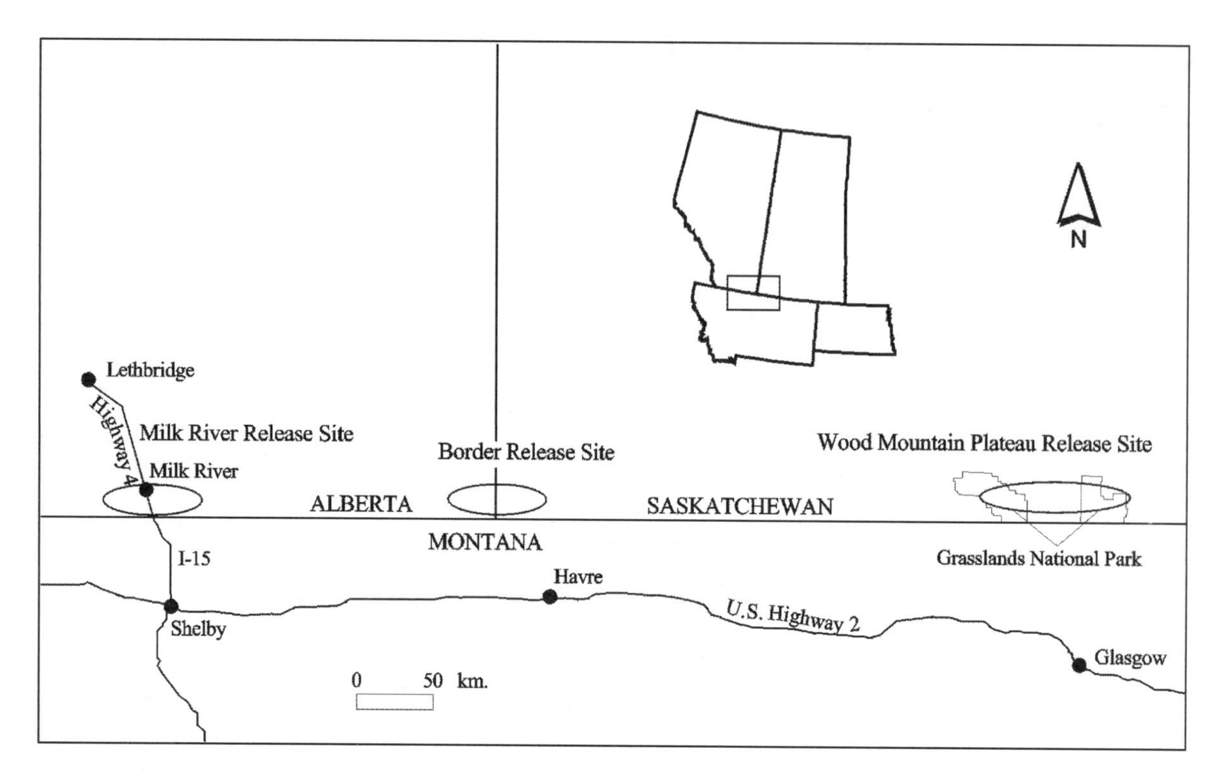

Figure 1. Release sites for the Canadian swift fox reintroduction program. The Milk River Ridge release site was located south of Lethbridge, Alberta. The Border release site was located along the borders of Alberta, Saskatchewan, and Montana. The Wood Mountain Plateau release site was located near Grasslands National Park in Saskatchewan.

Observations of swift foxes in portions of north-central Montana have been increasing in recent years, a result of the Canadian reintroduction program (Giddings and Knowles 1995). Swift foxes dispersed into north-central Montana from Canada throughout the late 1980s and early 1990s (Carbyn and Killaby 1989, Brechtel et al. 1993a). It was not known, however, if these dispersing foxes were surviving and reproducing in Montana.

Our goal was to verify and document the presence of the species in portions of north-central Montana. Furthermore, we wanted to document reproduction of swift foxes in Montana by locating rearing den sites. Additional study objectives included an investigation of relative distribution, home range size, reproductive success, survival, and food habits of radio-collared swift foxes.

Study Area and Methods

The study area was in north-central Montana, primarily in northern Blaine County, with secondary sites in northern Phillips and Valley counties (Fig. 2). The location of the primary study area was 48°45' to 49°00' North and 108°88' to 109°44' West. The landscape was gently rolling, glaciated short and mixed-grass prairie, and interspersed with dryland agricultural areas on upland sites. Moderate relief was created from creek drainages that cut through the prairies. Roads on the study area were either gravel or two-track trails (unimproved trails through grasslands) that experienced little human travel activity. Mean monthly temperatures ranged from -10.8° C to 20.6° C. Mean annual precipitation was 32 cm and ranged from 19 to 52 cm.

Dominant grasses on the study area were western wheatgrass (*Agropyron smithii*), junegrass (*Koeleria pyramidata*), blue grama (*Bouteloua gracilis*), needle-and-thread (*Stipa comata*), and Sandberg bluegrass (*Poa sandbergii*). Some localized areas were seeded to crested wheatgrass (*Agropyron cristatum*). The most common shrub was dwarf sagebrush (*Artemisia cana*). Other prominent plant species on the study area were American vetch (*Vicia americana*), scarlet globemallow (*Sphaeralcea* sp.), fringed sagewort (*Artemisia frigida*), white sage (*Artemisia ludoviciana*), toadflax (*Linaria* spp.), plains clubmoss (*Lycopodium* sp.), and prickly pear (*Opuntia* spp.). Soils on the study area had loam or clay-loam soil surfaces and clay or clay-loam subsoil types (National Cooperative Soil Survey 1976). Study area elevation ranged from 670 to 2,103 m. The Bureau of Land Management (BLM) managed 28% of the study area and 6% was managed by the state of Montana. Sixty-six percent of the study area was in private land ownership.

Capture and Handling

We systematically trapped 14 townships, each having an area of 92 km^2, using Tomahawk live traps (42 x 12 x 12 cm or 32 x 10 x 12 cm) placed in a block grid design

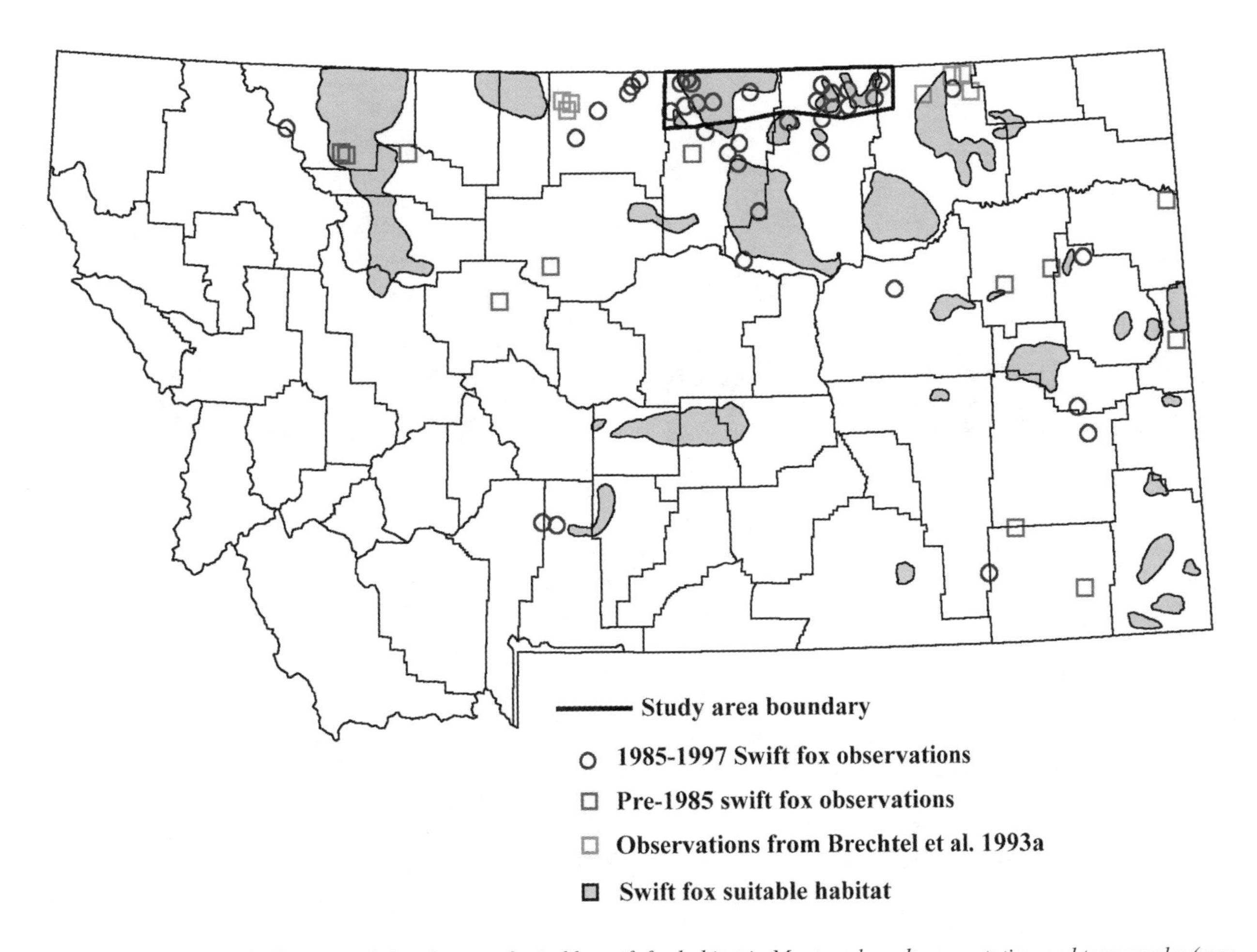

Figure 2. Location of study area and distribution of suitable swift fox habitat in Montana based on vegetation and topography (map modified from Montana Natural Heritage Program) with documented observations of swift foxes. Observations of swift foxes in Montana include pre-1985 observations, 1985–1997 observations, and observations of Canadian-released swift foxes (Brechtel et al. 1993).

(Fig. 3; Smith et al. 1975). Traps were placed on every corner of each 10-km² block in a township, resulting in a baseline number of 16 trap sites per township. This basic block design for trap placement was modified to increase the efficiency and ease of surveying a large area by placing traps as close to a pre-designated site as the road and trail system allowed. Loy (1981) found that placement of traps along roads or trails increased the ease of setting and checking traps and proved more successful for capturing swift foxes than a random placement of traps over the landscape. We trapped each township for 3 to 7 consecutive days, checking traps daily from 0530 to 1000 hr. Trapping was conducted from 12 August to 29 November 1996 and from 7 July to 15 August 1997. Trapping success was calculated as the number of foxes captured per 100 trap-nights of effort. We used beef tallow obtained from local grocery stores as bait, and a commercial trapping lure (Gary Jepson's Cow Country) was applied to a stick and placed on the top of each trap.

To handle captured swift foxes, we fitted a large mesh (2 x 2 cm) bag around the opening of the traps and foxes

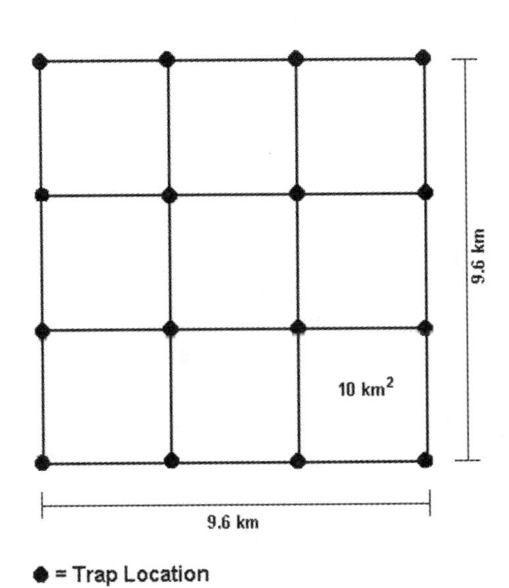

Figure 3. Example of grid design for trap placement within a 92-km² township. Traps were placed on the corners of each 10-km² area within a township, resulting in a baseline number of 16 trap sites per township.

ZIMMERMAN

Swift fox caught in trap.

ZIMMERMAN

Swift fox den site.

were transferred into the bag. Swift foxes were sedated using a 2:1 ketamine hydrochloride-xylazine hydrochloride mixture at a concentration of 10 mg/kg of estimated body weight (Kreeger 1996). Weight, sex, and age (juvenile or adult) were recorded for each captured fox. Ear length, hind foot length, tail length, and total lengths also were recorded. We tattooed the left ear with an individual number and radio-collared each fox. Radio-collars weighed 40–50 g and contained a mortality sensor that was activated after 4 (Wildlife Materials, Carbondale, Illinois) or 8 (Advanced Telemetry Systems [ATS], Isanti, Minnesota) hours of inactivity. Foxes were monitored during their recovery from the drug. Recovery times ranged from 20 to 40 minutes. Recovery times were faster for those foxes handled earlier in the morning (05:30–07:30 hr). As the foxes began to recover from the drug, movements were unsteady at first. However, muscle coordination returned quickly, within just a few minutes of moving away from the scene. We observed them with binoculars as they moved away from the site of capture until they stopped at a den site or traveled out of view.

Monitoring Swift Foxes

Foxes were relocated 4 times per week (or every other day) during the primary field seasons (fall 1996, summer 1997). We searched for foxes using aircraft every 10 to 14 days during winter 1996–1997 to document movements and survival. During the primary field seasons, we listened for signals while driving, using a truck-mounted bipolar antenna until a signal was heard. We stopped and obtained a bearing on the animal using a hand-held H-antenna. Regularly spaced radiotelemetry stations were used for stops when available. Painted and numbered wooden stakes designated radiotelemetry stations, which were placed along roads and trails in locations where signals were heard frequently. Telemetry stations were spaced 0.4 km apart along a road or trail. At least 3 bearings were obtained at different telemetry stations to determine a fox's location. Successive bearings for individual foxes were obtained within a maximum time of 4 minutes. Because swift foxes are mostly nocturnal (Hines and Case 1991), we located foxes from 2030 to 0200 hr and from 0300 to 0530 hr. We used 30 known locations of radio transmitters to determine the degree of precision of fox location bearings. The overall standard deviation of the bearing error was 13°.

We obtained telemetry locations from aircraft when a fox could not be located from the ground for 4 consecutive sampling nights, to locate rearing dens in the spring and summer of 1997, and during winter monitoring. An H-antenna was mounted to 1 wing strut of a Bellanca Scout 53819. Flights were made during the day, from 0700 to 1400 hr or from 1800 to 1930 hr. The accuracy of flight locations was not formally determined.

Home Range Estimation

We estimated locations using the program LOCATE II (Nams 1990) and home range size using the program CALHOME (Kie et al. 1994). We report the 100% minimum convex polygon (MCP; Mohr 1947) and the 95% adaptive kernel (Worton 1989) home range estimates. The minimum number of locations necessary to be included in the analysis was 30.

Survival Rate Estimation

We estimated an annual survival rate using the Kaplan-Meier estimation technique, modified for a staggered entry design (Pollock et al. 1989). SAS software was used for the analysis (SAS Institute 1997). When the mortality sensor in a radio-collar was activated, we retrieved the collar and/or carcass.

Rearing Den Characteristics and Reproduction

We located rearing dens of radio-collared swift foxes in the spring of 1997 using aerial and ground telemetry. Dens were observed on the ground from a distance of >100 m several times per week in the morning or evening, beginning in June and continuing until pups were seen outside the den. If pups were observed, we monitored rearing dens by ground twice per week through the summer of 1997 to count the number of pups present. All rearing dens were described and measured following dispersal of pups. Documentation included: number of entrances, dimensions of openings (width and length), opening exposure (N, S, E, W), slope, and distance to the nearest road.

Food Habits

We examined scats to evaluate swift fox food habits. Swift fox scats were collected in the study area during regular field activities; searching was concentrated along fence lines, trails, and gates. Scats also were collected from captured animals and at den sites. Scats were collected in every month from May to November. Individual scats were placed into separate, plastic zip-lock bags, labeled with the date, location, and swift fox number (when known), and frozen within 3 hours. We prepared scats for analysis by placing them into a strip of nylon hosiery and hand-washing them in a bucket (19 liters) with a tablespoon of laundry detergent (Smits et al. 1989). Scats were rinsed until the water appeared clear and were air-dried. Dried scats were placed on a sieve and separated into 7 categories: hair, bones, teeth, insect, feathers, vegetation, and other. Teeth were used to identify small mammals to genus in most cases. Insects were identified to order; feathers were simply classified as "bird," and plant material was labeled as vegetation. Food residues were reported by percent occurrence.

Results

Swift foxes were captured in 6 of 14 townships in which we trapped. In 1996, 9 swift foxes were captured in 1205 trap-nights (0.75 fox/100 trap-nights; Table 1). In 1997, 7 swift foxes were captured in 511 trap-nights (1.37 fox/100 trap-nights). Overall trapping success for the study was 0.93 fox/100 trap-nights.

The 9 swift foxes captured and collared in 1996 included 3 adult females, 3 adult males, and 3 juvenile males. In July of 1997, an adult male and female were captured and collared in northern Blaine County. We captured an adult male, an adult female and 2 female pups in northern Phillips County in 1997. These foxes were not radio-collared but were tattooed, measured, and released. An adult male swift fox captured in northwestern Valley County in 1997 had been previously marked with a tattoo ("VZ 25") in its left ear, identifying it as a fox released from the Canadian reintroduction project. This fox was born in May 1995 in the Valley Zoo in Edmonton, Alberta, and was released in the fall of 1995 in the east block of Grasslands National Park near Mankota, Saskatchewan (Fig. 1; Axel Moehrenschlager, Oxford University, personal communication). The capture location was about 26 km from the initial release site.

Home range estimates were calculated for 5 radio-collared swift foxes (Table 2). Four of the 9 swift foxes captured in the fall of 1996 were excluded from the home range analysis because of an inadequate number (11 to 20) of locations. Of the 4 foxes excluded from the analysis, 2 died early in the study period and 2 foxes left the study area and died at a later time. The MCP home range size ranged from 7.3 km^2 to 16.9 km^2 and averaged 10.4 km^2 (SE = 3.8 km^2). Adaptive kernel home range sizes ranged from 8.7 to 20.3 km^2 and averaged 12.3 km^2 (SE = 4.6 km^2).

We estimated an annual survival rate of 46% (SE = 13%). Our estimates were based on all 9 foxes captured in the fall of 1996 and the 2 adult foxes that were radio-collared in the summer of 1997. Eight of these foxes were adults and 3 were juveniles (young of the year). Of the 9

Table 1. Swift fox capture locations based on legal descriptions of townships in north -central Montana, 1996 –1997. Swift fox identification number, date, sex (M=Male, F=Female), age, and trapping success by year are included

	UTM				
Date	North	East	Swift Fox Number	Sex	Age
24 September 1996	5419000	624000	033	M	Adult
27 September 1996	5419000	624000	133	M	Juvenile
27 September 1996	5415900	620200	272	F	Adult
3 October 1996	5415900	620200	073	M	Adult
5 October 1996	5417300	620150	793	M	Juvenile
8 October 1996	5413000	627400	152	F	Adult
16 October 1996	5414680	630500	113	M	Juvenile
20 October 1996	5416260	630500	844	M	Adult
5 November 1996	5413200	638520	826	F	Adult
Total Trap Nights	1205				
Trap Success	0.75 foxes/100 trap nights				
10 July 1997	5401900	637250	834	M	Adult
14 July 1997	5403500	637200	350	F	Adult
4 August 1997	5406650	308250	Not collared	F	Juvenile
5 August 1997	5406650	308250	Not collared	M	Adult
5 August 1997	5409050	306700	Not collared	F	Adult
7 August 1997	5406650	308250	Not collared	F	Juvenile
13 August 1997	5406900	352050	VZ 25	M	Adult
Total Trap Nights	511				
Trap Success	1.37 foxes/100 trap nights				

swift foxes captured in the fall of 1996, 1 adult male, 1 adult female, and 2 juvenile male foxes died during winter and 1 juvenile male fox left the study area. Causes of death could not be determined for the foxes. One juvenile male, 1 adult male, and 2 adult females survived to mate and reproduce in the spring of 1997. Two foxes were added to the survival analysis in the summer of 1997. The mean number of days that foxes (n = 11) survived in 1 year was 212 days (SE = 140 days; Table 3). The number of days that foxes survived ranged from 28 to 399 days.

Eleven separate rearing dens were documented in the study. Two of the swift fox pairs and their pups occupied 4 different dens through the summer, and the third pair occupied 3 separate dens. All rearing dens were located less than 500 meters from a 2-track trail, and all rearing dens were located on slopes <15°. Most (73%) of rearing dens were located in areas with slopes of 0–5° and all rearing dens were surrounded by rangeland. Of the 11 rearing dens, 5 had 2 den entrances, 3 had 3 entrances, 2 had 4 entrances, and 1 had only 1 den entrance. The mean dimension (width, length) of den entrances was 24.6 cm x 42.0 cm (n = 34, SE = 6.8 cm x 17.8 cm). Over half (54%) of den entrances were exposed to a south/southeast direction, and 36% had a north/northeast exposure. One den site had an entrance toward the southwest, but no den entrances were exposed to the west or northwest.

Both pairs of radio-collared swift foxes that survived the winter of 1996–1997 produced offspring. One pair produced a litter of 5, and the other had a litter of 3. The swift fox pair captured in July of 1997 had produced a litter of 7 pups. Additionally, pups were captured in northern Phillips County, indicating that reproduction also had occurred there. Observations of rearing dens did not cause 2 pairs of foxes to move pups, but the third pair of foxes was particularly weary of researchers and often moved pups to different locations following visits near the den.

Mammals were the most common item in the diet of swift foxes, occurring in 91% (*n* = 59) of the 65 scats analyzed; followed by insects (66%; *n* = 43), vegetation (48%; *n* = 31), and birds (32%; *n* = 21; Table 4). Eleven percent of 65 scats contained unidentified mammal material.

The most frequently occurring mammal in scats was voles (*Cricetidae*, 68%, *n* = 44), followed by deer mice (*Peromyscus* spp., 18%, *n* = 12), Richardson's ground squirrel (*Spermophilus richardsonii*, 9%, *n* = 6), pocket mice (*Perognathus* sp., 3%, *n* = 2); 1 sample contained shrew remains (*Sorex* sp., 2%).

Insect material was mainly composed of beetles (Coleoptera, 49%, *n* = 32) and grasshoppers (Orthoptera, 43%, *n* = 28). Seven different families of beetle were identified, including Cicindelidae (tiger beetle), Curculionidae (weevil), Byrrhidae (pill beetle), Scarabaeidae (dung beetle), Carabidae (ground beetle), Tenebrionidae (darkling beetle), and Silphidae (carrion beetle). The 2 families of grasshopper identified were Acrididae (grasshopper) and Gryllidae (cricket). Ants (Formicidae), of the order Hymenoptera, occurred in 14% of the scats. A few scats contained remains of stink bugs (Hemiptera), caterpillars (Lepidoptera), and damselfly adults (Odonata).

Birds occurred in 32% of the scats examined. Two of the 65 samples had eggshell present. Feathers found in scats were too small to identify easily. However, mallard (*Anas platyrhynchos*) and horned lark (*Eremophila alpestris*) remains were found at den sites in the summer of 1997. On one occasion, a willet (*Catoptrophorus semipalmatus*) was found lying beside the trap when a juvenile female swift fox was captured south of Whitewater Lake (Phillips County). Vegetation or berries were present in 48% of the scats.

Mammals were represented evenly throughout the 3 seasons (Table 4). Insects occurred in all 3 seasons, peaking in the fall at 75% occurrence. Vegetation occurrence was highest in the summer (65%) but was present in feces in all 3 seasons. Bird occurrence was highest in the spring (60%), was present in feces through the summer (39%), and declined in occurrence in the fall (20%).

Discussion

A small percentage of the potential swift fox habitat in north-central Montana was surveyed by trapping in this study. There are about 688,000 ha of prairie in north-central Montana, primarily in Blaine, Phillips, and Valley counties (C. Knowles, FaunaWest Wildlife Consultants, Montana, personal communication). The total area surveyed in this study was 129,024 ha, or 19% of the total estimated area of prairie in north-central Montana.

Home range information was rather limited in this study. Our inability to detect signals from >1.6 km was insufficient for the difficult and remote access that was characteristic of the study area. There were circumstances when individual foxes could not be found during a sampling night. Furthermore, all winter locations were done by air during the early morning or late evening hours, but never during the night. Thus, no winter nocturnal locations were collected. The home ranges found in this study were probably underestimated. Kitchen et al. (1999) reported a mean home range estimate of 7.6 km^2 for swift foxes in Colorado using the 95% adaptive kernel method. Our 95% adaptive kernel home range estimates were comparable to those of Kitchen et al. (1999). Hines and Case (1991) reported larger 100% MCP home range sizes of up to 32.3 km^2 in Nebraska. It was speculated that the large home range sizes were the result of low population densities and/or low availability of prey. In Colorado, Andersen et al. (2003) reported a mean 100% MCP home range size of 29.0 km^2 (*n* = 5).

The estimated annual survival rate for swift foxes in this study—46%—was comparable to reported survival

Table 2. Home range size estimated by the Minimum Convex Polygon method (MCP; Mohr 1947) and Adaptive Kernel method (Worton 1989) and time periods of data collection of five swift fox in north -central Montana, 1996 –1997

Swift Fox Number	Sex	Age	Period radio-tracked	No. of locations	MCP (km^2)	Adaptive Kernel (km^2)
133	M	Juvenile	21 Oct. 1996–13 Sept. 1997	58	7.4	11.3
826	F	Adult	6 Nov. 1996–9 Oct. 1997	63	16.9	20.3
152	F	Adult	8 Oct. 1996–9 Oct. 1997	72	8.1	8.7
844	M	Adult	21 Oct. 1996–11 Sept. 1997	70	9.4	9.6
272	F	Adult	8 Oct. 1996–7 April 1997	30	10.1	11.3

Table 3. Dates captured, fates, and number of days surviving for 11 swift foxes during 1996 –1997 in north-central Montana. Sex: M=Male, F=Female; Age: A=Adult, J=Juvenile; Fate: D=Dead, C=Right censored

Swift Fox Number	Sex	Age	Beginning Date	End Date	Fate	Total Days Monitored
033	M	A	24 Sept. 1996	17 Feb. 1997	D	147
133	M	J	27 Sept. 1996	30 Oct. 1997	C	399
272	F	A	27 Sept. 1996	7 April 1997	D	193
073	M	A	3 Oct. 1996	19 Jan. 1997	D	109
793	M	J	5 Oct. 1996	4 March 1997	C	151
152	F	A	8 Oct. 1996	30 Oct. 1997	C	388
113	M	J	16 Oct. 1996	31 Dec. 1996	D	77
844	M	A	20 Oct. 1996	30 Oct. 1996	C	376
826	F	A	5 Nov. 1996	30 Oct. 1996	C	360
834	M	A	10 July 1997	23 Oct. 1997	C	106
350	F	A	14 July 1997	10 Aug. 1997	C	28
					Mean	212
					SE	140

Table 4. Percent occurrence of food items found in 65 swift fox scats and number of scats containing food items (in parentheses). Scats were collected in the spring, summer, and fall in north -central Montana, 1996–1997

	Spring: 1 May–21 June	Summer: 22 June–21 Sept.	Fall: 22 Sept.–30 Nov.	Total
Total No. of Samples	10	23	32	65
Mammals:	90 (9)	91 (21)	91 (29)	91 (59)
Cricetidae	90 (9)	48 (11)	84 (27)	72 (47)
Microtus, *Lagurus*, or *Clethrionomys*	80 (8)	43 (10)	81 (26)	68 (44)
Peromyscus spp.	40 (4)	4 (1)	22 (7)	18 (12)
Sciuridae	10 (1)	22 (5)	0	9 (6)
Spermophilus sp.	10 (1)	22 (5)	0	9 (6)
Heteromyidae	10 (1)	0	3 (1)	3 (2)
Perognathus sp.	10 (1)	0	3 (1)	3 (2)
Soricidae	0	4 (1)	0	2 (1)
Sorex spp.	0	4 (1)	0	2 (1)
Unidentified Mammal	0	22 (5)	6 (2)	11 (7)
Insects	50 (5)	61 (14)	75 (24)	66 (43)
Orthoptera	10 (1)	17 (4)	72 (23)	43 (28)
Coleoptera	50 (5)	57 (13)	44 (14)	49 (32)
Lepidoptera	10 (1)	41 (1)	3 (1)	5 (3)
Hemiptera	0	9 (2)	0	3 (2)
Hymenoptera	10 (1)	22 (5)	9 (3)	14 (9)
Odonata	0	0	3 (1)	2 (1)
Vegetation	60 (6)	65 (15)	31 (10)	48 (31)
Bird	60 (6)	39 (9)	19 (6)	32 (21)

Table 5. Small mammals occurring in north -central Montana. Information on small mammal distribution in Montana was obtained from Allen et al. (1994) and Hoffman and Pattie (1968).

Common Name	Scientific Name
Bushy-tailed woodrat	*Neotoma cinereus*
Northern grasshopper mouse	*Onocomys leucogaster*
Deer mouse	*Peromyscus maniculatus*
White-footed mouse	*Peromyscus leucopus*
Western harvest mouse	*Reithrodontomys megalotis*
Red-backed vole	*Clethrionomys gapperi*
Sagebrush vole	*Lagurus curtatus*
Meadow vole	*Microtus pennsylvanicus*
Prairie vole	*Microtus ochrogaster*
Long-tailed vole	*Microtus longicaudus*
Masked shrew	*Sorex cinereus*
Pygmy shrew	*Sorex hoyi*
Merriam's shrew	*Sorex merriami*
Dwarf shrew	*Sorex nanus*
Northern water shrew	*Sorex palustris*
Preble's shrew	*Sorex preblei*
Vagrant shrew	*Sorex vagrans*
Least chipmunk	*Tamias minimus*
Thirteen-lined ground squirrel	*Spermophilus tridecemlineatus*
Richardson's ground squirrel	*Spermophilus richardsonii*
Ord's kangaroo rat	*Dipodomys ordii*
Olive-backed pocket mouse	*Perognathus fasciatus*
Western jumping mouse	*Zapus princeps*
Northern pocket gopher	*Thomomys talpoides*
Black-tailed prairie dog	*Cynomus ludovicianus*
White-tailed jackrabbit	*Lepus townsendii*

estimates. Sovada et al. (1998) reported an annual survival rate of 43% for 40 radio-collared adult swift foxes and 33% (6-month period) for 24 juvenile swift foxes in Kansas. A 50% annual survival rate among 14 radio-marked swift foxes was reported in South Dakota (Sharps and Whitcher 1984). Zumbaugh (1984) also reported a mean survival rate of 50% for adult swift foxes based on 2 years of harvest data in Kansas. Minimum annual survival rates for wild born Canadian swift foxes have been estimated at 46% for adults and 36% for juveniles during their first year (Brechtel et al. 1993b). In this study, adults and juveniles (young of the year) were included to obtain the annual survival rate estimate. One might expect that the annual survival rate estimate was lower as a result of including juveniles in the overall analysis. If this was the case, then the overall survival rate indicates that swift foxes are surviving rather well on the northern Montana prairie.

Rearing den sites were located in open, relatively level areas, often near the top of a small rise or ridge. Other studies have indicated a preference for areas with <15° slope (Hillman and Sharps 1978, Loy 1981, Jackson and Choate 2000). Additional common characteristics included selection for an east or south exposure, a loamy soil, and areas of sparse vegetation (Kilgore 1969, Hillman and Sharps 1978, Loy 1981, Uresk and Sharps 1981, Hines and Case 1991). These characteristics also were common to den sites in this study.

Locating evidence of reproduction was one of the main goals of this study, and we identified litter sizes of 3, 5, and 7. Reports of litter sizes in the Canadian reintroduced population have ranged from 1 to 7 pups, with a mean litter size of 3.9 (Brechtel et al. 1993b). A record litter of 8 was reported for a pair of foxes on the Canadian prairie in the spring of 1997 (A. Moehrenschlager, Oxford University, personal communication). Covell (1992) reported that the mean litter size for swift foxes in Colorado was 2.4 pups for pairs and 4.2 pups for pairs with a helper fox. Hillman and Sharps (1978) and Kilgore (1969) reported larger mean litter sizes of 4 and 5 for swift foxes, respectively.

Mammals were the most frequent food item found in the swift fox scats collected. Uresk and Sharps (1981) also found mammals to be the most frequent food item (49%) in the diets of swift foxes in South Dakota, followed by insects (27%), plants (13%), and birds (6%). We found rodents, especially voles, to be the most common food item of swift foxes in north-central Montana. Hines and Case (1991) found the prairie vole (*Microtus ochrogaster*) to be the most common mammal species found in 52 scat samples in Nebraska. Cameron (1984) reported that rodents were important food items in spring (75%, $n = 4$) and summer (82%, $n = 22$) on the Pawnee National Grassland in Colorado. See Table 5 for a list of small mammal species that commonly occur in north-central Montana.

Cottontail rabbits and jackrabbits (*Leporidae*) were not detected in any of the 65 scat samples in this study, even though white-tailed jackrabbits (*Lepus townsendii*) seemed to be quite abundant on the study area. Hair analysis was not included in the methods for scat analysis. An examination of the hair in individual scats may have revealed the presence of jackrabbits. Cottontail rabbits and jackrabbits have been found to be common food items in other studies, especially in winter (Kilgore 1969, Cameron 1984, Zumbaugh et al. 1985, Hines and Case 1991). Kilgore (1969) documented that cottontail rabbits and jackrabbits decreased in scat occurrence in the spring but seemed important in the winter and early spring. The scats we collected were from late spring, summer, and fall.

A high percentage of scats examined had at least some insect material present. In a study in Nebraska, Hines and Case (1991) found that 56% of the swift fox scats had insect remains, most of which consisted of grasshoppers (37%) and beetles (25%). This result was quite similar to our result of 66% occurrence of insect material in scats. In South Dakota, Uresk and Sharps (1981) reported that insects were found in 27% of scats.

Vegetation found in scats was mainly grass and was usually in trace amounts, suggesting vegetation may have been ingested incidentally. However, 3 scats were entirely composed of vegetation. One was composed of grass, 1 was unidentified, and 1 was composed of berries, probably currants or gooseberries (*Ribes* sp.). Other vegetation found in scats included spike moss (*Selaginella* spp.) and nutlets from the family *Boraginaceae*. Hines and Case (1991) found that 54% of scats contained vegetation. Uresk and Sharps (1981) found vegetation in 13% of scats.

It is not known whether swift foxes ingest vegetation as they consume other prey or whether vegetation is consumed purposely. Zumbaugh et al. (1985) reported that only 1 stomach sample contained enough grass to indicate that it had been ingested purposely. Hines and Case (1991) concluded that vegetation was eaten intentionally in some instances. The 3 scats composed entirely of vegetation that we examined suggest that vegetation materials were ingested intentionally in our study area.

Birds occurred in 32% of the scats examined. Most feathers were found in scats deposited in late spring and into the summer. This was consistent with studies reporting that both bird and insect remains increased from winter to summer as their availability increased (Hines and Case 1991).

Management Implications

The swift fox has proved to be an adaptable and opportunistic species. However, it has been negatively affected by past human activities such as landscape changes, unregulated trapping, predator shooting, and poisoning programs. Thus, the degree to which we impose human activities on the landscape today will ultimately determine whether this recently established swift fox population in north-central Montana will persist. Potential threats to the viability of the swift fox population in north-central Montana include poisoning (coyotes and ground squirrels), the loss of native grassland prairie to cultivation, indiscriminate predator shooting, road deaths, and accidental trappings. Grassland habitat loss was one of the factors that led to the extirpation of swift foxes in Montana (Egoscue 1979, Allardyce and Sovada 2003). Currently in north-central Montana, there is no widespread trend toward converting prairie into agricultural lands. Historically, there was large-scale conversion of prairie to cultivated areas in some portions of north-central Montana. There is currently some small-scale conversion of prairie to cropland in northwestern Blaine County, along the Hill/Blaine County line (Jody Peters, Bureau of Land Management, personal communication). One positive aspect concerning the habitat in north-central Montana is that much of the native prairie today exists on public lands. Poisoning programs are not currently in use by Wildlife Services (Animal Damage Control) in north-central Montana. Wildlife Services is primarily involved in coyote trapping. The BLM has recently restricted the use of M44 poison baits and snaring near known swift fox locations on BLM-managed lands in northern Blaine County (Jody Peters, Bureau of Land Management, personal communication). Road deaths, however, may become an important concern as oil and gas development increases on public lands in north-central Montana. As more oil and gas structures are created, more

ZIMMERMAN

Free-ranging swift fox.

roads are developed to service those structures, which may increase the possibility of road deaths for swift foxes (Kilgore 1969, Matlack et al. 2000).

The influence or importance of any of these threats and habitat changes is unknown because of insufficient information on the distribution of swift foxes in Montana. Thus, additional surveys are needed to determine the extent of swift fox distribution and population density in Montana. Our experience indicated that using live-trapping and radio telemetry would be an expensive and ultimately inefficient means of surveying for swift fox distribution in Montana. Two alternate possibilities would be surveys of landowners and trappers and systematic track searches by townships. Surveys of landowners and trappers would be cheaper, but these surveys may be ineffective due to low response and unreliability of information. Our experience suggests that track searches would be reliable, quick, and less expensive. A track search method was used in Kansas and Oklahoma as a means of detection and proved to be quite successful (Christiane Roy, Kansas Department of Wildlife and Parks, personal communication).

After distribution has been defined for swift foxes in Montana, other management steps might be taken. Our data suggest that the population is currently reproducing and has similar survival rates as other swift fox populations in the U.S. This indicates occupied areas are suitable for swift foxes. Specific studies contrasting occupied with unoccupied areas that emphasize relative land cover, human activity type/intensity, and densities of coyotes would help determine if the present distribution represents full occupancy of suitable habitat or just early stages of colonization. Swift fox dispersal capabilities and the connectivity of suitable habitat should be studied to determine the probability of successful colonization into unoccupied areas. Population dynamics studies and perhaps augmentation efforts may be needed in the future to ensure the stability of the swift fox population.

It appears that swift foxes have begun to reestablish a population in north-central Montana. The area of potential habitat for further colonization is large (688,000 ha of native prairie primarily in Blaine, Phillips, and Valley counties). Much of the areas that swift foxes currently occupy in Montana are lands managed by the Bureau of Land Management (BLM) or the state of Montana. Public lands biologists, state biologists, and land managers will now have the opportunity to actively participate in the recovery of swift foxes in Montana by cooperating with private landowners and by considering the impacts of different land management strategies on the habitat requirements of swift foxes.

Acknowledgments

This project was funded by Montana Fish, Wildlife, and Parks and supported by Montana State University. Thanks to the following for technical support: Kristie Allen, Ludwig Carbyn, Alisa Gallant, Dan Gustafson, Melissa Hart, Craig Knowles, Bob Moore, Jay Rotella, Christiane Roy, Kevin Salsbery, Cathy Siebert, Marsha Sovada, Mark Taper, and Bo Wilmer. Thanks to the following agency members: Jody Peters (BLM), Kent Gilge, Dan Hughbanks, and Al Rosgaard (Montana Fish, Wildlife, and Parks). Thanks to the leasees and landowners of our study area for their cooperation. Thanks to Buttrey's of Havre and Media Works of Bozeman. Thanks to our technicians Jim Zimmerman and Heather Marstall.

Literature Cited

Allardyce, D., and M.A. Sovada. 2003. A review of the ecology, distribution, and status of swift foxes in the United States. Pp. 3–18 *in* M.A. Sovada and L.N. Carbyn, editors. The Swift Fox: ecology and conservation of swift foxes in a changing world. Canadian Plains Research Center, Regina, Saskatchewan.

Allen, K.L., T. Weaver, and D. Flath. 1994. Small mammals in northern rocky mountain ecosystems. Report to the Bureau of Land Management and U.S. Forest Service, Bozeman, Montana.

Andersen, D.E., T.R. Laurion, J.R. Cary, R.S. Sikes, M.A. McLeod, and E.M. Gese. 2003. Aspects of swift fox ecology in southeastern Colorado. Pp. 139–148 *in* M.A. Sovada and L.N. Carbyn, editors. The Swift Fox: ecology and conservation of swift foxes in a changing world. Canadian Plains Research Center, Regina, Saskatchewan.

Armstrong, D.M. 1972. Distribution of mammals in Colorado. University of Kansas Museum Natural History Publication, Lawrence, Kansas.

Banfield, A.W.F. 1974. The mammals of Canada. University of Toronto Press, Toronto, Ontario.

Bee, J.W., G.E. Glass, R.S. Hoffmann, and R.P. Patterson. 1981. Mammals in Kansas. University of Kansas Publication, Museum of Natural History Publication, Lawrence, Kansas.

Brechtel, S.H., L.N. Carbyn, D. Hjertaas, and C. Mamo. 1993a. Status of the swift fox recovery project. Swift fox recovery team final report for 3-year study from 1989–1992.

——. 1993b. Canadian swift fox reintroduction feasibility study: 1989 to 1992. Scientific and Technical Documents Division, Canadian Wildlife Service, Ottawa, Ontario.

Brechtel, S.H., L.N. Carbyn, G. Erikson, D. Hjertaas, C. Mamo, and P. McDougall. 1996. National recovery plan for the swift fox. Report No. 15. Ottawa: Recovery of Endangered Wildlife Committee. Scientific and Technical Documents Division, Canadian Wildlife Service, Ottawa, Ontario.

Cameron, M.W. 1984. The swift fox (*Vulpes velox*) on the Pawnee National Grassland: its food habits, population dynamics, and ecology. Thesis, University of Northern Colorado, Greeley, Colorado.

Carbyn, L.N., and M. Killaby. 1989. Status of the swift fox in Saskatchewan. Blue Jay 47:41–52.

Carbyn, L.N., Brechtel, D. Hjertaas, and C. Mamo. 1993. An update on the swift fox reintroduction program in Canada. Pp. 366–372 *in* G.L. Holroyd, H.L. Dickson, M. Regnier, and H.C. Smith, editors. Proceedings of the 3rd prairie conservation and endangered species workshop. Provincial Museum of Alberta Natural History Occasional Paper No. 19.

Cotterill, S.E. 1997. Population census of swift fox (*Vulpes velox*) in Canada: winter 1996–1997. Unpublished report for the swift fox national recovery team. Alberta Environmental Protection (Natural Resources Service), Edmonton, Alberta.

Covell, D.F. 1992. Ecology of the swift fox (*Vulpes velox*) in southeastern Colorado. Thesis, University of Wisconsin, Madison, Wisconsin.

Egoscue, H.J. 1979. *Vulpes velox*. Mammalian Species Number 122:1–5.

Giddings, B., and C.J. Knowles. 1995. The current status of swift fox in Montana. Montana Department of Fish, Wildlife and Parks and FaunaWest Wildlife Consultants, Helena, Montana.

Hall, E.R. 1981. The mammals of North America. John Wiley and Sons, New York, New York.

Hillman, C.N., and J.C. Sharps. 1978. Return of the swift fox to northern great plains. Proceedings of the South Dakota Academy of Science 57:154–162.

Hines, T.D., and R.M. Case. 1991. Diet, home range, movements, and activity periods of swift fox in Nebraska. Prairie Naturalist 23:131–138.

Hoffman, R.S., and D.L. Pattie. 1968. A guide to Montana mammals. University of Montana Press, Missoula, Montana.

Hoffman, R.S., P.L. Wright, and R.E. Newby. 1969. The distribution of some mammals in Montana. I. Mammals other than bats. Journal of Mammalogy 50:579–604.

Jackson, V.L., and J.R. Choate. 2000. Dens and den sites of the swift fox, *Vulpes velox*. Southwestern Naturalist 45:212–220.

Johnson, D.R. 1969. Returns of the American fur company, 1835-1839. Journal of Mammalogy 50:836–839.

Kie, J.G., J.A. Baldwin, and C.J. Evans. 1994. CALHOME: Home range analysis program, user's manual. U.S. Forest Service, Fresno, California.

Kilgore, D.L., Jr. 1969. An ecological study of swift fox (*Vulpes velox*) in the Oklahoma panhandle. American Midland Naturalist 81:512–534.

Kitchen, A.M, E.M. Gese, and E.R. Schauster. 1999. Resource partitioning between coyotes and swift foxes: space, time, and diet. Canadian Journal of Zoology 77:1645–1656.

Kreeger, T.J. 1996. Handbook of Wildlife Chemical Immobilization. International Wildlife Veterinary Services, Incorporated Publication, Laramie, Wyoming.

Loy, R.R. 1981. An ecological investigation of the swift fox (*Vulpes velox*) on the Pawnee National Grasslands, Colorado. Thesis, University of Northern Colorado, Greeley, Colorado.

Matlack, R.S., P.S. Gipson, and D.W. Kaufman. 2000. The swift fox in rangeland and cropland in western Kansas: relative abundance, mortality, and body size. Southwestern Naturalist 45:221–225.

Mohr, C.O. 1947. Table of equivalent populations of North American mammals. American Midland Naturalist 37: 223–249.

Moore, R.E., and N.S. Martin. 1980. A recent record of swift fox (*Vulpes velox*) in Montana. Journal of Mammalogy 61:161.

Nams, V.O. 1990. Locate II: user's guide. Pacer Publishers, Truro, Nova Scotia.

National Cooperative Soil Survey. 1976. Soil survey of Blaine County and part of Phillips County Montana. United States Department of Agriculture, Soil Conservation Service.

Pollock, K., S.R. Winterstein, C.M. Bunck, and P.D. Curtis. 1989. Survival analysis in telemetry studies: the staggered entry design. Journal of Wildlife Management 53:7–15.

SAS Institute. 1997. SAS SAS/STAT user's guide, version 6.0. SAS Institute, Cary, North Carolina.

Scott-Brown, J.M., S. Herrero, and J. Reynolds. 1987. Swift fox. Pp. 433–441 *in* M. Novak, J.A. Baker, M.E. Obbard, and B. Malloch, editors. Wild furbearer management and conservation in North America. Ontario Trappers Association, North Bay, Ontario.

Sharps, J.C., and M.F. Whitcher. 1984. Swift fox reintroduction techniques. South Dakota Department of Game, Fish, and Parks, Rapid City, South Dakota.

Smith, M.H., R.H. Cardner, J.B. Gentry, D.W. Kaufman, and M.H. O'Farrell. 1975. Density estimations of small mammal populations. Pp. 25–53 *in* S.F.B. Golley, K. Petrusewicz, and L. Ryszkowski, editors. Small Mammals: their productivity and population dynamics. Cambridge University Press, Cambridge, UK.

Smits, C.M., B.G. Sough, and C.A. Yasui. 1989. Summer food habits of sympatric arctic foxes, *Alopex lagopus*, and red foxes, *Vulpes vulpes*, in the northern Yukon Territory. Canadian Field-Naturalist 103:363–367.

Sovada, M.A., C.C. Roy, J.B. Bright, and J.R. Gillis. 1998. Causes and rates of mortality of swift foxes in western Kansas. Journal of Wildlife Management 62:1300–1306.

Uresk, D.W., and J.C. Sharps. 1981. Denning habitat and diet of the swift fox in western South Dakota. Great Basin Naturalist 46:249–253.

Warren, E.R. 1942. The mammals of Colorado. The Knickerbocker Press, New York, New York.

Worton, B.J. 1989. Kernel methods for estimating the utilization distribution in home range studies. Ecology 70:164–168.

Young, S.P. 1944. The wolves of North America Part 1. Their history, life habits, economic status, and control. Pp. 1–385 *in* S.P. Young and E.A. Goldman, editors. The wolves of North America, Part 1. Dover, New York, New York, American Wildlife Institute, Washington, D.C.

Zumbaugh, D.M. 1984. Natural history of foxes in Kansas. Thesis, Fort Hays State University, Hays, Kansas.

Zumbaugh, D.M., and J.R. Choate. 1985. Historical biogeography of foxes in Kansas. Transactions of the Kansas Academy of Science 88:1–13.

Zumbaugh, D.M., and J.R. Choate, and L.B. Fox. 1985. Winter food habits of the swift fox on the central high plains. Prairie Naturalist 17:41–47.

Swift Fox Detection Methods and Distribution in the Oklahoma Panhandle

■ Michael J. Shaughnessy, Jr.

Abstract: Baited tracking plates, infra-red triggered cameras, and spotlighting surveys were used to determine the distribution of swift foxes in the Oklahoma panhandle during January 1995 to July 1997. Tracking plates were the most cost and time effective method for detecting carnivores in the panhandle during the study. Cameras were valuable for track verification at tracking plates but were expensive to operate and spotlighting was not time effective. Swift foxes were distributed unevenly but widely across the Oklahoma panhandle. Highest concentrations occurred in the westernmost portion of the panhandle. Swift foxes were detected at highest rates in habitats with lowest coyote detection rates. Areas with large coyote populations supported few swift foxes.

The first records of swift fox (*Vulpes velox*) in Oklahoma date from 1888, when a specimen was obtained from what is now the Oklahoma panhandle (Caire et al. 1989). Historically, the swift fox was reported to range throughout the Oklahoma panhandle and Woodward County in northwestern Oklahoma (Caire et al. 1989).

The most comprehensive and recent study of swift foxes in Oklahoma was completed by Kilgore (1969). He examined denning habits, breeding and reproduction, food preferences, and parasites of swift foxes in Beaver County, from 1965 to 1966. Since 1969, most studies of the Oklahoma swift fox population have been completed in conjunction with swift fox investigations in neighboring states (e.g. Zumbaugh and Choate 1985). The U.S. Fish and Wildlife Service's decision to include the swift fox as a candidate species for federal listing under the Endangered Species Act (Federal Register 1995) prompted renewed interest in the distribution of swift foxes in Oklahoma. Three primary objectives were identified as the focus of the swift fox study in Oklahoma. The first objective was to evaluate the efficacy of scent-post surveys, spotlighting, and infra-red triggered cameras as carnivore detection techniques. Since few recent studies have been conducted on swift fox, there was no single preferred technique for investigating this particular carnivore. A determination of which techniques worked best in Oklahoma seemed vital to the success of the project. The second objective was to determine the current range of the swift fox in Oklahoma. Again, no recent range or population estimations existed for the swift fox in Oklahoma prior to 1989 (Caire et al. 1989). The most recent estimations of swift fox range and population status in Oklahoma were based upon data collected from neighboring states and from foxes collected in Oklahoma prior to 1970. A current status for the species in Oklahoma needed to be determined. The third objective was to investigate habitat affinities and interspecific associations (with other wild canids) of the swift fox.

Agonistic interactions between larger and smaller carnivores, particularly among canids, has been well documented (Polis et al. 1989, White et al. 1994, Peterson 1995, Ralls and White 1995, White et al. 1995, Johnson et al. 1996). Larger canids aggressively harass and sometimes kill smaller canids, excluding them from the ecosystem locally but not regionally. A major goal of this project was to determine if this interaction was present among panhandle canids. The panhandle potentially supports four canid species (swift fox, red fox [*Vulpes vulpes*], gray fox [*Urocyon cinereoargenteus*], and coyote [*Canis latrans*]). Detecting interactions of this kind between these species would not only further document these established interspecific interactions, but could also explain patterns in occurrence among the species.

Study Area

The Oklahoma panhandle is a strip of land approximately 267 kilometers long (east–west) and 55 kilometers wide (north–south) and is the most northwesterly portion of the state. Three counties comprise the panhandle: (east to west) Beaver County (approximately 470,172 ha); Texas County (approximately 527,855 ha); and Cimarron County (approximately 475,506 ha).

Historically, the Oklahoma panhandle was an extensive shortgrass prairie dominated by blue grama (*Bouteloua gracilis*), buffalograss (*Buchloe dactyloides*), and prairie three-awn (*Aristida oligantha*; Duck and Fletcher 1943). Much of the area was occupied by prairie dog towns, which occurred throughout all habitat types (Shackford et al. 1989, Shackford and Tyler 1991) and were reported to have stretched for miles (Shackford et al. 1989, Shackford and Tyler 1991). In addition, several major riparian areas cut through the landscape. These were dominated (and can be identified during dry times) by the presence of large eastern cottonwoods (*Populus deltoides*) and taller grasses. The extreme northwest corner of the panhandle is a mesa habitat, dominated by sand sagebrush (*Artemisia filifolia*), juniper (*Juniperus scopulorum*), and two-needle pine (*Pinus edulis*).

During the past 100 years, the panhandle landscape has been altered by agriculture, livestock grazing, and fossil fuel extraction. While historically dominant habitat types still persist, the quality and quantity of these habitats has changed. Grassland, mesa, and riparian areas are now almost entirely grazed by domestic cattle. The severity of grazing impact varies among habitats. Prairie dog towns have been greatly reduced in both number and size by the combined effects of periodic plague (*Yersinia pestis*) episodes and concentrated eradication efforts. Finally, another habitat has been added to this environment. Agriculture now predominates parts of the panhandle. Extensive grain monocultures have replaced tracts of shortgrass prairie.

Panhandle habitats were identified during this study by vegetation, geological/hydrological features, and/or land use. Four major habitats—mesa, riparian, agriculture, and range—were recognized in the Oklahoma panhandle. The mesa area, in the extreme northwestern section of the panhandle, encompasses approximately 74,290 hectares of continuous land and is the only single, continuous habitat in the panhandle. There are 4 major riparian corridors, of varying size, moving west to east in the Oklahoma panhandle. These corridors, along with their associated tributaries, drainages, and soils, were identified as panhandle riparian areas and account for approximately 133,881 hectares of land. These areas contained water sometime during the year but were usually also dry at times. Agricultural land is not a natural habitat formation, but was included as a habitat because so much of the prairie ecosystem had been altered to agricultural uses. Agricultural land was defined as any plowed and/or planted field, any field with central pivot irrigation, or any bare or stubble field. Agricultural land encompassed approximately 421,053 hectares of land in the panhandle. Finally, range habitat was defined as any land not primarily used for agriculture, and usually grazed at some time during the year. Range areas were typically characterized by a variety of naturally occurring and/or invading grasses and shrubs from 0.1–0.75 meters high. The overall composition of range areas was highly variable across the panhandle. Range areas accounted for approximately 844,292 hectares of land in the panhandle.

The Oklahoma panhandle supports 17 species of carnivores from 5 families (Caire et al. 1989; Table 1). Four of these (gray fox, western spotted skunk [*Spilogale gracilis*], hog-nosed skunk [*Conepatus mesoleucus*], and ringtail [*Bassariscus astutus*]) have been restricted to the mesa ecosystem in the extreme northwestern corner of the panhandle (Caire et al. 1989). The others are found throughout the panhandle.

Table 1. Carnivores of the Oklahoma panhandle

Scientific Name	Common Name
Family Mustelidae	
Taxidea taxus	Badger
Mustela nigripes	Black-footed ferret
Mustela frenata	Long-tailed weasel
Mephitis mephitis	Striped skunk
Spilogale putorius	Eastern spotted skunk
*Spilogale gracilis**	Western spotted skunk
*Conepatus mesoleucus**	Hog-nosed skunk
Family Canidae	
Vulpes velox	Swift fox
Vulpes vulpes	Red fox
*Urocyon cinereoargenteus**	Gray fox
Canis latrans	Coyote
Canis lupus	Wolf
Family Felidae	
Lynx rufus	Bobcat
Felis concolor	Cougar
Family Procyonidae	
Procyon lotor	Raccoon
*Bassariscus astutus**	Ringtail
Family Ursidae	
Ursus americanus	Black bear

*Restricted to mesa habitats

Methods

Track Plates

I determined the presence and distribution of carnivores with baited-tracking plates placed at pre-established tracking stations. An approximately 1m x 1m 26-gauge stainless steel tracking plate was placed at each tracking station and sprayed with a mixture of isopropyl alcohol and carpenter's chalk. The alcohol served as a dispersant, evaporated, and left a thick, uniform coating of chalk on the plate. Each plate had a 1-inch hole drilled through its center, allowing it to be placed directly over a stake which permanently marked the tracking station. Each plate cost approximately $7 U.S. to construct. Bait was then placed in the middle of the plate or on the stake (Fig. 1). Baits used during this study were beef scraps and canned mackerel. After 3 nights of operation, each plate was checked for tracks and moved (Pocatello Supply Depot progress report 1981). Our methods were similar to those developed by Egoscue (1956), Hatcher (1978), Orloff et al. (1986), and Paveglio and Clifton (1988).

Ninety permanent tracking stations were established throughout the panhandle using a stratified random sampling design (Fig. 2). The proportional area covered by counties and habitats in the panhandle was computed, and the ninety tracking stations were divided among first counties and then habitats according to these proportions (Table 2). Stations were also established near county and state lines due to the availability of well-maintained roads at these areas as well as the foreknowledge that swift fox populations occcurred in states adjacent to the panhandle.

PHOTO BY G.A. SMITH

Figure 1. Photo of tracking plate setup used during the swift fox study in the Oklahoma panhandle including plate, bait, and infra-red triggered camera in the Oklahoma panhandle.

In very small or excessively large habitats, numbers of stations were set to ensure an adequate sample size (i.e., minimum of 14 stations per habitat). Lastly, the specific locations of the stations were determined according to land accessibility and distance from other established stations. A linear distance of ≥4.8 km was maintained between all tracking stations and most were located within 1.0 km of a county road. Four complete surveys covering each season were conducted in the panhandle between January 1995 and July 1997.

Sampling effort for counties and habitats was computed in terms of functional plate nights. A functional plate night was defined as a night that a baited plate covered with sufficient chalk to record tracks was at a station. Plates were checked only once every 3 nights. Therefore, plates were recorded as operating 0 or 3 functional-plate nights per session depending upon the status of the plate at the time of its removal.

Table 2. Number of functional -plate nights*, by county and habitat type, used to determine distribution of swift foxes in the Oklahoma panhandle, 1995 –1997

	Cimarron Co.	Texas Co.	Beaver Co.	Total	# Plates
Range	104	132	140	376	41
Mesa	136	0	0	136	14
Agriculture	13	108	80	201	21
Riparian	43	51	43	137	14
Total	296	291	263	850	90
# Plates	31	33	26	90	

*Number of operational track plates/number of nights the plates were set up

I used functional plate nights to compute an index of tracking success at tracking stations and to standardize sampling effort. This index was computed by dividing the number of detections at a particular station by the number of functional plate nights of a particular habitat. While tracking stations were assigned proportionally among counties and habitats, variation in functional plate nights was the result of random rain events across counties and habitats.

Cameras

Infra-red triggered cameras were used to evaluate tracking plates. Cameras were of 2 types, but of similar design. Both types consisted of an infra-red detection unit connected to a camera housing and automatic shutter release. An auto-focus, auto-wind compact camera was placed in the camera housing and attached to the automatic shutter release. Cameras were set up at tracking-plate stations so that the infra-red trigger and the camera pointed at the center of the station. Infra-red triggers were set to a 3-minute delay between first exposure and next possible exposure in order to allow the camera sufficient time to advance the film. Cameras were triggered by detection of the thermal radiation given off by endotherms on the tracking plates which electronically tripped the shutter.

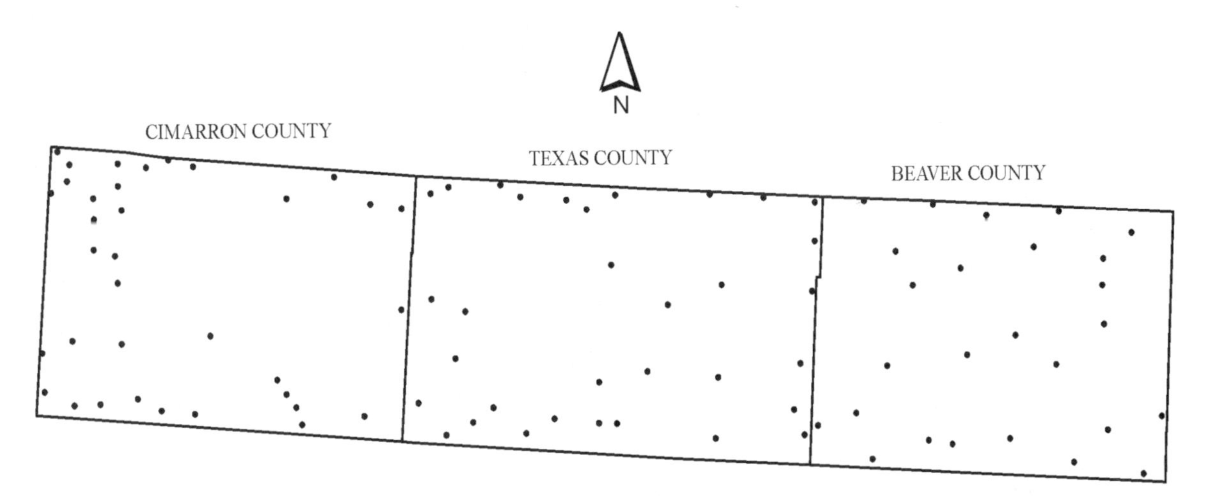

Figure 2. Approximate locations of the 90 permanent tracking stations of the swift fox study in the Oklahoma panhandle, 1995–1997.

A maximum of 10 cameras were operated during each sampling session. Cameras were placed at tracking stations based upon previous tracking station success. Priority was given to tracking stations where tracks on track plates suggested the presence of swift foxes. Secondary priority was given to tracking stations that seemed to be in optimal locations for detecting swift fox (or other carnivores), but for some reason (weather, insufficient chalk covering, etc.) tracks of carnivores were not detected.

Spotlighting

Spotlighting and predator calling was conducted opportunistically during the first two sampling sessions in the Oklahoma panhandle. Spotlighting and predator calling were conducted in all habitats. An 8.0-km section of road through the desired habitat was selected for spotlighting activity. Predator calling was conducted at 1.6-km intervals along the road, followed by a waiting period of 15 minutes at the site. The area was then illuminated with a minimum 500,000 candlepower spotlight and any carnivores were observed and recorded. If no carnivores were present, the procedure was repeated once more before moving to the next point along the transect.

Statistical Methods

The 3 counties of the panhandle divide the panhandle conveniently into west, central, and east sections. Using these divisions, I performed a chi-square goodness of fit test on the swift fox tracking plate survey data to determine if swift fox were distributed evenly among the west, central, and eastern extremes of the panhandle. I computed expected frequencies for the test based on sampling effort conducted in each county.

Additionally, analysis of swift fox distributions across the entire panhandle were conducted by further categorizing plates according to their location with respect to large scale panhandle habitat features. Track plates were placed into 1 of 7 categories based upon geological/hydrological features of the panhandle and habitat similarities, irrespective of county boundaries. This permitted evaluation of how swift fox detections might vary across the entire panhandle, independent of county lines.

To examine swift fox distributions between habitats, I conducted a chi-square goodness of fit test designed to test the null hypothesis of no difference in swift fox occurrence among the 4 habitats. Once again, counts of swift foxes from the tracking stations were used for this analysis.

Pseudo-replication of detection data was a potential problem in the analysis of the track plate data. In order to account for the potential that some detections between sampling periods were of the same individuals, county and habitat chi-square analyses were conducted 2 ways. First, all detections were considered novel detections, regardless of when or where they occurred, and analyzed using chi-square. In this approach, the actual detections were the statistical measurement. The second approach counted plates only once for each species detection, regardless of how many times the plate was visited by that species over the course of the entire study. In this approach, the tracking plate (whether or not it ever recorded a track, regardless of how many tracks it recorded) was the unit of statistical measurement. The second approach is a more conservative analysis of the data, and while it does not completely control for the possibility of pseudoreplication in the data, it lessens its effect. The outcomes of both analyses are compared.

I tested for interspecific associations of the swift fox using a chi-square contingency table analysis on data from the track plate surveys. The analysis compared detections of swift foxes in specific habitats to detections of coyotes in the same habitats.

Results

Sampling Effort

Sampling effort for the swift fox project in Oklahoma is presented in Table 2. I recorded 850 functional plate nights during the study (Table 2). Plate nights per county ranged from a low of 263 in Beaver County to 296 in Cimarron County (Table 2). Plate nights in habitats ranged from 136 in the mesa to 376 in range areas (Table 2).

Track Plates

Tracking plates had distinct advantages over other survey methods. First, tracking plates required less effort than other methods examined. One person could set 40 plates per day, depending on weather conditions. Second, operating plates was inexpensive after the initial purchase of the plates, stakes and sprayer. Costs for alcohol, bait and carpenter's chalk per sampling session were typically under $100 (not including mileage costs). I could also detect visits of multiple species even after bait had been taken. Carnivores frequently defecated on tracking plates, leaving further evidence of their presence and identity. Finally, carpenter's chalk sprayed on plates typically yielded clear, distinct tracks that were, in most cases, readily identifiable.

The principal disadvantage of tracking plates was that rain usually destroyed tracks. One sampling period was severely affected by rain. A second disadvantage of tracking plates was that I could not distinguish between individuals of the same carnivore species. This resulted in the statistical problem of pseudoreplication of the data, since one individual could potentially be responsible for tracks at a particular tracking plate during multiple sampling sessions.

Cameras

Infra-red triggered cameras enabled us to verify the presence of swift fox at stations and provided a photographic record of carnivore visitation (Fig. 3). Cameras could also detect swift foxes that visited tracking stations but did not step on tracking plates. Cameras functioned properly in the rain as well. During periods of rain, data from tracking stations that had infra-red triggered cameras could be salvaged even though rain had washed the track evidence away. A third advantage of the cameras was detection of multiple individuals of the same species at a single plate. Two individuals could be recorded together at a tracking station on the same exposure, verifying multiple visits at a tracking station.

The major disadvantage of the cameras was cost. Each camera unit, including infra-red trigger unit and compact camera, cost approximately $190. Film, batteries, and film processing for each sampling session cost approximately $25 per camera. The cost of operating the cameras for the duration of the study was approximately $1000. Using cameras, I was able to detect 2 carnivores (1 swift fox and 1 bobcat) which were not detected using the tracking plates. Clearly, cameras would not have been cost effective if used solely for novel detections. Cameras were also insensitive to endotherm size, mammals as small as mice would trigger the shutter switch. A mouse sitting on the tracking plate, eating the bait, had the potential to expose several frames of film before leaving. However, as a track verification tool, I felt the cost of cameras was offset by confirmation of track identifications.

Cameras also malfunctioned frequently. Problems with the cameras included drained batteries, improper film advancement, and poor exposures.

Spotlight Surveys

The obvious advantage to the spotlighting surveys was the visual records of carnivores in the habitat being investigated. Any carnivore observed during spotlighting could be recorded as positively occurring in the particular habitat. Spotlighting was the least expensive method used in our determination of swift fox occurrence.

Spotlighting was also the least effective method used for detecting swift fox. No foxes were detected through the spotlighting efforts. This may be attributable, at least in part, to the numbers of observers present during spotlighting surveys. During most spotlighting sessions, only one observer was present and no carnivores were detected during any of these sessions. However, during a spotlight ing session with 3 observers, multiple (6) coyotes were detected and one bobcat (*Lynx rufus*) was detected. Clearly, spotlighting effectiveness increased with the number of people present. For this reason, spotlighting was considered an ineffective use of time and was discontinued after the second survey session.

Analyses

The chi-square analysis of swift fox distributions among counties (detections) was significant ($\chi^2 = 29.61$, df = 2, $p < 0.001$). Swift fox occurrence was higher than expected in Cimarron County and less than expected in both Texas and Beaver counties (Fig. 4). When analyzed using the more conservative plate approach, differences in swift fox detections across counties remained significant ($\chi^2 = 9.228$, df = 2, $p < 0.01$). Swift fox detections were also not distributed evenly across the 7 large scale habitat designations ($\chi^2 = 24.18$, df = 6, $p < 0.001$), supporting the results of the county based chi-square analysis.

Swift foxes were not detected in habitats in proportion to survey effort ($\chi^2 = 12.51$, df = 3, $p < 0.01$). Swift foxes occurred more frequently than expected in the mesa habitat and less frequently than expected in the riparian habitat (Fig. 5). In range and agricultural habitats, swift foxes

PHOTO BY G.A. SMITH

Figure 3. Infra-red triggered camera photo of a swift fox visiting a tracking plate in the Oklahoma panhandle. Photos were used as track and visit verification tools.

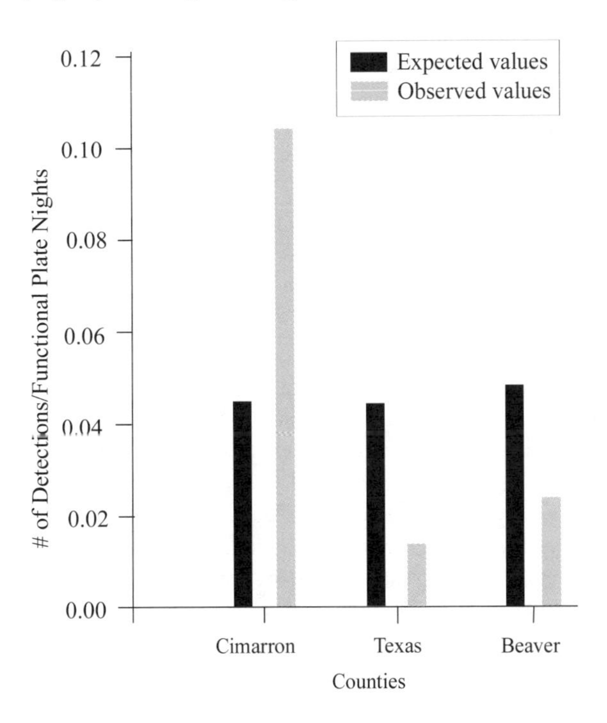

Figure 4. Results of chi-square analysis for swift fox occurrences in counties located in the panhandle of Oklahoma, 1995–1997.

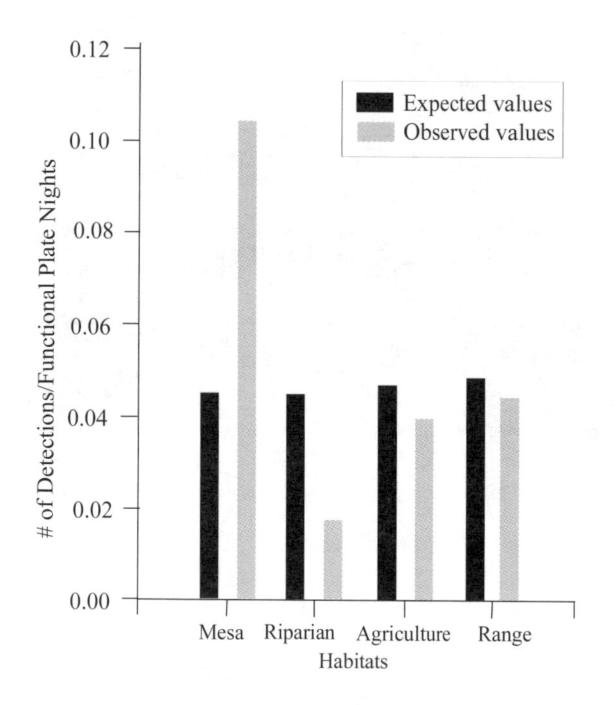

Figure 5. Results of chi-square analysis of swift fox occurrences across habitats in the Oklahoma panhandle, 1995–1997.

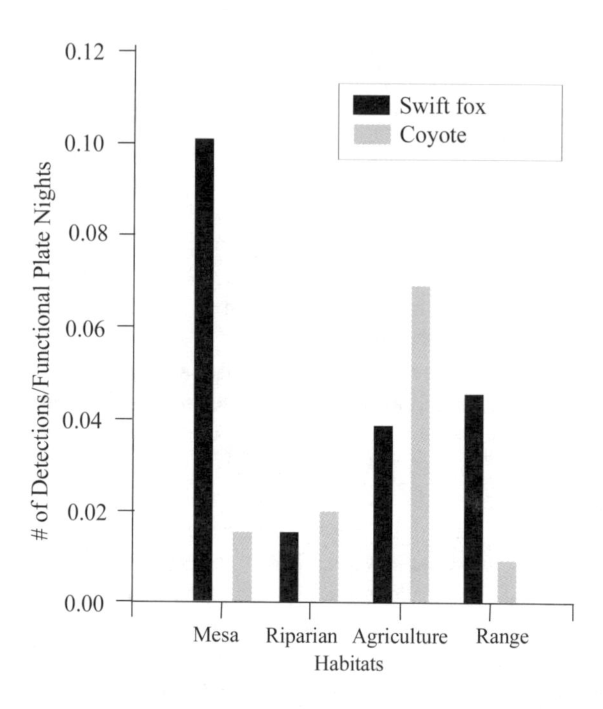

Figure 6. Swift fox and coyote occurrences in habitats of the Oklahoma panhandle, 1995–1997.

occurred slightly less often than expected (Fig. 5). Habitat differences in swift fox distribution using the more conservative approach, however, were not significant ($\chi^2 = 2.77$, df = 3, $0.50 > p > 0.25$). Power analysis was conducted on the data in this approach. Power for this test was low (w = 0.3557, u = 3, power = 0.3279).

I found differences ($\chi^2 = 13.61$, df = 4, $p < 0.01$) between the distributions of swift fox and coyotes among habitats. In the presence of coyotes, swift foxes were detected more frequently than expected in mesa and range areas but less frequently than expected in agricultural and riparian areas (Fig. 6). Coyotes were detected more frequently than expected in agricultural and riparian areas, but less frequently than expected in mesa and range areas.

Discussion

Sampling effort between track plates, cameras, and spotlighting was not even. Within 2 sampling sessions, the effectiveness of the various methods had become apparent. While 1 person could set and operate numerous track plates at one time, 1 person could conduct only 2 or 3 spotlighting sessions per night. Track plates also produced far more evidence (per unit effort) of carnivores than spotlighting during the first 2 sampling sessions. As a result, spotlighting was discontinued after the second sampling session.

Furthermore, carnivores tend to scent mark areas they visit, particularly when there is some new structure in that area (e.g., a scent station stake). Some carnivores are also curious about new scents. As a result, if a carnivore marked a scent station stake or plate, that plate was in effect "rebaited" (Conner et al. 1983). Due to this behavior, track plates were probably able to remain operational even after the bait had been taken.

Cameras were effective at detecting and recording carnivores, but their cost prohibited their widespread use in the panhandle. Track plates were as effective at carnivore detection, but more could be operated at one time and at lower cost than the cameras. Consequently, track plates were determined to be the most overall effective method for detection of swift foxes and other carnivores in the Oklahoma panhandle.

Additionally, cameras might have been helpful in identifying individuals at tracking plates during different sampling sessions, thereby addressing issues of track plate pseudoreplication. However, cameras were not used repetitively at the same sites because priority was given to the track verification ability of cameras. As a result, cameras were not useful in addressing issues of pseudoreplication at tracking plates.

Three primary areas of the Oklahoma panhandle support swift fox (Fig. 7). The highest concentrations of swift fox appear to be in the westernmost part of the panhandle (Cimarron County), with very regular detections in the mesa region of that county (extreme northwestern portion). Fewer detections of swift fox occurred in both southern Texas and southern Beaver counties, suggesting lower densities. Swift foxes were detected infrequently in other areas of Texas and Beaver counties (Fig. 7).

Cimarron County, in the western third of the panhandle, is the least human-populated county. The mesa area, in

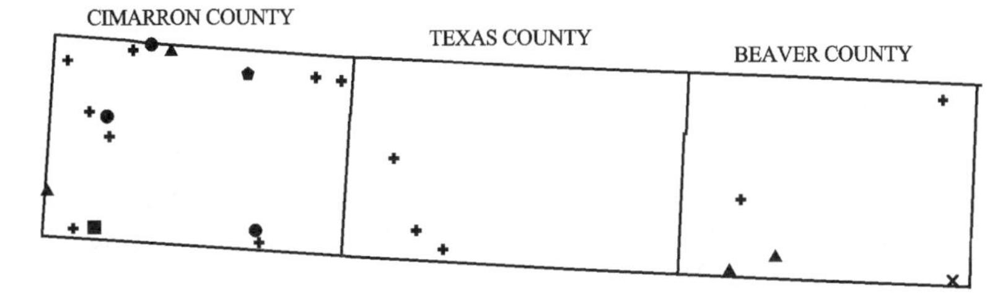

Figure 7. Swift foxes detection frequency (detections/functional plate nights) using baited tracking plates in the Oklahoma panhandle.

particular, supports a very low human population. This translates to larger tracts of unbroken range and possibly higher quality range than in the other 2 counties. Agriculture is also a very small component of land use in Cimarron County. Swift fox were not detected as often in agricultural areas (as discussed below). Land use in Texas County is primarily agriculture. Additionally, Texas County has recently undergone a major industry boom in pig farming and agriculture. This might explain the low numbers of swift fox detections in this county. The extreme southern portion of Texas County, however, remains committed to cattle production, which encourages range management practices. This land use pattern may explain why swift foxes were much less common in the northern sections of Texas County. Swift foxes were detected in Beaver County slightly more often than in Texas County, but still much less often than in Cimarron County. Beaver County land use is not skewed towards agriculture as much as Texas County, but still has a much larger agricultural component than Cimarron County.

Comparisons of habitat data during this project assumed that swift foxes (and coyotes) were equally detectable in all habitats. This assumption has been questioned by some authors because habitat biases that could affect detectability generally persist over time, and cannot be controlled by replication (Sargeant et al. 1998). The habitat classifications during this project, however, were considered to be defined broadly enough to minimize the impact of any detection biases associated with more specific habitat types.

The analysis of swift fox occurrence in habitats suggests that swift fox are more common in some habitats than others. In contrast, when track plates were used as the detection unit, instead of individual detections, the results were not significant. Power analysis suggests that this statistical test lacked sufficient power to confidently retain the null hypothesis (Thomas and Juanes 1996). However, this analysis was also the more conservative of the 2 analyses and reduced (but did not eliminate) the pseudo-replicative effect of possible multiple detections at a single plate over the course of the study. As a result, while habitat interpretations of these data are still valuable based upon the results of the first approach, interpretations should be viewed with caution due to the ambiguous results of the second approach.

Swift foxes occurred regularly in range and were detected more often than expected in mesa areas. Swift foxes occurred in agricultural areas slightly less often than expected. Swift foxes were rare in riparian areas (Fig. 5).

Mesa areas are apparently good habitats for swift foxes. The mesa was the only habitat which produced consistent swift fox detections. Coyotes were rarely detected in mesa habitat. Agricultural areas may be a substandard habitat for swift foxes. Swift foxes were not consistently detected in agricultural areas. Riparian areas apparently do not support swift fox populations. This may be due to a preference for these areas by coyotes.

The negative association between swift foxes and coyotes support the established theory that larger carnivores negatively impact populations of smaller carnivores through competition and possibly predation (Carbyn 1982, Sargeant et al. 1987, Harrison et al. 1989, White et al. 1994, Ralls and White, 1995, White et al. 1995). This interaction appears to be particularly pronounced between more closely related species (Carbyn 1982, Rudzinski et al. 1982, Bailey 1992). Although gray and red fox are more closely related to swift fox, they were not examined because they were not detected in the study area. Other comparisons of this type, considering swift fox distributions with other tracked carnivores, are planned for the future.

These interaction results could also be used to interpret the results of the habitat analysis. In general, swift fox may be avoiding or occurring in lower numbers in certain habitats (riparian and agricultural areas) because of the high numbers of coyotes in those habitats. Areas that are coyote depauparate (mesa) seem to be more suitable for swift fox.

Coyotes are identified by the ranchers as major

predators on stock, particularly during the calving season. As a result, ranchers employ a continuous coyote control plan in the Oklahoma panhandle which consists of opportunistically shooting coyotes when they see them (L. Green, Cimarron County game warden, personal communication; R. Apple, landowner, personal communication). This control procedure occurs most heavily in the major cattle production habitats. Since cattle spend so little time proportionally in agricultural and riparian areas, the killing of coyotes is less likely in these areas. The proximity of agricultural areas to dwellings and the low visibility in riparian areas may also be responsible for the lack of coyote control in these areas.

Swift fox are not considered to be detrimental to stock by ranchers. In fact, ranchers' attitudes toward the swift fox are generally favorable in Oklahoma (I. Labrier, landowner, personal communication; L. Green, Cimarron County game warden, personal communication). As a result, swift fox may not be subjected to control when they are seen by ranchers. While the control has not eliminated the coyote population in the panhandle, it may have kept coyote numbers relatively low in cattle production habitats (range and mesa).

The effects of human activities on swift fox behavioral patterns and home ranges are not entirely understood. In the Oklahoma panhandle though, swift foxes do not appear to be strongly affected by human activity associated with habitation and often occupy burrows near or under human habitations (S. Sparkman, landowner, personal communication). Human agricultural activities, though, do seem to affect swift foxes at least indirectly.

In their attempts to protect their stock from coyote predation, along with the unpredictable activity cycles on ranches, ranchers in the Oklahoma panhandle may be creating more suitable habitat for swift fox. Ranching activities and coyote control on ranches may at least partially explain why, in areas of primary cattle production (range and mesa habitats), swift fox are detected in relatively high numbers while coyotes are detected relatively infrequently.

Summary

The most effective method for detecting swift fox and other carnivores in the Oklahoma panhandle was baited tracking plates placed at permanent tracking stations and coated with isopropyl alcohol and chalk. The least effective method was spotlighting. Infra-red triggered cameras were valuable for verifying track identifications but were not cost effective or reliable enough to serve as an independent detection technique.

Swift fox were not evenly or randomly distributed throughout the Oklahoma panhandle. Swift fox occurred most often in the extreme western section of the panhandle (Cimarron County), particularly in the mesa region. Secondary concentrations of swift fox occurred in both southern Texas and southern Beaver counties.

Swift fox were not evenly distributed among habitats. Swift fox seemed to prefer the mesa habitat and avoided riparian areas. Swift fox also occurred in range and agricultural areas. However, relative numbers of occurrence in these habitats seemed to suggest that swift fox prefer range areas over agricultural areas.

A distinct negative interaction exists between swift fox and coyote in the Oklahoma panhandle. Swift fox were not detected in abundance in any habitats where coyotes were detected in abundance. This potential interaction among canids could explain the low detection numbers for swift fox in agricultural and riparian areas. Both of these habitats recorded high numbers of coyote detections.

Acknowledgments

I would like to express my sincere thanks to the Oklahoma Department of Wildlife Conservation who, through Section 6 funding, provided the funds which made this research possible. I would also like to thank all of the landowners in the Oklahoma panhandle who allowed me to work on their land. Particularly, I would like to thank Ina K. Labrier, the Apple family, Don Prather, Scott Sparkman, Robert Thrash, and John Fain. Also, I would like to thank Larry Green, the game warden of Cimarron County, who was instrumental in getting the project started. Thanks are also due to Mark Lomolino, Dave Perault, Greg Smith, and Tabatha Franklin for their contributions to this paper. Finally, I would like to extend thanks to Dr. David Legates and Dr. Larry Toothaker, who were instrumental in the planning and design of much of the statistical analyses.

Literature Cited

Bailey, E.P. 1992. Red foxes, *Vulpes vulpes*, as biological control agents for introduced arctic foxes, *Alopex lagopus*, on Alaskan Islands. Canadian Field-Naturalist 106:200–205.

Caire, W., J.D. Tyler, B.P. Glass, and M.A. Mares. 1989. Mammals of Oklahoma. University of Oklahoma Press, Norman, Oklahoma.

Carbyn, L.N. 1982. Coyote population fluctuations and spatial distribution in relation to wolf territories in Riding Mountain National Park, Manitoba. Canadian Field-Naturalist 96: 176–183.

Conner, M.C., R.F. Labisky, and D.R. Progulske. 1983. Scent-station indices as measures of population abundance for bobcats, raccoons, gray foxes, and opossums. Wildlife Society Bulletin 11:146–152.

Duck, L.G., and J.B. Fletcher. 1943. A game type map of Oklahoma. State of Oklahoma Game and Fish Department, Division of Wildlife Restoration. Oklahoma City, Oklahoma.

Egoscue, H.J. 1956. Preliminary studies of the kit fox in Utah. Journal of Mammalogy 37:351–357.

Federal Register. 1995. Endangered and threatened wildlife and plants: 12-month finding for a petition to list the swift fox as endangered. Federal Register 60:31663–31666.

Harrison, D.J., J.A. Bissonette, and J.A. Sherburne. 1989. Spatial relationships between coyotes and red foxes in eastern Maine. Journal of Wildlife Management 53:181–186.

Hatcher, R.T. 1978. Survey of the Red Fox (*Vulpes vulpes*) in Oklahoma. Oklahoma Cooperative Wildlife Research Unit, Biannual Progress Report, 31:35–36.

Johnson, W.E., T.K. Fuller, and W.L. Franklin. 1996. Sympatry in Canids: a review and assessment. *In* J.L. Gittleman editor. Carnivore behavior, ecology, and evolution. Volume 2. Cornell University Press, Ithaca, New York.

Kilgore, D.L., Jr. 1969. An ecological study of the swift fox (*Vulpes velox*) in the Oklahoma panhandle. American Midland Naturalist 8:512–534.

Orloff, S., F. Hall, and L. Spiegel. 1986. The distribution and habitat requirements of the San Joaquin kit fox in the northern extreme of their range. Transactions of the Western Section Wildlife Society 22:60–70.

Paveglio, F.L., and S.D. Clifton. 1988. Selenium accumulation and ecology of the San Joaquin kit fox in the Kesterson National Wildlife Refuge Area. Resource publication of the U.S. Fish and Wildlife Service, October 1988.

Peterson, R.O. 1995. Wolves as interspecific predators in canid ecology. *In* L.N. Carbyn, S.H. Fritts, and D.R. Seip, editors. Ecology and conservation of wolves in a changing world. Canadian Circumpolar Institute, Occasional Publication No. 35.

Pocatello Supply Depot. 1981. Indices of predator abundance in the western United States. Publication of the U.S. Fish and Wildlife Service.

Polis, G.A., C.A. Meyers, and R.D. Holt. 1989. The ecology and evolution of intraguild predation: potential competitors that eat each other. Annual Review of Ecology and Systematics 20:297–330.

Ralls, K., and P.J. White. 1995. Predation on San Joaquin kit foxes by larger canids. Journal of Mammalogy 76:723–729.

Rudzinski, D.R., H.B. Graves, A.B. Sargeant, and G.L. Storm. 1982. Behavioral interactions of penned red and arctic foxes. Journal of Wildlife Management, 46:877–884.

Sargeant, A.B., S.H. Allen, and J.O. Hastings. 1987. Spatial relationships between sympatric coyotes and red foxes in North Dakota. Journal of Wildlife Management 51:285–293.

Sargeant, G.A., D.H. Johnson, and W.E. Berg. 1998. Interpreting carnivore scent-station surveys. Journal of Wildlife Management 62:1235–1245.

Shackford, J.S., and J.D. Tyler. 1991. Vertebrates associated with black-tailed prairie dog colonies in Oklahoma. Publication of the nongame program, Oklahoma Department of Wildlife Conservation, Oklahoma City, Oklahoma.

Shackford, J.S., J.D. Tyler, and L.L. Choate. 1989. A survey of the black-tailed prairie dog in Oklahoma. Final report, nongame program, Oklahoma Department of Wildlife Conservation, Oklahoma City, Oklahoma.

Thomas, L., and F. Juanes. 1996. The importance of statistical power analysis: an example from *Animal Behavior*. Animal Behavior 52:856–859.

White, P.J., K. Ralls, and R.A. Garrott. 1994. Coyote-kit fox interactions as revealed by telemetry. Canadian Journal of Zoology 72.1831–1836.

White, P.J., K. Ralls, and C.A. Vanderbilt White. 1995. Overlap in habitat and food use between coyotes and San Joaquin kit foxes. Southwestern Naturalist 40:342–349.

Zumbaugh, D.M., and J.R. Choate. 1985. Historical biogeography of foxes in Kansas. Transactions of the Kansas Academy of Science 88:1–13.

Current Swift Fox Distribution and Habitat Selection Within Areas of Historical Occurrence in New Mexico

■ **Robert L. Harrison and C. Gregory Schmitt**

Abstract: Candidacy of the swift fox for protection under the Endangered Species Act highlighted the lack of detailed current knowledge of swift fox distribution and habitat selection in New Mexico. We surveyed the presence of swift fox within their historical range in New Mexico with scent-station and spotlight surveys from October 1996 through May 1997. We also collected specimens and examined New Mexico Department of Game and Fish fur harvest records and U.S. Department of Agriculture Wildlife Services incidental capture records. Swift fox presently occur throughout their historical range in New Mexico, with the exception of areas developed for cropland in eastern Curry and Roosevelt counties. We also did not find swift fox in southeastern Quay County where grass was longer and shrubs were more abundant than in the rest of the surveyed area. During winter and spring, swift fox in New Mexico prefer Bouteloua *rangeland with low shrub density and grass length less than 30 cm. Trends in agricultural development, fur harvest, and incidental capture by Wildlife Services are discussed.*

The swift fox (*Vulpes velox*) historically occurred in the shortgrass prairie of eastern New Mexico (Egoscue 1979) and was temporarily a candidate for endangered species listing by the United States Fish and Wildlife Service (Potter 1982, Clark 2001). In New Mexico, the swift fox is classified as a protected furbearer and may be legally harvested. Candidacy of the swift fox under the Endangered Species Act highlighted the general lack of current knowledge of swift fox status and biology. Regional population status and county-level distribution in New Mexico were reviewed by Hubbard (1994), using existing museum specimens and unpublished New Mexico Department of Game and Fish (NMDGF) fur harvest records. These data indicated the presence of swift fox throughout shortgrass prairie areas of eastern New Mexico, although no specimens were collected from large areas of potential habitat (Findley et al. 1975, Hubbard 1994; Fig. 1). Prior to this study, the last collection of a museum specimen was in 1982 (Hubbard 1994). Based upon these observations and swift fox habitat descriptions from other states (Egoscue 1979, Scott-Brown et al. 1987), we assume that the historical range of swift fox in New Mexico coincided with the shortgrass prairie (Fig. 1).

As a first step toward determining the current status of swift fox within areas of historical occurrence in New Mexico, we examined the current distribution and habitat selection of swift fox by scent-station and spotlight surveys. We also collected specimens and examined recent NMDGF fur harvest records and U.S. Department of Agriculture Wildlife Services (USDA-WS) incidental capture records.

Study Area

Surveys were conducted throughout areas of shortgrass prairie in eastern New Mexico (Fig. 1). Shortgrass prairie in New Mexico is described by Dick-Peddie (1993), who refers to it as plains-mesa grassland. Cropland habitats within the shortgrass prairie were also surveyed (Fig. 2). Topography is primarily flat or rolling hills. Land use is primarily rangeland, with cropland and Conservation Reserve Program (CRP) land dominating some areas adjacent to the Texas border. Cultivated crops within the study area were primarily winter wheat (*Triticum*) and milo (*Sorghum*). The majority of CRP lands were planted with

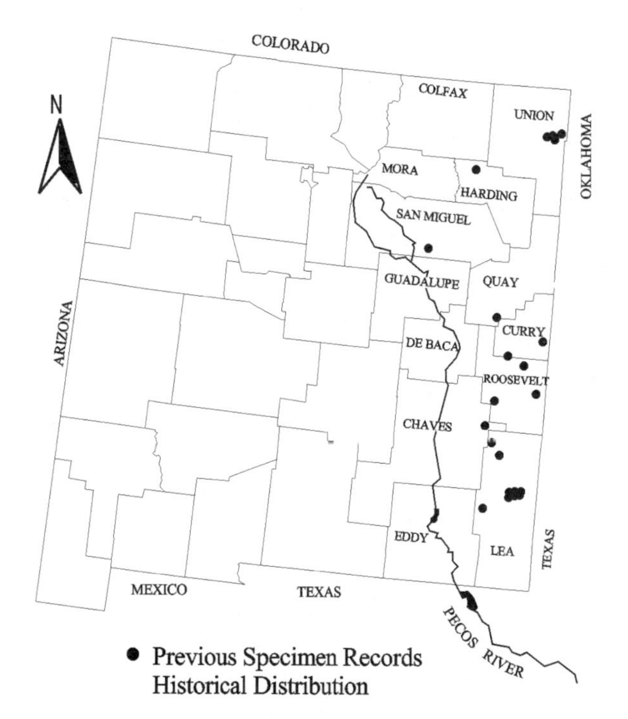

Figure 1. Historical distribution of swift fox in New Mexico. Shaded area represents shortgrass prairie east of the Pecos River. The latest historical museum specimen was collected in 1982.

lovegrass (*Eragrostis*) or little bluestem (*Schizachyrium*, formerly *Andropogon*; Allred 1997).

Swift fox are very similar to kit fox (*V. macrotis*), which occur in central and western New Mexico. The line of division between the ranges of the two species is indistinct, but generally follows the Pecos River (Rohwer and Kilgore 1973, Mercure et al. 1993). The species may be distinguished by skull and ear measurements (Dragoo et al. 1990, Hubbard 1994), but these measurements were not available to this study. Thus, we did not consider records or specimens potentially collected within the range of kit fox. Three counties (Chaves, De Baca, and Guadalupe) potentially have both swift and kit foxes.

Methods

Scent-station surveys were conducted from November 1996 through May 1997. We created scent stations (Linhart and Knowlton 1975, Conner et al. 1983) by clearing vegetation from 0.7 x 0.7 m areas, placing a plaster-of-paris tablet (Pocatello Supply Depot, USDA, Pocatello, Idaho) soaked in a mixture of mackerel and cod liver oil (Trailing Scent, On Target ADC, Dekalb, Illinois) in the center of each of the cleared areas, and sifting a 1:32 mixture of mineral oil and dried plaster sand over the areas and tablets. We secured tablets to the ground with nails inserted through previously drilled holes to prevent removal by rodents. We covered tablets with a thin layer of sand to prevent removal by birds. Scent stations were placed in transects of 10 stations, with stations separated by 1.6 km. Transects were located along public roadways and were separated by at least 8 km, which represents the diameter of the maximum reported home range size of swift fox (Hines and Case 1991).

At each scent station we recorded the land use (rangeland, cropland, or CRP land), genera of grasses and shrubs, road type (paved or unpaved), and number of fences and powerlines within 50 m of the station. CRP lands were identified by the presence of uniform stands of *Eragrostis* or *Schizachyrium*. CRP lands without such stands may occasionally have been misclassified as rangeland. We also estimated the average grass length within 4 categories (<15 cm, 15–30 cm, 30–45 cm, and >45 cm) and the average nearest neighbor distance between shrubs. If no shrubs were present, we assigned a value of 500 m to the average nearest neighbor distance.

Stations were examined on the day following setting. We identified species visiting scent stations by their tracks (Murie 1974, Halfpenny 1986). If swift fox tracks were not found on a transect after the first night, all the stations within the transect were reset with a new tablet and additional sand and observed after a second night. Visits by swift foxes to more than one station within a single transect were considered one observation of a swift fox. Swift fox tracks may be easily distinguished from those of gray fox (*Urocyon cinereoargenteus*), other canids, and domestic cats (*Felis catus*, Orloff et al. 1993), but not from those of kit fox. Presence of rodent tracks on stations was recorded but rodents were not identified as to species.

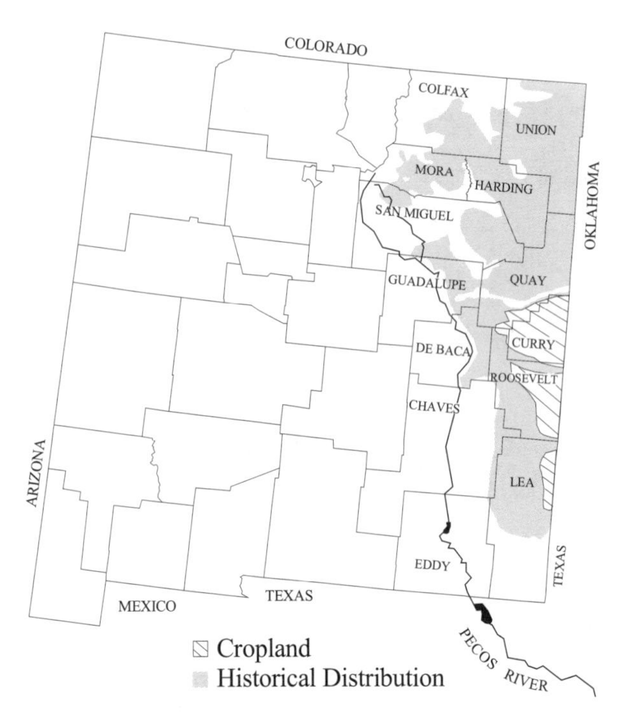

Figure 2. Location of cropland areas within the historical distribution of swift fox in New Mexico, following Dick-Peddie (1993). Minor cropland areas are not shown.

We conducted spotlight surveys from November 1996 through May 1997 by visually searching for swift fox with 1 or 2 1,000,000 candlepower spotlights while driving along public roadways. We searched for 3 to 4 hours per night, beginning at dusk. If a swift fox was seen less than 8 kilometers from a scent station where swift fox tracks were recorded, we counted the 2 observations as a single observation of a swift fox.

We collected swift fox specimens from USDA-WS, private trappers, and roadkills. Specimens will be given to the collection at the Museum of Southwestern Biology, University of New Mexico, Albuquerque, New Mexico.

We examined the effect of grass length, grass and shrub genera, roadway type, and number of fences and powerlines on swift fox visitation with log-likelihood ratio goodness of fit (G) tests (Zar 1984), using data from all stations to calculate expected visitation rates. We used *t*-tests (Zar 1984) to compare means of average nearest neighbor distance between shrubs between stations visited by foxes and stations not visited by foxes and to compare means of the number of stations per transect visited by rodents between transects visited by swift foxes and transects that were not visited by swift foxes.

Results

We recorded visits by swift foxes at 39 stations on 22

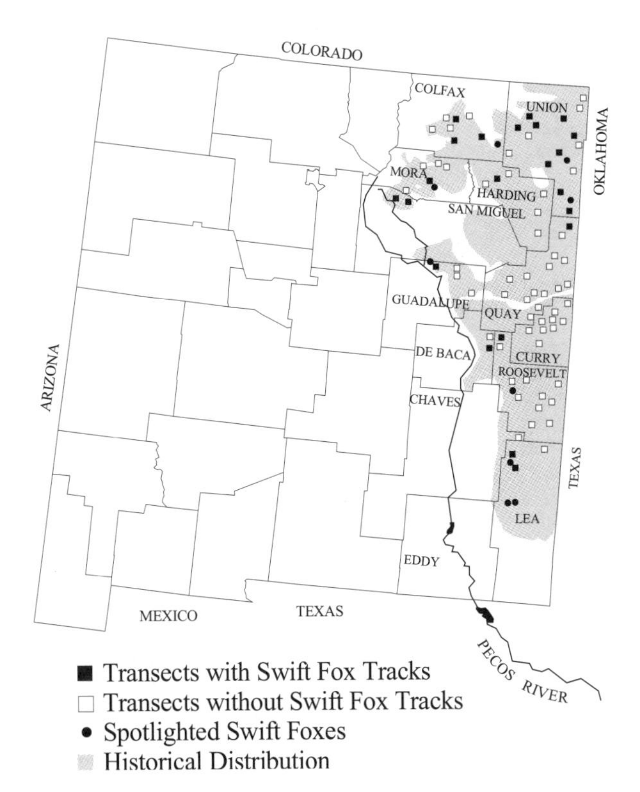

Figure 3. Locations of scent-station transects and spotlighted swift foxes from October 1996 through May 1997, in New Mexico. Shaded area represents shortgrass prairie east of the Pecos River.

of 80 transects and observed 9 swift foxes by spotlighting, achieving a total of 27 independent observations within 10 of the 12 counties surveyed (Table 1, Fig. 3). We found no swift foxes in Chaves or Curry counties. Area of swift fox habitat within counties estimated from Dick-Peddie (1993) correlated well with number of transects (Pearson $r = 0.812$, $P = 0.001$) and kilometers of spotlighting (Pearson $r = 0.778$, $P = 0.003$).

Collected specimens and NMDGF and USDA-WS records indicated the recent presence of swift fox in all counties of their historical range except San Miguel (Table 1, Fig. 4, Appendix A). NMDGF and USDA-WS records indicate swift or kit fox killed in Chaves, De Baca, and Guadalupe counties. However, these records list the locations of animals killed by county only. Thus, whether the foxes were captured within the ranges of swift or kit fox was not known. Swift fox killed by USDA-WS in Quay County were trapped in the northeastern section of the county (R. McCoy, USDA-WS, personal communication).

Habitat data were available from 670 stations. Swift fox tracks were located only in rangeland, and not in cropland or CRP land (Table 2). Swift foxes visited stations in areas of <15 cm and 15–30 cm grass length more than expected and visited stations in areas of grass length 30–45 cm or >45 cm less than expected (Table 2). Swift fox visited stations in areas of *Bouteloua* (grama) more than stations in areas of *Schizachyrium* or *Aristida* (three-awn) and visited

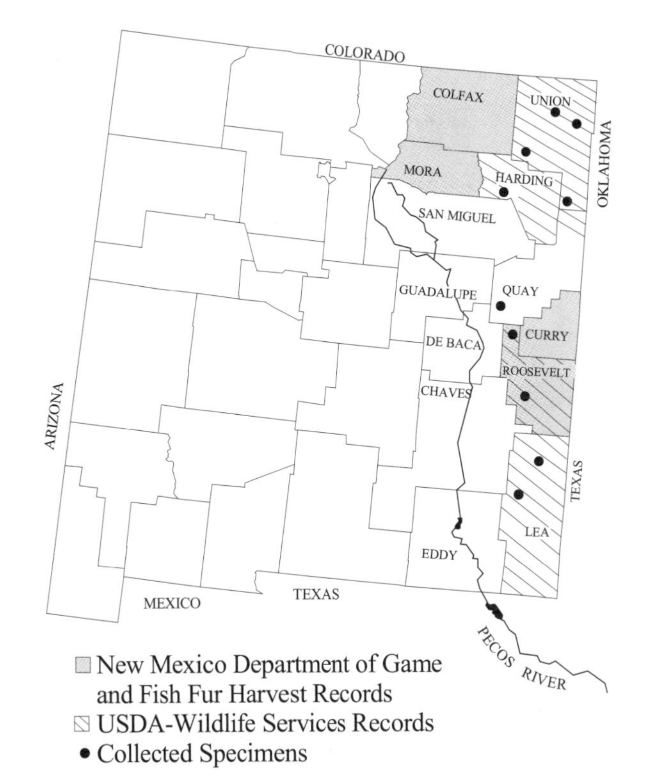

Figure 4. Swift fox reports from New Mexico Department of Game and Fish fur harvest records (1992-1996), U.S. Department of Agriculture Wildlife Services incidental capture records (1991-1997), and recent specimen collections (1996-1997). See Table 1 for numbers of foxes.

stations in areas without shrubs more than stations in areas with shrubs (Table 2). Roadway surface and number of fences and powerlines did not produce significant differences in visitation rates by swift fox (Table 2).

The average nearest neighbor distance between shrubs was greater at stations visited by foxes ($\bar{x} = 324.7$ m, SD = 232.3 m, $n = 35$) than at stations not visited by foxes ($\bar{x} = 194.5$ m, SD = 230.5, $n = 635$, $t = -3.251$, df = 669, $P = 0.001$). The average number of stations visited by rodents was less on transects that were visited by foxes ($\bar{x} = 2.7$ stations, SD = 3.0, $n = 19$) than on transects not visited by foxes ($\bar{x} = 6.4$ stations, SD = 2.5, n = 48, $t = 5.130$, df = 65, $P < 0.001$).

No evidence of swift fox was found in rangeland areas of southeastern Quay County. In SE Quay, grass was longer and shrubs, *Schizachyrium* and *Aristida*, were more abundant than in the rest of the study area (Table 3). The average nearest neighbor distance between shrubs at stations was less in SE Quay ($\bar{x} = 51.1$ m, SD = 128.5, $n = 120$) than in the rest of the study area ($\bar{x} = 234.0$ m, SD = 10.1, $n = 550$, $t = 8.200$, df = 668, $P < 0.001$). Rodents were more abundant in SE Quay ($\bar{x} = 7.7$ stations/transect visited by rodents, SD = 1.6, $n = 12$ transects) than in the rest of the study area ($\bar{x} = 4.8$, SD = 3.1, $n = 55$, $t = -3.064$, df = 65, $P = 0.003$).

Table 1. Evidence of swift fox by county within their historical range in New Mexico

County	Km^2 of swift fox habitat	Scent Station		Spotlight		Total survey observations	Specimens collected	NMDGF fur harvest records	USDA-WS records
		Number of transects	Swift fox visits to transects	Kilometres spotlighting	Swift foxes spotlighted				
Chaves	392	1		73		0			
Colfax	3898	7	3	605	1	4	1	1	
Curry	3646	9		504		0		3	
De Baca	1805	2	1	87		1			
Guadalupe	3141	4	1	304	1	1			
Harding	5221	6	1	429		1	9		15
Lea	5685	3	2	232	3	4	11		98
Mora	2250	5	1	353	1	1		1	
Quay	7444	15	1	869		1	1		3
Roosevelt	6353	11	1	533	1	2	4	15	2
San Miguel	4269	2	2	205		2			
Union	8928	15	9	747	2	10	9		1
Total	53032	80	22	4941	9	27	35	20	119

Scent-station and spotlight surveys were conducted from October 1996 through May 1997. Total survey observations = number of scent-station visits and spotlight observations separated by at least 8 km. Specimens were collected during 1996-1998. New Mexico Department of Game and Fish (NMDGF) fur harvest records and USDA Wildlife Services (USDA-WS) incidental capture records indicate number of animals reported killed during 1991–1996 and 1991–1997, respectively. No specimens collected were included in USDA-WS records, with the exception of 1 fox from Roosevelt County.

Table 2. Habitat characteristics of scent stations placed within the historical range of swift fox in New Mexico from January through May 1997

	Percentage of all stations (n=670)	Percentage of stations visited by foxes (n=35)	G	df	P
Land management					
Rangeland	87.2	100.0			
Cropland	8.4	0.0			
CRP land	4.4	0.0			
Average grass length					
<15 cm	29.0	34.3	10.689	2	0.005*
15-30 cm	29.1	42.9			
30-45 cm	31.2	22.8			
>45 cm	10.7	0.0			
Grass genera present					
Bouteloua	72.5	94.3	15.763	2	<0.001*
Schizachyrium	24.6	11.4			
Aristida	11.9	11.4			
Sporobolus	5.1	0.0			
Eragrostis	5.1	0.0			
Triticum	4.5	0.0			
Shrub genera present					
Yucca	48.5	25.7	11.070	2	0.004*
No shrubs	36.4	60.0			
Opuntia	15.5	14.3			
Prosopis	15.7	2.9			
Roadway type					
Gravel	60.4	65.7	0.222	1	0.665
Paved	39.6	34.3			
Number of fences					
2	62.5	60.0	1.352	2	0.511
1	18.5	25.7			
0	19.0	14.3			
Number of powerlines					
2	3.4	5.7	0.662	2	0.723
1	49.0	48.6			
0	50.6	45.7			

Categories with <5 stations were combined for analysis. G = log-likelihood ratio. * $P < 0.05$.

Table 3. Comparison of habitat characteristics of scent stations placed within and outside of southeastern Quay County, New Mexico, from January through May 1997

	Percentage of non -SE Quay stations (n=580)	Percentage of SE Quay (n=90)	G	df	P
Land managem ent					
Rangeland	86.4	92.2	2.539	1	0.118
Cropland/CRP land	13.6	7.8			
Average grass length					
<15 cm	32.6	6.7	46.159	3	<0.001*
15-30 cm	26.4	46.7			
30-45 cm	29.5	41.1			
>45 cm	11.5	5.5			
Grass genera present					
Bouteloua	71.4	80.0	4.552	1	0.035*
Schizachyrium/Aristida	31.9	54.4			
Shrub genera present					
Yucca	46.0	64.4	198.630	3	<0.001*
No shrubs	11.0	44.4			
Opuntia	8.3	63.3			
Prosopis	41.2	5.6			

Categories with <5 stations were combined for analysis. G = log-likelihood ratio. * $P < 0.05$.

Discussion

On a statewide scale, swift fox presently occur throughout their historical range in New Mexico, although significant gaps in Curry, Quay, and Roosevelt counties may be present (Fig. 3 and 4). In winter and spring, swift fox in New Mexico prefer *Bouteloua* rangeland free of shrubs with grass length <30 cm. Roadway type and number of fences and power lines are not important factors to swift fox distribution.

We did not find evidence of swift fox in the cropland areas of eastern Curry or Roosevelt counties, despite extra sampling (Fig. 2 and 3), with the exception of 1 specimen collected in an area of mixed cropland and rangeland near the western edge of cropland development in Roosevelt County (Fig. 4). Previous museum specimens from these areas were collected in 1957 and 1968, respectively (Hubbard 1994). Recent NMDGF and USDA-WS records do indicate swift fox in Curry and Roosevelt counties, but do not specify locations (Fig. 4). Whether swift fox populations in these areas have been reduced by agricultural development or related factors is not known. Swift fox have been found in areas of mixed agricultural use (Kilgore 1969). Cropland is often suitable habitat for red fox (*Vulpes vulpes*; Sheldon 1992). Red fox may exclude swift fox (Hines and Case 1991), and have been found to kill kit fox (Ralls and White 1995) and exclude arctic fox (*Alopex lagopus*; Bailey 1992). We collected 3 red fox specimens from cropland areas in eastern Roosevelt County. One red fox specimen was collected within 13 km of the swift fox specimen mentioned above that was collected in or near cropland. NMDGF records indicate the recent presence of red fox in all counties of eastern New Mexico. Swift fox did occur historically in cropland in northern Texas (Cutter 1958a) and in the Oklahoma panhandle (Kilgore 1969) and are currently present in cropland in Kansas (Sovada et al. 1998). Swift fox have not been found recently in cropland areas of western Texas (Mote 1996), where it is likely that red fox have expanded their range following introduction into central Texas (Davis and Schmidly 1994). Red fox occurrence is not known to overlap that of swift fox in Kansas (M. Sovada, U.S. Geological Survey, personal communication). The preference of red fox for diverse terrain may limit their use of open plains (Ables 1975), which are more suitable for swift fox. In areas of mixed rangeland and other habitats, red fox and swift fox may occur in close proximity. Red fox may be responsible for the absence of swift fox in eastern Curry and Roosevelt counties, but further research is needed.

We found no evidence of swift fox in Chaves County. However, only a small portion of Chaves County enters the historical range, and only 1 transect was set in that county.

No evidence for the presence of swift fox has been reported from southeastern Quay County, which is primarily rangeland. Southeastern Quay County had longer grass, more *Schizachyrium* and *Aristida*, and more shrubs than the rest of the surveyed area. Availability of suitable den sites may limit swift fox distribution (Egoscue 1979), but soils in southeastern Quay County do not appear to be obviously different from soils in other areas where swift fox were found (Maker et al. 1974). Absence of swift fox from an area with habitat different from the rest of the surveyed area supports the general habitat description given above. Reanalysis of habitat data with southeastern Quay County stations excluded did not change the conclusions reached.

Swift fox were found less frequently in areas with high shrub densities. Our data suggest that swift fox habitat use is unaffected by shrub density if the average nearest neighbor distance between shrubs is greater than 15 m. Below 15 m, swift fox visitation at stations decreased as average nearest neighbor distance decreased. No stations with average nearest neighbor distances less than 5 m were visited (n = 147 stations observed). Swift fox were not found in dense stands of mesquite (*Prosopis*), sagebrush (*Artemesia*), or shin oak (*Quercus havardii*), but were found in stands of Yucca at average nearest neighbor distance between 5 and 15 m. The reason that swift fox avoid dense stands of shrubs is unknown, but avoidance of predation by coyotes (*Canis latrans*) is the most likely explanation.

Rodent visits to stations were depressed in areas with swift fox, possibly because of predation by foxes. It is unknown whether rodent density along roads is indicative of rodent density away from roads. Also, the importance in swift fox diet of rodent species which visited stations is unknown. The diet of swift fox in New Mexico has not been studied, but in other areas many rodent species are consumed (Cutter 1958b, Kilgore 1969, Uresk and Sharps 1986, Hines and Case 1991).

Swift fox presently are widespread in New Mexico, but no population or demographic estimates are available. The stability and viability of the swift fox population in New Mexico remains to be examined. Anthropogenic factors potentially threatening the swift fox include habitat loss, fur harvesting, and incidental mortality by USDA-WS. Total crop and CRP land increased by an annual average of 0.26% in counties not overlapping kit fox range from 1970 to 1995 (Anon. 1970–1995). As of 1995, approximately 15% of potential swift fox habitat had been lost to crop and CRP land in counties not overlapping kit fox range. Significant habitat loss is unlikely in the near future, as funding for addition of new land to CRP programs has decreased and existing CRP land would be placed into production before new areas are developed (R. Lansford, New Mexico State University, personal communication).

Land management affects shrub density through overgrazing, control of shrubs, and fire suppression. The significance of the actions of individual managers is often demonstrated by the variation of shrub densities across fencelines, which can vary from dense shrub cover to no shrubs on opposite sides. Also, sharp contrasts of grass height often clearly separate CRP lands. The rate of habitat loss through these mechanisms remains to be determined.

Total estimated annual sport and commercial fur harvest of swift fox in counties not overlapping kit fox range peaked at 962 in 1985–1986, decreasing to an annual average of 19 from 1990 to 1995 (NMDGF unpublished records). Relatively large harvests in the 1980s did not appear to harm swift fox populations (Hubbard 1994). Although fur prices are currently low, advertising of fur garments has increased in recent years (personal observation), suggesting that fur harvests may increase in the future.

The average total number of swift fox killed annually from 1991 to 1997 by USDA-WS was 20 in counties not overlapping kit fox range (Table 1). However, incidental mortality was concentrated in Harding and Lea counties (Table 1), indicating that swift fox populations there may be in decline due to this factor. Trapping effort by USDA-WS is largely determined by the number of complaints of coyote depredations, which is unlikely to change significantly in the near future, given the relative stability of habitat and land use in eastern New Mexico. However, pan tension devices are effective for reducing incidental take of swift fox (Phillips and Gruver 1996; L. Kilgo, USDA-WS, personal communication), and should be used within swift fox range.

Other factors which are more difficult to assess may also affect the future of the swift fox population in New Mexico, such as predation by coyotes, competition with red fox, vehicle strikes, and an increase of shrubs in response to global warming or land management. Much further research is needed to predict the outlook for swift fox in New Mexico.

Acknowledgments

We thank G. Littauer for providing USDA-WS records. Portions of this study were funded under NMDGF professional services contract 97-516.57.

Literature Cited

Ables, E.D. 1975. Ecology of the red fox in America. Pp. 216–236 *in* M.W. Fox, editor. The wild canids. Krieger, Malabar, Florida.

Allred, K.W. 1997. A field guide to the grasses of New Mexico. Agricultural Experiment Station, New Mexico State University, Las Cruces, New Mexico.

Anonymous. 1970–1995. Sources of Irrigation Water and Irrigated and Dry Cropland Acreages in New Mexico, by County. Agricultural Experiment Station, New Mexico State University, Las Cruces, New Mexico.

Bailey, E.P. 1992. Red foxes, *Vulpes vulpes*, as biological control agents for introduced arctic foxes, *Alopex lagopus*, on Alaskan islands. Canadian Field-Naturalist 106:200–205.

Clark, J.R. 2001. Endangered and threatened wildlife and plants; Annual notice of findings on recycled petitions. Federal Register 66 (5):1295–1300.

Conner, M.C., R.F. Labisky, and D.R. Progulske, Jr. 1983. Scent-station indices as measures of population abundance for bobcats, raccoons, gray foxes, and opossums. Wildlife Society Bulletin 11:146–152.

Cutter, W.L. 1958a. Denning of the swift fox in northern Texas. Journal of Mammalogy 39:70–74.

——. 1958b. Food habits of the swift fox in northern Texas. Journal of Mammalogy 39:527–532.

Davis, W.B., and D.J. Schmidly. 1994. The mammals of Texas. Texas Parks and Wildlife, Austin, Texas.

Dick-Peddie, W.A. 1993. New Mexico vegetation, past, present, and future. University of New Mexico Press, Albuquerque, New Mexico.

Dragoo, J.W., J.R. Choate, T.L. Yates, and T.P. O'Farrell. 1990. Evolutionary and taxonomic relationships among North American arid-land foxes. Journal of Mammalogy 71:318–332.

Egoscue, H.J. 1979. *Vulpes velox*. Mammalian Species 122:1–5.

Findley, J.A., A.H. Harris, D.E. Wilson, and C. Jones. 1975. Mammals of New Mexico. University of New Mexico Press, Albuquerque, New Mexico.

Halfpenny, J. 1986. A field guide to mammal tracking in western America. Johnson Books, Boulder, Colorado.

Hines, T.D., and R.M. Case. 1991. Diet, home range, movements, and activity periods of swift fox in Nebraska. Prairie Naturalist 23:131–138.

Hubbard, J.P. 1994. The status of the swift fox in New Mexico. Unpublished report. New Mexico Department of Game and Fish, Santa Fe, New Mexico.

Kilgore, D.L., Jr. 1969. An ecological study of the swift fox (*Vulpes velox*) in the Oklahoma panhandle. American Midland Naturalist 81:512–534.

Linhart, S.B., and F.F. Knowlton. 1975. Determining the relative abundance of coyotes by scent station lines. Wildlife Society Bulletin 3:119–124.

Maker, H.J., H.E. Dregne, V.G. Link, and J.U. Anderson. 1974. Soils of New Mexico. New Mexico State University, Agricultural Experiment Station, Las Cruces, New Mexico. Research Report 285.

Mercure, A., K. Ralls, K.P. Koepfli, and R.K. Wayne. 1993. Genetic subdivisions among small canids: mitochondrial DNA differentiation of swift, kit, and arctic foxes. Evolution 47:131–138.

Mote, K. 1996. Swift fox investigations in Texas. Pp. 50–52 *in* B. Luce and F. Lindzey, editors. Unpublished annual report of the swift fox conservation team.

Murie, O.J. 1974. A field guide to animal tracks. Second edition. Houghton Mifflin, Boston, Massachusetts.

Orloff, S.G., A.W. Flannery, and K.C. Belt. 1993. Identification of San Joaquin kit fox (*Vulpes macrotis mutica*) tracks on aluminum tracking plates. California Fish and Game 79:45–53.

Phillips, R.L., and K.S. Gruver. 1996. Performance of the Paws-I-Trip™ pan tension device on 3 types of traps. Wildlife Society Bulletin 24:119–122.

Potter, J.C. 1982. Endangered and threatened wildlife and plants; review of vertebrate wildlife for listing as endangered or threatened species. Federal Register 47 (251): 58454–58460.

Ralls, K., and P.J. White. 1995. Predation on San Joaquin kit foxes by larger canids. Journal of Mammalogy 76:723–729.

Rohwer, S.A., and D.L. Kilgore, Jr. 1973. Interbreeding in the arid-land foxes, *Vulpes velox* and *Vulpes macrotis*. Systematic Zoology 22:157–165.

Scott-Brown, J.M., S. Herrero, and J. Reynolds. 1987. Swift fox. Pp. 433–441 *in* M. Novak, J.A. Baker, M.E. Obbard, and B. Malloch, editors. Wild Furbearer Management and Conservation in North America. Ministry of Natural Resources, Ontario.

Sheldon, J.W. 1992. Wild dogs. Academic Press, San Diego, California.

Sovada, M.A., C.C. Roy, J.B. Bright, and J.R. Gillis. 1998. Causes and rates of mortality of swift foxes in western Kansas. Journal of Wildlife Management 62:1300–1306.

Uresk, D.W., and J.C. Sharps. 1986. Denning habitat and diet of the swift fox in western South Dakota. Great Basin Naturalist 46:249–253.

Zar, J.H. 1984. Biostatistical Analysis. Second edition. Prentice-Hall, Englewood Cliffs, New Jersey.

Appendix A
*Locations of Swift Fox Specimens Collected in New Mexico 1996–1998**

Colfax County: 7.2 km S of Abbott. Harding County: 3.2 km S, 4.8 km E of Roy; 3.2 km N of Roy (35°59′26″N, 104°11′27″W); 2.4 km S of Roy; 7.7 km N, 1.9 km W of Roy; 1.6 km W of Roy; 8 km E of Roy (2 specimens); 3.2 km E of Mosquero; 6 km N, 1 km W of Mosquero. Lea County: 24 km N, 13.1 km E of Maljamar (33°04′59″N, 103°37′26″W); 7.7 km N, 2.6 km E of Maljamar (32°55′43″N, 103°44′17″W); 24 km N, 15.2 km E of Maljamar (33°04′04″N, 103°36′40″W); 25.9 km N, 8.5 km E of Maljamar (33°04′57″N, 103°40′37″W); 25.4 km N, 11.2 km E of Maljamar (33°04′59″N, 103°38′50″W); 25.4 km N, 13.1 km E of Maljamar (33°04′59″N, 103°37′26″W); 25.4 km N, 6.2 km E of Maljamar (33°05′00″N, 103°42′08″W); 19.5 km N, 5.4 km E of Maljamar (30°01′35″N, 103°42′16″W); 25.4 km N, 13.9 km E of Maljamar (33°04′58″N, 103°37′15″W); 21.3 km N of Maljamar (33°02′43″N, 103°46′00″W); 12.8 km SW of Tatum. Quay County: 9.6 km W of Jordan. Roosevelt County: 6.4 km S, 6.4 km E of Elida ; 8 km S, 9.6 km E of Elida (T5S, R33E, S14); 13.6 km S, 2.4 km E of House (T4N, R29E, S24); 14.0 km S, 11.0 km E of House (T3N, R30E, S3). Union County: 3.2 km S of Seneca; 30.9 km N, 12.8 km E of Mt. Dora (T30N, R34E, NE¼S29); 30.9 km N, 13.1 km E of Mt. Dora (T30N, R34E, NW¼S28); 20.2 km E of Gladstone (36°18′03″N, 103°44′33″W); 19.2 km S, 5.9 km W of Amistad (35°44′43″N, 103°12′49″W); 19.2 km S, 3.0 km W of Amistad (35°44′42″N, 103°11′14″W); 19.2 km S, 5.9 km W of Amistad (35°44′43″N, 103°12′49″W); 19.2 km S, 4.5 km W of Amistad (35°44′41″N, 103°12′17″W); 13.0 km E of Gladstone.

*Specimens will be given to the collection at the Museum of Southwestern Biology, University of New Mexico, Albuquerque, New Mexico. Unless otherwise noted, each location indicates one specimen collected. Latitude/longitude coordinates were determined with Global Positioning System (GPS) units by collectors. Distances to nearest town were based upon latitude/longitude coordinates and township/range locations provided by collectors or were estimated in the field by collectors not carrying GPS units.

A Literature Review of Swift Fox Habitat and Den-Site Selection

■ Robert L. Harrison and Julianne Whitaker-Hoagland

Abstract: We reviewed information about swift fox habitat and den-site selection published in journals, theses, and government agency reports. Information is presented on topography, vegetation, land use, soils, climate, water, and human disturbance. Swift foxes are primarily denizens of shortgrass and mixed-grass prairies, but they occasionally occur in landscapes dominated by shrubland and cropland. They are typically found in areas of level to gently rolling topography, and prefer short vegetation and low shrub density, but they have been found in a wide range of habitat conditions. There is limited evidence that swift foxes select areas of low slope, close proximity to roads, and loamy soils. Conversion of native prairie to cropland may result in exclusion of swift foxes, but the response of swift foxes to land use changes varies geographically and the mechanism of habitat selection has not been determined.

Swift foxes (*Vulpes velox*) inhabit level to gently rolling topography within the North American shortgrass and mixed-grass prairies (Egoscue 1979, Scott-Brown et al. 1987). Historically, swift foxes occurred in the Great Plains east of the Rocky Mountains from southern Alberta, Saskatchewan, and Manitoba to southeastern New Mexico and western Texas (Hall and Kelson 1959, Banfield 1974, Egoscue 1979). Our objective was to review and summarize literature describing swift fox habitat associations, habitat selection, and den-site selection. We reviewed articles in scientific journals, theses, and government agency reports available by December 2000. The review is organized into 5 sections: (1) descriptions of study areas from reviewed studies; (2) a review of habitat evaluations presented in the individual studies; (3) a review of studies on swift fox habitat selection at large scales; (4) a review of studies describing and analyzing habitat characteristics of den sites; and (5) a synthesis and discussion of the reviewed information. Online comprehensive swift fox bibliographies are available at http://www.wildlifer.com and http://www.npwrc.usgs.gov.

Descriptions of Study Areas from Reviewed Studies

In this section we first provide locations of studies we reviewed (Fig. 1), and then their habitat descriptions. The section is organized by location, topography, vegetation, soils, land use, and climate. Scientific names are from the Integrated Taxonomic Information System (http://www.itis.usda.gov/index.html).

Location

Cutter (1958a, b) conducted studies of swift fox dens and food habits in Hansford County in the Texas Panhandle. Harrison and Schmitt (2003) used scent stations to survey swift foxes and to gather habitat selection information in the eastern third of New Mexico. Kilgore's (1969) swift fox den investigations were conducted in southwestern Beaver County in the Oklahoma Panhandle. The Piñon Canyon Maneuver Site (PCMS) in Las Animas County, Colorado, was the site for studies of swift fox ecology (Rongstad et al. 1989, Covell 1992), and swift fox-coyote (*Canis latrans*) interactions (Kitchen et al. 1999). Several studies of the general ecology of swift foxes were conducted on the U.S. Forest Service's Pawnee National Grassland (PNG) and the Central Plains Experimental Range Station (CPER) in northeastern Colorado (Loy 1981, Cameron 1984, Roell 1999).

Studies of swift fox dens (Jackson 1996, Jackson and Choate 2000) and swift fox ecology (Sovada et al. 1998, Matlack et al. 2000) were conducted in cropland and rangeland landscapes located in Sherman and Wallace counties, Kansas. Studies of general swift fox ecology were conducted in grazed shortgrass prairie in Sioux County, Nebraska (Hines 1980, Hines and Case 1991). Floyd and Stromberg (1981) examined locations of swift fox specimens taken by trappers in Laramie County, Wyoming. In Albany County, Wyoming, Olson (2000) studied swift foxes ecology and Pechacek (2000) examined intraspecific interactions. Swift fox dens were examined in grasslands and river bottoms in southwestern Shannon County, South Dakota (Hillman and Sharps 1978, Uresk and Sharps 1986). Uresk and Sharps (1986) also investigated swift fox den habitat in Haakon County, South Dakota. Zimmerman (1998) studied general ecology and distribution of swift foxes recolonizing north central Montana from reintroduced foxes released in Canada. In southern Alberta and Saskatchewan, natal den sites were evaluated (Pruss 1994, 1999) and the effects of vegetation structure and snow depth on prey availability for swift foxes were examined (Klausz et al. 1996, Klausz 1997, Almási-Klausz and Carbyn 1999).

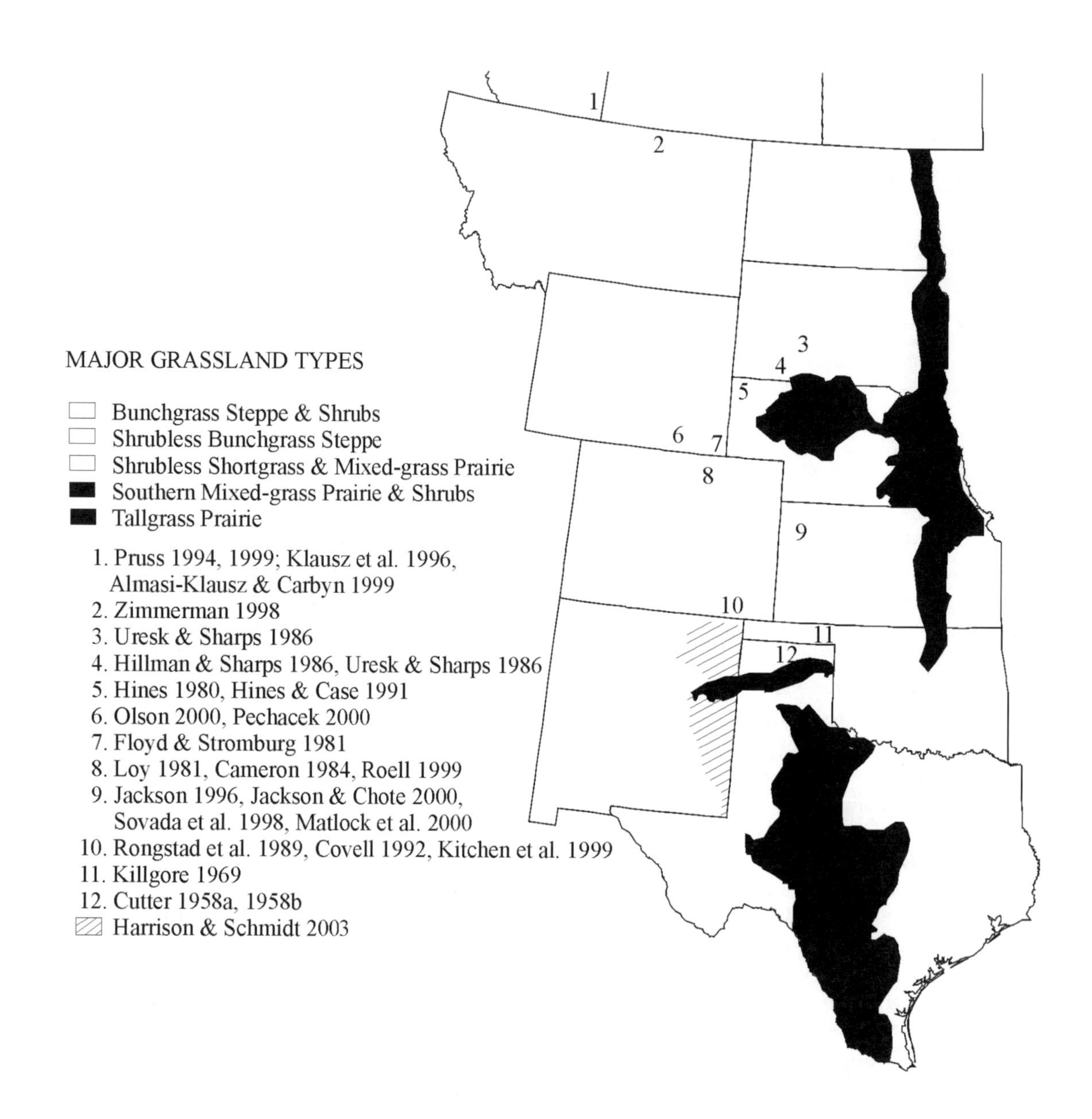

Figure 1. Locations of swift fox research and areas with published descriptions of study areas. Major grassland types were modified from Risser et al. (1981).

Topography

The topography in swift fox study areas has been most frequently described as flat to gently rolling hills (Floyd and Stromberg 1981, Loy 1981, Uresk and Sharps 1986, Hines and Case 1991, Sovada et al. 1998, Zimmerman 1998, Pruss 1999, Olson 2000, Pechacek 2000). Other topographical features reported include river canyons (Kitchen et al. 1999), floodplains (Uresk and Sharps 1986), drainages (Loy 1981), and intermittently wet streams and lakebeds (Loy 1981, Pechacek 2000). Swift foxes were observed incidentally in badland topography by Uresk and Sharps (1986) and by Woolley et al. (1995).

Vegetation

Vegetation in the Oklahoma Panhandle was originally dominated by buffalograss (*Buchloe dactyloides*) and blue grama (*Bouteloua gracilis*), but in some areas little bluestem (*Schizachyrium scoparius*, formerly *Andropogon scoparius*), side-oats grama (*Bouteloua curtipendula*), and three-awn (*Aristida* spp.) occurred (Kilgore 1969). The area was extensively cultivated and the original vegetation was replaced with Russian thistle (*Salsola tragus*), cocklebur (*Xanthium strumarium*), common sunflower (*Helianthus annuus*), and other weeds. Grasses present at scent stations placed in eastern New Mexico were primarily grama (*Bouteloua* spp.), bluestem (*Schizachyrium* spp.), and three-awn (Harrison and Schmitt 2003). Yucca (*Yucca* spp.), cholla (*Opuntia* spp.), and mesquite (*Prosopis* spp.) were the shrubs most commonly present.

Two vegetation types dominated the PCMS: shortgrass prairie and piñon pine (*Pinus edulis*)-one seed juniper (*Juniperus monosperma*) communities (Kitchen et al. 1999). Shortgrass prairie was dominated by blue grama, galleta (*Pleuraphis jamesii*), western wheatgrass (*Pascopyrum smithii*), candelabra cholla (*Opuntia imbricata*), and yucca (Rongstad et al. 1989). Piñon pine, one seed juniper, skunkbush sumac (*Rhus trilobata*), and mountain mahogany (*Cercocarpus montanus*) dominated river canyons and hills (Rongstad et al. 1989).

Two short-grass species, blue grama and buffalograss, dominated the vegetation on the PNG in northern Colorado (Cameron 1984). Other grasses present were crested wheatgrass (*Agropyron cristatum*), needle-and-thread grass (*Hesperostipa comata*), salt grass (*Distichlis spicata*), three-awn grass (*Aristida purpurea*), and western wheatgrass. Crested wheatgrass had been introduced in many locations by livestock management programs (Loy 1981). Cultivated land, primarily winter wheat, was interspersed with prairie (Roell 1999).

Rangeland vegetation in the Kansas study area consisted of short and midgrasses, especially blue grama, buffalograss, hairy grama (*Bouteloua hirsuta*), Japanese brome (*Bromus japonicus*), and smooth brome (*Bromus inermis*, Jackson 1996, Sovada et al. 1998). Crops in western Kansas were primarily wheat, corn, milo, sorghum, and sunflowers and most fields were fallow every other year (Sovada et al. 1998).

Dominant plants in the Nebraska study area were needleleaf sedge (*Carex duriuscula*), needle-and-thread grass, and blue grama (Hines 1980). Vegetation in Laramie County, Wyoming, was dominated by blue grama and buffalograss (Floyd and Stromberg 1981). In Albany county, Wyoming, swift foxes occurred in a sagebrush-grassland community dominated by grasses interspersed with low-growing sagebrush (*Artemisia tridentata*), greasewood (*Sarcobatus vermiculatus*), and saltbush (*Atriplex gardneri*; Olson 2000). Primary grasses were buffalograss, blue grama, needle-and-thread, western wheatgrass, and prairie junegrass (*Koeleria macrantha*).

The dominant vegetation on the Haakon and Shannon counties, South Dakota, study areas consisted of blue grama, buffalograss, needleleaf sedge, and western wheatgrass (Uresk and Sharps 1986). Dominant grasses in north-central Montana were western wheatgrass, prairie junegrass (*Koeleria pyramidata*), blue grama, and Sandberg bluegrass (*Poa secunda*; Zimmerman 1998). Some localized areas were seeded to crested wheatgrass. Silver sagebrush (*Artemisia cana*) was the dominant shrub.

The most common vegetation type in upland habitat in southern Alberta and Saskatchewan was the *Stipa-Bouteloua-Agropyron* association, consisting mostly of needle-and-thread, blue grama, and northern and western wheatgrasses (Klausz 1997, Pruss 1999). Shrubs present were winterfat (*Krascheninnikovia lanata*) and sagebrushes (*Artemisia frigida*, *Artemisia cana*).

Soils

The soils on the PNG and CPER were low in humic content and generally classified as well drained and sandy loams to clay loams (Roell 1999). In the Nebraska study area, soils were generally sandy loam (Hines 1980). The Albany County, Wyoming, study area contained saline upland, impervious clay, shallow clay, sandy, shallow loamy, loamy, and saline loamy soils (Olson 2000). Badlands, characterized by bare soil, were intermingled with uplands and mesas of clayey and loamy soils in South Dakota (Hillman and Sharps 1978). The soil type at the Haakon County study area was primarily clay to clay-loam (Uresk and Sharps 1986). Soils in north central Montana were characterized by loam or clay loam soil surfaces and clay or clay loam subsoil types (Zimmerman 1998).

Land Use

Cattle grazing was the predominant land use where many swift fox investigations were conducted (Hines 1980, Loy 1981, Uresk and Sharps 1986, Covell 1992, Klausz 1997, Sovada et al. 1998, Pechacek 2000, Harrison and Schmitt 2003). Cutter (1958b) reported that most of his study area was cultivated or overgrazed. Since federal acquisition in 1982 of the PCMS, cattle grazing has been discontinued on the area (Covell 1992). Some study areas also contained sheep grazing and center-pivot and dry-land farming (Kilgore 1969, Hines 1980, Floyd and Stromberg 1981, Loy 1981, Uresk and Sharps 1986, Sovada et al. 1998, Finley 1999, Harrison and Schmitt 2003). Harrison and Schmitt (2003) did not find swift foxes in cropland landscapes in eastern New Mexico.

Climate

The climate in Beaver County, Oklahoma, was characterized by limited and irregular precipitation, a high rate of evaporation, low relative humidity, high average wind velocity, hot summer days followed by cool nights, and moderate winters with occasional severe cold spells of short duration (Kilgore 1969). The mean annual precipitation was 47.2 cm, much of which fell as sudden torrential rains during late spring and summer, resulting in heavy run-off (Kilgore 1969). The average annual temperature was 14.5°C (range -16.1°C in February to 39.4°C in July). The climate on the PCMS was semi-arid with annual precipitation ranging from 26 to 38 cm and mean monthly temperatures ranging from 1°C in January to 23°C in July (Kitchen et al. 1999). The climate on the PNG and CPER was considered semi-arid with an average annual precipitation of 28–33 cm, most falling between April and September (Roell 1999). The winters on the PNG were mild, with an average temperature of -1.7°C. Average annual snowfall was 101.6 cm and there were usually 18 days each year with 2.5 cm or more of snow cover (Roell 1999). Summers were hot, with daytime temperatures often above 37.7°C (Cameron 1984).

The western Kansas study area received an average of 46.2 cm of precipitation per year, primarily in spring and summer (Sovada et al. 1998). Mean January temperature was -2.6°C and mean July temperature was 23.1°C. Average annual precipitation in the Nebraska study was 46.4 cm, falling primarily in April through July (Hines 1980). Annual snowfall averaged 142 cm. Average temperatures ranged from -6°C in January to 21°C in July (Hines 1980). Average annual precipitation at the Albany County, Wyoming, study area was 26 cm, including 98 cm of snow (Olson 2000). Mean monthly temperatures ranged from -12°C in January to 28°C in July. In South Dakota, precipitation on the Shannon County study area averaged 41.4 cm of rain, 78% of which occurred between April and September, and 78.7 cm of snowfall (Hillman and Sharps 1978). The Haakon County study area had an annual average precipitation of 43 cm of rain and 30 cm of snow (Uresk and Sharps 1986). In north-central Montana, mean annual precipitation was 32 cm (range 19 to 52 cm) and mean monthly temperatures ranged from -1.8°C to 13.9°C (Zimmerman 1998). In southern Alberta and Saskatchewan, average annual precipitation was 34.0 cm and mean daily summer temperatures ranged from 14.5°C to 16.0°C (Pruss 1999). Low levels of precipitation combined with strong winds, high insolation, and high summer temperatures resulted in high evapotranspiration.

Habitat Assessments of Local Study Sites

In this section we review published evaluations of habitat characteristics within swift fox study areas. The section is organized by topography, vegetation, soils, and proximity to roads and water.

Topography

Olson (2000) located swift foxes in areas of slopes of ≤3% more than expected, and in areas of slopes >3% less than expected. The use of areas with steeper slopes was greater when foxes were not rearing pups, perhaps reflecting a need to forage in areas not exploited during pup rearing.

Vegetation

Harrison and Schmitt (2003) found that swift foxes visited scent stations in areas with grass lengths averaging ≤30 cm more than expected, and visited stations in areas of grass length >30 cm less than expected. Swift foxes visited stations in habitats dominated by *Bouteloua* spp. more than stations in areas of *Schizachyrium* spp. or *Aristida* spp. Swift foxes were found in mesquite grassland, but foxes visited scent stations more in areas without shrubs than in areas with shrubs. Swift foxes were not found in dense stands of sagebrush, mesquite, or shin oak (*Quercus havardii*), but were found in stands of yucca where the average nearest-neighbor distance between shrubs was from 5–15 m. If the average distance between shrubs exceeded 15 m, occurrence of swift foxes was not affected. Harrison and Schmitt (2003) did not detect swift foxes in a large portion of eastern New Mexico where shrub density was greater than in the other surveyed portion of eastern New Mexico where swift foxes were detected.

Finley (1999) systematically live-trapped swift foxes in the eastern plains of Colorado and used Geographic Analysis Project (GAP) land cover maps to assess swift fox distribution and habitat use. He had greater capture success on 31-km^2 trapping grids containing ≥75% shortgrass prairie ($\bar{x}$ = 4.6 captures) than on grids containing 50–74% shortgrass prairie ($\bar{x}$ = 2.1 captures) or <50% shortgrass prairie ($\bar{x}$ = 1.5 captures). No difference was detected in captures rates between grids with 50-75% shortgrass prairie and grids with <50% shortgrass prairie. Ninety-two percent of swift fox captures occurred in 3 large blocks of relatively contiguous shortgrass prairie.

In Wyoming, Olson (2000) found that horizontal cover varied between years, but was greatest in greasewood, decreasing through sagebrush, sagebrush/grassland, grassland, and playa lake vegetation types. Grass height did not vary between vegetation types, but did vary between years. He found that swift fox home ranges contained sagebrush, sagebrush/grassland, grassland, playa lake, and greasewood vegetation types in proportion to availability in the overall study area throughout the year, except that home ranges of fox pairs during pup-rearing (April–August) included less greasewood than expected. The proportion of swift fox home ranges that contained greasewood, bare/rocky and riparian vegetation types was lower than simulated home ranges within the study area. With minor exceptions, the home ranges of males and females, and foxes with and without young, had similar vegetation types. However, among home ranges of fox pairs there was much variation in patterns of vegetation types. For example, combining all seasons and fox pairs (4 seasons averaging 6 pairs each), sagebrush was included more than expected 11 times and less than expected 6 times. Within home ranges, fox pairs did not exhibit selection for vegetation types during dispersal and pair formation periods, but during pup rearing in 1998 sagebrush was used less than expected, and in 1999 playa lake was used less than expected.

Composite rankings of relative prey abundance during pup rearing in Albany County, Wyoming, were highest in sagebrush, followed by greasewood, sagebrush/grassland, grassland, and playa lake (Olson 2000). Despite relatively high prey abundance, greasewood had the highest horizontal cover and swift foxes used this vegetation type less than expected. Higher abundance of food in sagebrush apparently outweighed risk of hunting in sagebrush. Higher abundance of prey in sagebrush may have also led to larger observed numbers of young in home ranges of fox pairs that included higher proportions of sagebrush vegetation. Olson (2000) found a positive correlation between proportion of sagebrush vegetation within home ranges and the number of young observed at dens and a negative correlation between the proportion of grassland vegetation within home ranges and the number of young observed at dens.

Soils

In Wyoming, fox home ranges during pup-rearing had a greater proportion of loamy soil and a lesser proportion of saline upland, impervious clay, shallow clay, and saline loamy than expected, relative to available soils (Olson 2000). Fox pairs used areas with less shallow clay than expected during dispersal and pair formation periods, and included impervious clay soils in home ranges less than expected during pair formation. Proportions of soil types at fox locations were not different from those available during dispersal and pair formation periods.

Proximity to Roads and Water

Olson (2000) found that the mean distance to roads of swift fox locations was 215 m, significantly lower than the mean distance of random points (1156 m). He also found that mean distances from fox locations to water sources were significantly lower than distances from random points to water. Swift fox locations were farthest from water sources in winter, when snow was available.

Studies of Swift Fox Habitat Selection at Large Scales

In this section we describe 2 studies which considered swift fox habitat selection at the scales of states and townships.

Lindberg (1986) used a mail survey of biologists and trappers to determine the distribution of swift foxes in Wyoming. He partitioned the state into 3 areas (east, central, west) based upon past swift fox sightings, habitat suitability, and geographic location in relation to swift fox populations outside Wyoming. The eastern area was within the Great Plains and consisted mainly of rangeland and cropland habitat types. Nearly all historic sightings of swift foxes in Wyoming occurred within this area (Lindberg 1986). Swift fox habitat within the central and western areas of the state consisted mainly of mixed rangeland and shrub-rangeland habitat. Mail survey results indicated 80% of swift fox observations were in the eastern area, 17% were in the central area, and 3% were in the western area of Wyoming. Eighty-two percent of all reported swift fox observations occurred within shortgrass, mixed-grass, and prairie/sagebrush habitat types (46.6%, 8.8%, and 26.8%, respectively). Roadside observations accounted for 6.8% of all swift fox observations, 4.0% were in fallow fields, 3.3% in alfalfa or small grains, and about 2% near buildings. Lindberg (1986) stated that the relatively low numbers of swift fox observations in agricultural habitat types in Wyoming was probably because of the relatively small portion of land under cultivation.

Zimmerman (1998) classified habitat in three counties of Montana as native grassland or cropland at the township level, and found no difference between the habitat composition of townships where swift foxes were captured and townships where they were not captured. She also found no difference in topographic roughness between townships where swift foxes were captured and townships where they were not.

Review of Studies Describing and Analyzing Habitat Characteristics of Den Sites

Swift foxes use dens throughout the year for protection, rearing of young, and predator avoidance (Egoscue 1979, Pruss 1999). Usually swift fox pairs have numerous burrows within their home ranges. Hillman and Sharps (1978) reported use of 13 different dens by a pair throughout a year. Swift foxes spend most of the day in or near a den (Egoscue 1979, Hines and Case 1991, Pruss 1994). Thus, a suitable den site is considered a critical habitat requirement for swift foxes (Snow 1973). In this section we review descriptions and analyses of habitat surrounding den sites. The section is organized by topography, vegetation, soils, land use, climate, water, and proximity to human disturbance.

Topography

All 25 dens observed by Cutter (1958a) were on sloping plains, hilltops, or other well-drained areas. Six of the 15 dens observed by Kilgore (1969) in shortgrass pastures were placed in what he termed "blow-ridges" (the result of extensive wind erosion and deposition that occurred during the 1930s) and generally bordered playa lakes and depressions in pastures. The rest were located in level areas. One such flatland den was occupied through 2 periods of flood, indicating that drainage was not the critical factor in den-site selection as Cutter (1958a) had suggested (Kilgore 1969).

Dens on the PNG were located most frequently along ridges or slopes above drainages or depressions (Loy 1981). The slope at 77% of 69 den sites was <3° (Loy 1981). Of 41 dens observed by Cameron (1984), 28 were located on flat terrain ≥30 m from the nearest altitude change, 3 were located at the base of a hill 3 m in height, 3 were located midslope, 5 were located at the edge of the top of a hill 3m in height, and 2 were located on the tops of hills.

In Kansas, average slope around 33 rangeland den sites

(2.9%) was not different from average slope around 27 cropland den sites (3.5%, Jackson 1996). Hines (1980) found dens in flatland to gently rolling hills with 3–6% slopes. Half of the 40 dens Hines (1980) observed were on flat terrain. Most dens from hilly areas were in valleys near the base of hills. Hines (1980) speculated that it might be impossible for swift foxes to construct dens on steep slopes because of shallow soils.

Olson (2000) found dens on slopes of 3% more than expected, and dens on slopes of 3–6% less than expected, but he did not explicitly state where these locations were in relation to hilltops. Uresk and Sharps (1986) noted that dens were generally at or near tops of ridges, possibly for better drainage.

All 11 natal dens found by Zimmerman (1998) were in areas with slopes <15°, and 73% were in areas of slopes from 0–5°. She commented that dens were often near tops of small ridges or rises. Pruss (1999) found that occupied dens were closer to the tops of hills than unoccupied dens, but the slope at occupied dens was not significantly different from that at unoccupied dens.

On the PNG, 69% of 107 den entrances faced either south or east. However, entrances were found facing all 4 principal exposures (Loy 1981). Dens faced all 4 cardinal directions in both cropland and rangeland in Kansas (Jackson and Choate 2000). There was no difference between number of dens with openings that received high and low solar radiation. Twelve of 14 dens found by Hines (1980) faced east or west, which was more than expected by chance. Uresk and Sharps (1986) observed that swift fox dens tended to have eastern exposures. Of 11 natal dens found by Zimmerman (1998), 54% faced south/southeast, and 36% faced north/northeast. No entrances were exposed to the west or northwest. Pruss (1999) found that neither occupied nor unoccupied sites faced a predominant direction.

Jackson and Choate (2000) reported that surface roughness (a subjective measure of the extent of furrowing ≤9 m away from dens) in cropland was greater around 15 dens than around 15 random points, and surface roughness in rangeland was not different between 15 den sites and 15 random points. Surface roughness was significantly greater in cropland than rangeland. Jackson and Choate (2000) also found that surface ruggedness (Beasom et al. 1983) was not significantly different between den sites and random points in either cropland or rangeland, but ruggedness in rangeland was greater than in cropland. Since no appreciable differences existed between ruggedness of areas surrounding den sites and randomly selected areas within a habitat type, Jackson and Choate (2000) suggested that swift foxes were not selecting for any particular ruggedness, but were rather exploiting available resources.

Vegetation

All 25 occupied dens found by Cutter (1958a) were in sparsely vegetated habitats. Cutter (1958a) suggested that swift foxes choose the most barren areas, devoid of any bushes or tall plants, to construct dens. The vegetation type at 77% (53) of 68 dens on the PNG was blue grama/buffalograss community and 23% of dens were located in shortgrass or mixed-grass communities with little or no tall vegetation near the den sites (Loy 1981). Almost all dens observed by Cameron (1984) were associated with surrounding areas of short vegetation allowing swift foxes long sight-lines. Cameron (1984), however, observed that den entrances might be located in isolated patches of taller grass.

Jackson and Choate (2000) found that swift foxes were not selecting areas of low percent cover, but rather exploiting what was available to them. No vegetation height differences were found between randomly selected sites and den sites.

Hines and Case (1991) reported that the percent occurrence of bare soil, litter, and live plants around 13 swift fox dens averaged 14%, 68.8%, and 17.2%, respectively. The species composition consisted principally of blue grama (40.7%), needle-and-thread grass (29.3%) and needleleaf sedge (22.2%). Olson (2000) found dens only within sagebrush, sagebrush/grassland, and grassland vegetation types; there was no preference among these types, despite variations in horizontal cover and prey density. They did not den in greasewood, playa lake, bare/rocky terrain, riparian, or saltbush vegetation types.

Grasses and sedges comprised 76% and 55% of the vegetation composition, while forbs constituted 23% and 43% at dens in Shannon and Haakon counties, South Dakota, respectively (Uresk and Sharps 1986). Shrubs were sparse at both sites, constituting only 1% and 2% of the composition. Pruss (1999) found that old grass remaining from the previous growing season was higher at occupied dens ($\bar{x} = 27.1$ cm) than at unoccupied dens ($\bar{x} = 22.1$ cm) in summer. New grass from the current growing season was not higher at occupied sites ($\bar{x} = 18.9$ cm) than at unoccupied sites ($\bar{x} = 17.0$ cm).

Soils

Swift fox dens found by Kilgore (1969) were located in Richfield clay loam and Ulysses-Richfield complex in cultivated fields and in silt loam in shortgrass pastures and artificial habitats. Jackson and Choate (2000) reported that swift fox dens were constructed in soils that were well drained. In rangeland, 8 of 14 den sites were in Ulysses soil; the remainder were in Keith, Colby, and Colby-Ulysses soils. Most den sites in cropland were in Colby or Ulysses soils. Texture of the soils used for den construction was either silt-loam or loam (Jackson and Choate 2000).

Hines (1980) observed swift fox dens in areas where soil classes ranged from loamy sand to loam, the majority being sandy loam. Seventy-nine percent of 19 observed dens were in sandy loam while 21% were in loamy sand. Large rocks (>10 cm) were found at only 1 den which had not been renovated. Hines (1980) speculated that swift foxes may be excluded from rocky areas and that dens are probably limited to loose, friable soils which do not contain high percentages of clay or sand, allowing easy digging and maintenance of den structure.

Olson (2000) found dens in impervious clay, loam, saline loam, saline upland, sandy, shallow clay, and shallow loamy soil classes, but not in clay, rocky, saline lowland, saline sub-irrigated, shallow sandy, or shallow soil classes. Foxes denned in loamy soils more than expected, and denned in impervious clay, saline loam, and sandy soils less than expected.

Soils were variable among the 7 den sites investigated by Uresk and Sharps (1986). Den sites in Shannon County contained soils that were loam, clay-loam, and sandy-clay-loam; and soils at dens found in Haakon County were mostly clay. Uresk and Sharps (1986) did not believe that swift foxes were selecting for a particular soil type for the construction of dens.

Land Use

Cutter (1958a) and Kilgore (1969) found swift fox dens in cultivated fields. Kilgore (1969) believed swift foxes did not show a preference for native pasture versus cultivated fields. Hines (1980) did not observe any dens in hay or cultivated fields, but these habitat types comprised a small percentage of his study area. Because dens in cultivated fields observed in the spring had disappeared by the fall of the same year, Kilgore (1969) speculated that dens in cultivated fields were temporary and rarely reopened. However, Jackson and Choate (2000) reported that several dens disturbed by farm implements were reopened. They also determined that construction and maintenance of dens by swift foxes was not different between cultivated fields and native rangeland. Kilgore (1969) stated that dens in shortgrass pastures tended to have more entrances than those in cultivated fields, perhaps indicating prolonged occupancy. However, Jackson and Choate (2000) found no difference between the number of den openings in rangeland and cropland.

Water

All 69 dens observed by Loy (1981) were located within 1 km of available water sources such as ponds, seeps, and cattle tanks, but the overall availability of water in his study area was not quantified. Because none of the 40 swift fox dens observed by Hines (1980) were within 4 km of permanent available water sources, he suggested that water was not a limiting factor for den site selection for swift foxes. Olson (2000) found that dens were located closer to water than were random points. Pruss (1999) found that the distance of occupied dens from water ($\bar{x}$ = 1014 m) was not different from that of unoccupied sites ($\bar{x}$ = 870 m). She speculated that occupied sites might be located farther from water to avoid coyotes or for better drainage.

Proximity to Human Disturbance

Distances to the nearest road ranged from 33 to 2400 m for 40 dens located by Hines (1980), however, 68% of the dens were within 200 m of roads. Sixty-six percent of telemetry locations were <1 km from roads, but only 30% of the study area was <1 km from roads. Thus, Hines and Case (1991) suggested that swift foxes select roads or roadway right-of-ways. All 69 dens located by Loy (1981) were <1 km from roads, with 1 den located 5 m from a heavily traveled highway. Olson (2000) found that dens were located closer to roads and fences than random points. All the swift fox dens observed by Hillman and Sharps (1978) were <1.6 km from roads. Loy (1981) and Hillman and Sharps (1978) searched for dens from roads, potentially biasing their observations. All 11 natal dens found by Zimmerman (1998) were within 500 m of a road or 2-track trail. Eighty-one percent of occupied dens found by Pruss (1999) were <200 m of roads, and occupied dens averaged significantly closer to roads than unoccupied dens (267 m v. 419 m). She suggested that swift fox denned closer to roads to avoid coyotes, to obtain vehicle-killed carrion, or to use roads as travel corridors.

Cutter (1958a) found 3 dens <100 m from human residences and 6 <100 m from windmills that were probably visited by people weekly. Hillman and Sharps (1978) found several dens near human occupied residences and 2 swift fox burrows beneath abandoned farm buildings. Dens located by Hines (1980) averaged 2 km (range 0.5 to 5.0 km) from occupied residences, and he believed that swift foxes were relatively tolerant of human activity.

Cutter (1958a) found 4 occupied swift fox dens along fence rows; none were believed to be natal dens. Kilgore (1969) found 1 den in a culvert, and 2 dens in a cemetery. He stated that culverts were used in drought years and when debris accumulated over the ends, providing a darkened enclosure. In the cemetery, den entrances were excavated at the edges of concrete caps covering 2 of the graves. One of the cemetery dens was used for whelping (Kilgore 1969).

Synthesis and Discussion of Reviewed Information

In general, swift foxes are most commonly found in rangeland areas of low slope, short vegetation, and loamy soil. They avoid areas of dense shrubs. Swift foxes are also found in what are considered atypical habitats for swift foxes to use, such as cropland and mixed shrub and grassland. Swift foxes have been found to exhibit preferences of slope, vegetation, soil, land use, and proximity to human disturbances. They are clearly able to use a wide variety of habitat configurations. Swift foxes are absent from some grassland areas, and there have been no investigations into the reasons for their absence.

Land use is the habitat variable of most obvious impact upon swift foxes. Swift foxes are absent from some, but not all, areas where native prairie has been converted to cropland. Swift foxes may be excluded from cropland by farming practices, such as flood irrigation, frequent plowing, or herbicide application, which disturb prey communities. If such disturbances are not widespread, foxes may easily move to nearby undisturbed areas. Dens covered by plowing are sometimes reused (Kilgore 1969, Jackson and Choate 2000). Swift foxes avoid areas with high vegetation, and thus fields of tall crops may not be used. However, Jackson and Choate (2000) stated that agricultural activities in themselves do not appear to prevent swift foxes from finding suitable habitat for denning. There has been no suggestion that availability of suitable den sites is limited in cropland. Sovada et al. (1998) and Matlack et al. (2000) did not find differences in survival between rangeland and cropland landscapes. Cropland under some conditions may be useful habitat for swift foxes.

The intensity of agricultural practices may be significant. Fallow fields are used by swift foxes in Kansas. The availability of fallow fields is reduced when fields are converted from dryland to irrigated agriculture. Irrigated agriculture typically involves much more activity, such as harvesting and chemical application, than dryland farming. To date there have been no comparisons of swift fox ecology between areas of dryland and irrigated or center-pivot farming, although study areas in Colorado, Kansas, and Nebraska had both types. Comparisons may be difficult due to intermixing of several cropland types as well as native prairie.

There is consistent circumstantial evidence that recent changes in the canid community as a result of conversion of prairie to cropland may be significant to swift foxes (Ralls and White 1995, Sovada et al. 1998). As discussed by Harrison and Schmitt (2003), cropland dominated landscapes are often occupied by red foxes (*Vulpes vulpes*; Sheldon 1992:194). Red foxes may exclude swift foxes (Hines and Case 1991), and have been found to kill kit foxes (Ralls and White 1995) and exclude arctic foxes (*Alopex lagopus*; Bailey 1992). Red foxes may be responsible for the absence of swift foxes from some cropland areas. Red foxes were not present in the Kansas study areas (Sovada et al. 1998). Red foxes were not present in Hansford County, Texas, during Cutter's (1958a, b) studies (Packard and Bowers 1969), nor were they present in Beaver County, Oklahoma, during Kilgore's (1969) study (Glass and Halloran 1960, Hatcher 1982). They are present in cropland areas of eastern New Mexico where Harrison and Schmitt (2003) did not find swift foxes. Red foxes are also present in areas of western Texas (Davis and Schmidly 1994) where an unpublished study was unable to locate swift foxes (Mote 1996).

In general, cattle grazing does not conflict with swift fox presence. Grazing may help maintain the shorter vegetation that foxes prefer (Olson 2000). However, overgrazing, fire suppression, and elimination of black-tailed prairie dogs (*Cynomys ludovicianus*) may lead to increased shrub density (Weltzin et al. 1997) and reduced swift fox density (Warrick and Cypher 1998). Coyote control to reduce depredation on livestock might benefit swift foxes, if it is done in a manner that does not harm swift foxes.

Topography reported in published studies is similar to the overall topography of the Great Plains. There is consistent, although limited, evidence that swift foxes select areas of low slope, especially for den sites. Swift foxes have only rarely been reported in areas of rough topography, such as badlands or canyons. Though many authors have observed, often without quantification, that swift foxes choose den sites close to the tops of hills, there is only weak evidence for this trend. Pruss's (1999) study was the only published study found that modeled den-site characteristics. Pruss (1999) found position on hills to be the strongest discriminating characteristic. Her model correctly classified 88% of occupied sites as occupied and 79% of unoccupied sites as unoccupied. However, Pruss (1999) compared occupied fox den sites with unoccupied den sites, which were primarily badger (*Taxidea taxus*) burrows. Swift foxes often modify badger, prairie dog (*Cynomys* spp.), or other holes for their own use (Hillman and Sharps 1978, Carbyn et al. 1994), thus unoccupied sites represent potential den sites rejected by swift foxes. However, badger holes are probably not distributed randomly across landscapes, and thus the comparison is questionable. It is clear that swift foxes will den in areas other than hilltops. It is logical that foxes would avoid sites that might flood, but as Kilgore (1969) observed anecdotally, flooded sites may be reoccupied. To what extent low-lying dens would flood in semi-arid habitats has not been examined.

There is no strong evidence for consistent selection for exposure of den sites. Only 1 study with limited sample size (Zimmerman 1998) reported absence of den exposures in specific directions. Prevailing winds may affect den exposure in localized areas.

Climate has been found to be an important factor in kit fox population dynamics (Cypher et al. 2000), and is probably important for swift foxes as well. In a 15-year study, Cypher et al. (2000) found correlations between the previous year's precipitation, prey density, and population density of kit foxes. They reported much variation in these measurements over the course of the study. It is likely that swift fox populations may respond in a similar manner. It would be helpful for swift fox researchers to report not only the average precipitation in their study areas, but also recent precipitation levels and the range of precipitation known to occur. The relationship between annual and growing season, or effective, precipitation should also be examined (Cypher et al. 2000). This information would enhance comparisons between study areas and years, and would be particularly useful for comparisons of population density, prey density, and habitat quality.

Swift foxes occur predominantly in areas of short vegetation, and avoid areas of thick shrubs. Results from studies of den sites conflicted about whether or not swift foxes select sites with vegetation height different than average, with the strongest evidence (Jackson and Choate 2000) suggesting no preference. Differences in grass height between den sites and non-den sites are likely to be small. The biological significance of differences of average vegetation heights of only a few centimeters is questionable, but may be related to abundance of insect prey (Pruss 1999). One study (Harrison and Schmitt 2003) found an association of fox locations with specific grass communities within the shortgrass and mixed-grass ecosystems, but otherwise the issue has not been examined, probably due to the superficial uniformity of most grassland study areas.

Swift fox mortality may be higher in canyons or areas with trees or dense shrubs than in grasslands. Olson (2000) found that swift foxes killed by predators were found more often than expected in sagebrush and less than expected in sagebrush/grassland. Swift foxes may avoid such areas because they are unable to see long distances and the shrubs provide cover for larger canids (Kitchen et al. 1999, Harrison and Schmitt 2003).

Many authors have reported soil types and made the logical suggestion that swift foxes prefer areas with soils that are easy to dig and maintain good den structure. Only Olson (2000) compared soil types at dens with soils at random points. He confirmed the suggestion that foxes select for loamy soils and against clayey soils, and found that fox selection for soil types was more consistent than selection for vegetation types. Olson (2000) suggested that soil type might be a better predictor of swift fox habitat suitability than vegetation type. Although it has been reported that swift foxes sometimes occupy dens dug by other species, to what extent swift foxes rely upon other species to dig dens has not been studied.

The importance of water to swift foxes is not clear. The strongest evidence (Olson 2000) suggests that swift foxes use areas and select den sites close to water, but other authors have suggested that dens are located either close to, far from, or in a way unrelated to water. Flaherty and Plakke (1986) tested water stress in swift foxes and found that they can remain in water balance with food alone. The closely related kit fox (*Vulpes macrotis*) has low physiological demand for water and has behavioral strategies that conserve water (Golightly and Ohmart 1984). Swift fox habitat is generally semi-arid and free-standing water is often undependable. Water sources in many areas are artificial. Thus, swift foxes can probably survive without free-standing water.

If not harassed, swift foxes may be quite tolerant of human proximity and residences, as has been found in other western foxes (Harrison 1997, Cypher and Frost 1999). There is limited, but consistent, evidence that swift foxes are attracted to roads or the vicinity of roads. No study has found that swift foxes avoided roads. No published study has considered the density of roads or the effects of observing animals initially trapped by roads. Small mammal abundance may be higher near roads and the disturbed areas associated with them (Schwartz et al. 1994, Olson 2000). Klausz et al. (1996) stated that linear habitats such as roadsides often contain higher vegetation cover than adjacent fields grazed by cattle and often serve as protective havens for small mammals. Roadsides and coulees in southern Canada had higher abundance of small mammals than uplands in late winter (Klausz 1997, Almási-Klausz and Carbyn 1999). Roadways may also serve as travel corridors or sources of carrion. Topography and soil type may influence placement of roads as well as dens, confounding the relation between dens and roads (Olson 2000).

The available literature provides a basic, but limited, understanding of swift fox habitat and den-site selection. The most important topic for future research is the absence of swift foxes from some cropland and grassland areas that appear to be suitable. Also, we need more information on the degree to which swift foxes can tolerate fragmentation of prairie habitat, a factor important primarily in areas of mixed cropland and rangeland (Finley 1999). A long-term study examining the relationships between climate, prey density, and swift fox population size, similar to Cypher et al.'s (2000) study of kit foxes, would provide valuable insights into the mechanisms that regulate population dynamics. Other topics for future study are den-site selection in relation to soils and position on hillsides, and habitat selection in relation to road density, water sources, and vegetation height and species composition. Studies that use point locations should use random points for comparisons.

Literature Cited

Almási-Klausz, E.E., and L.N. Carbyn. 1999. Winter abundance and distribution of small mammals in the Canadian mixed-grass prairies and implications for the swift fox. Pp. 206–209 *in* J. Thorpe, T.A. Steeves, and M. Gollop, editors. Proceedings of the Fifth Prairie Conservation and Endangered Species Conference. Provincial Museum of Alberta Natural History Occasional Paper No. 24. Edmonton, Alberta.

Bailey, E.P. 1992. Red foxes, *Vulpes vulpes*, as biological control agents for introduced arctic foxes, *Alopex lagopus*, on Alaskan islands. Canadian Field-Naturalist 106:200–205.

Banfield, A.W.F. 1974. The mammals of Canada. University of Toronto Press, Toronto, Ontario.

Beasom, L.S., E.P. Wiggers, and J.R. Giardino. 1983. A technique for assessing land surface ruggedness. Journal of Wildlife Management 47:1163–1166.

Cameron, M.W. 1984. The swift fox (*Vulpes velox*) on the Pawnee National Grassland: its food habits, population dynamics and ecology. Thesis, University of Northern Colorado, Greeley, Colorado.

Carbyn, L.N. , H.J. Armbruster, and C. Mamo. 1994. The swift fox reintroduction program in Canada from 1983 to 1992. Pp. 247–271 *in* M.L. Bowles and C.J. Whelan, editors. Restoration of an endangered species: conceptual issues, planning and implementation. Cambridge University Press, Cambridge, UK.

Covell, D.F. 1992. Ecology of the swift fox (*Vulpes velox*) in southeastern Colorado. Thesis, University of Wisconsin, Madison, Wisconsin.

Cutter, W.L. 1958a. Denning of the swift fox in northern Texas. Journal of Mammalogy 39:70–74.

——. 1958b. Food habits of the swift fox in northern Texas. Journal of Mammalogy 39:527–532.

Cypher, B.L., and N. Frost. 1999. Condition of San Joaquin kit foxes in urban and exurban habitats. Journal of Wildlife Management 63:930–938.

Cypher, B.L., G.D. Warrick, M.R.M. Otten, T.P. O'Farrell, W.H. Berry, C.E. Harris, T.T. Kato, P.M. McCue, J.H. Scrivner, and B.W. Zoellick. 2000. Population dynamics of San Joaquin kit foxes at the Naval Petroleum Reserves in California. Wildlife Monographs 145:1–43.

Davis, W.B., and D.J. Schmidly. 1994. The Mammals of Texas. Texas Parks and Wildlife, Austin, Texas.

Egoscue, H.J. 1979. *Vulpes velox*. Mammalian Species 122:1–5.

Finley, D.J. 1999. Distribution of the swift fox (*Vulpes velox*) on the eastern plains of Colorado. Thesis, University of Northern Colorado, Greeley, Colorado.

Flaherty, M., and R. Plakke. 1986. Response of the swift fox, *V. velox*, to water stress. Journal of the Colorado–Wyoming Academy of Science 18:51. Abstract only.

Floyd, B.L., and M.R. Stromberg. 1981. New records of the swift fox (*Vulpes velox*) in Wyoming. Journal of Mammalogy 62:650–651.

Glass, B.P., and A.H. Halloran. 1960. Status and distribution of the red fox (*Vulpes vulpes*) in Oklahoma. Southwestern Naturalist 5:71–74.

Golightly, R.T., Jr., and R.D. Ohmart. 1984. Water economy of two desert canids: coyote and kit fox. Journal of Mammalogy 65:51–58.

Hall, E.R., and K.R. Kelson. 1959. The mammals of North America. John Wiley and Sons, New York, New York.

Harrison, R.L. 1997. A comparison of gray fox ecology between residential and undeveloped rural landscapes. Journal of Wildlife Management 61:112–122.

Harrison, R.L., and C.G. Schmitt. 2003. Current swift fox distribution and habitat selection within area of historical occurrence in New Mexico. Pp. 71–77 *in* L.N. Carbyn and M.A. Sovada, editors. Proceedings of the Swift Fox Symposium: ecology and conservation of swift foxes in a changing world. Canadian Plains Research Center, Regina, Saskatchewan.

Hatcher, R.T. 1982. Distribution and status of red foxes (Canidae) in Oklahoma. Southwestern Naturalist 27:183–186.

Hillman, C.N. and J.C. Sharps. 1978. Return of swift fox to northern Great Plains. Proceedings of the South Dakota Academy of Science 57:154–162.

Hines, T.D. 1980. An ecological study of *Vulpes velox* in Nebraska. Thesis, University of Nebraska, Lincoln, Nebraska.

Hines, T.D., and R.M. Case. 1991. Diet, home range, movements, and activity periods of swift fox in Nebraska. Prairie Naturalist 23:131–138.

Jackson, V.L. 1996. Denning ecology of swift foxes (*Vulpes velox*) in western Kansas. Thesis, Fort Hays State University, Hays, Kansas.

Jackson, V.L., and J.R. Choate. 2000. Dens and den sites of the swift fox, *Vulpes velox*. Southwestern Naturalist 45:212–220.

Kilgore, D.L., Jr. 1969. An ecological study of the swift fox (*Vulpes velox*) in the Oklahoma panhandle. American Midland Naturalist 81:512–534.

Kitchen, A.M., E.M. Gese, and E.R. Schauster. 1999. Resource partitioning between coyotes and swift foxes: space, time, and diet. Canadian Journal of Zoology 77:1645–1656.

Klausz, E.E. 1997. Small mammal winter abundance and distribution in the Canadian mixed grass prairies and implications for the swift fox. Thesis, University of Alberta, Edmonton, Alberta.

Klausz, E.E., R.W. Wein, and L.N. Carbyn. 1996. Interaction of vegetation structure and snow conditions on prey availability for swift fox in the northern mixed-grass prairies: some hypotheses. Pp. 281–286 *in* W. D. Willms and J. F. Dormaar, editors. Proceedings of the Fourth Prairie Conservation and Endangered Species Workshop. Provincial Museum of Alberta Natural History Occasional Paper No. 23.

Lindberg, M. 1986. Swift fox distribution in Wyoming: a biogeographical study. Thesis, University of Wyoming, Laramie, Wyoming.

Loy, R.R. 1981. An ecological investigation of the swift fox (*Vulpes velox*) on the Pawnee National Grasslands, Colorado. Thesis, University of Northern Colorado, Greeley, Colorado.

Matlack, R.S., P S. Gipson, and D.W. Kaufman. 2000. The swift fox in rangeland and cropland in western Kansas: relative abundance, mortality, and body size. Southwestern Naturalist 45:221–225.

Mote, K. 1996. Swift fox investigations in Texas, 1996. Pp. 50–52 *in* B. Luce and F. Lindzey, editors. Annual Report of the Swift Fox Conservation Team 1996. Unpublished report. Wyoming Game and Fish Department, Lander, Wyoming.

Olson, T.L. 2000. Population characteristics, habitat selection patterns, and diet of swift foxes in southeast Wyoming. Thesis, University of Wyoming, Laramie, Wyoming.

Packard, R.L., and J.H. Bowers. 1969. Distributional notes on some foxes from western Texas and eastern New Mexico. Southwestern Naturalist 14:450–451.

Pechacek, P. 2000. Activity radii and intraspecific interactions in the swift fox (*Vulpes velox*). Biologia 55:201–205.

Pruss, S.D. 1994. An observational natal den study of wild swift fox (*Vulpes velox*) on the Canadian prairie. Thesis, University of Calgary, Alberta.

——. 1999. Selection of natal dens by the swift fox (*Vulpes velox*) on the Canadian prairies. Canadian Journal of Zoology 77:646–652.

Ralls, K., and P.J. White. 1995. Predation on San Joaquin kit foxes by larger canids. Journal of Mammalogy 76:723–729.

Roell, B.J. 1999. Demography and spatial use of swift fox (*Vulpes velox*) in northeastern Colorado. Thesis, University of Northern Colorado, Greeley, Colorado.

Rongstad, O.J., T.R. Laurion, and D.E. Anderson. 1989. Ecology of swift fox on the Piñón Canyon Maneuver Site, Colorado. Final report to the U.S. Army, Directorate of Engineering and Housing, Fort Carson, Colorado.

Schwartz, A.O., A.M. Vivas, A. Orris, and C.J. Miller. 1994. Small mammal species associations in three types of roadside habitats in Iowa. Prairie Naturalist 26:45–52.

Scott-Brown, J.M., S. Herrero, and J. Reynolds. 1987. Swift fox. Pp. 433–441 *in* M. Novak, J.A. Baker, M.E. Obbard, and B. Malloch, editors. Wild Furbearer Management and Conservation in North America. Ministry of Natural Resources, Ontario.

Sheldon, J.W. 1992. Wild dogs. Academic Press, San Diego, California.

Snow, C. 1973. San Joaquin kit fox, *Vulpes macrotis mutica*: related subspecies and the swift fox, *Vulpes velox*. Habitat management series for endangered species. Report Number 6. Bureau of Land Management, U.S. Department of Interior, Denver, Colorado.

Sovada, M.A., C.C. Roy, J.B. Bright, and J.R. Gillis. 1998. Causes and rates of mortality of swift foxes in western Kansas. Journal of Wildlife Management 62:1300–1306.

Uresk, D.W., and J.C. Sharps. 1986. Denning habitat and diet of the swift fox in western South Dakota. Great Basin Naturalist 46:249–253.

Warrick, G.D., and B.L. Cypher. 1998. Factors affecting the spatial distribution of San Joaquin kit foxes. Journal of Wildlife Management 62:707–717.

Weltzin, J.F., S. Archer, and R.K. Heitschmidt. 1997. Small-mammal regulation of vegetation structure in a temperate savanna. Ecology 78:751–763.

Woolley, T.P., F.G. Lindzey, and R. Rothwell. 1995. Swift fox surveys in Wyoming—annual report. Pp. 61–79 *in* S.H. Allen, J.W. Hoagland, and E.D. Stukel, editors. Report of the Swift Fox Conservation Team. Unpublished report. North Dakota Game and Fish Department. Bismarck, North Dakota.

Zimmerman, A.L. 1998. Reestablishment of swift fox in north central Montana. Thesis, Montana State University, Bozeman, Montana.

Part III – Censusing and Techniques

Swift Fox Detection Probability Using Tracking Plate Transects in Southeast Wyoming

■ Travis L. Olson, J. Scott Dieni, Frederick G. Lindzey and Stanley H. Anderson

Abstract: Monitoring the status of swift fox populations was one of the goals identified in the Conservation Assessment and Conservation Strategy for Swift Fox in the United States. An effective population monitoring program requires development of techniques capable of detecting swift fox presence with high probability. Tracking plate transects have been used as a method of detecting swift fox presence, but its effectiveness has not been determined. We conducted a study near Medicine Bow, Wyoming from March to September 1997 to estimate the probability of detecting swift foxes within areas known to be occupied by marked foxes using tracking plate transects. We estimated detection probability at 0.67 (95% CI = 0.30 - 0.93) in late June, 0.88 (95% CI = 0.47 - 1.0) in late August, and at 1.0 (95% CI = 0.66 - 1.0) near active dens in early July. Detection rate of radiocollared adults increased from late June to late August, despite 5 fewer adult foxes in August. Traditional methods of monitoring population trends usually employ indices, but the reliability of indices is dependent upon the relationship between the index and population size. These relationships are difficult to determine, especially for secretive nocturnal species. We propose using tracking plate transects to detect declines in swift fox populations.

The need to monitor swift fox (*Vulpes velox*) population status has come to the forefront of research on the smallest of North American canids, since proposal for listing as an endangered species in 1992. In June of 1995, the U.S. Fish and Wildlife Service concluded that listing of the swift fox as endangered was warranted but precluded, giving affected states the opportunity to gather additional data on the species (Federal Register 1995). The Swift Fox Conservation Team was formed to develop management objectives for the species, and the Conservation Assessment and Conservation Strategy (CACS; Swift Fox Conservation Team 1997) was developed to identify steps to be taken to ensure swift fox survival. Reliable techniques for monitoring swift fox populations are needed.

Researchers have used various techniques to detect carnivore presence, such as trapping, tracking plate scent stations, spotlighting, remote cameras, and scat surveys (Wood 1959, Linhart and Knowlton 1975, Hillman and Sharps 1978, Frederickson 1979, Barrett 1983, Orloff 1992, Orloff et al. 1993, Dieni et al. 1997, Foresman and Pearson 1998), but the probability of detecting carnivore presence with these methods has not been determined. Dieni et al. (1997) reported that tracking plate surveys for swift fox done in fall provided a greater detection frequency than scat or spotlight surveys. Tracking plates are relatively inexpensive and easy to use, making this technique an attractive detection method. Orloff et al. (1993) reported that tracks left on tracking plates by San Joaquin kit foxes (*Vulpes macrotis mutica*) were readily distinguishable from those left by other wild canids.

Our objective was to estimate the probability of detecting resident swift foxes with tracking plate transects. Transects were established within known swift fox home ranges, and the proportion of transects detecting marked swift fox presence was used as the detection probability estimate. Sampling design and inherent assumptions with the method are discussed.

Study Area

We conducted our study on a 220 km^2 area in northwestern Albany County, Wyoming (42°N, 106°07'W), near the town of Medicine Bow. Topography of the area was primarily flat with numerous dry lakebeds and several saline lakes. Elevation averaged 2075 m. The climate was characterized by long, cold, snowy winters, and warm dry summers. Precipitation averaged 26 cm, including 59 cm of snow annually. The habitat was primarily grass dominated, interspersed with patches of low-growing (<1 m) sagebrush (*Artemisia tridentata*) and areas of taller greasewood (*Sarcobatus vermiculatus*). Other predators present on the study area were badgers (*Taxidea taxus*), coyotes (*Canis latrans*), golden eagles (*Aquila chrysaetos*), and ferruginous hawks (*Buteo regalis*). White-tailed prairie dog (*Cynomys leucurus*) colonies of variable size were also found on the study area. Land ownership was private and the primary land use was cattle grazing. Human developments consisted of fences, windmills, stock ponds, and secondary roads.

Methods

We captured swift foxes between March and June 1997 using Tru-catch live traps baited with meat scraps. Each fox was ear-tagged, fitted with a radio-collar (Advanced Telemetry Systems Inc., Isanti, Minnesota), weighed, and released. Radio-collars were equipped with an 8-hour mortality delay switch and a 12-month battery. Traps were checked twice nightly to minimize the time trapped female foxes were separated from newborn pups. Foxes previously captured in 1996 and their mates were targeted for recapture before old radiotransmitter batteries failed. We considered foxes a mated pair if they were found to be using the same den.

Foxes were located at night using a combination of a vehicle mounted omni antenna and a hand held "H" antenna between 10 May and 15 June 1997. We triangulated fox locations from roads, using at least 3 intersecting azimuths per location. Foxes were also located at dens during daylight hours. To delineate swift fox pair home ranges we plotted circles centered on each pair's natal den site with a radius of 2,119 m. This radius figure was the mean (plus 1 SD) dispersion distance from the center of activity for male foxes on our study area during the winter of 1996–97 as estimated by P. Pechacek (Wyoming Cooperative Fish and Wildlife Research Unit, Laramie, Wyoming, personal communication). We also plotted nocturnal and diurnal fox locations taken during the spring of 1997, and all of these locations fell within estimated circular home ranges. We then identified areas of least overlap with home ranges of adjacent pairs.

To estimate the probability of detecting marked swift foxes within areas known to be occupied by marked foxes we conducted two 7-day test trials, 1 in early summer (27 June–3 July) and 1 in late summer (28 August–3 September). A 1-km transect with 4 plate stations spaced 0.3 km apart was placed within the estimated home range of each pair. Transects were placed in areas of least overlap with adjacent home ranges to minimize the number of adult foxes which would likely encounter each transect. Transects were situated in selected locations (e.g., along fencelines, 2-track road intersections) to increase the likelihood of fox visitation. Each station (Fig. 1) consisted of a 61 cm x 61 cm tracking plate (16-gauge sheet steel) and an infra-red remotely triggered camera/transmitter system (TrailMaster TM 1500, Goodson and Assoc. Inc. Lenexa, Kansas) located 40 cm from each track plate. Tracking plates were sprayed with a talcum powder-ethanol mixture (approximately 50 g talcum powder / 3.6 L ethanol), leaving a thin coat of talc on the plate, and baited with approximately 10 g of canned mackerel (*Trachurus murphyi*) in the center (Woolley et al. 1995). Mackerel was used both as an attractant and as an incentive for foxes to re-visit plates on subsequent nights. Cameras were triggered when an infra-red beam of light centered across the plates approximately 20 cm above the ground was broken, allowing us to identify foxes as marked (radio-collared; Fig. 2) or unmarked (non-collared; Fig. 3). Marked foxes were known to be residents. Tracking plates were checked each morning, and swift fox tracks were measured and recorded, but plates were not re-baited until early evening of that day.

The transect/fox pair was the sample unit, and the proportion of transects detecting presence of tracks from marked swift foxes during each trial was considered the detection probability estimate. We constructed approximate 95% confidence intervals for detection probability estimates ($n = 8$ is too small to use the normal approximation) as described by Johnson and Kotz (1969). We

Figure 1. Tracking plate/camera stations used to detect swift foxes near Medicine Bow, Wyoming, during late June and late August 1997.

Figure 2. Marked swift fox visiting tracking plate station in late June 1997 near Medicine Bow, Wyoming.

Figure 3. Unmarked swift fox visiting tracking plate station in late August 1997 near Medicine Bow, Wyoming.

Table 1. Proportion of tracking plate transects (1 km) detecting swift fox during late June (trial 1) and late August (trial 2) 1997, near Medicine Bow, Wyoming

	Proportion		Days[a]	
Evidence of swift fox presence	Trial 1 (n = 9 pairs)	Trial 2 (n = 8 pairs)[b]	Trial 1	Trial 2
Photograph of marked fox and track	0.33	0.88	3	6
Photograph of marked fox with or without track	0.67	1.00	6	6
Track with or without photograph of marked fox	0.67	0.88	7	6
Track or scat[c]	0.77	0.88	7	6
Track, scat[c], or photograph of marked fox	1.00	1.00	7	6

[a]Number of transect days required to achieve maximum proportion
[b]3 fox pairs were reduced to 1 member by trial 2
[c]Swift fox scat found near (<5 m) tracking plate stations

compared detection results between trials using a paired-samples Wilcoxon test (Agresti and Finlay 1986). Regression analysis was used to compare the shortest distances from tracking plate transects to active dens (den where 1 or both foxes of a pair were found during a trial), with number of days until we first recorded evidence (track, photograph, scat) that a swift fox had encountered a transect. Statistical analyses were conducted using Minitab (release 12.21) and SPSS (release 8.0).

We also estimated the probability of detecting swift fox near active dens using tracking plates during a 2-night trial (13–14 July). Tracking plates were placed at the ends of 2 perpendicular transects (4 plates total), centered on the active den site of each marked fox pair. Prior to running this test we observed pups to determine how far they ranged from their dens, and placed plates far enough from dens (100 m) to reduce the possibility of pups visiting plates. One transect was baited with mackerel and the other with bacon. After the first night both transects were rotated 45 degrees clockwise around the den to vary the placement of tracking plates.

Results

We captured and radio-collared both members of 9 swift fox pairs between 9 March and 6 June. Five foxes died between the first and second trials, eliminating 1 pair and leaving 3 single foxes.

Detection Probability

We operated 9 transects during trial 1 (27 June-3 July), and recorded swift fox tracks on 6 transects (67%) by day 7 of the trial (Table 1). Therefore, we estimated the probability of detecting swift foxes within known swift fox home ranges at 0.67 (95% CI = 0.30 - 0.93). Of the 6 transects where swift fox tracks were recorded, we detected marked foxes on 3 of the transects and were unable to determine the identity of foxes that stepped on the other 3 transects due to a lack of photographs. Although we detected tracks on only 6 transects, we found scat (<5 m from tracking plate station) on 1 additional transect and photographs of marked foxes at the 2 remaining transects, resulting in a fox encounter rate of 1.0. We recorded photographs of unmarked foxes on 2 transects where we also recorded photographs of marked foxes.

We ran only 8 of the 9 transects during trial 2 (28 August–3 September) because both members of 1 fox pair died between trials. We recorded swift fox tracks on 7 transects by day 6 of the trial (Table 1), resulting in a detection probability estimate of 0.88 (95% CI = 0.47 - 1.0). Photographs of marked foxes were taken on all 8 transects by day 6 of the trial, resulting in an encounter rate of 1.0. Tracks and scat were detected on 5 transects, and photographs of unmarked foxes (presumably pups) were recorded on 6 of 8 transects.

Trial Comparisons

The average number of nights detecting swift fox tracks on a transect increased significantly ($P = 0.05$; one-tailed, $n = 8$) from trial 1 ($\bar{x} = 2.25$, SE = 0.98) to trial 2 ($\bar{x} = 4.0$, SE = 0.85). The average number of stations on a transect which had swift fox tracks each night also increased significantly ($P = 0.03$; one-tailed, $n = 8$) from trial 1 ($\bar{x} = 0.68$, SE = 0.35) to trial 2 ($\bar{x} = 1.34$, SE = 0.33). The average number of stations on a transect which detected radio-collared swift foxes each night was similar ($P = 0.5$, $n = 8$) for the 2 trials (trial 1: $\bar{x} = 0.64$, SE = 0.35, trial 2: $\bar{x} = 0.61$, SE = 0.14). Combining results of the 2 trials, we detected tracks of radio-collared swift foxes on all 9 transects.

On 33 occasions (trials combined), we photographed swift foxes but did not detect them with track plates. Fifteen of those occasions were caused by 3 factors: poor tracking medium due to rain or dew (6); disturbance by cattle (6); missing bait (3). On 23 occasions, we identified swift fox tracks on tracking plates, but no photographs had been taken.

The mean shortest distance from tracking plate transects to active dens was 740 m ($n = 9$, SE = 130, range = 175–1,350) for trial 1 and 1,086 m ($n = 8$, SE = 384, range = 300–2,340) for trial 2. We did not detect a significant relationship (trial 1: $R^2 = 0.27$, $P = 0.08$; trial 2: $R^2 = 0.05$, $P = 0.28$) between the shortest distance from tracking plate transects to active dens, and the number of days before a swift fox had encountered a transect during either trial.

Detection Near Dens

We recorded swift fox tracks on at least 1 of the 4 tracking plates at all active den sites both nights of the trial, giving a detection probability point estimate of 1.0 (95% CI = 0.66 - 1.0). An average of 2.1 (SE = 0.45) tracking plates of the 4 at each den site detected tracks the first night and 2.5 (SE = 0.29) the second night. There was little difference in the number of plates visited that were baited with mackerel (22) and bacon (20). Bait was completely eaten on 17 plates baited with mackerel and 19 plates baited with bacon.

Discussion

Estimated detection probability of adults stepping on tracking plates increased from late June to late August even though 5 radio-collared foxes died between trials. The average number of nights that tracks were detected on each transect, and the average number of stations detecting tracks on a transect each night increased significantly from the first to second trial. We expected to detect more foxes in late summer as fox pups began foraging on their own, but the number of transects detecting radio-collared (adult) foxes increased as well. This suggests that adults may be more active or range farther from natal dens in late summer and fall than during early summer. These results corroborate those of Dieni et al. (1997), who found that detection frequencies with tracking plates were higher in the fall than in the spring.

Late summer is a preferable time to conduct surveys because the likelihood of detecting swift foxes is increased by the presence of fox pups. Much of the swift fox literature suggests that swift fox pups do not disperse until late summer or early fall (Kilgore 1969, Egoscue 1979, Hines 1980). Rongstad et al. (1989) reported that radio-collared swift fox pups did not disperse from their parent's home ranges until November and December. If juveniles do not disperse until late in the fall, then detection of juvenile tracks during late summer will indicate adult presence as well. Detection probability of adult swift foxes in the fall and winter may be similar to what we measured in late summer, but these seasons may be less desirable for conducting surveys due to unpredictable and possibly severe weather. Dieni et al. (1997) found that detection results with tracking plates were very sensitive to winter weather conditions. Precipitation is the main hindrance to using tracking plates, but selecting typically dry time periods (e.g., late June–mid September in Wyoming) to conduct surveys will help mitigate this problem. We recommend late summer as the best time to monitor.

The swift fox detection rate using tracking plates appeared a product of the likelihood that ≥1 fox would encounter ≥1 plate and leave identifiable tracks. We expected the number of tracking plate transects encountered by foxes, and thus the cumulative detection rate, to increase with time, and indeed, maximum detection rates were not reached until days 7 and 6 of the trials, respectively. This suggests that the detection rate may have been greater during trial 1 if we had run transects more than 7 days. The encounter rate (evidence of swift fox presence from scat, tracks, or photographs) also reached a maximum (1.0) on days 7 and 6 of the trials, respectively. However, the detection rate of marked foxes was lower than the encounter rate for both trials due to a lack of photographs of marked foxes, or a lack of identifiable tracks. Swift foxes may respond differently to tracking plates encountered in close proximity to active den sites than to more distant plates. It is possible that the camera flash occasionally scared a fox from the plate before it stepped on it, thus reducing the detection rate. Cameras and transmitters were situated far enough from plates that a fox could break the infra-red beam without stepping on the plate. If the camera flash did deter future visits by foxes, our estimates of detection probability are conservative. Decreased tracking plate spacing would likely increase the swift fox encounter rate and thus detection rate.

We failed to detect a relation between distances from transects to active dens and numbers of days until we first recorded evidence that swift foxes encountered transects. However, transects were intentionally placed within portions of home ranges least overlapped by home ranges of adjacent pairs. This approach likely caused transects to be located closer to active dens than might have occurred if transects were randomly placed within home ranges. Swift fox home range size may vary (across populations or temporally) from those we observed and affect encounter rates.

Estimates of swift fox detection probability using tracking plate transects need to be obtained in different swift fox populations to further establish reliability of this technique. Variations in swift fox behavior, home range size, or prey availability may influence detection probability. One possible bias in our results is that 8 of the marked foxes on our study area during the summer of 1997 were likely exposed to tracking plates during tests conducted as part of another study during 1996. Familiarity with tracking plates could have influenced their reactions to the plates, although different attractants were used between years (scent tabs in the previous study and mackerel in this study).

Our detection probability estimate did not depend upon the number of foxes detected or number of visits per transect, only on fox presence or absence. As transect length increases, the number of swift fox home ranges intersected will increase and thus the number of foxes exposed to a transect will increase. Therefore, when using presence/absence monitoring, transect length will directly influence the probability of detecting at least 1 fox on a transect. For example, if swift fox presence is detected on a 10 km transect in year 1 when there are 5 fox pairs present, and presence is still detected in year 2 when only 1 fox pair remains, the result for the transect would be the same even though there are fewer foxes. If short transects (1 km) are used, the number of foxes exposed to a transect will be minimized and presence/absence results for a large number of transects will be more reflective of actual population change. If foxes are not detected on a transect we would be fairly confident (given our estimate of the detection probability) that the transect no longer intersects a fox home range.

Our trials were designed to reflect how tracking-plate transects might be used in future swift fox population persistence monitoring. Therefore, we used short transect lengths (1 km) to minimize the number of swift foxes potentially exposed to each transect. The primary assumption of persistence monitoring is that if foxes are present initially, foxes should still be present in future years if the population is not declining, whether they are the same or different individuals. This should hold true in good quality swift fox habitat with stationary or increasing populations. Survey transects chosen for persistence monitoring should be located in good quality swift fox habitat rather than marginal habitat where populations may fluctuate and home ranges may not be filed each year.

Remotely triggered cameras detected marked swift foxes on more transects than did tracking plates (6 during trial 1, 8 during trial 2), suggesting that cameras may be a better detection tool than tracking plates (Foresman and Pearson 1998). They can be used during rainy weather with no adverse effects, whereas tracking plates can only be used effectively during dry weather. Cameras also eliminate the need to identify tracks and may yield a greater detection rate even when used without tracking plates. However, cameras were not infallible. On 23 occasions, swift fox tracks were detected when no photographs were taken. In most cases, the lack of a photograph could be explained by exposure of an entire roll of film due to cattle disturbance before a swift fox visited the tracking plate station. It was also possible for a fox to step on the tracking plate edge and not trigger the camera because of alignment of the infra-red beam across the center of the plate. Camera systems are also very expensive (about $600 US; plus cost of film, film development, and batteries), and may be subject to theft in the field.

Management Implications

Traditional methods of monitoring population trends usually employ indices because of the difficulty and expense of obtaining density estimates (Johnson and Pelton 1981). Despite the frequent use of scent station indices in monitoring carnivore populations, the relationship between an index and population abundance is usually not known (Linhart and Knowlton 1975, Roughton and Sweeney 1982, Conner et al. 1983, Sargeant et al. 1998). Changes in carnivore scent station visitation rates in the long term may reflect population change, but it will still be difficult to determine when management actions need to be taken (Sargeant et al. 1998). Perhaps a more efficient and realistic approach to monitoring is to emphasize the detection of declines in persistence over time rather than changes in visitation rates, which are confounded by variation in individual responses.

Our results will be useful in development of a Monte Carlo power simulation model used to determine the number of transects needed to detect a given level of decline (Beier and Cunningham 1996, Zielinski and Stauffer 1996). Zielinski and Stauffer (1996) conducted power simulations for monitoring fishers (Martes pennanti) and martens (M. americana) but were forced to include a bias adjustment for the possibility of failing to detect presence with the survey method (tracking plates). Our work presents an estimate of this detection error (1-detection probability) for detecting swift foxes with tracking plate transects. Incorporation of a detection error estimate in power simulations rather than a bias adjustment should improve accuracy of power calculations. If the detection error with a given survey method is low (i.e. high detection probability), the number of survey transects needed to detect a given level of decline will be minimized.

Ideally, one would like to be able to detect changes in populations (increases or decreases) with a monitoring program, but the survey effort needed to accomplish this may be prohibitive (Zielinski and Stauffer 1996). Monitoring to detect only declines will greatly reduce the field effort required, and would provide managers with needed information, especially when dealing with a candidate species for threatened status such as the swift fox.

Acknowledgments

Funding for this project was provided by the Wyoming Game and Fish Department. We would like to thank B. Luce and the nongame program for providing housing in the field. We especially thank P. Jeschke and C. Graff for allowing us to conduct research on their properties; without their support this study would not have been possible. We sincerely thank T. Woolley and D. Wroe for their advice and assistance, and we especially thank G.A. Sargeant for a very thorough review of the manuscript.

Literature Cited

Agresti, A., and B. Finlay. 1986. Statistical methods for the social sciences. Dellen Publishing, San Francisco, California.

Barret, R.H. 1983. Smoked aluminum track plots for determining furbearer distribution and relative abundance. California Fish and Game 69:188–190.

Beier, P., and S.C. Cunningham. 1996. Power of track surveys to detect changes in cougar populations. Wildlife Society Bulletin 24:540–546.

Conner, M.C., R.F. Labisky, and D.R. Progulske. 1983. Scent-station indicies as measures of population abundance for bobcats, raccoons, gray foxes, and opossums. Wildlife Society Bulletin 11:146–151.

Dieni, J.S., F.G. Lindzey, T.P. Woolley, and S.H. Anderson. 1997. Swift fox investigations in Wyoming: annual report. Wyoming Cooperative Fish and Wildlife Research Unit, Laramie, Wyoming.

Egoscue, H.J. 1979. *Vulpes velox*. Mammalian Species 122:1–5.

Federal Register. 1995. Endangered and threatened wildlife and plants: 12-month finding for a petition to list the swift fox as endangered. Federal Register 60:31663–31666.

Foresman, K.R., and D.E. Pearson. 1998. Comparison of proposed survey procedures for detection of forest carnivores. Journal of Wildlife Management 62:1217–1226.

Frederickson, L. 1979. Furbearer population surveys: techniques, problems and uses in South Dakota. Pp. 62–70 *in* Proceedings of the Midwest furbearer conference. Cooperative Extension Service, Kansas State University, Manhattan, Kansas.

Hillman, C.N., and J.C. Sharps. 1978. Return of the swift fox to northern great plains. Procedings of the South Dakota Academy of Science 57:154–162.

Hines, T.D. 1980. An ecological study of *Vulpes velox* in Nebraska. Thesis, University of Nebraska, Lincoln, Nebraska.

Johnson, K.G., and M.R. Pelton. 1981. A survey of procedures to determine relative abundance of furbearers in the southeastern United States. Proceedings of the Annual Conference of the Southeastern Association of Fish and Wildlife Agencies 35:261–272.

Johnson, N.L., and S. Kotz. 1969. Discrete distributions. John Wiley and Sons, New York, New York.

Kilgore, D.L., Jr. 1969. An ecological study of the swift fox (*Vulpes velox*) in the Oklahoma panhandle. American Midland Naturalist 81:512–534.

Linhart, S.B., and F.F. Knowlton. 1975. Determining the relative abundance of coyotes by scent station lines. Wildlife Society Bulletin 3:119–124.

Orloff, S.G. 1992. Survey techniques for the San Joaquin kit fox (*Vulpes macrotis mutica*). Pp. 185–197 *in* D.F. Williams, S. Byrne, and T.A. Rado, editors. Endangered species of the San Joaquin Valley, California: their biology, management, and conservation. California Energy Commission, Sacramento, California.

Orloff, S.G., A.W. Flannery, and K.C. Belt. 1993. Identification of San Joaquin kit fox (*Vulpes macrotis mutica*) tracks on aluminum tracking plates. California Fish and Game 79:45–53.

Rongstad, O.J., T.R. Laurion, and D.E. Andersen. 1989. Ecology of swift fox on the Piñon Canyon maneuver site, Colorado. Final report to Directorate of Engineering and Housing, Fort Carlson, Colorado.

Roughton, R.D., and M.W. Sweeny. 1982. Refinements in scent-station methodology for assessing trends in carnivore populations. Journal of Wildlife Management 46:217–229.

Sargeant, G.A., D.H. Johnson, and W.E. Berg. 1998. Interpreting carnivore scent-station surveys. Journal of Wildlife Management 62:1235–1245.

Swift Fox Conservation Team. 1997. Conservation assessment and conservation strategy for swift fox in the United States. R. Kahn, L. Fox, P. Horner, B. Giddings, and C. Roy, technical editors. Montana Fish, Wildlife, and Parks, Helena, Montana.

Wood, J.E. 1959. Relative estimates of fox population levels. Journal of Wildlife Management 23:53–63.

Woolley, T.P., F.G. Lindzey, and R. Rothwell. 1995. Swift fox surveys in Wyoming—annual report. Pp. 61–76 *in* S.H. Allen, J. Whitaker Hoagland, and E. Dowd Stukel, editors. Report of the swift fox conservation team 1995. North Dakota Game and Fish Department, Bismarck, North Dakota.

Zielinski, W.J., and H.B. Stauffer. 1996. Monitoring Martes populations in California: survey design and power analysis. Ecological Applications 6:1254–1267.

Scent-Station Survey Techniques for Swift and Kit Foxes

■ **Glen A. Sargeant, P.J. White, Marsha A. Sovada and Brian L. Cypher**

Abstract: We compared scent-station survey techniques for monitoring the distribution and relative abundance of swift and kit foxes. We used data collected at Camp Roberts, California, during 1988-97 (kit fox); the U.S. Department of Energy's Naval Petroleum Reserves, California, during 1984-96 (kit fox); and in Sherman and Wallace counties, Kansas, during 1996 (swift fox). Principal results included the following: 1) Scent-station surveys were not cost-effective for determining the distribution of swift foxes. 2) Monthly visitation rates of swift foxes declined from April to August, then increased eight-fold in October, after juveniles began visiting stations. 3) Spring and summer surveys, but not autumn surveys, detected a sustained decline in kit fox abundance. 4) Swift foxes visited scent stations more frequently during the first night of each monthly survey than during subsequent nights. 5) Repeated operation of the same scent stations yielded less information about abundance than could have been obtained by establishing new stations. 6) Swift foxes visited stations with a sand-and-mineral oil substrate 2.4 times as frequently as track plates. These results suggest intuitive perceptions are frequently incorrect and emphasize the need for objective, experimental comparisons of scent-station survey techniques for monitoring swift and kit foxes.

Scent-station surveys have become a popular method for monitoring the distribution and relative abundance of swift foxes (*Vulpes velox*) and kit foxes (*V. macrotis*). However, scent stations frequently fail to detect canids that are present (Griffith et al. 1982, Sargeant et al. 1998) and efforts to link visitation rates to the abundance of swift and kit foxes have been inconclusive (e.g., Harris 1987). Thus, consensus regarding the usefulness of scent-station surveys for either purpose has not been achieved.

Investigators who rely on scent-station surveys of swift or kit foxes choose different techniques of data collection on an ad hoc basis, guided principally by intuition. Some differences among surveys result because the 2 ultimate objectives of scent-station surveys–assessing distributions and monitoring changes in population–dictate different proximate objectives. Methods for assessing distributions should minimize errors of omission (failing to detect resident foxes) and commission (detecting foxes where they are not resident), whereas methods for monitoring relative abundance should result in precise estimates of visitation rates and control for confounding factors. Investigators with similar goals, however, also choose different techniques. This reflects general disagreement regarding the best time to conduct surveys, the most effective means of attracting foxes to stations, the most reliable technique for detecting visits, and the most efficient means of allocating sampling effort. Few techniques have been compared experimentally, survey results are largely unpublished, and available information has not been synthesized.

Our objective was to evaluate scent-station survey techniques for swift and kit foxes, based on an exploratory analysis of 3 data sets and a synthesis of published and unpublished reports. We compared scent-station surveys with other methods for determining the distribution of swift foxes, tested trends in scent-station visitation rates of a rapidly declining kit fox population, examined seasonal variation in visitation rates of a swift fox population, compared replication with repeated operation of stations, and compared 2 types of tracking surfaces (track plates and sand) and 2 types of lures (fatty-acid scent [FAS] and food). In this paper, we discuss the utility of scent stations for surveys of distribution and abundance, relate choices of techniques to proximate objectives of surveys, and compare our results with contemporary perceptions of scent-station survey techniques for swift and kit foxes.

Study Areas

Our data were collected at the Camp Roberts Army National Guard Training Site, California (35° 45' N., 120° 45' W.), during 1988-97 (kit fox); at the U.S. Department of Energy's Naval Petroleum Reserves (NPRC), California (119° 30' N., 35° 15' W.), during 1984-96 (kit fox); and at 2 areas in Sherman and Wallace counties, western Kansas (39° 14' N., 101° 31' W. and 39° 05' N., 101° 33' W.), during 1996 (swift fox). These 4 areas represented a diversity of habitats, wildlife communities, and land management practices encountered within the ranges of swift and kit foxes.

Camp Roberts, California, encompassed 172 km^2 of rolling hills between the Salinas River floodplain and the Santa Lucia Mountains. Dominant vegetation included grassland, oak (*Quercus* spp.) woodland, and mixed chaparral: kit foxes, however, occurred primarily in grassland and low- to medium-density oak woodland (Reese et al. 1992). The NPRC was located 42 km southwest of Bakersfield, in Kern County, and comprised 323 km^2 of rolling hills dissected by steep draws and dry stream channels. Vegetation consisted of xerophytic shrubs and patchy herbaceous cover dominated by exotic annual grasses and forbs. Study areas in Kansas encompassed 259 km^2 each. One comprised relatively flat cropland devoted primarily to production of winter wheat, corn, milo, sunflowers and sorghum. The other included rolling hills of moderately to heavily grazed pastures with a few cultivated areas interspersed.

Temperatures averaged 14°C in winter and 23°C in summer at Camp Roberts, with mean annual rainfall of 28.5 cm, and 9°C in winter and 29°C in summer at the NPRC, where mean annual precipitation was only 12.5 cm. In Kansas, temperatures averaged 10°C in winter and 26°C in summer, with mean annual precipitation of 46.2 cm.

At Camp Roberts and the NPRC, potential predators or competitors of kit foxes included coyotes (*Canis latrans*), red foxes (*Vulpes vulpes*), gray foxes (*Urocyon cinereoargenteus*), bobcats (*Lynx rufus*), badgers (*Taxidea taxus*), and golden eagles (*Aquila chrysaetos*). Coyotes, badgers, and domestic dogs (*Canis familiaris*) were the only potential predators or competitors observed on study areas in Kansas.

Camp Roberts was used primarily for military training and for grazing by sheep and cattle. The NPRC was devoted primarily to the production of petroleum products. Associated activities included the construction of roads, well pads, and other facilities. Farming and grazing were the dominant uses of study areas in Kansas.

Materials and Methods

Data Collection

At Camp Roberts and the NPRC, we patterned scent-station surveys after Linhart and Knowlton (1975), as modified by Roughton and Sweeny (1982). Each station consisted of a 0.9-m diameter circle of smoothed earth or sand with an FAS disk (Pocatello Supply Depot, USDA) placed in the center. We placed stations on alternate sides of unpaved roads at 480-m intervals, in lines of 10. We placed lines at least 1 km apart and distributed them as regularly as possible, subject to the availability of roads and, at Camp Roberts, to access restrictions in ordinance impact areas. At Camp Roberts, we conducted surveys during autumn (September–November) of 1988; spring (March–May) of 1997; summer (June–August) and autumn of 1989, 1993 and 1995; and spring, summer, and autumn of 1990–92, 1994, and 1996. At the NPRC, we conducted surveys during summer and autumn of 1984; spring, summer and autumn of 1985–91; and spring of 1992–96. At both sites, we operated lines for 1 night per survey.

In Kansas, we constructed stations of 0.37-m^2 (61x61-cm) sheets of 18-ga galvanized steel plate coated with carpenters chalk (track plates). We attached a blank scent disk (Pocatello Supply Depot, USDA) soaked in commercially processed liquefied mackerel in cod-liver oil to the center of each track plate. During the October survey, we supplemented ~60% of track plates with a station composed of sand, which we mixed with mineral oil (approximately 16:1 ratio) to enhance the clarity and durability of tracks. Sand-and-oil stations were placed ~5 m from track plates and had the same dimensions and lure. We placed stations along unpaved roads at 500-m intervals, in lines of 17–19, with a minimum distance of 1.6 km between lines. We conducted surveys monthly during April-August and in October, 1996. During each monthly survey, we checked stations daily until useable data were collected for 3 nights. We considered data to be useable if weather conditions did not interfere with the identification of tracks. While conducting surveys, we also collected telemetry data that showed all stations were placed within home ranges of swift foxes (M.A. Sovada, U.S. Geological Survey, unpublished data).

Statistical Analysis

Stations, lines, and surveys are 3 common choices of experimental units for analyses of scent-station data (Sargeant et al. 1998). We opted for a conservative approach. We treated surveys as experimental units when we considered several simultaneously (i.e., when testing for association and for seasonal differences in visitation) and treated lines as experimental units when we considered surveys individually (i.e., when comparing tracking media). In all cases, we used daily visitation rates (number of visits/number of station-nights) as our population index. SAS (SAS Institute, Inc. 1988) and S-PLUS (MathSoft, Inc. 1997) were used to perform analyses.

To test for an association between abundance and daily visitation rates, we assumed a monotonic decline in kit fox numbers at Camp Roberts and used Spearman's rank-order correlation coefficient (ρ; Daniel 1990). We applied the same method to data from Kansas to test for a seasonal trend in daily visitation rates during April-August. We used analysis of variance with year as a blocking factor (Sokal and Rohlf 1981) to test for seasonal differences in visitation at Camp Roberts and the NPRC.

To compare tracking media, we used a one-tailed t-test with October data from Kansas, which we paired by station and date within line (Sokal and Rohlf 1981). We used a one-tailed test because we anticipated higher daily visitation rates to sand stations than to track plates. To determine whether observed visits were equally likely to occur on the first, second, and third nights of surveys, we used chi-square tests of homogeneity (Daniel 1990) with data from Kansas. Our null hypothesis was $P_1 = P_2 = P_3$ where P_1, P_2, and P_3 were the respective proportions of visits expected on the first, second, and third nights of surveys. We conducted separate tests for April-August, when most visits were attributable to adult foxes, and October, when juveniles were probably responsible for most visits.

To determine whether stations should be operated for more than 1 night in succession and to assess the value of repeating surveys, we resampled empirical distribution functions (ED's; Efron and Tibshirani 1993:31-35) comprising subsets of data from Kansas and the NPRC. In Kansas, where we operated stations for 3 nights in succession during each monthly survey, results of each survey could be organized as an $n \times 3$ matrix. The first column of each matrix constituted an ED for stations operated 1 night each (i.e., replicate stations). The set of n rows, however, constituted an ED for the joint trivariate distribution sampled when data were collected by operating each station for 3 nights in succession (i.e., repeated stations). Similarly, results of spring, summer, and autumn surveys at the NPRC could be envisioned as a set of three $n \times 1$ vectors, each of which was a seasonal ED, or as a set of n 1×3 vectors, an ED for repetition across seasons. From these distributions, we calculated estimates of mean daily visitation rates and their standard errors by resampling. We used means and standard errors to estimate coefficients of variation resulting from replication (establishment of new stations) and repetition (repeated operation of the same stations). We recommend methods of allocating survey effort that resulted in smaller coefficients of variation.

Results

Capture records (J. Eliason, U.S. Army National Guard, Camp Roberts, unpublished data) suggested a rapid, sustained decline in the Camp Roberts fox population. From 1988 to 1996, the number of individuals captured annually declined from 103 to 9, despite identical trapping techniques and similar trapping effort. This decline in population was evident from spring ($\rho = 0.89$, $P = 0.009$) and summer scent-station surveys ($\rho = 0.71$, $P = 0.02$), but not from autumn surveys ($\rho = 0.43$, $P = 0.13$; Fig. 1). Declines in spring and summer visitation rates were of much smaller magnitude than changes in the number of foxes captured. The disparity suggests individual foxes were more likely to visit stations when fox density was low than when it was high.

In Kansas, monthly visitation rates for track plates declined steadily from April to August, then increased nearly eight-fold in October (Fig. 2), after juveniles began traveling away from natal dens (M.A. Sovada, U.S. Geological Survey, unpublished data). During April-August, foxes habituated to track plates operated for more than 1 night. Habituation resulted in higher visitation rates ($\chi^2_2 = 11.97$, $P = 0.003$) on the first night (2.6%) stations were operated successfully than on the second (0.9%) or third (0.9%). In October, visitation rates were higher ($\chi^2_2 = 6.77$, $P = 0.034$) for track plates, but not for sand-and-oil stations ($\chi^2_2 = 0.095$, $P = 0.95$), on the first night of successful operation (plates = 15.1%; sand = 27.0%) than on the second (plates = 11.9%, sand = 25.2%) and third (plates = 6.9%, sand = 26.4%). We found no evidence of consistent seasonal variation in visitation rates at Camp Roberts ($F_{2,12} = 0.01$, $P = 0.99$; Fig. 1) or the NPRC ($F_{2,13} = 2.64$, $P = 0.11$; Fig. 3).

Swift foxes were not readily detected with track plates during spring or summer in Kansas. Although all stations were placed within fox home ranges, we observed a daily visitation rate of only 1.5% and monthly track plate surveys frequently failed to detect foxes that were known to be present. During October, foxes in Kansas visited sand-

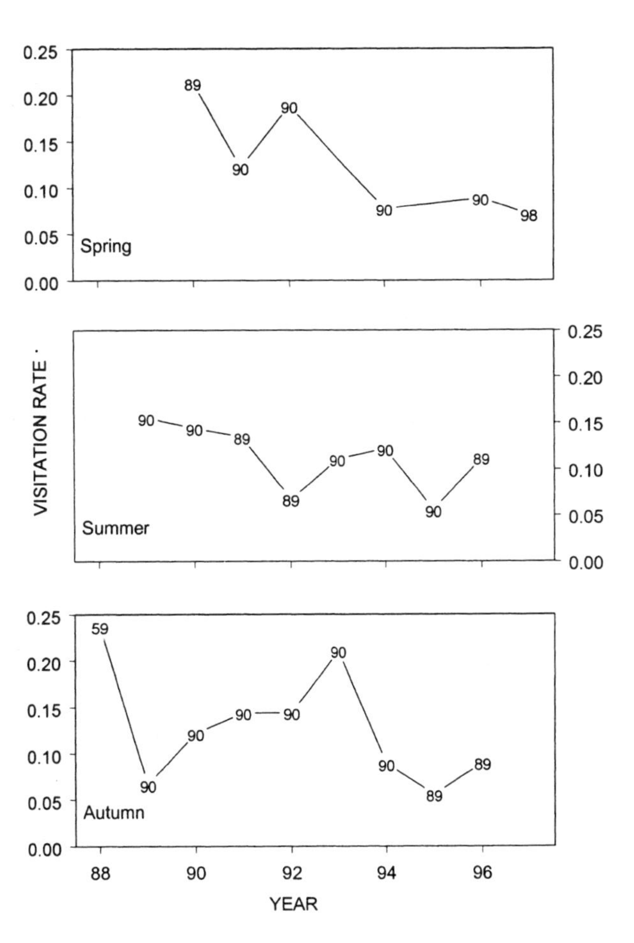

Figure 1. Scent-station visitation rates for kit fox at Camp Roberts, California, 1988-96. Plotting symbols are numbers of stations operated.

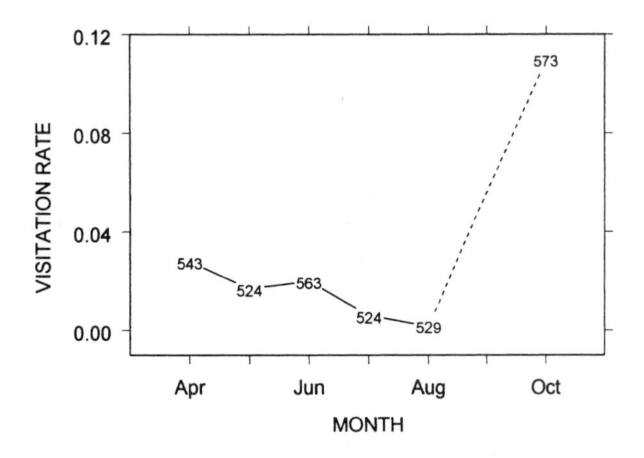

Figure 2. Scent-station visitation rates for swift fox in Sherman and Wallace counties, Kansas, during April-October, 1996. Plotting symbols are numbers of stations operated.

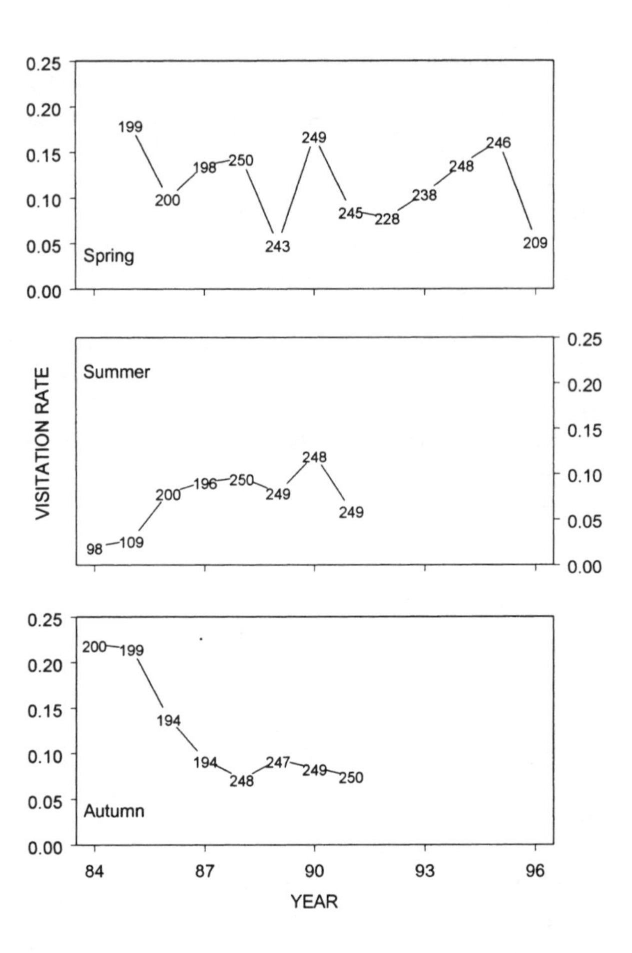

Figure 3. Scent-station visitation rates for kit fox at the Naval Petroleum Reserves in Kern County, California, 1984–96. Plotting symbols are numbers of stations operated.

and-oil stations (26.2%) more readily (t_6 = 2.34, P = 0.029) than track plates (11.2%).

Because track plates were visited at higher rates on the first night stations were operated successfully than on the second or third, operating stations for more than 1 night unexpectedly increased coefficients of variation for 5 of 6 monthly surveys conducted in Kansas. At the NPRC, where the same stations were monitored 3 times annually, establishing new stations each season would have produced a smaller coefficient of variation for all 7 years where a comparison was possible.

Discussion

Surveys of Distribution

Low detection rates, such as we observed in Kansas, are apparently the rule for swift foxes and other canids. In Wyoming, swift foxes were photographed visiting track plates without leaving identifiable tracks in 50% of cases in 1 trial and in 12% of cases in another (T.L. Olson, Wyoming Cooperative Wildlife Research Unit, unpublished data). In Utah, Griffith et al. (1981) were able to detect tracks at only 28.8% of stations where coyotes left identifiable tracks within 10 m. In Minnesota, Sargeant et al. (1993) found evidence of red foxes in 96.4% of quarter sections they searched. Foxes were nevertheless detected at only 63.3% of 4800 m lines of 10 scent stations operated concurrently in the same counties (W.E. Berg, Minnesota Department of Natural Resources, unpublished data). At typical levels of sampling effort, scent stations may not be effective for detecting the presence of swift or kit foxes. They clearly cannot be used to determine where swift foxes are absent.

Although detection rates may be increased by sampling intensively and conducting fall surveys, other methods of detection are likely to be more cost effective. Because we conducted 6 monthly surveys, each survey line in Kansas eventually detected foxes. However, the time and expense of searching legal quarter-sections (0.65 km^2; Sargeant et al. 1993) for evidence of swift fox was less than required to conduct repeated scent-station surveys. Moreover, probabilities of detecting foxes were higher (M.A. Sovada, U.S. Geological Survey, unpublished data). Similarly, C.C. Roy (Kansas Department of Wildlife and Parks, unpublished data) detected evidence of swift foxes in approximately 40% of townships she surveyed in western Kansas, simply by spending 2 hours identifying and searching likely swift fox habitat in each township.

Surveys of Relative Abundance

Estimates of carnivore abundance are usually imprecise because most carnivores are cryptic, secretive, neophobic, and occur at low densities. Hence, short-term variation in abundance is seldom of sufficient magnitude for conclusive validation of population indices. Because manipulation of populations on a scale sufficient for index validation is neither feasible nor acceptable, scent-station population indices for swift and kit foxes have not been experimentally validated or calibrated. Results from Camp Roberts, where spring and summer surveys detected a decline in the kit fox population, are important in this context. However, it is noteworthy that visitation rates at Camp Roberts declined proportionately less than population size. Thus, individuals visited stations at comparatively high rates when population density was low. Negative relations between individual visitation and abundance, if present, reduce the sensitivity of scent-station indices (Smith et al. 1994).

Despite limited information for swift and kit foxes, circumstantial evidence suggests a correspondence between scent-station visitation rates and canid abundance (Sargeant et al. 1998). Prevailing evidence thus suggests long-term trends in spring and summer visitation rates reflect gross changes in swift and kit fox populations when samples are of sufficient number and surveys are properly designed. Scent-station surveys have low spatial resolution, however, because reliable inferences require large sample sizes (Zielinski and Stauffer 1996, Sargeant et al. 2003), kit and swift foxes have home ranges of 4-20 km^2 (Zoellick et al. 1987, White and Ralls 1993, M.A. Sovada, unpublished data), and observational units (stations or groups of stations) should be spaced sufficiently to ensure sampling of different individuals (Zielinski and Stauffer 1996).

Survey Timing

Most scent-station surveys of canids are conducted wholly or partly in autumn (e.g., Wood 1959, Linhart and Knowlton 1975, Morrison et al. 1981, Linscombe et al. 1983). Our results suggest presumed benefits of autumn surveys, which include favorable weather (Linhart and Knowlton 1975) and high visitation rates (Beltrán et al. 1991) that maximize statistical power for detecting changes in visitation (Roughton and Sweeny 1982), are outweighed by other considerations. Although spring and summer surveys tracked changes in the abundance of kit foxes at Camp Roberts, autumn surveys did not. At the NPRC, spring, summer, and autumn surveys were suggestive of different population trends (Fig. 3).

The persistence of canid populations depends on the distribution and abundance of adults in spring, when reproduction occurs. We believe juveniles are responsible for increases in visitation rates that are often observed in autumn. Thus, autumn surveys are appropriate only if the distribution and abundance of juveniles in autumn correspond closely with the distribution and abundance of adults in spring. Available evidence suggests they may not.

Details of dispersal are not well-known for swift and kit fox, but juvenile canids frequently disperse great distances in autumn (Storm et al. 1976, Gese and Mech 1991, Harrison 1992). Thus, surveys conducted after dispersal occurs in autumn may very well detect swift or kit foxes in areas where they are not resident. Density-independent factors strongly influence reproduction of swift and kit foxes. Reproductive rates of kit foxes are associated with prey density (White and Garrott 1998), which fluctuates markedly with precipitation (White and Ralls 1993). Neonatal survival rates of kit foxes are variable and are also controlled by rainfall through effects on prey availability (Spiegel and Torn 1996). Most juvenile swift and kit foxes do not survive from autumn to spring (Fitzgerald and Roell 1995, Ralls and White 1995, M.A. Sovada, unpublished data). Further, swift and kit foxes are similar, in many demographic respects, to coyotes and red foxes. Litter sizes of coyotes may decline when prey is scarce (Todd and Keith 1983), and reproductive output of coyotes and red foxes shows evidence of density-dependent limitation (Windberg 1995). For these reasons, we expect weak relations between autumn visitation rates and spring population densities.

Replication and Repetition

For efficiency, surveys should balance information gained by establishing new stations (replication) with effort saved by operating stations repeatedly (repetition). Replication is advisable when the act of data collection biases the outcome of future survey efforts (i.e., when foxes habituate to stations) and when repetition produces redundant data (i.e., when successive operation of the same stations produces essentially the same results).

It seems counterintuitive that repetition reduced the sensitivity of monthly surveys conducted in Kansas, but this surprising result has a simple explanation. Swift foxes in Kansas habituated to stations and daily visitation rates decreased after the first night. This caused proportionately greater reductions in means than in their standard errors and increased coefficients of variation. Thus, our attempt to increase the sensitivity of indices by obtaining three nights data for each station was counterproductive.

Conversely, we expected to find that repetition of surveys was less efficient than establishment of new stations at the NPRC. We suspected redundancy of results from different surveys due to variation in individual behavior. Our suspicion is supported by Kahn and Beck (1996), who photographed marked swift foxes at scent stations and noted that some individuals consistently traveled roads and visited a number of stations in succession.

Repeated operation of permanent stations is an element of most scent-station surveys. If surveys are repeated, steps that reduce habituation of foxes will improve the precision and reduce the bias of survey results. When costs of replication and repetition are similar, we recommend replication.

Lures

Swift foxes in Kansas quickly habituated to stations baited with mackerel in cod-liver oil. We believe visitation rates declined after the first night of operation and with successive monthly surveys, until autumn when juveniles began visiting stations, because foxes had visited stations without obtaining food they anticipated. Had we rewarded foxes with food, we might have faced the opposite problem: visitation rates that increased with habituation. Seasonal variation in the attractiveness of food may further complicate matters. Foxes in Wyoming always consumed mackerel baits when visiting stations in early spring, but not in summer when food was more abundant (Woolley et al. 1995). Effects of seasonal differences in food availability can be controlled via survey design, but effects of annual differences cannot. Annual differences in attractiveness of food baits may therefore reduce the effectiveness of food-based lures for surveys of relative abundance because such lures may provoke different responses on different survey occasions.

FAS is an obvious alternative to food-based lures. Swift and kit foxes, however, may avoid FAS because they associate it with coyotes or other canids (Orloff 1992, Hoagland 1995). Most evidence of differences in the performance of lures is anecdotal, however, and reasons given for hypothesized differences are speculative. Although coyotes were the principal source of kit fox mortality on both our California study areas (Cypher and Scrivner 1992, Standley et al. 1992), foxes visited stations baited with FAS at relatively high rates.

Tracking Media

Previous efforts to maximize visitation rates of swift foxes have focused on the use of various baits and lures (Allen et al. 1995, Luce and Lindzey 1996). Our results, however, suggest the choice of tracking medium may be much more important. Visitation rates to stations composed of natural materials were much higher than for track plates. Although natural materials may require greater care in track identification, this disadvantage is outweighed by benefits of higher detection rates. Moreover, mixing sand with mineral oil greatly facilitated identification of swift fox tracks in Kansas.

Conclusions

Responses of carnivores to scent stations may vary locally. Our results nonetheless suggest contemporary perceptions of scent-station techniques are frequently inaccurate and sometimes lead to inefficient survey designs. We question the use of scent stations for surveys of distribution; acknowledge potential of the method for assessments of relative abundance; discourage autumn surveys; show that repeated operation of stations can lead to habituation and reduce, rather than increase, precision of results; demonstrate a marked reduction in visitation rate resulting from wariness of swift foxes toward track plates; and prescribe cautious extension of our results to different populations.

For the most part, our insights were gained through study of data collected for other purposes. We hope they inspire controlled experimental comparisons of scent-station survey techniques for swift and kit foxes. Given current interest in the precarious population status of these smallest North American canids and widespread use of scent-station surveys, such comparisons seem overdue.

Acknowledgments

We thank our cooperators for generously providing data that made this paper possible: Enterprise Advisory Services, Inc.; the United States Department of Energy; Chevron, U.S.A; EG&G Energy Measurements, Inc.; the California Army National Guard, Camp Roberts, California; the Kansas Department of Fish, Wildlife and Parks; the North Dakota Game and Fish Department; and Northern Prairie Wildlife Research Center, United States Geological Survey. Special thanks are due to Bill Berry, Julie Eliason, Michael Hanson, Greg Warrick, and Christiane Roy for assistance with data collection and processing.

Literature Cited

Allen, S.H., J.W. Hoagland, and E.D. Stukel, editors. 1995. Report of the swift fox conservation team. North Dakota Game and Fish Department, Bismarck, North Dakota.

Beltrán, J.F., M. Delibes, and J.R. Rau. 1991. Methods of censusing red fox (*Vulpes vulpes*) populations. Hystrix 3:199–214.

Cypher, B.L., and J. H. Scrivner. 1992. Coyote control to protect endangered San Joaquin kit foxes at the Naval Petroleum Reserves, California. Pp. 42–47 *in* J.E. Borrecco and R.E. Marsh, editors. Proceedings of the 15th Vertebrate Pest Conference. Newport Beach, California.

Daniel, W.W. 1990. Applied nonparametric statistics. Second edition. PWS-Kent, Boston, Massachusetts.

Efron, B., and R.J. Tibshirani. 1993. An introduction to the bootstrap. Chapman and Hall, London, England.

Fitzgerald, J.P., and B. Roell. 1995. Preliminary results of ecological investigations of the swift fox (*Vulpes velox*) in northern Weld County, Colorado, October 1994–September 1995. Pp. 56–60 *in* S. H. Allen, J. W. Hoagland, and E. D. Stukel, editors. Report of the swift fox conservation team. North Dakota Game and Fish Department, Bismarck, North Dakota.

Gese, E.M., and L.D. Mech. 1991. Dispersal of wolves (*Canis lupus*) in northeastern Minnesota, 1969–1989. Canadian Journal of Zoology 69:2946–2955.

Griffith, B., H.M. Wight, W.S. Overton, and E.C. Meslow. 1981. Seasonal properties of the coyote scent station index. Pp. 197-220 *in* F.L. Miller, A. Gunn, and S.R. Hieb, editors. Symposium on census and inventory methods for populations and habitats. Forest, Wildlife and Range Experiment Station, University of Idaho, Moscow, Idaho. 1980 Symposium proceedings, Banff, Alberta.

Harris, C.E. 1987. An assessment of techniques for monitoring San Joaquin kit fox population abundance on Naval Petroleum Reserve #1, Kern County, California. U.S. Department of Energy Topical Report Number EGG 10282-2159, EG&G/EM Santa Barbara Operations. National Technical Information Service, Springfield, Virginia.

Harrison, D.J. 1992. Dispersal characteristics of juvenile coyotes in Maine. The Journal of Wildlife Management 56:128–138.

Hoagland, J.W. 1995. Distribution and ecology of swift fox in Oklahoma. Pp. 33-38 *in* S. H. Allen, J.W. Hoagland, and E.D. Stukel, editors. Report of the swift fox conservation team. North Dakota Game and Fish Department, Bismarck, North Dakota.

Kahn, R., and T. Beck. 1996. Swift fox investigations in Colorado, 1996. Pp. 10–15 *in* B. Luce and F. Lindzey, editors.

Annual Report of the Swift Fox Conservation Team. Wyoming Game and Fish Department, Lander, Wyoming.

Linhart, S.B., and F.F. Knowlton. 1975. Determining the relative abundance of coyotes by scent station lines. Wildlife Society Bulletin 3:119–124.

Linscombe, G., N. Kinler, and V. Wright. 1983. An analysis of scent station response in Louisiana. Proceedings of the Annual Conference of the Southeastern Association of Fish and Wildlife Agencies 37:190–200.

Luce, B., and F. Lindzey, editors. 1996. Annual report of the swift fox conservation team. Wyoming Game and Fish Department, Lander, Wyoming.

Mathsoft, Inc. 1997. S-PLUS 4 for Windows. Data Analysis Products Division, MathSoft, Seattle, WA.

Morrison, D.W., R.M. Edmunds, G. Linscombe, and J.W. Goertz. 1981. Evaluation of specific scent station variables in Northcentral Louisiana. Proceedings of the Annual Conference of the Southeastern Association of Fish and Wildlife Agencies 35:281–291.

Orloff, S.G. 1992. Survey techniques for the San Joaquin kit fox (*Vulpes macrotis mutica*). Pp. 185–197 *in* D. F. Williams, S. Byrne, and T. A. Rado, editors. Endangered species of the San Joaquin Valley, California: their biology, management, and conservation. California Energy Commission, Sacramento, CA.

Ralls, K., and P.J. White. 1995. Predation on San Joaquin kit foxes by larger canids. Journal of Mammalogy 76:723–729.

Reese, E.A., W.G. Standley, and W.H. Berry. 1992. Habitat, soils, and den use of San Joaquin kit fox (*Vulpes velox macrotis*) at Camp Roberts Army National Guard Training Site, California. U.S. Department of Energy Topical Report Number EGG 10617-2156, EG&G/EM Santa Barbara Operations. National Technical Information Service, Springfield, Virginia.

Roughton, R.D., and M.D. Sweeny. 1982. Refinements in scent-station methodology for assessing trends in carnivore populations. The Journal of Wildlife Management 46:217–229.

Sargeant, A.B., R.J. Greenwood, M.A. Sovada, and T.L. Shaffer. 1993. Distribution and abundance of predators that affect duck production—Prairie Pothole Region. U.S. Fish and Wildlife Service Resource Publication 194.

Sargeant, G.A., D.H. Johnson, and W.E. Berg. 1998. Interpreting carnivore scent-station surveys. The Journal of Wildlife Management 62:1235-1245.

——. 2003. Sampling designs for carnivore scent-station surveys. The Journal of Wildlife Management 67:289–299.

SAS Institute, Inc. 1988. SAS language guide for personal computers, Release 6.03 Edition. SAS Institute, Cary, North Carolina.

Smith, W.P., D.L. Borden, and K.M. Endres. 1994. Scent-station visits as an index to abundance of raccoons: an experimental manipulation. Journal of Mammalogy 75:637–647.

Sokal, R.R., and F.J. Rohlf. 1981. Biometry. Second edition. W. H. Freeman, New York, New York.

Spiegel, L.K., and J. Torn. 1996. Reproduction of San Joaquin kit fox in undeveloped and oil-developed habitats of Kern County, California. Pp. 53–69 *in* Studies of the San Joaquin kit fox in undeveloped and oil-developed areas. Staff Report, California Energy Commission.

Standley, W.G., W.H. Berry, T.P. O'Farell, and T.T. Kato. 1992. Mortality of San Joaquin kit fox (*Vulpes velox macrotis*) at Camp Roberts Army National Guard Training Site, California. U.S. Department of Energy Topical Report Number EGG 10617-2157, EG&G/EM Santa Barbara Operations. National Technical Information Service, Springfield, Virginia.

Storm, G.L., R.D. Andrews, R.L. Phillips, R.A. Bishop, D.B. Siniff, and J.R. Tester. 1976. Morphology, reproduction, dispersal, and mortality of midwestern red fox populations. Wildlife Monograph 49.

Todd, A.W., and L.B. Keith. 1983. Coyote demography during a snowshoe hare decline in Alberta. The Journal of Wildlife Management 47:394–404.

Windberg, L.A. 1995. Demography of a high-density coyote population. Canadian Journal of Zoology 73:942–954.

White, P.J., and K. Ralls. 1993. Reproduction and spacing patterns of kit foxes relative to changing prey availability. The Journal of Wildlife Management 57:861–867.

White, P.J., and R.A. Garrott. 1998. Factors regulating kit fox populations. Canadian Journal of Zoology 76:1982–1988.

Wood, J.E. 1959. Relative estimates of fox population levels. The Journal of Wildlife Management 23:53–63.

Woolley, T P., F.G. Lindzey, and R. Rothwell. 1995. Swift fox surveys in Wyoming—annual report. Pp. 61–80 *in* S.H. Allen, J.W. Hoagland, and E.D. Stukel, editors. Report of the swift fox conservation team. North Dakota Game and Fish Department, Bismarck, ND.

Zielinski, W.J., and H.B. Stauffer. 1996. Monitoring Martes populations in California: survey design and power analysis. Ecological Applications 6:1254–1267.

Zoellick, B.W., T.P. O'Farrell, and T.T. Kato. 1987. Movements and home range of San Joaquin kit foxes on the Naval Petroleum Reserves, Kern County, California. U.S. Department of Energy Topical Report Number EGG 10282-2184, EG&G/EM Santa Barbara Operations. National Technical Information Service, Springfield, Virginia.

Reducing Capture-Related Injuries and Radio-Collaring Effects on Swift Foxes

■ **Axel Moehrenschlager, David W. Macdonald and Cynthia Moehrenschlager**

Abstract: Capture, handling, and radio-collaring effects should be minimized for the welfare of endangered canids and to ensure that data from monitored individuals are representative of study populations. While investigating the reintroduced and endangered swift fox (Vulpes velox) population in Canada we: (1) identified factors that reduced swift fox capture injuries; (2) compared handling times of pups, juveniles, and adults; and (3) examined changes in swift fox neck sizes over time to improve age-specific radiocollaring specifications. In southern Alberta and Saskatchewan, 125 swift foxes were trapped for a total of 273 captures from January 1995 to February 1998. No capture-related deaths or major injuries occurred, but 8.3% of initial captures and 1.1% of recaptures resulted in minor injuries. Injury rates were independent of sex, age, season, and fox origin, but frequent trap checks and the lining of box traps with hardboard reduced injuries. Minor injuries did not affect the likelihood of recapture, the number of recaptures, survival, or the cause of death. Handling times were shorter for pups than juveniles and adults. Neck circumference was correlated to fox weight, neck sizes increased as pups developed into juveniles, but those of juveniles did not increase as adults. To minimize potential radiocollar effects on growing swift fox pups, we recommend that: (1) applications for telemetry data should be identified; (2) if pups are collared, trapping should be conducted to re-fit radiocollars by the time the foxes are juveniles; and (3) if pups are not recaptured, the potential effects of collar tightness on body condition and survival should be monitored and addressed.

Most questions regarding behavior, population dynamics, and predator-prey interactions can be seriously addressed only through the tagging and subsequent monitoring of subject animals. Consequent effects on the subject need to be understood since: (1) welfare considerations dictate that the intervention should not cause significant distress to the study animal; and (2) scientific applicability requires the subject to act in an unbiased fashion relative to unmarked individuals in the population (Macdonald and Amlaner 1980). If handled animals are significantly affected, conclusions based on biased data may be misleading and subsequent conservation measures could be misguided.

The trapping and marking of wild mammals can cause decreases in fitness by causing stress, physical injury, immunosuppression, or a change in scent that may subsequently affect survival (White and Knowlton 1972, Goldberg and Haas 1978) or reproduction (Larsen and Gauthier 1989). Marking or radiocollaring can also affect activity patterns (Pouliquen et al. 1989), behaviour (Daly et al. 1992), physiology (Clute and Ozoga 1983, MacArthur et al. 1986), and survival (Garshelis and Siniff 1983, Wolton and Trowbridge 1985, Swenson et al. 1999). Such effects have been documented for canids. In North America, box and leg-hold trapping or darting from helicopters has occasionally resulted in the overstress, injury, or death of coyotes (*Canis latrans*), red foxes (*Vulpes vulpes*), and wolves (*Canis lupus*; Andelt 1980, Van Ballenberghe 1984, Olsen et al. 1986, White et al. 1991). Although causal linkages are lacking, capture stress has been implicated as one factor that may have contributed to a rabies outbreak in African wild dogs (*Lycaon pictus*) which caused their extirpation from the Serengeti (reviewed by Woodruffe et al. 1997).

Radio-collars have been found to affect the body weight and survival of San Joaquin kit foxes (*Vulpes macrotis mutica*) in the first 30 days after collaring (Cypher 1997). Tight fitting radio-collars can cause neck-abrasions (Cypher 1997) and occasionally swift or kit foxes die after a paw is trapped in a loosely fitted collar (Cypher 1997; S. Black, Calgary Zoological Society, Calgary, Alberta, personal communication). This dichotomy presents a dilemma for dealing with young animals whose neck sizes will grow over time. Being cautious, Koopman et al. (1997) only collared half the pups from kit fox litters. In some carnivores, this problem has been tackled by using expandable collars (Weber and Meia 1992), harnesses (Jackson et al. 1985), or transmitter implants (Reid et al. 1986, Koehler et al. 1987, Van Vuren 1989), but these techniques have not been attempted with swift or kit foxes.

Swift foxes were extirpated from Canada in the 1930s (Scott-Brown et al. 1987), but reintroduced since 1983 (Carbyn et al. 1994); the current population of approximately 300 individuals is still highly endangered (Cotterill 1997, Moehrenschlager and Moehrenschlager 1999). Since swift fox conservation planning is largely contingent on capture and telemetry data, our objectives are threefold: (1) to identify factors that reduce swift fox capture injuries; (2) to compare handling times of pups, juveniles, and adults; and (3) to enhance age-specific radio-collaring strategies by examining changes in swift fox neck sizes over time. We evaluate trapping results of 125 foxes that were captured 273 times from January 1995 until February 1998 in southern Alberta and Saskatchewan.

Study Area

Swift fox trapping was conducted on the prairies of southern Alberta and Saskatchewan (49°00.0'N–49°28.8'N; 109°44.2'W–110°36.0'W), where the altitude varies from 850 to 1050 m and average annual precipitation is 327 mm (Smoliak 1985). The main vegetation types are representative of the Mixed Prairie Association, which is characterized by mid- and short-grasses, numerous forbs, and few scrubs (Smoliak 1985). Human habitation was sparse although areas were used for cultivation and cattle ranching. Temperatures ranged from -58°C to +38°C.

Methods

Swift fox trapping was conducted from January 1995 to February 1998 twice per year. Winter trapping was conducted in: (1) 10–26 January 1995, (2) 16 December 1995–15 February 1996, (3) 11 November 1996–15 February 1997, and (4) 6–14 February 1998. Fall trapping spanned the following time periods: (1) 11 August–22 October 1995; (2) 16 August–8 September 1996; and (3) 6 August–29 September 1997. Radio-collars were removed during the fall of 1997 and the winter of 1998. Age classifications were based on recaptures of previously tattooed pups and on the size, color, and wear of teeth. Ralls et al. (1990) classed foxes that were less than 1 year old as juveniles and older foxes as adults. We classified foxes into 3 age classes: pups were 3–6 months old, juveniles were 6–9.5 months, and adults were >15 months. Foxes were not trapped between 9.5 and 15 months of age because this period fell within the breeding, gestation, and pup-rearing stages, which span from 15 February to 1 August in Canada (Moehrenschlager 2000).

Trap types were variable but consistent among trap sessions. We used 109-cm x 39-cm x 39-cm Tomahawk (Tomahawk Live Trap Company, Tomahawk, Wisconsin) double-door, 83-cm x 31-cm x 31-cm Tomahawk single door, and 83-cm x 26-cm x 26-cm Havaharts (Havahart, Pennsylvania 17543). After 26 January 1995, trap bottoms and corners were lined with 3-mm hard board to reduce the likelihood of jaw, canine, or paw injuries. Trap placement was systematic in January 1995 and February 1996, random from November 1996 to January 1997, and subjective near dens or core-use areas during other time periods. Trapping was conducted almost exclusively at night to coincide with fox activity, to avoid heat-stress, and to prevent disturbance of foxes by people. Traps were generally set between 1800 and 2000 hr, checked between 2400 and 0200 hr, and closed following a second check between 0600 and 0800 hr. Fox capture durations were calculated as the maximum time span in the trap; this was calculated by subtracting the trap set from the trap check time. Maximum capture times were compared between injured and uninjured foxes with an unpaired t-test (Sokal and Rohlf 1981). Respective upper and lower temperature limits for conducting the trapping were +25°C and -20°C, but traps were also closed within this range depending on snow, rain, and wind conditions.

Fox handling methods changed following the first winter trapping session. Like previous Canadian trappers (Brechtel et al. 1993), we used a noose to remove foxes from traps in January 1995; thereafter a denim handling bag was used (see O'Farrell et al. 1986, Ralls et al. 1990, Covell 1992). While foxes resisted removal using the noose by bracing against the trap, biting the cord, or biting the trap, the animals entered the handling bag without physical restraint.

Foxes were handled by 2 field workers. The first positioned the animal on his/her lap to shelter it from the wind, one hand restrained the head and covered the eyes, and the second hand restrained the body. The second field worker conducted parasite counts, canine measurements, and body condition assessments. The neck circumference was measured to determine neck-size differences between age classes. To ensure that changes in seasonal fur thickness did not confound neck size comparisons, two neck circumference measurements were taken during the final trapping session in February 1998. First the tape measure was placed on top of the fur, while the second measurement was taken after moving hairs aside under the tape measure. Neck-size comparisons were tested relative to body weight using simple regression, between age classes by one-way ANOVA followed by Tukey multiple comparisons and between time periods among individuals using a paired t-test (Zar 1984; SPSS 9, SPSS Inc., Chicago, Illinois).

To minimize stress on the animal, foxes were only handled once in each trap session. Foxes were sexed, ear-tattooed, treated with Ivermectin (0.2 mg/kg) to combat parasite infestation, and radiocollared. Foxes were vaccinated against canine distemper, adenovirus, parainfluenza, parvovirus, and leptospirosis with Duramune (Wyeth-Ayerst, St. Davids, Pennsylvania; 10 mg/ml), and Imrab (Wyeth-Ayerst, St. Davids, Pennsylvania; 10 mg/ml) provided vaccination against rabies. Radio-collars weighed 48 g, had a mortality sensor, and had an antenna 15 to 20 cm long (Lotek Inc., Newmarket, Ontario). Handling was timed to the nearest minute and times were compared between age classes by one-way ANOVA and subsequent Tukey multiple comparisons (Zar 1984).

Objectively classifying injuries is complicated by a variety of factors pertaining to specific situations and animals (Kirkwood et al. 1994, Sainsbury et al. 1994). To simplify this procedure, we divided captures into 3 subjective categories: (1) no injury, (2) minor injury, and (3) major injury. A thorough examination of animal condition was conducted only when animals were trapped and handled; recaptured foxes were examined for obvious wounds and only re-handled if an injury was apparent. Animals were classified as having no injury if no external wounds, limb irregularities, or tooth breakages could be detected. Minor injuries were external wounds that would not obviously decrease fitness through a deterioration of physical condition or reproductive potential. This category included the occurrence of chipped teeth, a broken canine, broken claw sheaths, external lacerations <0.5 cm, and superficial abrasions. Major injuries were ones that were thought to affect fitness if untreated, such as external lacerations ≥0.5 cm in length, jaw or skull fractures, and the breakage or dislocation of limbs.

A hierarchical, saturated loglinear model using backwards elimination at the alpha = 0.05 level was used to explore the role of age, sex, season (winter vs. fall), and fox origin (captive-bred, translocated from Wyoming, or wild-born in Canada) in predicting the likelihood of injury occurrence. Recapture numbers and the minimum number of post-trapping survival days were tested between injured and uninjured foxes using unpaired *t*-tests. Numbers of dead and surviving foxes were compared between injured and uninjured treatments using Fisher's exact test. Variance measures throughout are standard deviations and x^2 values from likelihood ratio tests (Sokal and Rohlf 1981).

Results

One hundred and twenty-five individual swift foxes were trapped in 273 captures (Table 1); foxes were handled on 181 occasions. Fifty-three radiocollared foxes died during the study.

No mortalities or major injuries resulted from trapping or handling. Handling revealed minor injuries on 15 (8.3%) occasions, which were not unique to individual trap sessions or age classes (Table 1). Foxes waited in traps for a maximum of 5.70 ± 1.93 hours. Despite the small variation in trap check frequencies, the maximum amount of time that injured foxes were in traps was longer (6.82 ± 1.55 hrs.) than that of uninjured animals (5.70 ± 1.83 hrs.; t = 1.94, df = 8.5, P = 0.087).

The minor injuries consisted of 8 cases with single canine chips, 2 cases with 2 chipped canines, 1 instance of a fractured canine, 1 occurrence of a canine chip and a minor paw laceration, 2 cases of broken claw sheaths, and 1 case of a bloody ear during tattooing. Of the 15 animals with minor injuries, 11 were radiocollared while 3 pups, which were too small, and 1 adult, which was caught in the last trap session, were not. Of the collared injured foxes, 9 died during the study, whereas 2 survived its duration. This proportion was similar to that of radiocollared, uninjured foxes (Fisher's exact, P = 0.65) since 32 died, 7 survived, and radio-contact was lost with 3. The minimum number of days that injured foxes survived (302.1 ± 202.2) was similar to that of uninjured animals (320.1 ± 308.7; t = 0.23, df = 23.8, P = 0.82). Known mortality causes of previously injured and uninjured foxes were similar (x^2 = 2.44, df = 2, P = 0.30; Table 2).

Age and season-specific captures are outlined in Table 1. Of the 125 captured foxes, 115 were born in the wild in Canada, 4 had been captive-bred and released, and 6 had been translocated from Wyoming. Sex, age, origin, season, and related interactions were not significant predictors of injury occurrence in the final loglinear model. However, the first trap session, when traps were not lined with wood and foxes were removed from traps using a noose, had proportionately more injuries than the other sessions combined (Table 1; x^2 = 3.51, df = 1, P = 0.061).

Recaptured animals were immediately released in 92 instances (n = 48) and only 1 fox (1.1%) showed signs of injury (small abrasion on paw). We recaptured 48% (60/125) of foxes; 21.6% (27/125) were re-trapped only within the trap session of their initial capture while 26.4% (33/125) were caught in more than 1 trapping session. Foxes were re-caught as frequently as 6 times in one trapping session and 14 times over the course of the study. The proportion of injured and uninjured foxes that were

Table 1. Number of swift fox captures, with no injury and with minor injury in southern Alberta and Saskatchewan from January 1995 to February 1998.

		Winter 1994–95		Fall 1995		Winter 1995–96		Fall 1996		Winter 1996–97		Fall 1997		Winter 1997–98		Total	
Age Class	Treatment	No injury	Minor injury	No injury	Minor injury	No injury	Minor injury	No injury	Minor injury	No injury	Minor injury	No injury	Minor injury	No injury	Minor injury	No injury	Minor injury
Adult	Caught and handled	15	3	7	0	19	0	5	1	24	3	5	0	4	1	***79***	***8***
	Recaptured	11	1	3	0	9	0	4	0	11	0	2	0	2	0	***42***	***1***
Juvenile	Caught and handled	6	1			6	2			26	0			10	0	***48***	***3***
	Recaptured	4	0			4	0			12	0			1	0	***21***	***0***
Pup	Caught and handled			10	2			17	0			13	1			***40***	***3***
	Recaptured			2	0			6	0			20	0			***28***	***0***
Total		***36***	***5***	***22***	***2***	***38***	***2***	***32***	***1***	***73***	***3***	***40***	***2***	***17***	***1***	258	15

Table 2. Comparative causes of mortality for radio -collared foxes in southern Alberta and Saskatchewan between January 1995 and February 1998 that experienced minor injuries versus those that did not.

Mortality Cause	Percentage of Injured Foxes (n = 9)	Percentage of Uninjured Foxes (n = 32)
Predators	44.4	78.1
Human	22.2	9.4
Starvation	0	6.3
Unknown	33.3	6.3

recaptured within trapping sessions was similar ($x^2 = 0.30$, df = 1, $P = 0.59$). The number of recaptures per animal was also similar for injured (1.25 ± 0.50 captures) and uninjured (1.71 ± 1.14 captures) animals within trapping sessions ($t = 1.54$, df = 5.9, $P = 0.18$).

On average, captured swift foxes required 20.0 ± 6.4 minutes to handle. Handling times were similar between foxes that were radiocollared and those that were not ($t = 0.90$, df = 92.5, $P = 0.37$). Handling times differed with fox age (F = 14.4, $P < 0.001$). The Tukey HSD test revealed that handling times for pups (15.6 ± 3.0 min., n = 33) were shorter than for juveniles (22.8 ± 8.0 min., n = 30) and adults (21.2 ± 5.4 min., n = 49, $P < 0.001$).

Neck circumference was correlated to body weight ($r^2 = 0.58$, $P < 0.001$, Figure 1). When first-captured, neck sizes of adults (15.7 ± 1.15 cm, n = 46) and juveniles (15.2 ± 1.08 cm, n = 38) were larger than pup neck sizes (13.3 ± 0.96 cm n = 44; F = 61.7, $P < 0.001$) but there was no difference between adults and juveniles ($P = < 0.05$). The neck sizes of 8 swift fox pups that were recaptured as juveniles after 155.7 ± 21.1 days increased significantly (max. = 2.4 cm., mean difference = 1.2 cm; $t = 4.57$, df = 8, $P = 0.002$). These results were not confounded by seasonal changes in neck fur thickness since winter measurements with or without fur under the tape measure yielded similar results ($t = 0.25$, df = 19.9, $P = 0.81$). To avoid potentially adverse effects associated with neck growth, pups were not radiocollared whereas juveniles and adults were.

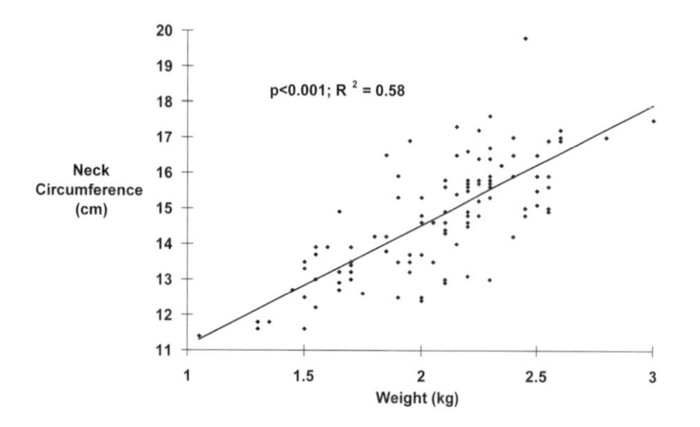

Figure 1. Relationship between body weight and neck circumference for swift foxes trapped in southern Alberta and Saskatchewan between January 1995 and February 1998.

Discussion

Trapping and handling methods were successful in meeting 2 prime considerations. In terms of animal welfare, foxes did not sustain any major injuries, minor injuries were infrequent, and injured foxes were in traps for short time periods. Comparing the injured and uninjured foxes, their similar recapture frequencies, survival rates, and causes of mortality suggest that incurred injuries did not affect swift fox fitness.

Minor injuries were noted among 8.3% of handled foxes. Of 23 papers or reports that involved swift or kit fox captures (Golightly and Ohmart 1983, 1984: Zoellick and Smith 1986; Scrivner et al. 1987; Zoellick et al. 1989; Ralls et al. 1990; Womer and Richards 1990; Covell 1992; Zoellick and Smith 1992; Brechtel et al. 1993; EG&G/EM 1993; White and Ralls 1993; Carbyn et al. 1994; Mamo 1994; White et al. 1994, Carbyn and Klausz 1995; Ralls and White 1995; White et al. 1996; Cypher 1997; Koopman et al. 1997; Warrick and Cypher 1998; Sovada et al. 1998; Kitchen et al. 1999), 8 mentioned capture-related injuries or deaths. Injuries described in Cypher (1997) refer to collar-related incidents but not to capture-related occurrences. Although individual injuries are not outlined, O'Farrell et al. (1986) recorded 5 trap-related deaths of kit foxes after 928 captures between 1980 and 1986. Of 63 dead foxes that were recovered in Kern County California, 1 died during an attempt to capture it in a pipe (EG&G/EM 1993). Ralls et al. (1990) noted that some foxes broke teeth in custom-made traps. Fifteen captures involving 13 foxes in Alberta resulted in 4 minor injuries (Mamo 1994). Of 300 captures, Covell (1992) noted that "occasionally, canines were lost and in 1 instance a jaw was broken on the steel mesh of traditional box traps." Of 47 foxes captured in Wyoming from 1994 to 1996, 7 (14.9%) sustained minor injuries (chipped canines) and 1 (2.1%) had a major injury (broken mandible; Carbyn and Klausz 1995; L.N. Carbyn, Canadian Wildlife Service, Edmonton, Alberta, personal communication). In a 2000/2001 census that evaluated the distribution and abundance of reintroduced swift foxes in Canada and Montana, trapping techniques similar to those presented in this paper were employed by one experienced and five previously untrained trap-teams (Moehrenschlager and Moehrenschlager, 2001). Of 149 captured individuals, 2 (1.3%) had major injuries in the form of jaw breaks and 1 fox was successfully released after surgery; the second was euthanized due to complications that arose during treatment. The rate of minor injuries was even lower than those presented in this paper as only 8 (5.4%) of foxes had minor injuries (7 cases of one chipped canine or one chipped premolar, 1 case of a small cut on one paw).

After the first capture period, box traps were lined with hardboard and injuries decreased. Researchers have reduced box-trap mesh sizes to reduce injuries (Ralls et al. 1990, Sovada et al. 1998) and wood-lining should be used to complement such mesh modifications in the future.

Injured foxes had been in traps longer than uninjured individuals. Individuals may have been more likely to injure themselves if stress levels increased over time. In red foxes, the cumulative amount of time foxes spent in box traps was correlated to rising cortisol levels (White et al. 1991). While frequent trap checks seem to reduce swift fox injuries, they also yield other advantages. During the pup-rearing season, mother and pup separation times may be reduced by checking traps twice during the night (Olson et al. 1997). Moreover, coyotes or dogs can kill foxes that are captured in box or leg-hold traps (Covell 1992, EG&G/EM 1993, Moehrenschlager and List 1996), but frequent trap-checks might reduce this possibility.

Handling times were shorter for swift fox pups, which were relatively docile, than adults. Since handling duration correlates to the subject's stress levels (Kirkwood et al. 1994), pups may be less stressed by handling than adults.

Swift fox neck sizes increased significantly during the pup (<6 months old) to juvenile (6–9.5 months old) transition, whereas juveniles had similar neck-sizes to adults. This raises the concern that tight collars could have negative effects on radio-collared pups as they grow. The retrieval of 63 radiocollars, including 40 fox necropsies, showed no collar-related injuries in Canada (Moehrenschlager 2000), swift fox pups that were collared at 5 months of age and recaptured 5 months later in Kansas were uninjured (M.A. Sovada, Northern Prairie Wildlife Research Center, Jamestown, North Dakota, personal communication), and San Joaquin kit foxes rarely had collar-related injuries or mortalities (Cypher 1997). Nevertheless, radio-collars can have some adverse effects. Behavior of captive adult swift foxes was altered by collars, and short-term weight gain might have been affected (Moehrenschlager 2000). Weight loss of collared San Joaquin kit fox adults and pups was greater than among controls. Moreover, the body condition and survival of juvenile dispersers may have been affected by radio-collars. Consequently, Cypher (1997) recommended that radio-collaring should be avoided during periods of increased vulnerability or stress.

If radio-collars can affect swift or kit foxes in some cases, neck constriction by radio-collars could particularly impact the body weight, body condition, or survival of growing swift fox pups. However, telemetry data of this age group might be essential to understanding population demographics and implement informed conservation actions. The following recommendations attempt to balance these concerns: (1) pups can be collared if an applicable need for telemetry data has been identified; (2) if pups are collared, trapping should be conducted to re-fit radiocollars by the time the foxes are juveniles; and (3) if pups are not recaptured, the potential effects of collar tightness on the body condition and survival of these animals should be monitored and addressed.

Since injuries did not affect recapture and survival rates in this study, the classification of the observed injuries as 'minor' seems appropriate. Due to their absence we could not assess the consequences of major injuries, which likely would affect swift fox fitness. Minor injuries were reduced through the lining of traps with wood and through frequent trap checks. To further increase swift or kit fox welfare, we suggest that future comparisons of trapping, handling, and radio-collaring methodologies should incorporate stress hormone analyses.

Acknowledgments

We thank our dedicated field staff, Jeff Johnson, Jasper Michie, Jörn Scharlemann, Clint Stokke, and Ian Welsh. We would also like to thank the landowners that graciously gave us permission to work on their land, such as the Buchanan, Heydlauff, Kusler, Petrowski, Saville, Stokke, Trumpour, Walburger, and Weisgerber families. The manuscript benefited greatly from the comments of Sandie Black and one anonymous reviewer. This study was supported by Alberta Environmental Protection, the Alberta Sport Recreation Parks and Wildlife Foundation, the Canadian Wildlife Service, Express Pipelines, Green Plan International, Nature Saskatchewan, Parks Canada, the People's Trust for Endangered Species, the Rocky Mountain Elk Foundation, Saskatchewan Environment and Resource Management, the Swift Fox Conservation Society, the University of Alberta Biodiversity Fund, Wildlife Preservation Trust Canada, and World Wildlife Fund Canada.

Literature Cited

Andelt, W.F. 1980. Capturing coyotes for studies of their social organization. Wildlife Society Bulletin 8:252–254.

Brechtel, S.H., L.N. Carbyn, D. Hjertaas, and C. Mamo. 1993. Canadian swift fox reintroduction feasibility study: 1989 to 1992. Alberta Fish and Wildlife Report, Edmonton, Alberta.

Carbyn, L.N., H. Armbruster, and C. Mamo. 1994. The swift fox reintroduction program in Canada from 1983 to 1992. Pp. 247–271 *in* M.L. Bowles and C.J. Whelan, editors. Restoration of endangered species: conceptual issues, planning and implementation. Cambridge University Press, Cambridge, UK.

Carbyn, L.N., and E.Klausz. 1995. Live trapping swift foxes in south-eastern Wyoming and survival upon release in southern Alberta. Canadian Wildlife Service Report, Edmonton, Alberta.

Clute, R.K., and J.J. Ozoga. 1983. Icing of transmitter collars on white-tailed deer fawns. Wildlife Society Bulletin 11:70–71.

Cotterill, S.E. 1997. Population census of the swift fox (*Vuples velox*) in Canada: winter 1996–1997. Prepared for the Swift Fox National Recovery Team. Alberta Environmental Protection, Natural Resources Service, Wildlife Management Division. Edmonton, Alberta.

Covell, D.F. 1992. Ecology of the swift fox (*Vulpes velox*) in southeastern Colorado. Thesis, University of Wisconsin-Madison, Madison, Wisconsin.

Cypher, B.L. 1997. Effects of radiocollars on San Joaquin kit foxes. Journal of Wildlife Management 81:1412–1423.

Daly, M., M.I. Wilson, P.R. Behrends, and L.F. Jacobs. 1992. Sexually differentiated effects of radio transmitters on predation risk and behaviour in kangaroo rats *Dipodomys merriami*. Canadian Journal of Zoology 70:1851–1855.

EG&G/EM Incorporated. 1993. Effects of supplemental feeding on survivorship, reproduction, and dispersal in San Joaquin kit foxes. U.S. Department of Energy Topical Report, EG&G/EM Santa Barbara Operations Report No. EGG 11265–2059.

Garshelis, D.L., and D.B. Siniff. 1983. Evaluation of radio-transmitter attachments for sea otters. Wildlife Society Bulletin 11:378–383.

Goldberg, J.S., and W. Hass. 1978. Interactions between mule deer dams and their radio-collared and unmarked fawns. Journal of Wildlife Management 42:422–425.

Golightly, R.T., and R.D. Ohmart. 1983. Metabolism and body temperatures of two desert canids: coyotes and kit foxes. Journal of Mammalogy 64:624–635.

——. 1984. Water economy of two desert canids: coyote and kit fox. Journal of Mammalogy 65:51–58.

Jackson, D.H., L.S. Jackson, and W.K. Seitz. 1985. An expandable drop-off transmitter harness for young bobcats. Journal of Wildlife Management 49:46–49.

Kirkwood, J.K., A.W. Sainsbury, and P.M. Bennett. 1994. The welfare of free-living wild animals: methods of assessment. Animal Welfare 3:257–273.

Kitchen, A.M., E.M. Gese, and E.R. Schauster. 1999. Resource partitioning between coyotes and swift foxes: space, time, and diet. Canadian Journal of Zoology 77:1645–1656.

Koehler, D.K., T.D. Reynolds, and S.H. Anderson. 1987. Radio-transmitter implants in 4 species of small mammals. Journal of Wildlife Management 51:105–108.

Koopman, M.E., J.H. Scrivner, and T.T. Kato. 1997. Patterns of den use by San Joaquin kit foxes. Journal of Wildlife Management 62:373–379.

Larsen, D.G., and D.A. Gauthier. 1989. Effects of capturing pregnant moose and calves on calf survivorship. Journal of Wildlife Management 53:564–567.

MacArthur, R.A., V. Geist, and R.H. Johnston. 1986. Cardiac responses of bighorn sheep to trapping and radio instrumentation. Canadian Journal of Zoology 64:1197–1200.

Macdonald, D.W., and C.J. Amlaner, Jr. 1980. A practical guide to radio tracking. Pp. 143–159 *in* C.J. Amlaner, Jr. and D.W. Macdonald, editors. A handbook on biotelemetry and radio tracking. Pergamon Press, Oxford, UK.

Mamo, C.C. 1994. Swift fox (*Vulpes velox*) population survey assessment. Report prepared for Alberta Fish and Wildlife Services and the Swift Fox Conservation Society. Alberta Environmental Protection, Edmonton, Alberta.

Moehrenschlager, A. 2000. Effects of ecological and human factors on the behaviour and population dynamics of reintroduced Canadian swift foxes (*Vulpes velox*). D.Phil Dissertation, University of Oxford, Oxford, UK.

Moehrenschlager, A., and R. List. 1996. Comparative ecology of North American prairie foxes—conservation through collaboration. Pp. 22–28 *in* D. W. Macdonald and F. H. Tattersall, editors. The WildCRU review: the tenth anniversary report of the Wildlife Conservation Research Unit at Oxford University, George Street Press Limited, Stafford, UK.

Moehrenschlager, A., and C. Moehrenschlager. 2001. Census of swift foxes in Canada and northern Montana. Alberta Environmental Protection Report, Edmonton, Alberta.

Moehrenschlager, C., and A. Moehrenschlager. 1999. Canadian swift fox population assessment: Winter, 1999. Alberta Environmental Protection Report, Edmonton, Alberta.

O'Farrell, T.P., C.E. Harris, T.T. Kato, and P.M. McCue. 1986. Biological assessment of the effects of petroleum production at maximum efficient rate, Naval Petroleum Reserve 1 (Elk Hills), Kern County, California, on the endangered San Joaquin kit fox (*Vulpes macrotis mutica*). U.S. Department of Energy Topical Report, EG&G/EM Santa Barbara Operations Report No. EGG10282-2107.

Olsen, G.H., S.B. Linhart, R.A. Holmes, G.J. Dasch, and C.B. Male. 1986. Injuries to coyotes caught in padded and unpadded steel foothold traps. Wildlife Society Bulletin 14:219–223.

Olson, T.L., J.S. Dieni, and F.G. Lindzey. 1997. Swift fox survey evaluation, productivity, and survivorship in southeast Wyoming. Wyoming Cooperative Fish and Wildlife Research Unit, Laramie, Wyoming.

Pouliquen, O., M. Leishman, and T.D. Redhead. 1990. Effects of radio-collars on wild mice, *Mus domesticus*. Canadian Journal of Zoology 68:1607–1609.

Ralls, K., and P.J. White. 1995. Predation on San Joaquin kit foxes by larger canids. Journal of Mammalogy 76:723–729.

Ralls, K., P.J. White, J. Cochran, and D.B. Siniff. 1990. Kit-fox coyote relationships in the Carrizo Plain Natural Area. Annual Report to the U.S. Fish and Wildlife Service.

Reid, D.G., W.E. Melquist, J.D. Woolington, and J.M. Noll. 1986. Reproductive effects of intraperitoneal transmitter implants in river otters. Journal of Wildlife Management 50:92–94.

Sainsbury, A.W., P.M. Bennett, and J.K. Kirkwood. 1994. The welfare of free-living animals in Europe: harm caused by human activities. Animal Welfare 4:183–206.

Scott-Brown, J.M., T.P. O'Farrell, and K.L. Hammer. 1987. Swift fox. *In* M. Novak, J.A. Baker, M.E. Obbard, and B. Malloch, editors. Wild furbearer management and conservation in North America. Ontario Ministry of Natural Resources, Ontario.

Scrivner, J.H., T.P. OFarrell, and T.T. Kato. 1987. Dispersal of San Joaquin kit foxes, (*Vulpes macrotis mutica*), on Naval Petroleum Reserve #1 Kern County, California. EG&G Energy Measurements Rep. EGG 10282-2190.

Smoliak, S. 1985. Flora of the Manyberries Research Substation. Lethbridge Research Station Contribution No. 6. Lethbridge, Alberta.

Sokal, R.R., and F.J. Rohlf. 1981. Biometry, 2nd edition W.H. Freeman & Company, San Francisco, California.

Sovada, M.A., C.C. Roy, J.B. Bright, and J.R. Gillis. 1998. Causes and rates of mortality of swift foxes in western Kansas. Journal of Wildlife Management 62:1300–1306.

Swenson, J. E., K. Wallin, G. Ericsson, G. Cederlund, and F. Sandergren. 1999. Effects of ear-tagging with radiotransmitters on survival of moose calves. Journal of Wildlife Management 63(1):354–358.

Van Ballenberghe, V. 1984. Injuries to wolves sustained during live-capture. Journal of Wildlife Management 48:1425–1429.

Van Vuren, D. 1989. Effects of intraperitoneal transmitter implants on yellow-bellied marmots. Journal of Wildlife Management 53:320–323.

Warrick, G.D., and B.L. Cypher. 1998. Factors affecting the spatial distribution of San Joaquin kit foxes. Journal of Wildlife Management 62:707–717.

Weber, J.M., and J.S. Meia. 1992. The use of expandable radio collars for radio-tracking fox cubs. Pp. 698–700 *in* I.G. Priede and S.M. Swift, editors. Wildlife telemetry: remote monitoring and tracking of animals. Ellis Horwood, New York, New York.

White, M., and F.F. Knowlton. 1972. Effects of dam-newborn fawn behaviour on capture and myopathy. Journal of Wildlife Management 36:897–906.

White, P.J., T.J. Kreeger, U.S. Seal, and J.R. Tester. 1991.

Pathological responses of red foxes to capture in box traps. Journal of Wildlife Management 55:75–80.

White, P.J., and K. Ralls. 1993. Reproduction and spacing patterns of kit foxes relative to changing prey availability. Journal of Wildlife Management 57:861–867.

White, P.J., K. Ralls, and R.A. Garrott. 1994. Coyote—kit fox interactions as revealed by telemetry. Canadian Journal of Zoology 72:1831–1836.

White, P.J., C.A. Vanderbilt White, and K. Ralls. 1996. Functional and numerical responses of kit foxes to a short-term decline in mammalian prey. Journal of Mammalogy 77: 370–376.

Wolton, R.J., and B.J. Trowbridge. 1985. The effects of radio-collars on wood mice, *Apodemus sylvaticus*. Journal of Zoology 206:222–224.

Womer, D.E., and E.A. Richards. 1990. Cardiac and respiratory parameters of the swift fox. Journal of Wildlife Management 54(3):418–419.

Woodruffe, R., J.R. Ginsberg, D.W. Macdonald and the IUCN/SSC Canid Specialist Group. 1997. The African wild dog—status survey and conservation action plan. IUCN, Gland, Switzerland.

Zar, J.H. 1984. Biostatistical analysis. Prentice-Hall, Inc., Englewood Cliffs, New Jersey.

Zoellick, B.W., and N.S. Smith. 1986. Capturing desert kit foxes at dens with box traps. Wildlife Society Bulletin 14:286–288.

——. 1992. Size and spatial organization of home ranges of kit foxes in Arizona. Journal of Mammalogy 73:83–88.

Zoellick, B.W., N.S. Smith, and R.S. Henry. 1989. Habitat use and movements of desert kit foxes in western Arizona. Journal of Wildlife Management 53:955–961.

Part IV – Population Ecology

A Review of Small Canid Reproduction

■ Cheryl S. Asa and Carolina Valdespino

Abstract: There are many aspects of reproduction that, although unusual among mammals, are common across canid species that have been studied. These include, for example, monogamy, paternal care, monestrum with exceptionally long proestrous and diestrous phases, a copulatory tie, incorporation of adult offspring into the social group, behavioral suppression of mating in these offspring, obligate pseudopregnancy in subordinate females, and alloparental care. However, this pattern is based predominately on observations of larger species such as the gray wolf, coyote, and African wild dog. Although data from smaller species, save the red fox, are sparse, these more limited reports suggest some interesting differences. For example, smaller canid species may be more likely than larger ones to have a second monestrous, ovulatory cycle per year. There are also apparent differences in the relationships between adult body size, neonate size, and gestation length. It is clear that more research on the smaller canids is needed, rather than relying on data from the larger species, especially when conservation decisions involve calculations of the reproductive potential of populations.

In reviews of canid mating systems, small canids have been considered less strictly monogamous with smaller, more precocial litters, when compared to larger species (e.g., Macdonald and Moehlman 1983, Moehlman 1986, Moehlman 1989, Geffen et al. 1996). However, reviews of canid reproductive physiology have been few and have relied more on data from large species such as the gray wolf (*Canis lupus*; see Asa 1997, Asa 1998, Asa and Valdespino 1998). In this chapter, we present information on parameters of reproduction in small canids and compare it to data from the larger species. Using the categories of Macdonald and Moehlman (1983), small canids are those <6 kg, medium are 6 to 13 kg (e.g., jackals, *C. mesomelas*, *C. adustus*, *C. aureus*, and coyotes, *C. latrans*), and large species are >13 kg (e.g., gray wolf and wild dog, *Lycaon pictus*).

Puberty

Comparisons between small and medium canids show complete overlap, typically ranging from 8 to 12 months (Table 1). However, puberty has been reported to occur about one year later in the larger African wild dog and gray wolf. The earlier age of puberty of fennec foxes, *Fennecus zerda*, relative to the other small canids may well be due to the measure used, i.e., semen samples contained sperm at first collection by electroejaculation at 6 months of age (Asa and Valdespino, unpublished). Time of puberty is often calculated by subtracting gestation length from the date of first parturition, which is an imprecise measure of physiological puberty, i.e., first ovulation in females or first production of sperm in males. Initial ovarian cycles are often infertile, and young males are often precluded from mating because of social immaturity even though spermatogenesis has commenced (see Asa, 1996). There also have been reports of early puberty (10 months) in gray wolves, e.g., in captivity, perhaps related to unlimited availability of food or to altered social circumstances (Medjo and Mech 1976). However, 22 to 24 months is typical for the gray wolf and wild dog, so when similar measures are used, small and medium-sized canids do appear to reach puberty earlier.

Table 1. Comparisons of time of puberty and length of proestrus and estrus among small, medium and large canids [a]

Species	Puberty (mo)	Proestrus	Estrus (d)
Alopex lagopus	9-10		3-5
Fennecus zerda	≤6[b]		1-2
Nyctereutes procyonoides	8-11	2-14 d	4-6
Speothos venaticus	10-12		4
Urocyon cinereoargenteus	10-12		
Vulpes corsac	9-10		
V. vulpes	9-12	10-19 d	3-8
Canis aureus	10-11		3-4
C. latrans	9-10	2-3 mo	10
C. lupus	22	1.5-2 mo[c]	7-9
Lycaon pictus	23-24	14 d	20

[a]From Asdell's Patterns of Mammalian Repr oduction, unless otherwise noted
[b]Asa and Valdespino, unpublished
[c]from Asa et al. (1986).

Reproductive Cycle

The phases of the typical reproductive cycle include proestrus, estrus, diestrus (or luteal phase) and anestrus (Conaway 1991, Asa 1996). Proestrus is the period when increasing estradiol makes the female attractive to the male and stimulates courtship. The ensuing estrous phase in canids follows a rise in progesterone just before ovulation. Combined estradiol and progesterone make the female receptive to mating (in most other mammals, estradiol alone is sufficient to induce mating behavior). The luteal or diestrous phase following estrus is characterized by elevated progesterone and lower, fluctuating levels of estradiol. In canids the length of the diestrous phase of non-pregnant females is remarkable in that it is roughly equivalent in length to pregnancy. It is not associated with courtship or mating, but seems to prime the female to behave maternally (Asa 1996, 1997, 1998; Asa and Valdespino 1998). Anestrus is the period of relative gonadal quiescence that separates the breeding seasons.

The type of reproductive cycle reported for all canids for which there are data is monestrum (that is, only one period of estrus and ovulation per breeding season, a reproductive pattern that is quite unusual for mammals) (Asa 1996, 1998). Polyestrous species, in contrast, have repeated opportunities to conceive during successive cycles that are not interrupted by periods of anestrus (the typical non-rodent mammalian cycle of polyestrous species is two to four weeks). The breeding seasons for both types of reproductive patterns are separated by a period of anestrus.

That monestrum has been selected for in canids (especially in the more social species) may be related to the interplay of aspects of their social and reproductive systems (see Asa 1997, Asa 1998, Asa and Valdespino 1998). Briefly, monestrum may minimize social disruption in the pack by limiting periods of estrus to one per adult female. If canids were polyestrous, and it was still likely that only the dominant pair copulated, the subordinate females in a social group would continue having estrous cycles through the end of the breeding season, and the dominant pair would repeatedly need to prevent them from mating.

Monestrum is physiologically linked to the spontaneous, extended diestrous phase or pseudopregnancy that occurs in all female canids that ovulate but fail to conceive. The sustained elevation in progesterone during pseudopregnancy prevents further ovulatory cycles. An additional benefit of pseudopregnancy is that the hormonal milieu, which is virtually indistinguishable from pregnancy, primes the subordinate females to behave maternally toward the dominant female's pups. However, this explanation is based primarily on data from medium and large canids, especially the gray wolf (see Asa 1997, 1998; Asa and Valdespino 1998).

Among small canids, the length of the proestrous phase (Table 1) has only been reported for the raccoon dog, *Nyctereutes procyonoides* (Valtonen et al. 1977, Valtonen and Makela1980) and red fox, *Vulpes vulpes*, (see Hayssen et al. 1993). The values reported (from 2 to 19 days) contrast markedly with the 1.5 to 3 months that can occur in coyotes and the gray wolf (Table 1). As with measures of puberty, the discrepancies may be due to methodological differences. For most species, proestrus must be described behaviorally as the period when the female is attractive to and courted by the male but not receptive to copulation. In the genus *Canis*, however, the sanguinous uterine discharge can often be seen on the vulvar area, or has been more accurately detected with vaginal smears. It would be useful to have more reports for the small canids, especially because the interactions during the proestrous period probably facilitate the formation or maintenance of the pair bond (see Rothman and Mech 1979, Asa et al. 1986). If so, a comparison of length of proestrus and stability of the pair bond across years would be revealing.

Estrus, in contrast to proestrus, is typically characterized by female receptivity to copulation. But, as with proestrus, the length of the estrous phase and frequency of copulation may be related to pair-bond dynamics. Fortunately, more data exist for the length of estrus (Table 1), probably because copulation is more notable and easily observable than the courtship behavior that defines proestrus. The larger canids and also the coyote, again, have the longest estrus, whereas the smaller species are fairly uniform, with about four days being the mode. The fennec fox is the exception with a typical estrus of only one day (Valdespino 1999).

Copulation is accompanied by a copulatory lock in all canids for which there are reports except perhaps the wild dog (Kleiman 1967). This phenomenon appears to be restricted to the Canidae and probably serves the dual functions of mate guarding and facilitation of sperm transport (see Asa 1997). The longest reported is for the fennec fox, with a mean of almost 2 hours (Valdespino 1999, Valdespino et al. 2002).

Body Weight, Gestation, and Litter Size

Among small canids, gestation length ranges from about 50 to 60 days for all species except the bat-eared fox, *Otocyon megalotis*, and bush dog, *Speothos venaticus* (Table 2), whose pregnancies are comparable in length to those of the large and medium canids at ≥60 days. Neonates of bat-eared foxes and bush dogs also have the highest body weights among the small canids. For the small canids, in general, mean neonatal body weight is correlated with gestation length ($r = 0.89$, $P < 0.01$) but not with adult body weight ($r = 0.56$, $P = 0.12$), and appears independent of litter size. Thus, small canid species with longer gestations tend to give birth to larger pups. For the larger canids, gestation lengths of 60 to 66 days (similar to

Table 2. Comparisons of adult to neonatal body weight (BW), litter size, and gestation length among small canids

Species	Adult BW (kg)[a]	Neonatal BW (g)[b]	Neonatal/adult BW[c]	Litter size[b]	Gestation (d)[b]
Alopex lagopus	1.4-9	60-90	0 .012-0.017	1-25	46-58
Fennecus zerda	1-1.5	24-30	0.02-0.024	1-4	50-52
Nyctereutes procyonoides	4-10	60-110	0.008-0.016	3-11	51-70
Otocyon megalotis	3-5.3	100-140	0.024-0.034	2-5	60-75
Speothos venaticus	5-7	130-190	0.022-0.032	3-6	65-75
Urocyon cinereoargenteus	2.5-7	85-115	0.018-0.024	2-6	51-63
Vulpes bengal	1.8-3.2	52-65	0.021-0.026	4	50-51
V. macrotis	1.9-2.2			2-7	49-55[d]
V. velox	1.8-3	55-60[e]	0.023-0.025	3-7	51
V. vulpes	4-5.5	50-150	0.011-0.032	1-6	50-60

[a] From Nowak and Paradiso (1983)
[b] From Hayssen et al. (1993)
[c] Range of neonatal BW/median female adult BW
[d]Egoscue (1956)
[e]M. Sovada, Northern Prairie Wildlife Research Center, Jamestown, ND, personal communication.

Table 3. Comparisons of adult to neonatal body weight (BW), litter size, and gestation length between medium and large canids

Species	Adult BW (kg)[a]	Neonatal BW (g)[b]	Neonatal/adult BW[c]	Litter size[b]	Gestation (d)[b]
Canis aureus	7-15	201-214	0.018-0.019	1-5	60-63
C. latrans	7-20	200-284	0.014-0.021	2-15	60-64
C. lupus	20-80	300-500	0.006-0.01	1-12	60-66
Lycaon pictus	17-36	~300	0.011	2-16	60-80

[a] From Nowak and Paradiso (1983)
[b] From Hayssen et al. (1993)
[c] Range of neonatal BW/median adult BW

those of the bat-eared fox and bush dog) are associated with even larger pups and extremely variable intraspecific litter sizes (Table 3). However, for medium and large canids (Table 3) mean neonatal body weight is correlated with adult weight, not with gestation length ($r = 0.99$; $P < 0.01$).

To examine the degree of pre- versus post-natal investment, the ratio of neonatal to adult body weight might be used as a rough measure of helplessness, i.e., neonates that are smaller relative to the adult body weight they later achieve might be considered more altricial and thus require more parental investment post-partum than larger pups. In comparing small to medium- and large-sized canids, it is interesting that the overlap in these ratios is extensive (Tables 2 and 3). Gray wolves appear to have the most altricial pups, but arctic foxes (*Alopex lagopus*) and raccoon dogs can have pups with a similarly low ratio of neonate to adult weight. At the high end of the ratios, bat-eared foxes and bush dogs appear to have the most precocial pups.

A limitation of this approach to describing pre- and post-natal investment is the need to use means and ranges of variables used in the analysis (weights, litter sizes, and gestations), since some may be unrepresentative (i.e., exceptional) and thus misleading. Unfortunately, median or modal values have not been established.

Litter size is especially variable; thus it is difficult to generalize about its relationship to the other parameters (Tables 2 and 3). Arctic foxes have the greatest range in litter size (1–25, Table 2), whereas the next largest reported ranges are for the wild dog (2–16) and coyote (2–15). (Of course, species with the highest maximum litter size will also have the greatest range, since 1 or 2 is typically the reported minimum.) Variability in litter size may be related more to ecological variability, e.g., food availability and climate throughout the species range, than to allometry. For example, arctic fox populations with the highest recorded litter sizes were those with the most fluctuating population sizes, thought to be a function of immediate resource levels and the degree of resource predictability (Tannerfeldt and Angerbjorn 1998). Kit fox litter sizes have also been reported to be larger in years with more food abundance (Spiegel and Tom 1996, Cypher et al. 2000).

Seasonality

Most canids are reported to breed seasonally. Species in temperate latitudes are presumed to respond to changes in photoperiod, whereas those with more equatorial distribution may be constrained by seasonal changes in rainfall and/or food availability. Evidence for a photoperiodic effect comes from observation that red foxes, maned wolves (*Chrysocyon brachyurus*) and wild dogs translocated across the equator shift breeding cycles by 6 months (Ewer 1973, Cunningham 1905).

An examination of breeding activity for species with an extensive latitudinal range provides additional information on seasonal control. The most extensive data set, from records of red foxes in North America, the British Isles,

Table 4. Mating dates by latitude for the red fox, *Vulpes vulpes*

Latitude	Location	Time of mating	Reference
56°N	Round Island, Alaska	End of March	Zabel and Taggart 1989
46-49°N	North Dakota	Mid-February	Sargent et al.1981
47°N	North Dakota	First week of February	From Storm et al 1976
43-46°N	South Dakota	Beginning of February	Sargent et al. 1981
43-44°N	Saratoga Springs, NY	February	Pearson and Enders 1943
43°N	Saratoga, NY	End of February	Schoonmaker 1938
40-42°N	Iowa & Illinois	Third week of Janu ary	Storm et al. 1976
58°N	Northern Scotland	Mid-February	Lever 1963
57°N	Braemer, Scotland	Early February	Douglas 1965
55°N	Ireland	Last week of January	Fairley 1970
50-53°N	Southern England	Mid-January	Lever 1963
43°N	Pisa, Italy	Early February -end of March	Cavallini 1996
37°N	Spain	December	Travaini et al. 1993

Table 5. Dates of pup emergence and calculated date of mating for swift foxes in southern Saskatchewan [a], southern Colorado [b], and northern Mexico [a]

Location	Year	Pup Emergence	Calculated time of mating [c]
Southern Saskatchewan	1995	25 May	13 March
	1996	9 June	28 March
	1997	4 June	23 March
Southern Colorado	1989–90	12 May	1 March
	1995	25 April	11 February
Northern Mexico	1996	26 April	12 February

[a]From Moehrenschlager et al., this volume
[b]Covel 1992
[c]Calculation assumes that mating occurred 73 days be fore first emergence (52-day gestation + 21 -day denning pre -emergence = 73 days).

and Europe, reveals a clear trend for reproduction (based on dates of observed matings) to occur earlier at lower latitudes (Table 4). Another analysis of red fox reproductive seasonality relied on post-mortem examination of testes and epididymides for spermatogenesis and on ovaries and uteri for evidence of ovulation or presence of embryos, respectively (Lloyd and Englund 1973). Although this method was not as precise as observations of actual matings, the results were similar. Their data, obtained from Switzerland, the British Isles, and Scandinavia (47 to 68° North latitude), revealed a pattern of later reproductive activity at higher latitudes.

More limited data from swift foxes (dates of pup emergence from dens) reveal a similar trend (Table 5). Surprisingly, though, Kilgore (1969) observed swift foxes in Oklahoma mating from late December through early January, much earlier than reported for nearby southern Colorado or even for northern Mexico (see Table 5). Kit foxes also were reported to mate as early as December in southern California (Spiegel and Tom 1996). However, altitude was not provided in these studies, which could have been a confounding effect.

The regularity of the breeding season, its apparent association with latitude, and the 6-month phase-shift in canids translocated across the equator suggest photoperiodic control. In general, temperate zone mammals can be divided into long- and short-day breeders (Ortavant et al.1985), depending on whether their breeding season begins during spring when the hours of daylight are increasing (long days) or during fall when daylight is decreasing (short days). Because the breeding season for most temperate zone canids is winter or early spring, when daylight hours are increasing, they might be thought to be long-day breeders. However, as mentioned, some populations of red (Travaini et al. 1993), swift (Kilgore 1969) and kit (Spiegel and Tom 1996) foxes have been reported to mate as early as December, before the winter solstice, whereas other populations of the same species may not mate until March. The spermatogenic cycle (time required for sperm production) is about 6 weeks for domestic dogs (Foote et al. 1972) and coyotes (Kennelly 1972), and likely for other canids. Thus, the testosterone increase that stimulated spermatogenesis for a December mating had to have occurred at least as early as mid-November, when daylength was decreasing. However, for March mating, spermatogenesis might not have commenced until January, a time when daylength was increasing. Because at least some species may mate before or after the winter solstice, the typical long-day/short-day designation cannot be easily assigned to canids.

Lloyd and Englund (1973) evaluated various aspects of photoperiod and daylength in their study of red foxes and failed to find a consistent explanation for even the latitudinal gradient effect. However, Forsberg et al. (1989) were able to extend or advance the period of spermatogenesis in silver foxes (color variant of the red fox) by exposure to artificially short daylength, indicating that short days are indeed stimulatory in that species. A similar experiment was also successful in extending the breeding period in arctic foxes (Kuznetsov 1979, cited in Forsberg et al. 1989). Still, the explanation for the natural period of estrus extending from December through March in temperate zone fox species remains unclear. Perhaps temperate zone canids are stimulated by short days to begin reproductive recrudescence, but other factors determine how quickly recrudescence progresses. The pineal gland is thought to mediate the response to changing photoperiod in other temperate latitude mammals (Hastings et al. 1985), yet neither pinealectomy nor superior cervical ganglionectomy (part of the retinal/pineal pathway) of gray wolves affected the occurrence or onset of seasonal reproduction (Asa et al. 1987). Such studies have not been performed with other canids, so their dependence on the pineal pathway is unknown.

Another aspect of seasonal breeding is reflected in the number of litters produced per year. In the wild, the only canids that have been reported to sometimes produce more than one litter annually are the wild dog and fennec fox, and then only if the first litter is lost (Frame et al. 1979, Gaultier-Pilters 1967). Although there are no comparable data from the wild, a second litter following loss of the first has been reported for bat-eared foxes in captivity (Rosenburg 1971). More interesting, some small canids in captivity (fennec: Asa and Valdespino 1998; bush dog: Porton et al. 1987; and crab-eating fox, *Cerdocyon thous*: Brady 1978) are known to sometimes produce two litters per year even if the first survives. The inter-estrus intervals, ranging from 8–10 months, are similar to the mean interval for the domestic dog (Christie and Bell 1971), which together with the lack of seasonality in these cycles suggests they are free-running, i.e., not being influenced by circannual cues. Although details on conditions of housing were not always provided, it is more common for small canids to be kept indoors and exposed to artificial lights, which are not likely to provide the same photoperiodic cues as would be present outdoors. Although the natural range of these three species, as well as the wild dog, is ≤30° latitude, where changes in photoperiod are not as pronounced, arctic foxes were also induced to produce two litters per year under artificial "short days" (Kuznetsov 1979, cited in Forsberg 1989).

Prolactin

The pituitary peptide hormone prolactin is best known for its role in stimulating milk production, but it increases seasonally in male as well as female mammals in temperate latitudes (e.g., Pelletier 1973, Mirarchi et al. 1978, Muduuli et al. 1979, Suttie 1984). Prolactin values are not available for many canids, but are reported to peak in spring in the gray wolf (April: Kreeger et al. 1991), red fox (Maurel et al. 1984), and arctic fox (Mondain-Monval et al. 1985, Smith et al. 1985), which is also the time when pups are born. Because prolactin has been implicated in the support of parental behavior, not only in females (Bridges and Ronsheim 1990), but also in males of some species (Dixson and George 1982, Gubernick and Nelson 1989), this seasonal increase may support or facilitate paternal behavior in these species.

The annual prolactin peak may also be implicated in the seasonal decline in male reproductive parameters. The seasonal decrease in testosterone occurs as prolactin is increasing, but it is not clear whether there is a causal relationship. However, high levels of prolactin have been shown to be antispermatogenic in domestic dogs (Shafik 1994). It would be especially interesting to know the annual pattern of prolactin secretion in those canids that sometimes reproduce more than once per year.

Conclusions

Although small canid reproduction is in most ways similar to that of the larger canid species, the major difference is for earlier puberty. Whether the potential for 2 litters per year, at least in some captive conditions, occurs more generally for small canids will require more systematic study. Data are sparse for most of the smaller species, which severely limits more meaningful comparisons, especially in terms of the influence of ecology and body size. More research is needed on life history and reproductive traits of these species, particularly because conservation decisions typically involve consideration of reproductive potential of populations.

Literature Cited

Asa, C.S. 1996. Reproductive physiology. Pp. 390–417 *in* D.G. Kleiman, M.E. Allen, K.V. Thompson, and S. Lumpkin, editors. Wild Mammals in Captivity, Volume 1: Principles and Techniques of Captive Management, Section A: Captive Propagation. University of Chicago Press, Chicago, Illinois.

——. 1997. Hormonal and experiential factors in the expression of social and parental behavior in canids. Pp. 129–149 *in* J. A. French and N. G. Solomon, editors, Cooperative Breeding in Mammals. Cambridge University Press, Cambridge, United Kingdom.

——. 1998. Dogs (*Canidae*). Pp. 80–87 *in* E. Knobil and J. D. Neill, editors, Encyclopedia of Reproduction. Academic Press, New York.

Asa, C.S., U.S. Seal, M.A. Letellier, and E.D. Plotka. 1987. Pinealectomy or superior cervical ganglionectomy do not alter reproduction in the wolf (*Canis lupus*). Biology of Reproduction 37:14–21.

Asa, C.S., U.S. Seal, E.D. Plotka, M.A. Letellier, and L.D. Mech. 1986. Effect of anosmia on reproduction in male and female wolves (*Canis lupus*). Behavioral and Neural Biology 46:272–284.

Asa, C.S., and C. Valdespino. 1998. Canid reproductive biology: integration of proximate mechanisms and ultimate causes. American Zoologist 38:251–259.

Brady, C.A. 1978. Reproduction, growth, and parental care in crab-eating foxes, *Cerdocyon thous*, at the National Zoological Park, Washington. International Zoo Yearbook 18:130–134.

Bridges, R.S., and P.M. Ronsheim. 1990. Prolactin (PRL) regulation of maternal behavior in rats: bromocriptine treatment delays and PRL promotes the rapid onset of behavior. Endocrinology 126:837–848.

Cavallini, P. 1996. Ranging behaviour of red foxes during the mating and breeding season. Ethology, Ecology and Evolution 8:57–65.

Christie, D.W. and E.T. Bell. 1971. Some observations on the seasonal incidence and frequency of oestrus in breeding bitches in Britain. Journal of Small Animal Practice 12:159–167.

Conaway, C.H. 1971. Ecological adaptation and mammalian reproduction. Biology of Reproduction 4:239–247.

Covell, D.F. 1992. Ecology of the swift fox (*Vulpes velox*) in southeastern Colorado. Thesis, University of Wisconsin-Madison, Madison, Wisconsin.

Cunningham, D.J. 1905. Cape hunting dogs (*Lycaon pictus*) in the gardens of the Royal Zoological Society of Ireland. Proceedings of the Royal Society of Edinburgh 25:843–848.

Cunningham, D.J., M.R.M. Otten, T.P. O'Farrell, W.H. Berry, C.E. Harris, T.T. Kato, P M. McCue, J.H. Scrivner, and B.W. Zoellick. Submitted. Population dynamics of San Joaquin kit foxes at the Naval Petroleum Reserves in California. Wildlife Monographs.

Dixson, A.F., and L. George. 1982. Prolactin and parental behavior in a male New World primate. Nature 299:551–553.

Douglas, M.J.W. 1965. Notes on red fox (*Vulpes vulpes*) near Braemar, Scotland. Journal of Zoology 147:228–233.

Ewer, R.E. 1973. The Carnivores. Cornell University Press, Ithaca, New York.

Fairley, J.S. 1970. The food, reproduction, form, growth and development of the fox *Vulpes vulpes* (L.) in northeast Ireland. Proceedings of the Royal Irish Academy 69B:103–137.

Foote, R.H., E.E. Swierstra, and W.L. Hunt. 1972. Spermatogenesis in the dog. Anatomical Record 173:341–352.

Forsberg, M., J.A. Fougner, M. Madej and E.J. Einarsson. 1989. Photoperiodic regulation of reproduction in the male silver fox (*Vulpes vulpes*). Journal of Reproduction and Fertility 87:115–123.

Frame, L.H., J.R. Malcolm, G.W. Frame, and H. Van Lawick. 1979. Social organization of African wild dogs (*Lycaon pictus*) on the Serengeti Plains, Tanzania 1967–1978. Zeitschrift fur Tierpsychologie 50:225–249.

Gaultier-Pilters, H. 1967. The Fennec. African Wildlife 21: 117–125.

Geffen, E., M.E. Gompper, J.L. Gittleman, H.K. Luh, D.W. Macdonald, and R.K. Wayne. 1996. Size, life history traits, and social organization in the Canidae: a reevaluation. American Naturalist 147:140–160.

Gubernick, D.J., and R.J. Nelson. 1989. Prolactin and paternal behavior of the biparental California mouse. Hormones and Behavior 23:203–210.

Hastings, M.H., J. Herbert, N.D. Martensz, and A.C. Roberts. 1985. Annual reproductive rhythms in mammals: mechanisms of l light synchronization. Pp. 182–204 *in* R.J. Wurtman, M.J. Baum, and J.T. Potts, editors, Medical and Biological Effects of Light. Annals of the New York Academy of Science.

Hayssen, V., A. van Tienhoven, and A. van Tienhoven. 1993. Asdell's Patterns of Mammalian Reproduction. Cornell University Press, Ithaca, New York.

Kennelly, J.J. 1972. Coyote reproduction. I. The duration of the spermatogenic cycle and epididymal sperm transport. Journal of Reproduction and Fertility 31:163–170.

Kilgore, D.L., Jr. 1969. An ecological study of the swift fox (*Vulpes velox*) in the Oklahoma panhandle. American Midland Naturalist 81:512–534.

Kleiman, D.G. 1967. Some aspects of social behavior in the Canidae. American Zoologist 7:365–72.

Kreeger, T.J., U.S. Seal, Y. Cohen, E.D. Plotka, and C.S. Asa. 1991. Characterization of prolactin secretion in gray wolves. Canadian Journal of Zoology 69:1366–1374.

Kuznetzov, G.A. 1979. Producing two litter per year from arctic foxes. Nauch. Trudy nauchnoissled. Insti. Pushnogo Zverovod Krolikovod. 20:5–10. *In* Forsberg et al. 1989.

Lever, R.A. 1963. Weights of fox cubs. Proceedings of the Zoological Society of London 140:337–338.

Lloyd, H.G., and J. Englund. 1973. The reproductive cycle of the red fox in Europe. Journal of Reproduction and Fertility, Supplement 19:119–130.

Macdonald, D.W., and P.D. Moehlman. 1983. Cooperation, altruism, and restraint in the reproduction of carnivores. Perspectives in Ethology 5:433–67.

Maurel, D., A. Lacroix, and J. Boissin. 1984. Seasonal reproductive endocrine profiles of two wild animals: the red fox (*Vulpes vulpes*) and the European badger (*Meles meles L.*) considered as short-day mammals. Acta Endocrinologica Copenhagen 105:130–138.

Medjo, D.C., and L.D. Mech. 1976. Reproductive activity in nine and ten month old wolves. Journal of Mammalogy 57:406–408.

Mirarchi, R.E., B.E. Howland, P.F. Scanlon, R.L. Kirkpatrick and L.M. Sanford. 1978. Seasonal variation in plasma LH, prolactin and testosterone in adult male white-tailed deer. Canadian Journal of Zoology 56:121–127.

Moehlman, P.D. 1986. Ecology of cooperation in canids. Pp. 64–86 *in* D.I. Rubenstein and R.W. Wrangham, editors. Ecological aspects of social evolution: birds and mammals. Princeton University Press, Princeton, New Jersey.

Moehlman, P.D. 1989. Intraspecific variation in canid social systems. Pp. 143–163 *in* J.L. Gittleman, editor. Carnivore behavior, ecology and evolution. Cornell University Press, Ithaca, New York.

Mondain-Monval, M., O.M. Moller, A.J. Smith, A.S. McNeilly, and R. Scholler. 1985. Seasonal variations of plasma prolactin and LH concentrations in the female blue fox (*Alopex lagopus*). Journal of Reproduction and Fertility 74:439–448.

Muduuli, D.S., L.M. Sanford, W.M. Palmer, and B.E. Howland. 1979. Secretory patterns and seasonal changes in LH, FSH, prolactin and testosterone in the male pygmy goat. Journal of Animal Science 49:543–553.

Nowak, R.M., and J.L. Paradiso. 1983. Walker's mammals of the world. The Johns Hopkins University Press, Baltimore, MD.

Ortavant, R., J. Pelletier, J.P. Ravault, J. Thimonier, and P. Volland-Nail. 1985. Photoperiod: main proximal and distal factor of the circannual cycle of reproduction in farm animals. Oxford Reviews of Reproductive Biology 7:305–345.

Pearson, O.P., and R.K. Enders. 1943. Ovulation, maturation and fertilization in the fox. Anatomical Record 85:69–83.

Pelletier, J. 1973. Evidence for photoperiodic control of prolactin release in rams. Journal of Reproduction and Fertility 35:143–147.

Porton, I.J., D.G. Kleiman, and M. Rodden. 1987. A seasonality of bush dog reproduction and the influence of social factors on the estrous cycle. Journal of Mammalogy 68:867–871.

Rosenburg, H. 1971. Breeding the bat-eared fox *Otocyon megalotis* at the Utica Zoo. International Zoo Yearbook 11:101–102.

Rothman, R.J., and L.D. Mech. 1979. Scent-marking in lone wolves and newly-formed pairs. Animal Behaviour 27:750–760.

Sargeant, A.B., S.H. Allen, and D.H. Johnson. 1981. Determination of age and whelping dates of live red fox pups. Journal of Wildlife Management 45:760–765.

Schoonmaker, W.J. 1938. Notes on mating and breeding habits of foxes in New York state. Journal of Mammalogy 19:375–376.

Shafik, A. 1994. Prolactin injection, a new contraceptive method: experimental study. Contraception 50:191–199.

Smith, A.J., M. Mondain-Monval, O.M. Moller, R. Scholler, and V. Hansson. 1985. Seasonal variations of LH, prolactin, androstenedione, testosterone and testicular FSH binding in the male blue fox (*Alopex lagopus*). Journal of Reproduction and Fertility 74:449–458.

Spiegel, L.K., and J. Tom. 1996. Reproduction of San Joaquin kit fox in undeveloped and oil-developed habitats of Kern County, California. Pp. 53–69 *in* Studies of the San Joaquin kit fox in undeveloped and oil-developed areas. Staff Report, California Energy Commission.

Storm, G.L., R.D. Andrews, R.L. Phillips, R.A. Bishop, D.B. Siniff, and J.R. Tester. 1976. Morphology, reproduction, dispersal and mortality of midwestern red fox populations. Wildlife Monographs 49:1–82.

Suttie, J.M., G.A. Lincoln, and R.N.B. Kay. 1984. The endocrine control of antler growth in red deer stages. Journal of Reproduction and Fertility 71:7–15.

Tannerfeldt, M., and A. Angerbjorn. 1998. Fluctuating resources and the evolution of litter size in the arctic fox. Oikos 83:545–559.

Travaini, A., J.J. Aldama, R. Laffitte, and M. Delibes. 1993. Home range and activity patterns of red fox *Vulpes vulpes* breeding females. Acta Theriologica 38:427–434.

Valdespino, C. 1999. The reproductive system of the fennec fox (*Vulpes Zerda*). Dissertation, University of Missouri-St. Louis, Missouri.

Valdespino, C., C.S. Asa, and J.E. Bauman. 2002. Ovarian cycles, copulation and pregnancy in the fennec fox (*Vulpes zerda*). Journal of Mammalogy 83:99–109.

Valtonen, M.H., and J.I. Makela. 1980. Reproduction and breeding of the raccoon dog. Scientifur 4:18–19.

Valtonen, M.H., E.J. Rajakoski, and J.I. Makela. 1977. Reproductive features in the female raccoon dog (*Nyctereutes procyonides*). Journal of Reproduction and Fertility 51:517–518.

Zabel, C.J., and S.J. Taggart. 1989. Shift in red fox, *Vulpes vulpes*, mating system associated with El Nino in the Bering Sea. Animal Behavior 38:830–838. Table 1. Comparisons of time of puberty and length of proestrus and estrus among small, medium and large canids.

Factors Influencing Populations of Endangered San Joaquin Kit Foxes: Implications for Conservation and Recovery

■ **Brian L. Cypher, Patrick A. Kelly and Daniel F. Williams**

Abstract: The endangered San Joaquin kit fox inhabits the San Joaquin and adjacent valleys in California, and is potentially affected by a number of population influences of both natural and anthropogenic origins. Among natural factors, food availability can vary both spatially and temporally, and this variation produces marked fluctuations in kit fox abundance. Larger predators and landscape physiography can affect local abundance of kit foxes. Among anthropogenic factors, habitat loss resulting from agricultural, industrial, and urban development has significantly reduced both the abundance and distribution of San Joaquin kit foxes, and is the primary reason that federal and state protections were implemented. Compatible land uses include oil and gas production, military training activities, and livestock grazing; kit foxes also can persist in some agricultural and urban areas if sufficient food and den sites are available. Pesticides, vehicles, illegal shooting, disease, and predator control programs do not currently appear to be important population influences. Habitat loss and alteration appear to be the most important factors influencing range-wide distribution and abundance of kit foxes while food availability is the most important factor influencing local population dynamics. Interactions between factors and temporal variation in environmental conditions increase the challenge of recovering San Joaquin kit foxes, but a number of ongoing conservation initiatives provide cause for cautious optimism.

To effectively conserve and recover populations of rare species, it is imperative to understand the factors that influence these populations. These factors can be either natural or anthropogenic in origin. Of particular importance are factors that influence long-term population trends and those that may cause profound short-term variations in abundance, because these factors have the greatest potential for increasing extinction risk (Caughley and Gunn 1996). Once the important factors influencing a population are identified and relationships between these factors and population dynamics are understood, then conservation and recovery strategies that maximize efficacy and cost-efficiency can be designed and implemented.

The San Joaquin kit fox (*Vulpes macrotis mutica*) occurs in the San Joaquin Valley and some adjacent valleys in central California (Fig. 1). The historical distribution and abundance of this small arid-land canid are not known. O'Farrell (1983) estimated that abundance may have been between 8,700 and 12,100 prior to 1930, and then may have declined to about 7,000 by 1975. The San Joaquin kit fox was listed as Federally Endangered in 1967 (Federal Register 1967) and as "rare" (changed to "threatened" in 1985) in California in 1971 (Morrell 1972). The primary reason for affording formal protection to this taxon was the profound loss of natural communities in the San Joaquin Valley, primarily due to agricultural conversion. Other factors potentially threatening San Joaquin kit foxes included predator control programs, secondary poisoning from rodenticides, and hunting and trapping (U.S. Fish and Wildlife Service 1998).

Due to its rare status, considerable research has been conducted on San Joaquin kit foxes, mostly in the past 15 years. This research has primarily focused on investigating basic biology and ecology, and also on assessing the influence of various natural and anthropogenic factors on foxes. Our objectives are to (1) summarize available data on the effects of various factors on populations of San Joaquin kit foxes, (2) identify those factors that appear to exert the greatest influence and therefore may be most important to address in conservation and recovery efforts, and (3) identify critical information needs for future research.

Food Availability

San Joaquin kit foxes exhibit considerable ecological plasticity with regards to use of food items, as is commonly observed among canids. Rodents and leporids are usually the primary prey, but a diversity of other items are consumed as secondary foods. Kangaroo rats (*Dipodomys* spp.) have been identified as primary prey of San Joaquin kit foxes in a number of locations in the San Joaquin Valley (Hawbecker 1943, Laughrin 1970, Morrell 1972, Spiegel et al. 1996). California ground squirrels (*Spermophilus beechyi*) have been identified as the primary prey in many other locations (Logan et al. 1972, Orloff et al. 1986, Cypher and Warrick 1993). Black-tailed jackrabbits (*Lepus californicus*) and desert cottontails (*Sylvilagus audubonii*) also can be the primary items in kit fox diets (Scrivner et al. 1987, Knapp and Chesemore 1992). Pocket mice (*Perognathus inornatus* and

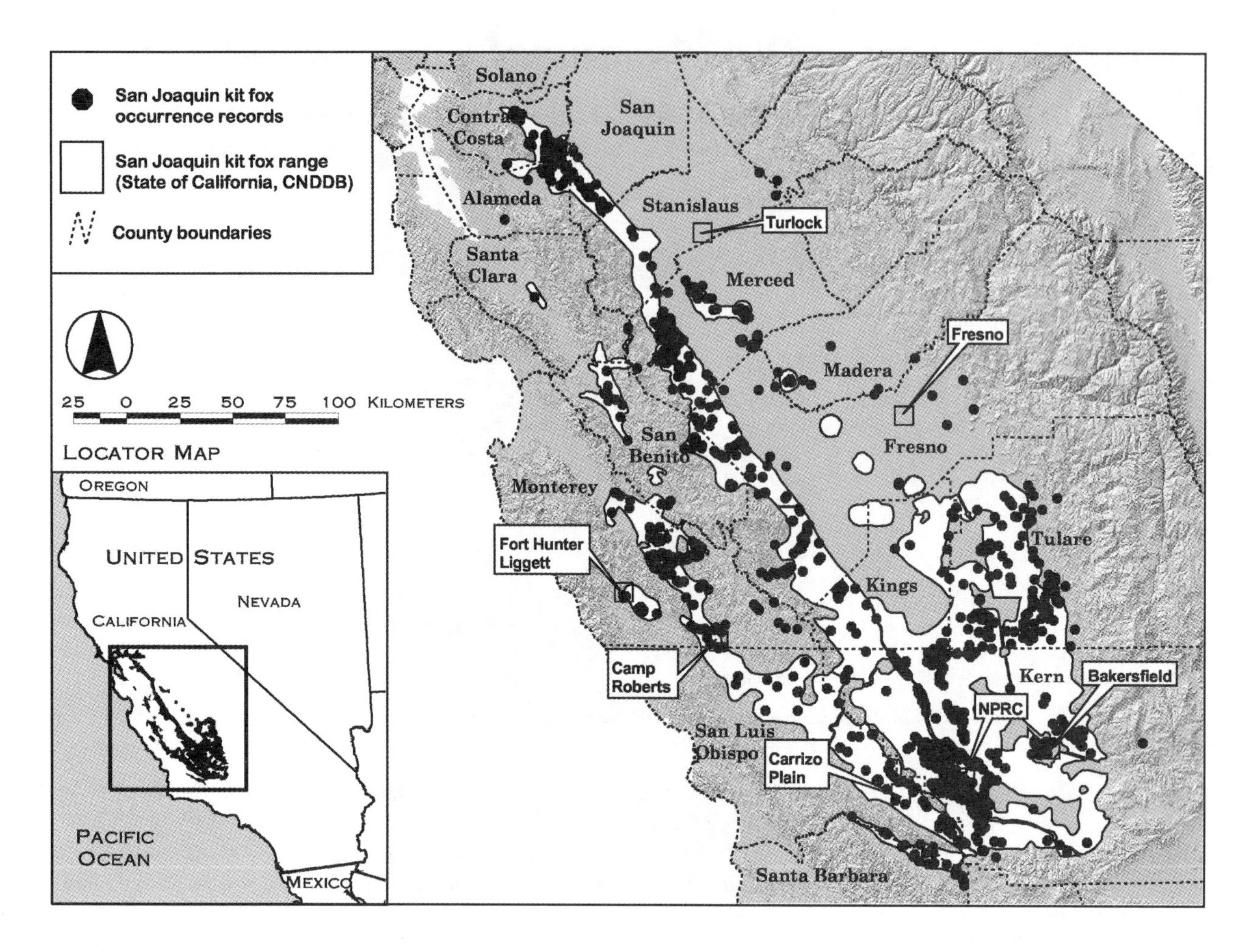

Figure 1. Current distribution of the San Joaquin kit fox in central California.

Chaetodipus californicus) and San Joaquin antelope squirrels (*Ammospermophilus nelsoni*) may be the primary items consumed when kangaroo rat and leporid abundance is low (White et al. 1996, Cypher, unpublished data). When sufficiently available, kangaroo rats may be used preferentially (e.g., Koopman 1995). Other items consumed by kit foxes include mice (*Peromysus* spp., *Mus musculus*, *Onychomys torridus*, *Reithrodontomys megalotis*), pocket gophers (*Thomomys bottae*), various birds and bird eggs, various snakes and lizards, grasshoppers (Acrididae), darkling beetles (*Eleodes* spp.), Jerusalem crickets (Gryllacrididae), scorpions (*Centruruoides* spp.), domestic sheep (consumed as carrion), and human refuse. Some vegetation, primarily grasses, is incidentally ingested along with prey.

Within the range of the San Joaquin kit fox, prey availability varies temporally and spatially. Spatial variation results from differences in habitat conditions among areas. Topography, soil types, fire history, grazing, and anthropogenic habitat disturbance all can alter vegetation structure and composition, which may affect prey availability. For example, intense or repeated range fires can destroy shrubs rendering areas less suitable for leporids (Warrick and Cypher 1998). Kangaroo rats may be less abundant in low areas subject to flooding (Williams 1985, U.S. Fish and Wildlife Service 1998), areas with a dense cover of exotic grasses (Goldingay et al. 1997), and areas of intensive petroleum production (Spiegel and Small 1996a). California ground squirrels are more common in grasslands and disturbed habitats (e.g., fallow agricultural lands, urban areas) than in shrublands (Grinnell and Dixon 1918). This spatial variation potentially influences local abundance of kit foxes.

Temporal variation in prey availability is primarily a function of annual changes in environmental conditions. These changes are largely a result of extensive variation in annual precipitation (Fig. 2) and consequential effects on primary productivity. The San Joaquin Valley is subject to extended periods (i.e., ≥2 years) of below-average precipitation (National Climatic Data Center 1996) during which primary productivity is relatively low (e.g., Otten and Holmstead 1996). Abundance of small mammals in general, and kangaroo rats in particular, track precipitation trends. Cypher et al. (2000) reported a significant relationship between small mammal abundance and the previous year's effective precipitation (i.e., precipitation occurring

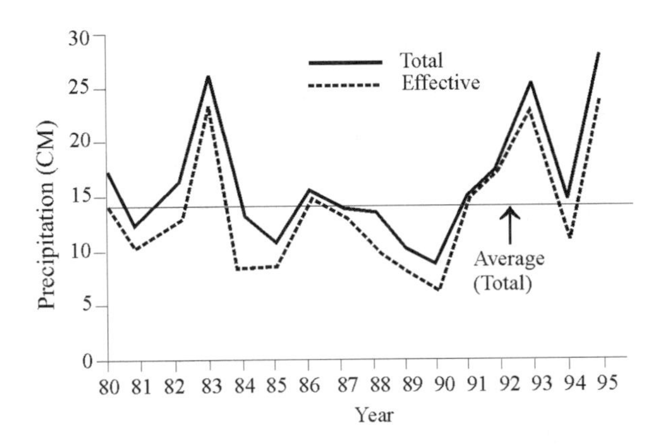

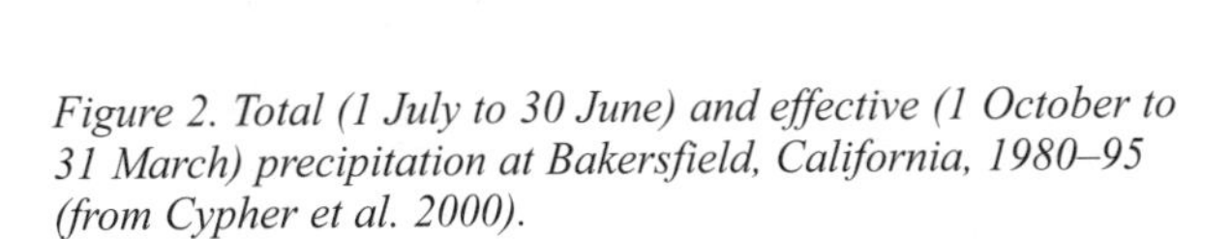
Figure 2. Total (1 July to 30 June) and effective (1 October to 31 March) precipitation at Bakersfield, California, 1980–95 (from Cypher et al. 2000).

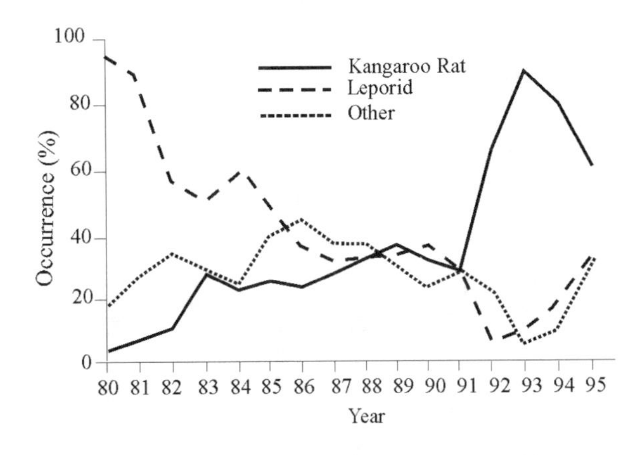

Figure 3. Frequency of occurrence of food items in scats of San Joaquin kit foxes at the Naval Petroleum Reserves in California, 1980–95. "Other" includes other rodents, birds, reptiles, insects, and refuse (from Cypher et al. 2000).

between 1 October and 31 March) indicating that there may be about a 1-year lag response by small mammals in favorable conditions, although pocket mouse populations can increase rapidly in response to favorable conditions (Otten and Holmstead 1996). High precipitation levels in conjunction with cool temperatures may negatively impact small mammals, particularly kangaroo rats. Such conditions during winter 1994–spring 1995 resulted in a dramatic decline in kangaroo rat abundance throughout the southern San Joaquin Valley (Single et al. 1996).

Leporid populations in the San Joaquin Valley also exhibit temporal variation. Precipitation can influence the abundance of leporids (Bronson and Tiemeir 1959), particularly cottontails (Ralls and Eberhardt 1997). Abundance of jackrabbits can be influenced by predation and also may exhibit population cycles of 5–10 years during which abundance can change significantly (Gross et al. 1974). The cause(s) of these cycles is not known.

San Joaquin kit foxes exhibit both functional and numerical responses to variation in food availability. The primary functional response observed is prey switching. At the Naval Petroleum Reserves in California (NPRC), kit foxes switched from leporids to kangaroo rats (Fig. 3) as the relative availability of each varied during a 16-year period (Cypher et al. 2000). When the availability of both items was relatively low, use of other items increased. Changes in local space use (e.g., home range size and location) in response to variations in prey availability have not been observed (White et al. 1996), but kit foxes may use home ranges of sufficient size to encompass adequate food resources during periods of low prey availability (von Schantz 1984).

Numerical responses by kit foxes to variation in food availability have been documented at several locations (Berry and Standley 1992, White et al. 1996, Cypher et al. 2000). These responses can be both dramatic and rapid. During 1983–95 at NPRC, the number of kit foxes captured on a 216-km^2 study area increased from the lowest number (46) in 1991 to the highest number (363) in 1994 in response to a rapidly increasing prey base, particularly kangaroo rats. By 1995, the number captured had declined by more than half in response to a crash in kangaroo rat abundance. Variation in prey availability may affect numerical responses in kit foxes through changes in both survival and reproductive success, but reproductive success may be the more important mechanism. Warrick et al. (1999) documented both increased survival and reproduction among kit foxes provided with supplemental food. Changes in survival rates attributable to variations in prey abundance have not been observed. However, a positive relationship between kit fox reproductive success and prey availability has been documented in several locations (Spencer et al. 1992, White and Ralls 1993, Spiegel and Tom 1996). Interestingly, Cypher et al. (2000) reported a negative relationship between leporid abundance and survival of adult kit foxes, possibly due to a concomitant increase in kit fox competitors (see "Interspecific Competition" below).

Interspecific Competition

A number of species potentially engage in competitive interactions with San Joaquin kit foxes. These include coyotes (*Canis latrans*), bobcats (*Lynx rufus*), badgers (*Taxidea taxus*), gray foxes (*Urocyon cinareoargenteus*), red foxes (*V. vulpes*), and large raptors such as golden eagles (*Aquila chrysaetos*), red-tailed hawks (*Buteo jamaicensis*), and great horned owls (*Bubo virginianus*), and large snakes such as Pacific rattlesnakes (*Crotalus viridis*) and gopher snakes (*Pituophis melanoleucus*).

These species may engage in either exploitative or interference competiton, or both, with kit foxes.

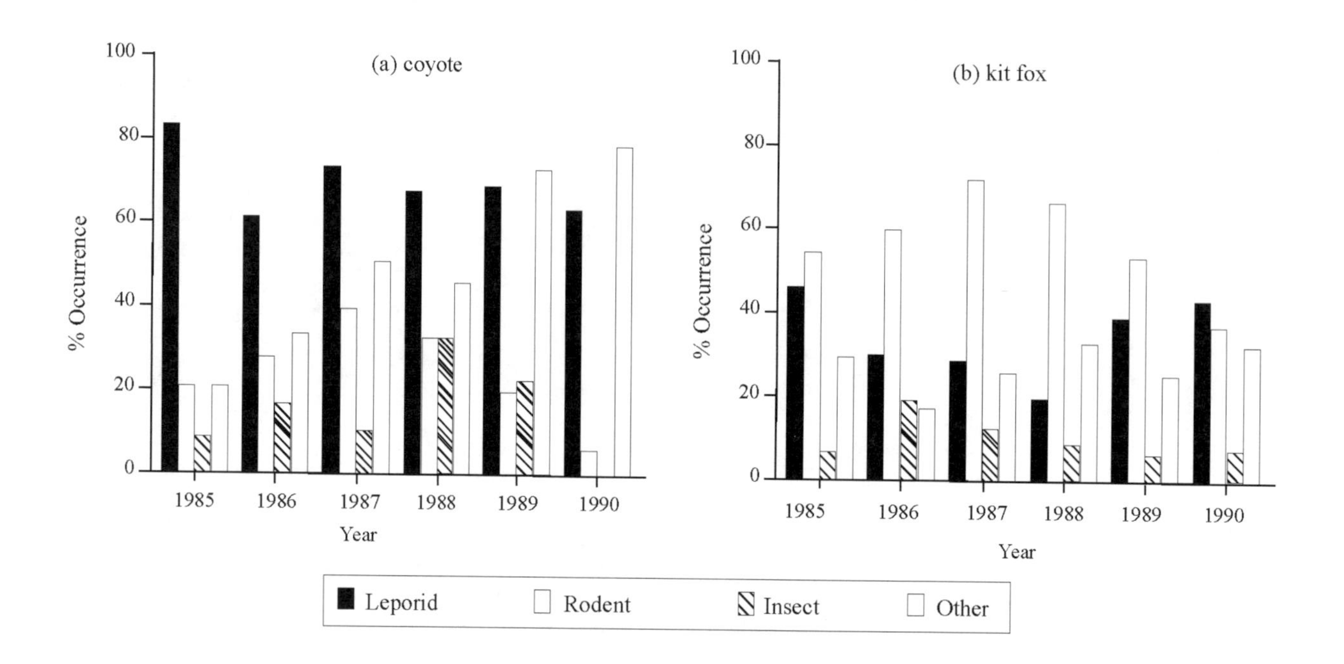

Figure 4. Use of food items by coyotes and San Joaquin kit foxes at the Naval Petroleum Reserves in California, 1985–90 (from Cypher and Spencer 1998).

Exploitative competition involves competition for resources, particularly food. All the species listed above consume the rodents and leporids that are the primary items in the diet of kit foxes, and therefore potentially reduce food availability for kit foxes. Kit foxes may compete with some species for other resources, such as with badgers for dens (K. Spencer, Fort Hunter Liggett, personal communication). In interference competition, kit foxes are excluded or killed by other species. This mortality may be a result of classic predation whereby kit foxes are killed for food by other species. Bobcats, golden eagles and red foxes occasionally kill and consume kit foxes (Balestreri 1981, Briden et al. 1992, Ralls and White 1995, Spiegel and Disney 1996). However, other species may kill kit foxes to reduce competition for resources. Coyotes kill kit foxes, but frequently do not consume the carcass, indicating that food acquisition was not the purpose of the kill (Disney and Spiegel 1992, Ralls and White 1995, Cypher and Spencer 1998). In a study on the Carrizo Plain Natural Area, coyotes consumed over half of the kit foxes they killed, but prey availability was considered low (Ralls and White 1995). In a study conducted by the California Energy Commission (CEC), most of the kit foxes killed by coyotes were not consumed (L. Spiegel, California Energy Commission, personal communication).

The effects of exploitative competition on kit foxes are unknown. Although dietary overlap occurs between kit foxes and other species, it is unknown whether this overlap significantly reduces food availability for kit foxes. No studies have addressed this question, probably because of the difficulty in manipulating or even quantifying the abundance of kit foxes, competitors, and prey. Thus, exploitative competition is usually just inferred based on use of common resources by kit foxes and competitors (e.g., Cypher and Spencer 1998).

Interference competition between kit foxes and various competitors has been well documented, but the effects on kit fox populations are poorly understood. Larger predators have long been recognized as a significant source of mortality for kit foxes (Seton 1925). Predators are usually identified as the primary mortality factor in kit fox survival studies, and reported proportions of deaths attributable to predators include 46% in Merced County (Briden et al. 1992), 75% at Camp Roberts (Standley et al. 1992), 78% on the Carrizo Plain Natural Area (Ralls and White 1995), 89% in the CEC study (Spiegel and Disney 1996), 58% for adults and 51% for juveniles at NPRC (Cypher et al. 2000), and 80% for juveniles in Alameda County (Hall 1983). Actual rates are likely higher as many deaths classified as unknown are probably due to predators as well (Cypher and Spencer 1998). Coyotes are almost always identified as the primary predator involved.

Declines in kit fox abundance have been associated with high predation rates (Ralls and White 1995), and with an increase in coyote abundance (Cypher et al. 2000). At NPRC, survival of adult kit foxes was inversely related to

PHOTO BY JASON STORLIE

Figure 5. San Joaquin kit fox in den.

coyote abundance (Cypher et al. 2000), and Cypher and Spencer (1998) reported that fox mortality from predators was at least partially additive. Also, coyotes may exclude kit foxes from some areas (Warrick and Cypher 1998), particularly marginal areas such as rugged terrain (see "Landscape Physiography" below). However, a coyote control program conducted at NPRC during 1985–90 resulted in the removal of 591 coyotes, but no increase in either abundance or survival of kit foxes was detected (Cypher and Scrivner 1992).

Kit foxes have evolved a variety of strategies that facilitate coexistence with competitors. Kit foxes exhibit considerable ecological plasticity which permits them to occupy a wide range of habitat conditions and to use a diversity of food resources, both of which reduce competitive pressure by increasing the potential for resource partitioning. Coyotes and bobcats may prefer more rugged terrain and areas with higher shrub densities whereas kit foxes may favor areas of more gentle terrain with lower shrub densities (White et al. 1995, Warrick and Cypher 1998). Also, leporids were the primary item in the diets of coyotes at NPRC during 1985–90 (Cypher et al. 1994) while kit foxes were consuming primarily rodents, particularly kangaroo rats (Fig. 4; Cypher and Spencer 1998). Leporids also tend to be the primary item in bobcat diets in the western United States (Gashwhiler et al. 1960, Jones and Smith 1979). Another factor facilitating coexistence with larger predators is year-round den use (Fig. 5) by kit foxes (White et al. 1995, Cypher and Spencer 1998). Dens constitute escape cover and allow kit foxes to elude pursuing predators. Koopman et al. (1998) estimated that 99–106 dens may be present in an average kit fox home range, although individual foxes use an average of about 12 dens each year.

Competition from red foxes may pose a significant threat to San Joaquin kit foxes. Red foxes are not native within the range of the San Joaquin kit fox. Red foxes from eastern North America apparently were introduced for hunting and some also may have escaped from fur farms (Lewis et al. 1993). Sightings of red foxes in the San Joaquin Valley have increased significantly in recent years (Lewis et al. 1993). Red foxes are larger than kit foxes and several kit fox deaths have been attributable to red foxes (Ralls and White 1995; G. Warrick, Endangered Species Recovery Program, personal communication). Also, red foxes are closer in size to kit foxes compared to coyotes and bobcats, and this may allow them to enter kit fox dens and also may increase the potential for dietary overlap. Finally, due to their close taxonomic relationship to kit foxes, red foxes may pose a threat as vectors for infectious diseases (see "Other Factors" below). Paradoxically, coyotes may effectively limit red fox abundance through competitive exclusion (Major and Sherburne 1987, Sargeant et al. 1987), and therefore it may be advantageous to kit foxes if at least some coyotes are present.

Landscape Physiography

Landscape physiography influences the relative quality of kit fox habitat. The 2 components of landscape physiography that are most important to kit foxes are terrain and vegetation characteristics. Terrain affects kit foxes through its influence on food availability, competitor abundance, and possibly the capacity of kit foxes to detect and elude predators. The availability of kit fox prey, particularly small mammals, can vary with terrain. Some species of kangaroo rats such as giant (*D. ingens*) and San Joaquin (*D. nitratoides*) kangaroo rats are more abundant in gentle terrain ($<5°$ slope), while others such as Heermann's kangaroo rat (*D. heermanni*) are more abundant in rugged terrain ($\geq 5°$slope) (Cypher 1995, Otten and Cypher 1999). Depending upon environmental conditions, kangaroo rat abundance can differ significantly between gentle and rugged terrain (Cypher 2001).

Certain kit fox competitors may be more abundant in rugged terrain. Almost all the bobcat captures on NPRC were located in rugged terrain (Warrick and Cypher 1998). Additionally, there was some evidence that coyotes may be more abundant in rugged terrain (Warrick and Cypher 1998). On NPRC, kit fox abundance declined in the early 1980s concomitant with an increase in coyote abundance. However, the decline in fox abundance was greater in rugged areas leading Warrick and Cypher (1998) to speculate that it may be more difficult for kit foxes to detect and elude predators in rugged terrain.

Vegetation characteristics, particularly community structure, also may affect kit foxes. Vegetation structure may influence the abundance of kit fox prey and competitors. Leporids appear to be more abundant where shrub cover is present (Warrick and Cypher 1998). Coyotes and bobcats also may be more abundant in areas with shrubs (Warrick and Cypher 1998), which may be a function of increased escape and resting cover, increased leporid abundance, or both. Furthermore, kit foxes may have a more difficult time detecting and eluding predators in dense shrub cover (Warrick and Cypher 1998).

The density of herbaceous ground cover also may affect kit foxes through effects on food availability. Invasive exotic grasses, particularly bromes (*Bromus* spp.), now dominate the herbaceous flora of the San Joaquin Valley (Heady 1977). These grasses form dense swards that may reduce habitat quality for certain kit fox prey species, such as kangaroo rats (Goldingay et al. 1997). Cypher (1995) reported a negative relationship between San Joaquin antelope squirrel abundance and ground cover density. These rodent species may have a difficult time locomoting and eluding predators in dense vegetation. Vegetation density varies with a variety of natural and anthropogenic factors. Soil type, southern exposures, low annual precipitation, burning, and grazing can all reduce the density of herbaceous ground cover.

Habitat Loss

Loss of native habitat was the primary reason that San Joaquin kit foxes were afforded formal protection. Extensive conversion of natural communities in the San Joaquin Valley (Fig. 6) began in the mid-1800s and has been facilitated by various federal and state initiatives designed to promote agricultural development (U.S. Fish and Wildlife Service 1998). By 1979, only about 7% of the original grassland and scrub communities south of Stanislaus County remained undeveloped (Brode et al. 1980).

Agricultural, industrial, and urban developments result in habitat loss, degradation, and fragmentation. Such development reduces overall kit fox abundance by decreasing the amount of habitat available to kit foxes and by reducing habitat quality (and thus, carrying capacity) on adjacent undeveloped areas. Destruction of habitat may result in direct mortality or displacement of individuals into marginal or already occupied areas. Fragmentation may result in areas that are too small to support viable fox populations or even individual family groups. Fragmentation also may restrict movements between habitat patches thereby affecting gene flow and reducing the potential for recolonization of patches in the event of local extirpations. Degradation can result from habitat alterations (e.g., changes in vegetation composition or structure) that reduce the abundance of food resources or escape cover, or that increase the abundance of competitors. Foxes in habitat patches near residential areas may be subject to increased disturbance from humans as well as domestic pets, which could result in reduced survival and reproductive success or avoidance of these areas.

Interestingly, due to their considerable adaptability and ecological plasticity, kit foxes are able to persist in certain areas converted to agricultural, industrial, and urban uses,

PHOTO BY DANIEL WILLIAMS

Figure 6. Agricultural development in the San Joaquin Valley, California.

and some of these developments may even provide benefits to foxes (see "Competing Land Uses" below). However, the ability of kit foxes to persist in such areas on a long-term basis is questionable.

The rate of habitat loss for San Joaquin kit foxes may be slowing somewhat. Most of the arable land in the San Joaquin Valley has already been cultivated, and the amount of land converted to agriculture in the future is likely to be minimal. Indeed, significant tracts of cultivated lands are now considered to be marginal or unsuitable for economic agricultural production. Many of these lands may soon be "retired" from agricultural uses, and at least some of these lands likely will be made available for restoration or natural rehabilitation to habitat suitable for kit foxes (U.S. Fish and Wildlife Service 1998). This will result in an increase in habitat available for kit foxes as well as other species. Industrial development will continue within the range of San Joaquin kit foxes, and the impacts of this development are unclear. The most significant types of industrial developments likely will be oil and gas production facilities and large factories. Urban development may constitute the biggest threat to remaining habitat. The human population in the San Joaquin Valley is growing rapidly, and is expected to double in the next 40 years (American Farmland Trust 1995). Consequently, urban development (e.g., construction of residential areas, small businesses, and infrastructure facilities) also is occurring at a rapid pace. In many parts of the San Joaquin Valley, much of the land being developed was formerly in agricultural production, but some previously undisturbed land also is being converted. Thus, loss of prime habitat for foxes continues.

Competing Land Uses

Oil and Gas Production

Crude oil and natural gas production occurs in many parts of the range of San Joaquin kit foxes, particularly the southern portion of the San Joaquin Valley. Production activities consist of creating well pads and drilling wells; constructing supporting facilities such as access roads, pipelines, powerlines, tank settings, compressor stations, anode beds, and waste water sumps; constructing processing facilities such as oil refineries and natural gas plants; and maintenance activities. Potential impacts from such activities include habitat loss, den loss, reduced prey availability, mortality during construction or operation activities, exposure to toxic materials, and disturbance. The establishment of regulations regarding oil field operations has helped reduce impacts through required mitigation actions such as minimizing habitat disturbance, reducing the frequency of spills and requiring rapid cleanup, and requiring that sumps be netted to exclude wildlife. However, impacts still occur on occasion. At NPRC, kit fox mortalities directly attributable to oil field activities included 35 foxes struck by vehicles, 1 fox entombed in a den, 3 foxes that drowned in spilled oil, 1 fox that drowned in a sump, 2 foxes entrapped in pipes, and 2 foxes entrapped in an oil well cellar (Cypher et al. 2000). Additionally, 2 kit fox pups were trapped in a well cellar, but were discovered and released unharmed (O'Farrell and Sauls 1987). In the CEC study, one kit fox died after becoming covered by oil (Spiegel and Disney 1996).

At NPRC, kit fox population trends were similar in both developed and undeveloped areas (Fig. 7) suggesting that the same factors were influencing foxes in both areas (Cypher et al. 2000). Fox abundance was generally lower in the developed areas, but these areas were mostly in more rugged terrain, which may be less optimal for reasons discussed earlier (see "Landscape Physiography" above). No evidence was found that oil field activities affected survival, reproduction, dispersal, space use, den use, food habits, or food availability. A small number of dens were destroyed (ca. 25), but over 1,000 dens were located on NPRC and dens were not considered to be a limiting factor. In the CEC study (Spiegel 1996), kit fox

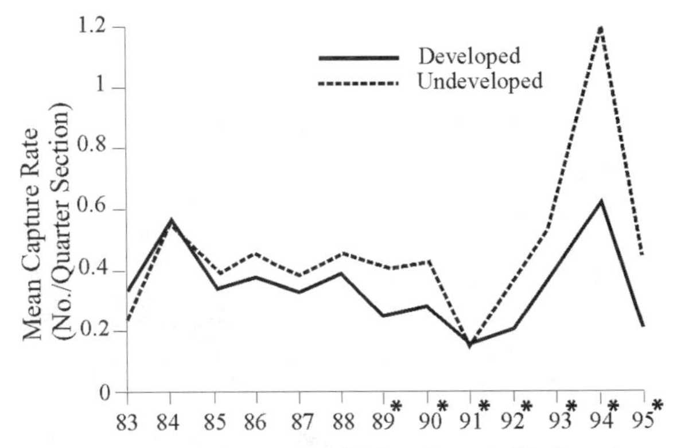

*Figure 7. Mean capture rates for San Joaquin kit foxes in developed (>15% habitat disturbance) and undeveloped (≤15% habitat disturbance) quarter-sections (65 ha) on the Naval Petroleum Reserves in California, 1981–95. An * indicates years in which rates differed between developed and undeveloped quarter-sections, based on* t-tests *(from Cypher et al. 2000).*

survival and reproduction were not affected by oil field activities. The types of foods available to foxes differed between developed and undeveloped areas, and consequently so did kit fox food habits. Interestingly, fox home ranges were smaller in developed areas, possibly due to the availability of anthropogenic foods. The most significant impact was that fox abundance in the developed area was about half that in the undeveloped area (Spiegel and Small 1996b). This difference was attributed to a reduction in carrying capacity associated with 70% habitat disturbance in the developed area.

Responsible oil field operations conducted at low to moderate levels do not appear to significantly impact kit foxes. Kit foxes are present in most oil fields in the southern and western San Joaquin Valley. The greatest impact from oil field operations appears to be the loss of some habitat and a concomitant reduction in carrying capacity. The presence of oil fields may even be beneficial to kit foxes in that the land may otherwise have been developed for a less compatible use.

Military Activities

Several military installations owned and operated by the U.S. Department of Defense occur within the range of the San Joaquin kit fox. The 2 properties of greatest significance are the Camp Roberts Army National Guard Training Site (17,200 ha) and Fort Hunter Liggett (67,400 ha) located in the Salinas Valley. Military activities occur year-round on both installations, but activity significantly increases during periodic training exercises which occasionally include firing live ammunition (Berry et al. 1992).

Potential impacts associated with military operations include direct mortality from live-fire exercises, mortality from on-road and off-road vehicle impacts, destruction of dens, and disturbance due to human activities. Additionally, non-military activities, such as grazing and sport hunting, are conducted on both installations and

PHOTO BY WILLIAM STANDLEY
Figure 8. San Joaquin kit fox at Camp Roberts Army National Guard Training Site, California.

these activities constitute competing land uses for kit foxes. From 1988 to 1991, four kit foxes died at Camp Roberts due to authorized activities. One fox died when the antenna on its radiocollar became entangled in concertina wire, 2 were struck by vehicles, and 1 was believed to have been shot by a hunter (Berry et al. 1992). Also, some kit fox dens at Camp Roberts were damaged during military training exercises, although no foxes were believed to be in the dens at the time and den availability was not considered to be a limiting factor (Berry et al. 1992).

Despite these occasional impacts, military properties appear to be beneficial for kit foxes. These properties are large and military activities, including training activities, only impact a relatively small percentage of the total area. Also, public access to these facilities is strictly limited, reducing disturbance to kit foxes. Finally, these installations have implemented natural resource management plans that include protective measures for endangered species such as the San Joaquin kit fox (Fig. 8).

Grazing

Livestock grazing by sheep or cattle occurs on many of the lands inhabited by San Joaquin kit foxes. The majority of grazing is conducted during winter and spring with livestock generally present during December–April. Duration of grazing in a given year and stocking rates are at least partly determined by annual precipitation with both duration and rates being reduced in years with less precipitation (U.S. Bureau of Land Management 1996).

Impacts to kit foxes associated with grazing activities are unknown. Potential negative impacts include reduced carrying capacity for prey species resulting in lower food availability for foxes, collapsed dens from livestock, and possibly mortality, injury, or disease transmission from sheep dogs (e.g., Balestreri 1981). Concern regarding these potential impacts has led to livestock grazing being prohibited on certain lands inhabited by foxes. For example, grazing is proscribed on lands owned by NPRC, CEC, and the California Department of Fish and Game as well as some private lands.

However, grazing has occurred on many lands that have consistently been occupied by kit foxes. Such areas include the Carrizo and Elkhorn Plains, most Bureau of Land Management lands, and large tracts of private lands inhabited by kit foxes. Thus, there is at least some compatibility between grazing and kit foxes. Indeed, there is a growing perception that grazing may be a valuable tool for reducing the dense cover of invasive exotic grasses that is present in many areas and that may be reducing prey availability for kit foxes (see "Landscape Physiography" above). Both Laughrin (1970) and Balestreri (1981) reported that grazing appeared to increase habitat suitability for kit foxes and their prey. Currently, the elimination of grazing from any lands occupied by kit foxes is strongly discouraged because doing so potentially could negatively impact kit foxes (e.g., U.S. Fish and Wildlife Service 1998). A number of studies are either in progress or planned to evaluate the effects of grazing on kit fox habitat and prey, and to determine the value of grazing as a habitat management tool.

Agriculture

The conversion of natural lands to agricultural uses is the primary source of habitat loss in the San Joaquin Valley (U.S. Fish and Wildlife Service 1998). The San Joaquin Valley is one of the most productive agricultural areas in the United States. Over 250 agricultural commodities are produced and are worth $13 billion annually (American Farmland Trust 1997). The varied crops include row crops (e.g., cotton, tomatoes, carrots), field crops (e.g., alfalfa, small grains, corn), orchards (e.g., citrus, nuts), and vineyards. The profound alterations and impacts associated with cultivated lands render most unsuitable for permanent habitation by kit foxes. These alterations and impacts include repeated tilling, minimization or elimination of natural vegetation, irrigation including flood irrigation, and use of chemicals such as pesticides and fertilizers. These activities usually result in low or no prey availability, low den site availability, and high potential for mortality or injury from farm equipment and exposure to chemicals. In a study of agricultural conversion of native habitat, Knapp (1978) determined that 3 kit foxes were entombed in their dens, 1 may have died after being struck by farm equipment, and a number of other foxes were displaced from the area and their fate was unknown. Additionally, agricultural lands may provide more favorable habitat conditions for non-native red foxes that potentially could engage in both interference and exploitative competition with kit foxes (U.S. Fish and Wildlife Service 1998).

Kit foxes may be able to use certain types of agricultural lands, at least to meet some life history requirements (e.g., Knapp 1978). Due to the high frequency of soil disturbance and irrigation, few types of agriculture provide suitable conditions for kit fox dens. However, foxes have been reported to forage in certain types of agricultural lands, such as alfalfa fields and orchards, where prey may be present. Also, kit foxes may even consume certain crops. Tomato seeds and cotton seeds were found in kit fox scats in quantities suggesting that these items were ingested as foods and not incidentally (S. Clifton, Endangered Species Recovery Program, personal communication). A study underway in the San Joaquin Valley is attempting to determine which types of agricultural lands are used by kit foxes, how frequently, and for what purpose(s).

Urbanization

The human population in the San Joaquin Valley is growing rapidly. Concomitantly, cities and towns in the region also are rapidly expanding, and an increasing amount of land is being converted to urban uses, including residential, commercial, and industrial developments (see "Habitat Loss" above). Furthermore, unlike almost all other land uses, urbanized lands very rarely revert to some other use that may be more compatible for natural communities. Urbanization usually results in the elimination of most natural habitat, reduction or elimination of natural foods, a profound increase in human disturbance, increased potential for vehicle impacts, potential for entombment during construction activities, increased exposure to chemicals (e.g., automotive fluid leaks, yard care products), and increased risk of injury or disease transmission from domestic animals.

However, kit foxes inhabit several urban areas in the San Joaquin Valley (Fig. 9) including the cities of Bakersfield (pop. ca. 260,000) and Taft (pop. ca. 15,000). A general perception was that kit foxes in urban areas represented remnant populations consisting of displaced or transient individuals with a low probability of persistence. But frequent sightings and observations of recognizable individuals occupying the same areas in multiple years suggested that urban populations may be both larger and more persistent than previously believed. Also, during periods of poor environmental conditions (e.g., drought, low prey availability) when kit foxes in non-urban areas exhibited low reproduction (White and Ralls 1993, Spiegel and Tom 1996, Cypher et al. 2000), kit foxes in urban areas successfully reproduced (B. Cypher, personal observation). Furthermore, preliminary results from an investigation of kit foxes in Bakersfield, indicated that survival of kit foxes in urban areas may be significantly higher than that of foxes in non-urban areas (B. Cypher, unpublished data). Kit foxes were observed to den in culverts, pipes, open fields, parks, school grounds, storm drainage basins, golf courses, and railroad right-of-ways (B. Cypher, unpublished data). Foxes in urban areas were found to consume a variety of natural as well as anthropogenic (e.g., refuse, pet food) foods (Cypher and Warrick 1993).

PHOTO BY JASON STORLIE

Figure 9. San Joaquin kit fox in Bakersfield, California.

Thus, kit foxes exhibit a strong capacity to adapt to and exploit urban environments. Den site and food availability seem sufficient to support foxes. Furthermore, food availability is likely more consistent in urban areas than in natural lands because of the presence of anthropogenic foods and because potential prey are plentiful due to the consistent abundance of water (primarily from landscape irrigation). Urban kit foxes appear to be adept at eluding both domestic animals and vehicles. Foxes in Bakersfield frequently were observed successfully crossing 6-lane highways, including freeways (J. Storlie, Endangered Species Recovery Program, personal communication). Based on the number of individuals trapped during a 9-month period, the population of urban foxes in Bakersfield is likely 200–400 (B. Cypher, unpublished data). A study of the ecology of urban kit foxes currently is being conducted in Bakersfield to determine the demographics and potential persistence of this population. Kit fox populations maintained in urban areas could serve as a buffer against catastrophes in natural habits, contribute to genetic diversity within the kit fox metapopulation, and serve as a source of stock for reintroduction efforts in areas where natural colonization is unlikely to occur in a timely manner.

Pesticides

Use of pesticides, particularly rodenticides, may have been a significant source of kit fox mortality in the past (e.g., Jensen 1972). Rodenticides were used extensively on agricultural and range lands to control rodents, particularly California ground squirrels, which are a significant agricultural pest within the range of the San Joaquin kit fox (Bell et al. 1994). Kit foxes may occasionally die from direct consumption of baits, but secondary poisoning from consumption of rodent carcasses is probably the greater threat. Rodenticides were identified as the cause of death for 1 fox that died in Bakersfield (Bell et al. 1994) and 2 foxes that died at Camp Roberts (Standley et al. 1992). Another concern is the potential reduction in prey availability, particularly in areas where ground squirrels constitute the primary food for kit foxes. Intensive rodent control programs may have contributed to the low occurrence of kit foxes in the northern part of their range (Orloff et al. 1986, Bell et al. 1994).

Stricter regulation of the use of rodenticides within the range of the kit fox may have significantly reduced impacts to foxes from these chemicals. These regulations dictate when and where rodenticides can be used, bait placement, and the collection and disposal of rodent carcasses (U.S. Environmental Protection Agency 1995). Also, a shift in agricultural practices away from intensive, wide-spread rodent control programs and toward focal treatments of problem areas has likely helped as well. Thus, although some mortality probably still occurs on occasion, rodenticides may currently constitute only a minor threat to kit foxes.

Other Factors

No evidence has been detected indicating that disease significantly impacts San Joaquin kit fox populations. Disease is difficult to detect in wildlife populations except in situations were epidemics of virulent pathogens result in the occurrence of large numbers of dead or moribund individuals. Such events have not been documented in San Joaquin kit foxes. Foxes with debilitating diseases are likely to die in dens or may be predisposed to another source of mortality making it difficult to find carcasses. Exposure to a number of pathogens has been documented in San Joaquin kit foxes including canine parvovirus, canine distemper, infectious canine hepatitis, rabies, leptospirosis, toxoplasmosis, brucellosis, tularemia, and coccidioidomycosis (McCue and O'Farrell 1988, Standley and McCue 1997). Two kit foxes died of rabies at Camp Roberts (Standley et al.1992). Otherwise, the occurrence of clinical disease has rarely been documented.

San Joaquin kit foxes used to be harvested for their fur. Such harvests may have been substantial at times and could have reduced fox numbers. Grinnell et al. (1937) report of 1 trapper who caught 100 kit foxes in 1 week in a 40-mi^2 area. Legal harvesting was discontinued in the 1960s when the taxon received formal federal and state protection. However, some illegal shooting still occasionally occurs (e.g., Briden et al. 1992, Standley et al. 1992, Cypher et al. 2000), but there is no evidence that this mortality affects kit fox abundance or population dynamics.

Kit foxes were killed in the past both as target and incidental species during predator control programs. Mortality from such programs may have been substantial (Grinnell et al. 1937). However, kit foxes are no longer a target species in such programs since being protected as an endangered species. Also, whenever predator control efforts are conducted in the San Joaquin Valley (usually for coyotes), the methods used greatly decrease the probability of killing or injuring kit foxes. Such methods include shooting coyotes from the ground or air, denning, and using pan-tension devices on leg-hold traps to exclude kit foxes (e.g., Cypher and Scrivner 1992).

Vehicles kill a number of kit foxes annually. The proportions of radiocollared kit foxes killed by vehicles during various studies have included 9% at NPRC (Cypher et al. 2000), 6% at Camp Roberts (Standley et al. 1992), and 2% in the CEC study (Spiegel and Disney 1996). The number of kit foxes killed by vehicles in a given year is probably density dependent. Mortality from vehicles is probably not of a sufficient magnitude to significantly influence kit fox abundance or population dynamics. Also, reducing this source of mortality would be difficult, although limiting vehicular speeds and the number of roads in areas where kit fox conservation is a primary objective might be considered.

Conclusions

Numerous factors, some natural and some anthropogenic, influence the distribution, abundance, and population dynamics of San Joaquin kit foxes. These factors result in highly dynamic demographics within the kit fox metapopulation. Local extirpations and recolonizations may be common on smaller habitat patches while kit fox abundance fluctuates markedly on larger patches (e.g., Cypher et al. 2000). Characterizing kit fox population dynamics and generating predictive models are difficult due to the number of factors, variability observed within factors, and complex interactions between factors including non-linear and possibly density-dependent relationships (e.g., White and Garrott 1997). All of this increases the challenge of conserving San Joaquin kit foxes.

Despite the difficulty in characterizing kit fox population dynamics, 2 generalizations seem reasonable. First, loss and alteration of habitat is the most important factor determining the range-wide distribution and abundance of San Joaquin kit foxes. Second, food availability is probably the most important factor influencing local population dynamics. The mechanism by which food availability affects these dynamics appears to be variation in reproductive success (White and Garrott 1997, Cypher et al. 2000); when food availability is low, reproductive success declines resulting in decreased abundance. Food availability is strongly influence by annual precipitation.

Despite the number of studies conducted to date on San Joaquin kit foxes, additional data are needed to increase the efficacy of conservation efforts and the probability of successfully recovering kit foxes. More data need to be collected on demographic parameters of kit foxes at multiple locations and across the full range of environmental conditions experienced by kit foxes. In particular, additional information is needed on the effects of various ecological factors on demographic parameters, and interactive effects between factors. All of the above information is necessary to assess metapopulation viability and extinction risk.

Further data are needed on interspecific competition between kit foxes and other predators, particularly non-native red foxes. Also, use of non-traditional habitats, such as agricultural and urban areas, by kit foxes needs to be examined further to determine the capacity of these habitats to support kit foxes, or to at least be used for specific purposes such as foraging or dispersal. Genetic variability and exchange within the metapopulation also need to be quantified. Finally, additional testing of various management techniques is needed to determine the most effective strategies for conserving kit foxes. In particular, evaluation is needed of techniques for enhancing habitat, creating dispersal corridors, and relocating/reintroducing foxes.

There is cause for cautious optimism regarding the future of San Joaquin kit foxes and the prospects for recovery. Kit foxes exhibit considerable ecological and behavioral plasticity which apparently allows them to adapt to altered environmental conditions including the use of agricultural lands and urban areas that once were believed to be completely incompatible for foxes. Foxes also have demonstrated an ability to use non-natural structures as dens, and to consume a variety of foods including anthropogenic foods. Furthermore, a number of ongoing or planned initiatives are addressing kit fox conservation. A draft recovery plan (U.S. Fish and Wildlife Service 1998) has been completed. Various habitat conservation plans resulting from consultations between private parties and the U.S. Fish and Wildlife Service (as required under the Federal Endangered Species Act) have either been implemented or are being drafted. As mentioned earlier, large tracts of agricultural land may be retired in the near future and may be available for conservation of rare species including kit foxes. Finally, a number of studies are underway or planned that will address information needs for kit foxes including providing data on conservation strategies. All of the above increase the potential for the successful conservation and recovery of San Joaquin kit foxes. Much of the information gathered on San Joaquin kit foxes is potentially useful in conservation efforts for swift foxes.

Acknowledgments

Preparation of this manuscript was supported by the U.S. Department of Energy, and by the U.S. Fish and Wildlife Service and the U.S. Bureau of Reclamation through the Endangered Species Recovery Program. We thank M. Lomolino and an anonymous reviewer who provided helpful comments for improving the manuscript.

Literature Cited

American Farmland Trust. 1995. Alternatives for future urban growth in California's Central Valley: the bottom line for agriculture and taxpayers. American Farmland Trust, Washington, DC.

——. 1997. Farming on the edge. American Farmland Trust, Washington, D.C.

Balestreri, A.N. 1981. Status of the San Joaquin kit fox at Camp Roberts, California, 1981. Department of the Army, Fort Ord, California.

Bell, H.M., J.A. Alvarez, L.L. Eberhardt, and K. Ralls. 1994. Distribution and abundance of San Joaquin kit fox. California Department of Fish and Game, Nongame Bird and Mammal Section, Sacramento, California.

Berry, W.H., and W.G. Standley. 1992. Population trends of San Joaquin kit fox at Camp Roberts Army National Guard Training Site, California. U.S. Department of Energy Topical Report No. EGG 10617-2155.

Berry, W.H., W.G. Standley, T.P. O'Farrell, and T.T. Kato. 1992. Effects of military-authorized activities on the San Joaquin kit fox (*Vulpes velox macrotis*) at Camp Roberts Army National Guard Training Site, California. U.S. Department of Energy Topical Report No. 10617-2159.

Briden, L.E., M. Archon, and D.L. Chesemore. 1992. Ecology of the San Joaquin kit fox in Western Merced County, California. Pp. 81–87 *in* D.F. Williams, S. Byrne, and T.A Rado, editors. Endangered and sensitive species of the San Joaquin Valley, California: their biology, management, and conservation. California Energy Commission, Sacramento, California.

Brode, J.M., D.P. Christenson, J. Lindell, T. Charmley, R.C. Long, P. Schempf, S. Montgomery, D. Johnson, and J. Boggs. 1980. Recovery plan for blunt-nosed leopard lizard. U.S. Fish and Wildlife Service, Portland, Oregon.

Bronson, F.H., and O.W. Tiemeier. 1959. The relationship of precipitation and black-tailed jackrabbit populations in Kansas. Ecology 40:194–198.

Caughley, G., and A. Gunn. 1996. Conservation biology in theory and practice. Blackwell Science, Cambridge, Massachusetts.

Cypher, B.L. 1995. Influence of physiography and vegetation on small mammals at the Naval Petroleum Reserves, California. Transactions of the Western Section of The Wildlife Society 31:45–52.

——. 2001. Spatiotemporal variation in rodent abundance in the San Joaquin Valley, California: implications for conserving rare species. Southwestern Naturalist 46:66–75.

Cypher, B.L., and J.H. Scrivner. 1992. Coyote control to protect endangered San Joaquin kit foxes at the Naval Petroleum Reserves in California. Pp. 42–47 *in* J.E. Borrecco, and R.E. Marsh, editors. Proceedings of the 15th vertebrate pest conference. University of California, Davis, California.

Cypher, B.L., and K.A. Spencer. 1998. Competitive interactions between coyotes and San Joaquin kit foxes. Journal of Mammalogy 79:204–214.

Cypher, B.L., K.A. Spencer, and J.H. Scrivner. 1994. Food-item use by coyotes at the Naval Petroleum Reserves in California. Southwestern Naturalist 39:91–95.

Cypher, B.L., and G.D. Warrick. 1993. Use of human-derived food items by urban kit foxes. Transactions of the Western Section of The Wildlife Society 29:34–37.

Cypher, B.L., G.D. Warrick, M.R.M. Otten, T.P. O'Farrell, W.H. Berry, C.E. Harris, T.T. Kato, P.M. McCue, J.H. Scrivner, and B.W. Zoellick. 2000. Population dynamics of San Joaquin kit foxes at the Naval Petroleum Reserves in California. Wildlife Monographs 45.

Disney, M., and L.K. Spiegel. 1992. Sources and rates of San Joaquin kit fox mortality in western Kern County, California. Transactions of the Western Section of The Wildlife Society 28:73–82.

Federal Register. 1967. Conservation of endangered species and other fish or wildlife. Federal Register 32:4001.

Gashwiler, J.S., W.L. Robinette, and O.W. Morris. 1960. Foods of bobcats in Utah and eastern Nevada. Journal of Wildlife Management 23:226–229.

Goldingay, R.L., P.A. Kelly, and D.F. Williams. 1997. The kangaroo rats of California: endemism and conservation of keystone species. Pacific Conservation Biology 3:47–60.

Grinnell, J., and J.S. Dixon. 1918. Natural history of the ground squirrels of California. Bulletin of the California State Commission on Horticulture 7:597–708.

Grinnell, J., J.S. Dixon, and J.M. Linsdale. 1937. Fur-bearing mammals of California, Volume 2. University of California Press, Berkeley, California.

Gross, J.E., L.C. Stoddart, and F.H. Wagner. 1974. Demographic analysis of a northern Utah jackrabbit population. Wildlife Monographs 40.

Hall, F.A., Jr. 1983. Status of the San Joaquin kit fox, *Vulpes macrotis mutica*, at the Bethany wind turbine generating (WTG) project sites, Alameda County, California. California Department of Fish and Game, Sacramento, California.

Hawbecker, A.C. 1943. Food of the San Joaquin kit fox. Journal of Mammalogy 24:499.

Heady, H.F. 1977. Valley grassland. Pp. 491–514 *in* M.B. Barbour, and J. Major, editors. Terrestrial vegetation of California. Wiley and Sons, New York, New York.

Jensen, C.C. 1972. San Joaquin kit fox distribution. U.S. Fish and Wildlife Service, Sacramento, California.

Jones, J.H., and N.S. Smith. 1979. Bobcat density and prey selection in central Arizona. Journal of Wildlife Management 43:666–672.

Knapp, D.K. 1978. Effects of agricultural development in Kern County, California, on the San Joaquin kit fox in 1977. California Department of Fish and Game, Non-game Wildlife Investigations Final Report, Project E-1-1, Job V-1.21.

Knapp, D.K., and D.L. Chesemore. 1992. Impact of agricultural development on San Joaquin kit foxes, Kern County, California. Page 378 *in* D.F. Williams, S. Byrne, and T.A Rado, editors. Endangered and sensitive species of the San Joaquin Valley, California: their biology, management, and conservation. California Energy Commission, Sacramento, California.

Koopman, M.E. 1995. Food habits, space use, and movements of the San Joaquin kit fox on the Elk Hills Naval Petroleum Reserves in California. Thesis, University of California, Berkeley, California.

Koopman, M.E., J.H. Scrivner, and T.T. Kato. 1998. Patterns of den use by San Joaquin kit foxes. Journal of Wildlife Management 62:373–379.

Laughrin, L. 1970. San Joaquin kit fox: its distribution and abundance. California Department of Fish and Game, Wildlife Management Branch, Administrative Report 70-2.

Lewis, J.C., K.L. Sallee, and R.T. Golightly, Jr. 1993. Introduced red fox in California. California Department of Fish and Game, Nongame Bird and Mammal Section Report 93-10.

Logan, C.G., W.H. Berry, W.G. Standley, and T.T. Kato. 1992. Prey abundance and food habits of San Joaquin kit fox at Camp Roberts Army National Guard Training Site, California. U.S. Department of Energy Topical Report No. EGG 10617-2158.

Major, J.T., and J.A. Sherburne. 1987. Interspecific relationships of coyotes, bobcats, and red foxes in western Maine. Journal of Wildlife Management 51:606–616.

McCue, P.M., and T.P. O'Farrell. 1988. Serologic survey for selected diseases in the endangered San Joaquin kit fox, *Vulpes macrotis mutica*. Journal of Wildlife Diseases 24:274–281.

Morrell, S. 1972. Life history of the San Joaquin kit fox. California Fish and Game 58:162–174.

National Climatic Data Center. 1996. Local climatological data, Bakersfield, California. National Clamatological Data Center, Asheville, North Carolina.

O'Farrell, T.P. 1983. San Joaquin kit fox recovery plan. U.S. Fish and Wildlife Service, Portland, Oregon.

O'Farrell, T.P., and M.L. Sauls. 1987. Biological survey of Naval Petroleum Reserve #2 (Buena Vista), Kern County, California. U.S. Department of Energy Topical Report No. EGG 10282-2166.

Orloff, S.G., F. Hall, and L. Spiegel. 1986. Distribution and habitat requirements of the San Joaquin kit fox in the northern extreme of their range. Transactions of the Western Section of The Wildlife Society 22:60–70.

Otten, M.R.M., and B.L. Cypher. 1999. Occurrence and prevalence of three rodent species on the Naval Petroleum Reserves: sampling implications. Transactions of the Western Section of The Wildlife Society 35:22–28.

Otten, M.R.M., and G.L. Holmstead. 1996. Effect of seeding burned lands on the abundance of rodents and leporids on Naval Petroleum Reserve No. 1, Kern County, California. Southwestern Naturalist 41:129–135.

Ralls, K., and L.L. Eberhardt. 1997. Assessment of abundance of San Joaquin kit foxes by spotlight surveys. Journal of Mammalogy 78:65–73.

Ralls, K., and P.J. White. 1995. Predation on San Joaquin kit foxes by larger canids. Journal of Mammalogy 76:723–729.

Sargeant. A.B., S.H. Allen, and J.O. Hastings. 1987. Spatial relations between sympatric coyotes and red foxes in North Dakota. Journal of Wildlife Management 51:285–293.

Scrivner, J.H., T.P. O'Farrell, and T.T. Kato. 1987. Diet of the San Joaquin kit fox, *Vulpes macrotis mutica*, on Naval Petroleum Reserve #1, Kern County, California, 1980–1984. U.S. Department of Energy Topical Report No. EGG 10282-2168.

Seton, E.T. 1925. Lives of game animals. Volume 1. Cats, wolves and foxes. Doubleday, Doran and Company, New York, New York.

Single, J.R., D.J. Germano, and M.H. Wolfe. 1996. Decline of kangaroo rats during a wet winter in the southern San Joaquin Valley, California. Transactions of the Western Section of The Wildlife Society 32:34–41.

Spencer, K.A., W.H. Berry, W.G. Standley, and T.P. O'Farrell. 1992. Reproduction of the San Joaquin kit fox on Camp Roberts Army National Guard Training Site, California. U.S. Department of Energy Topical Report No. EGG 10617-2154.

Spiegel, L.K. 1996. Studies of San Joaquin kit fox in undeveloped and oil-developed areas: an overview. Pp. 1–14 *in* L.K. Spiegel, editor. Studies of the San Joaquin kit fox in undeveloped and oil-developed areas. California Energy Commission, Sacramento, California.

Spiegel, L.K., B.L. Cypher, and T.C. Dao. 1996. Diet of the San Joaquin kit fox at three sites in western Kern County, California. Pp. 39–52 *in* L.K. Spiegel, editor. Studies of the San Joaquin kit fox in undeveloped and oil-developed areas. California Energy Commission, Sacramento, California.

Spiegel, L.K., and M. Disney. 1996. Mortality sources and survival rates of San Joaquin kit fox in oil-developed and undeveloped lands of southwestern Kern County, California. Pp. 71–92 *in* L.K. Spiegel, editor. Studies of the San Joaquin kit fox in undeveloped and oil-developed areas. California Energy Commission, Sacramento, California.

Spiegel, L.K., and M. Small. 1996a. Estimation of relative abundance, biomass, and diversity of small mammals between an undeveloped site and an oil-developed site in Kern County, California. Pp. 125–131 *in* L.K. Spiegel, editor. Studies of the San Joaquin kit fox in undeveloped and oil-developed areas. California Energy Commission, Sacramento, California.

——. 1996b. Estimation of relative abundance of San Joaquin kit foxes between an undeveloped site and an oil-developed site in Kern County, California. Pp. 115–124 *in* L.K. Spiegel, editor. Studies of the San Joaquin kit fox in undeveloped and oil-developed areas. California Energy Commission, Sacramento, California.

Spiegel, L.K., and J. Tom. 1996. Reproduction of San Joaquin kit fox in undeveloped and oil-developed habitats of Kern County, California. Pp. 53–69 *in* L.K. Spiegel, editor. Studies of the San Joaquin kit fox in undeveloped and oil-developed areas. California Energy Commission, Sacramento, California.

Standley, W.G., W.H. Berry, T.P. O'Farrell, and T.T. Kato. 1992. Mortality of San Joaquin kit fox at Camp Roberts Army National Guard Training Site, California. U.S. Department of Energy Topical Report No. EGG 10627-2157.

Standley, W.G., and P.M. McCue. 1997. Prevalence of antibodies against selected diseases in San Joaquin kit foxes at Camp Roberts, California. California Fish and Game 83:30–37.

U.S. Bureau of Land Management. 1996. Caliente resource management plan. U.S. Department of the Interior, Bureau of Land Management, Bakersfield, California.

U.S. Environmental Protection Agency. 1995. Protecting endangered species: interim measures for San Joaquin kit fox. U.S. Environmental Protection Agency, Pesticides and Toxic Substances Leaflet H-7506C, Sacramento, California.

U.S. Fish and Wildlife Service. 1998. Recovery plan for upland species of the San Joaquin Valley, California. U.S. Fish and Wildlife Service, Portland, Oregon.

von Schantz, T. 1984. Carnivore social behavior—does it need patches? Nature 307:389–390.

Warrick, G.D., and B.L. Cypher. 1998. Factors affecting the spatial distribution of a kit fox population. Journal of Wildlife Management 62:707–717.

Warrick, G.D., J.H. Scrivner, and T.P. O'Farrell. 1999. Demographic responses of kit foxes to supplemental feeding. Southwestern Naturalist 44:367–374.

White, P.J., and R.A. Garrott. 1997. Factors regulating kit fox populations. Canadian Journal of Zoology 75:1982–1988.

White, P.J., and K. Ralls. 1993. Reproduction and spacing patterns of kit foxes relative to changing prey availability. Journal of Wildlife Management 57:861–867.

White, P.J., K. Ralls, and C.A. Vanderbilt White. 1995. Overlap in habitat and food use between coyotes and San Joaquin kit foxes. Southwestern Naturalist 40:342–349.

White, P.J., C.A. Vanderbilt White, and K. Ralls. 1996. Functional and numerical responses of kit foxes to a short-term decline in mammalian prey. Journal of Mammalogy 77:370–376.

Williams, D.F. 1985. A review of the population status of the Tipton kangaroo rat, *Dipodomys nitratoides nitratoides*. U.S. Fish and Wildlife Service, Endangered Species Office Final Report, Sacramento, California.

Aspects of Swift Fox Ecology in Southeastern Colorado

■ David E. Andersen, Thomas R. Laurion, John R. Cary,
Robert S. Sikes, Mary A. McLeod and Eric M. Gese

Abstract: We studied the ecology of swift fox on the Piñon Canyon Maneuver Site (PCMS) in southeastern Colorado from March 1986 to September 1987. Forty-two foxes were captured 162 times; 23 were radiocollared. Mean minimum convex polygon home range size of 5 adult swift fox was 29.0 km² (range = 12.8 to 34.3 km²) and, although home ranges of adjacent social groups overlapped, core areas described by 50% harmonic means were almost entirely exclusive. Swift fox diet (as determined from scats) consisted primarily of small and medium-sized mammals (monthly mean % volume = 64%), arthropods ($\bar{x}$ = 19%), and small birds ($\bar{x}$ = 8%). Mean litter size (n = 5) was 3.4 (range = 2 to 5) and not all females produced litters. Kaplan-Meier estimates of annual survivorship were 0.45 for adults (n = 8) and 0.126 for juveniles (n = 14). Predation by coyotes was the primary cause (63%) of fox mortality. Fox carcasses collected off of the PCMS (where coyote hunting and trapping were permitted) indicated that juvenile mortality due to predation by coyotes was lower there than on the study site. We conclude that where coyotes are abundant, predation by coyotes is a significant source of mortality for swift fox and that den availability might be an important aspect of swift fox management.

Swift fox (*Vulpes velox*) formerly inhabited shortgrass and midgrass prairies of North America, from eastern New Mexico and northwestern Texas to southern Alberta and Saskatchewan, and from eastern Colorado, Wyoming, and Montana to western Iowa (Scott-Brown et al. 1987, Carbyn et al. 1994). By the mid-1950s, swift fox were uncommon in eastern and northern portions of their historic range, and rare or absent from other portions (Martin and Sternburg 1955, Glass 1956, Long 1965, Pfeifer and Hibbard 1970, Kerwin 1972, Hillman and Sharps 1978). Explanations for this range reduction include the loss of prairie habitat to agriculture (Chambers 1978, Russell and Scotter 1984) and both direct and indirect effects of poisoning campaigns directed primarily at the wolf (*Canis lupus*). Young (1944:336) noted that swift foxes often were the first to consume poisoned bait intended for wolves, and Carbyn (1986) suggested that exterminating the wolf in prairie habitats may have allowed coyote (*C. latrans*) densities to increase. As coyotes often prey upon swift and kit fox (*V. macrotis*) (Seton 1929:564, Kilgore 1969, Scott-Brown et al. 1986, O'Neal et al. 1987, Covell 1992, Cypher and Scrivner 1992, Disney and Spiegel 1992, Ralls and White 1995, Sovada et al. 1998), increases in coyote densities can in turn increase predation rates on swift and kit fox.

Swift fox are slowly becoming re-established in parts of their historical range, but populations are affected by a variety of human activities including hunting and trapping (Kilgore 1969, Linhardt and Robinson 1972, Loy 1981), indiscriminate shooting (Miller and McCoy 1965, Kilgore 1969, Hines 1980, Hines and Case 1991), poisoning programs for coyote control (Seton 1929, Bunker 1940, Hillman and Sharps 1978), and mortality caused by vehicles on roads (Cutter 1958, Hines 1980, Samuel and Nelson 1982, Scott-Brown et al. 1986, Hines and Case 1991). Additional sources of fox mortality include predation by golden eagles (*Aquila chrysaetos*) (Cameron 1984, Scott-Brown et al. 1986), American badgers (*Taxidea taxus*), red fox (*V. vulpes*), bobcats (*Lynx rufus*), domestic dogs (*C. familiaris*) (Scott-Brown et al. 1986, Disney and Spiegel 1992), and potentially great horned owls (*Bubo virginianus*) (Kilgore 1969).

Scott-Brown et al. (1987) reviewed available literature on swift fox and suggested the need for population studies and especially the need for information on rates and causes of mortality. These types of data, combined with information on general ecological patterns, are essential for species management. The current study was designed to provide information on swift fox ecology in southeastern Colorado pertinent to population management. We used radiocollared individuals to examine home range sizes and patterns of habitat use between neighboring individuals and also to assess mortality rates and causes of mortality. Additionally, regular observations of this population allowed us to gather data on reproduction and food habits.

Study Area

The 1040-km² Piñon Canyon Maneuver Site (PCMS) is located 52 km northeast of Trinidad in Las Animas County, Colorado (Fig. 1). The area was first settled in the late 1860s and has undergone 2 homesteading booms associated with cattle and sheep ranching. Cattle ranching has dominated in this area since the early 1950s (Friedman

1985). The PCMS was acquired by the U.S. Army in 1983 for use by mechanized infantry. All hunting, trapping, and predator control was prohibited on the area from 1983 until 1987. Beginning in January 1987 an experimental program of coyote population control to reduce coyote densities was initiated in the southwestern region of the PCMS, outside of the area of our intensive study site (Gese 1987). In areas surrounding the PCMS, coyotes have consistently been subjected to intense removal efforts by ranchers (Covell 1992). Because of the restrictions on many types of human use on the PCMS, this site provides an excellent opportunity to examine elements of swift fox ecology and population dynamics that are impacted by human activities elsewhere and to provide comparative data for managed populations.

Elevation on the PCMS ranged from 1,300 to 1,740 m and climate was semiarid with average annual precipitation ranging from 26 to 38 cm on different parts of the study site. Vegetation on the PCMS was composed of grasslands, shrublands, and woodlands (Shaw and Diersing 1990). Grasslands covered 55% of the total area (Firchow 1986) and were dominated by blue grama *(Bouteloua gracilis)*, western wheatgrass *(Agropyron smithii)*, and galleta *(Hilaria jamesii)*. Shrublands were composed of a grassland understory with an overstory of shrubs or succulents, including walking-stick cholla *(Opuntia imbricata)*, soapweed *(Yucca glauca)*, wolfberry *(Lycium pallidum)*, winterfat *(Ceratoides lanata)*, and bigelow sage *(Artemisia bigelovii)*. Woodlands were dominated by one-seed juniper *(Juniperus monosperma)* and piñón pine *(Pinus edulis)*, with a shrubby understory of wax current *(Ribes cereum)*, sumac *(Rhus trilobata)*, and true mountain mahogany *(Cercocarpus montanus)*.

A 75-km^2 intensive study area within the PCMS (Fig. 1) was selected in which we attempted to capture and radiocollar all swift fox. The intensive study area was outside of the area of coyote removal described above, and only the first removal effort (January 1987) occurred during the present study (Gese 1987). Habitat within the intensive study site was primarily short-grass prairie that graded into piñón/juniper shrub in association with limestone breaks or at the heads of canyons of tributaries to the Purgatoire River.

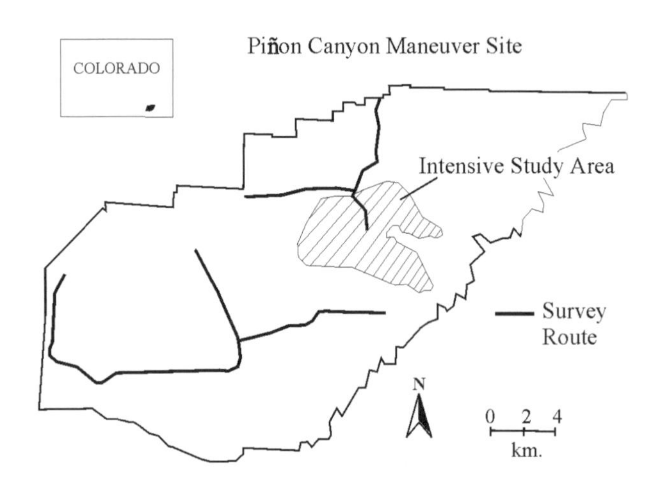

Figure 1. Map of the Piñon Canyon Maneuver Site located in southeastern Colorado. Boundaries of the intensive study area used to assess swift fox ecology and the 87-km truck survey route are indicated.

Methods

Trapping, Capture, and Radio Telemetry Monitoring

Swift foxes were captured using single- and double-door National live traps (61 x 24 x 24 cm and 81 x 24 x 24 cm, respectively) baited with chicken or pork. Traps were prebaited with a door wired open, and were set when the bait had been taken for 1–2 nights. Radiocollared foxes were recaptured by enclosing the entrance to their den with a small pen and an attached trap (Zoellick and Smith 1986). We usually recaptured radiocollared individuals in 1–2 nights. Beginning May 1986, we attempted to capture and radiocollar pups as soon as possible after they appeared above ground and weighed >700 g. Pups were captured in traps set next to natal dens; no prebaiting was necessary. We manually restrained all individuals and recorded sex, weight, age class, and standard body measurements (total length, and length of tail, hindfoot, and ear). Individuals were classified as pup or adult based on their development at the time of first capture. We checked all females for evidence of lactation or for pregnancy by abdominal palpation.

Radio-collars weighing between 35 and 50 g (<5% body weight, Eberhardt et al. 1982) and with a battery life of 150 to 200 days were affixed to all foxes captured in the intensive study site. Most radiocollars were equipped with mortality sensors. We used a portable receiver and hand-held 4-element Yagi antenna to locate animals during daylight hours ≥2 times/week. These daytime locations facilitated collection of scats and provided information on den use and date of death. Nighttime locations were obtained from simultaneous bearings recorded from 2 fixed-location receiving stations. Each station had a 13.7 m rotatable mast with paired 11-element Yagi antennas and a null-peak system (Mech 1983). Night tracking was conducted in 6-hour blocks, either 1800 to 2400 hours or 2400 to 0600 hours. Four to 12 locations were obtained on each animal during each 6-hour tracking period. The number of locations obtained for each animal depended upon weather conditions and the number of animals being tracked. Night tracking was conducted on 54 nights in 7 periods: from 15 July–7 August 1986; 23–30 September 1986; 23–30 October 1986; 3–7 January 1987; 17–22 January 1987; 11–21 February 1987; and 18–27 March 1987. Because night locations best represented home-range use during periods of fox activity, we performed all home range analyses using only nighttime locations.

Home Range Estimation and Core Areas of Activity

Prior to home range analyses, location data were screened for accuracy using a maximum error polygon size and rate of movement from a prior location. Error polygons are in part a function of azimuth precision, distance from the signal source to receiver, the number of simultaneous azimuths used to determine the location of the source of a signal, and the relation of azimuths to one another (White and Garrott 1990). Azimuth precision and the number of simultaneous azimuths used (2) were the same for all nighttime locations. However, because receiving stations were fixed, distance to individual animals and the relation of azimuths to one another varied among individuals, resulting in relatively large error polygons for some radiocollared foxes. Thus, we assessed location data for each fox separately. Error polygons for all but 2 foxes averaged <5 km^2, so for these individuals we excluded obvious outliers (<2% of locations) from further analyses. For the remaining 2 individuals we excluded all locations with error polygons >10 km^2. Five additional locations were removed from the overall data set based on calculated movement rates that seemed excessive (>0.20 m/sec compared to an average of 0.026 m/sec). In total, 46 locations were excluded from home range analyses based on excessive error polygons or consideration of rate of movement.

Home ranges were calculated from data that included at least 1 location from each nighttime tracking period. The number of tracking periods per fox ranged from 13 to 50. Home ranges were calculated using the minimum convex polygon method (100%; Mohr 1947, Southwood 1966) and we calculated 50% and 95% core activity areas using both harmonic mean (Dixon and Chapman 1980) and adaptive kernel (Worton 1989) estimators. To assess the extent to which home ranges might change over time, we calculated seasonal minimum convex polygon home ranges for the 3 adults (at time of capture) located most often across seasons (3 months comprised each season with spring beginning on 1 March).

Population Dynamics and Estimation of Survival Rates

To determine the relative abundance of swift fox on the PCMS, an 87-km survey route was driven in the morning after each new snowfall or after it ceased snowing in the winter of 1986–87. This route was driven 5 times from 2 December 1986 to 21 January 1987, but surveys were sometimes abbreviated if required by local weather conditions. The location of all swift fox tracks observed on the road were recorded. We compared sex and age composition of swift fox on the entire PCMS to foxes on the intensive study area by attempting to capture foxes in live traps placed near locations of tracks observed on the survey route.

We used cementum annuli of fox teeth (canines) to estimate the age structure of foxes trapped in areas adjacent to the PCMS where coyotes were not protected through hunting and trapping restrictions. Skulls of swift fox were obtained from trappers during the 1986–87 winter in Cheyenne Wells, Colorado, about 280 km northeast of the study site ($n = 43$), and Springfield, Colorado, about 160 km east of the study site ($n = 30$).

Deaths were recorded when indicated by the mortality sensor on radiocollars or when individuals remained motionless during nighttime tracking periods. Causes of death were determined from condition of carcasses and tracks and signs at the kill site. In a few cases, cause and date of death were recorded for untagged individuals that were found by investigators within the study area. Survival rates were calculated using Kaplan-Meier product limit estimators based on the staggered entry design described by Pollock et al. (1989). Annual (June 1986 through May 1987) survival rate was calculated for eight adults collared during this period. Survival rate of pups was calculated over an 11-month period (July 1986 through May 1987) and annual survival rate was estimated by extrapolation. Seasons for survival calculations were the same periods of time described for home range analyses.

Food Habits

We determined food habits from scats that were collected throughout the PCMS. Scats were collected most often during snow tracking, while obtaining day locations, or at den sites. Scats found around dens probably were pup scats; adults generally left the den site to defecate (determined from snow tracking). Scats were air dried and broken apart, and food remains were identified from reference materials collected locally. The percent volume of each item in individual scats was estimated visually (to the nearest 10%) and the mean of these estimates was calculated to give total percent volume for each month and season.

Results and Discussion

Radio Telemetry

Forty-two swift fox were captured 162 times on the PCMS. Twenty-three individuals (9 adults and 14 pups) captured within the 75-km^2 intensive study area were equipped with radiotransmitters (Table 1) and were located 995 times during the day and 1,539 times at night. Five pairs (or family groups when pups were present) were followed during 1986 and 1987.

Home Range Estimates

We had sufficient data to estimate home range size for five adult swift foxes (2 females and 3 males). Minimum convex polygon home ranges averaged 29.0 km^2 (range 12.8 to 34.3 km^2, Table 2) and are similar to previous reports for swift fox (Hines and Case 1991), but somewhat larger than those reported for kit foxes (Spiegel and Bradbury 1992, Zoellick et al. 1992). Minimum convex

Table 1. Summary data for swift foxes captured and equipped with radio transmitters on a 75 -km^2 intensive study area of the Piñon Canyon Maneuver Site in Colorado, March 1986 –November 1987. Table includes mortality data for 4 additional animals that were n ot radio-collared.

Animal No.	Sex	Date of Capture dd/mm/yy	No. Times Captured	No. Dens	Radio-days	Daytime Locations	Nighttime Locations	Cause of Death	Date of Death or Last Day Monitored dd/mm/yy
Adults									
1	M	30/03/86	6	4	89	17	0	coyote	27/06/86
2	F	25/0586	13	20	493	156	285		30/09/87
3	M	15/0586	17	23	503	151	312		30/09/87
4	F	04/06/86	2	5	56	18	18	coyote	30/07/86
5	M	28/05/86	5	15	282	104	217	coyote	06/03/87
18	M	21/12/86	5	8	210	74	94	eagle	19/07/87
23	F	09/02/87	6	7	233	67	34		30/09/87
36	M	12/05/87	3	11	141	40	0		30/09/87
37	F	03/06/87	4	5	119	27	0		30/09/87
Pups									
6	F	07/06/86	5	—	56	0	24	coyote	02/08/86
7	M	07/06/86	7	—	55	0	7	coyote	01/08/86
8	F	07/06/86	16	17	480	145	195		30/09/87
11	M	07/06/86	8	—	113	0	38	coyote	28/09/86
13	M	20/06/86	7	12	194	72	104	coyote	31/12/86
14	M	28/05/86	4	—	29	0	3	eagle	26/06/86
15	F	27/11/86	1	—	15	2	0	coyote	12/12/86
16	M	27/11/86	3	—	39	9	13	coyote	05/01/87
17	F	20/12/86	2	12	265	90	195	coyote	11/09/87
38	M	04/06/87	5	4	98	3	0	badger	10/09/87
39	F	04/06/87	5	2	118	2	0		30/09/87
40	F	06/06/87	6	4	97	8	0	coyote	11/09/87
41	F	01/07/87	1	4	71	4	0	coyote	10/09/87
42	M	04/07/87	1	4	88	6	0		30/09/87
Unmarked									
—	F (adult)							vehicle	21/04/86
—	M (pup)							suspected coyote	07/06/86
—	M (pup)							suspected coyote	07/06/86
—	M (pup)							unknown	01/07/86

Table 2. Minimum convex polygon (100% MCP) home range size estimates (km ^{2}) and harmonic mean and adaptive kernel estimates of core activity areas (km^2) from night locations for adult swift foxes radio -tracked on the Piñon Canyon Manuever Site, Colorado, 19 86–1987.

Fox number	Number of locations	Tracking nights	Months monitored	MCP	Core activity areas: Harmonic mean 50%	Harmonic mean 95%	Adaptive kernal 50%	Adaptive kernal 95%
Male								
3	312	50	16	34.3	3.6	18.5	3.5	16.4
5	217	43	10	32.3	4.8	23.8	5.5	23.4
18	94	21	7	33.5	4.9	23.3	6.3	30.1
Female								
2	285	48	16	31.4	1.9	24.9	1.9	23.9
23	34	13	7	12.8	1.7	14.0	3.5	20.1

polygon estimates of home ranges of adjacent individuals or family pairs overlapped appreciably (Fig. 2). Sizes of 50% core activity areas of individuals or families were similar regardless of whether the harmonic means or adaptive kernel estimates were used and ranged from 1.3 to 6.3 km^2 depending on the estimator (Table 2). The core activity areas of adjacent, same-sex adults were almost entirely exclusive of one another (Fig. 2), suggesting some degree of territoriality. These data are important because swift foxes previously were not believed to be territorial (Hines 1980, Samuel and Nelson 1982, Cameron 1984, Scott-Brown et al. 1987). Similarly, earlier studies of kit fox indicated no tendency toward territoriality (Morrel 1972, McGrew 1979) whereas recent studies using telemetry have demonstrated territoriality for kit fox (White et al. 1994).

The areas delineated by daytime locations (essentially den locations) and the 50% core activity areas were similar, indicating that swift fox spent most of their time in the vicinity of a den even during the active nocturnal period. That various areas within an individual's home range were used with different intensities is further demonstrated by the fact that 95% activity areas were 5–13 times larger than 50% activity areas (Table 2). The size of an individual's home range can vary among different areas and also temporally (Gittleman and Harvey 1982). In the present study, home ranges varied seasonally and were smallest during the summer for both sexes (Fig. 3). The summer range of female number 2 was especially restricted (Fig. 3A), as she was rearing a litter during this time. Winter ranges of all animals were by far the largest and included virtually the entire area of the overall minimum convex polygon calculated for each individual (Fig. 3).

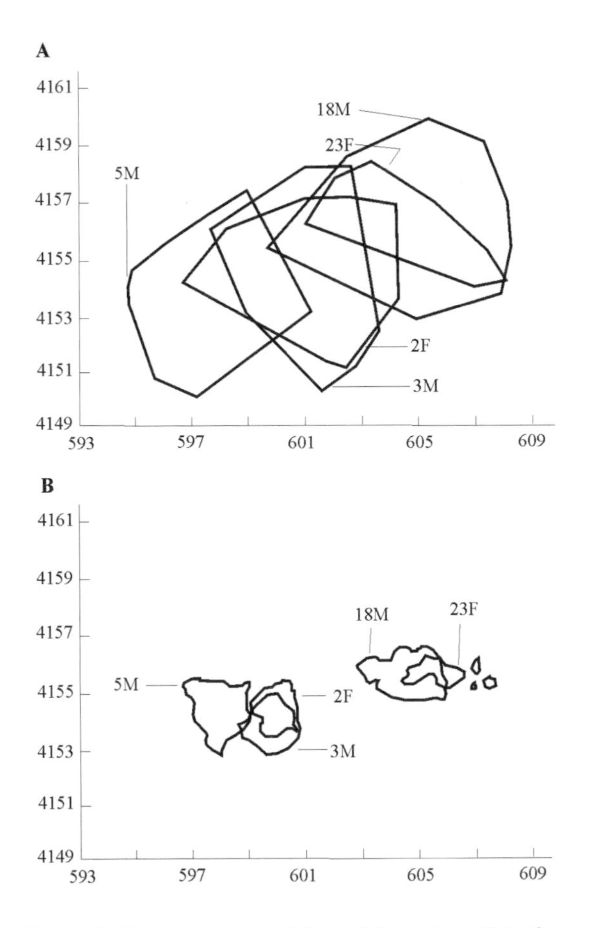

*Figure 2. Home ranges of adult swift foxes (*n = 5*) in the intensive study area of the Piñon Canyon Maneuver Site, Colorado, 1986–1987. A) Range boundaries calculated using the 100% minimum convex polygon method. B) Core activity areas using 50% harmonic mean estimators. Labels associated with each range boundary indicate the individual foxes' identification number and sex.*

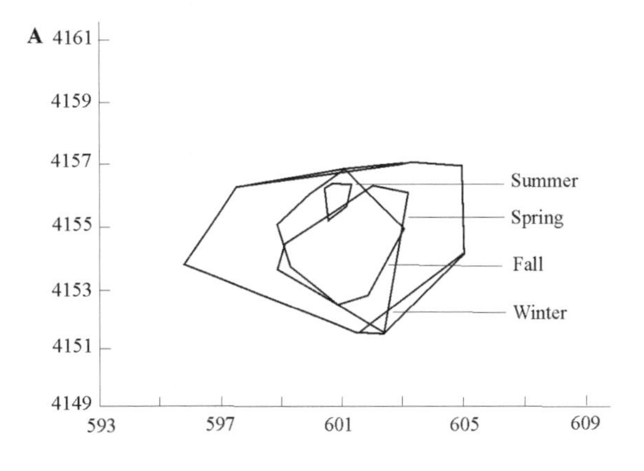

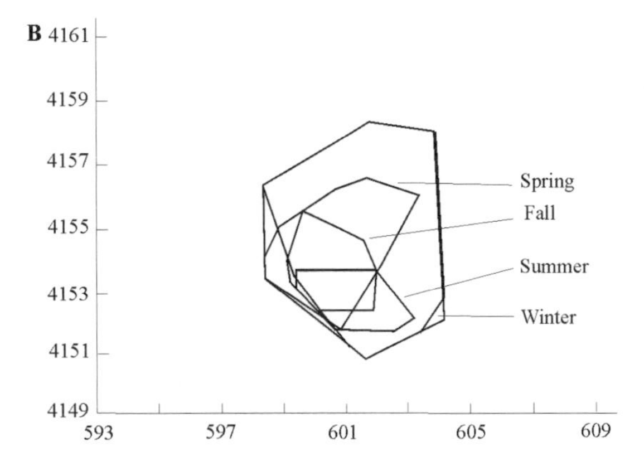

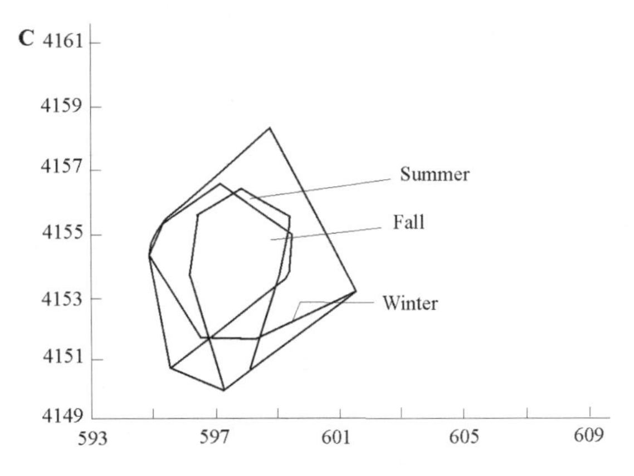

Figure 3. Seasonal[a] home ranges calculated by the 100% minimum convex polygon method for 3 adult swift fox in the intensive study area of the Piñon Canyon Maneuver Site, Colorado, 1986–1987. A) Female number 2; B) Male number 3; C) Male number 5. Outermost boundaries indicate the overall minimum convex polygon ranges for each individual.
[a] - Seasonal periods, Spring = March–May, Summer = June–August, Fall = September–November, and Winter = December–February.

Not surprisingly, mated pairs and family group members had home ranges that were similar. Pup movements initially were restricted to only a small portion of their parents' home range, but by September they appeared to be using most of their family group's range. Dispersal in this species has been reported as early as August (Kilgore 1969), but we observed no dispersal of pups until November and December.

Movement Data

Swift fox are assumed to be primarily monogamous, but some polygamy has been reported in both this species (Kilgore 1969) and in kit foxes (Egoscue 1962, 1975). If one mate dies, the surviving adult may move to another adult's home range, or stay and accept an ingressing mate. As a consequence, most movements outside of family group ranges probably are by the young of the year when they disperse or are forced out of their natal home range. We observed only 4 instances of ingress and egress; 3 of these involved pups and 1 involved an adult. Two juvenile females moved into the area and replaced mates lost to predation, and a single juvenile male presumably became the mate of a female that was occupying a territory with a male litter mate. One adult male moved outside of his previously known home range and into an adjacent territory when both his mate and the mate of the female in the adjoining territory were killed.

Population Dynamics and Mortality

Nineteen swift fox mortalities were recorded (Table 1); 12 fox (63%) were killed by coyotes, 2 were suspected of being killed by coyotes, 2 were killed by eagles, 1 was most likely killed by a badger (judging from tracks, fresh excavations, and fox remains at the den site), 1 was hit by a vehicle, and 1 pup died of unknown causes in a den. Pups suffered the highest mortality. Forty-two percent of the individuals that were classified as adults at initial capture survived at least 10 additional months (n = 9, SE = 0.21 months), whereas >50% of radiomarked pups were killed within 100 days of capture (Table 1). Estimated annual survival rate was 0.45 for adults, and the 11-month survival rate was 0.15 (a rate of 0.126 on an annual basis) for pups. Survivorship curves for both adults and juveniles beginning in June 1987 were similar to curves from 1986 (Fig. 4).

The high level of predation by coyotes that we documented is consistent with subsequent findings for swift fox in this same area in 1989–1991, when coyotes accounted for 85% of fox mortality (Covell 1992). In the closely related kit fox, Disney and Spiegel (1992) reported that coyotes and domestic dogs accounted for about 75% of all fox mortality on their developed study site. White et al. (1995) documented considerable overlap in habitat and food use by coyotes and kit foxes, and reported that 65% of all verified kit fox mortalities in their study were attributable to coyotes. Given the level of predation of coyotes on these arid-land foxes and the resource competition between coyotes and foxes (White et al. 1994), the potential exists for coyotes to suppress fox populations where densities of the former are high, especially in times of low prey availability.

In the present study, only 3 of 14 radiocollared pups were alive at the end of the study period. Eleven of 14 pup mortalities (including data for 3 pups that were not radiocollared) were caused or suspected of being caused by coyotes (Table 1). Death of the uncollared pups was inferred from the fact that they were not captured after their presumed father was killed by coyotes. Further, 7 of the 10 pups killed by coyotes in the study area were removed from the population before October; only 3 were killed in December and January. These data indicate that predation by coyotes was occurring well before hunting or trapping normally would have occurred had these activities been allowed on the study site.

In contrast to the low pup survival on the PCMS, 55 of 73 swift fox carcasses (75%) obtained from nearby areas outside the PCMS in southeastern Colorado after the 1986–87 winter were juveniles (Table 3). The preponderance of juveniles in this sample suggests much lower levels of predation by coyotes on juvenile foxes outside as compared to inside the PCMS, even if juvenile swift fox are substantially more vulnerable to harvest. As the density of the unexploited population of coyotes likely was higher on the PCMS than off the site, these circumstantial data underscore the negative correlation between coyote density and swift fox survivorship, especially for juvenile foxes.

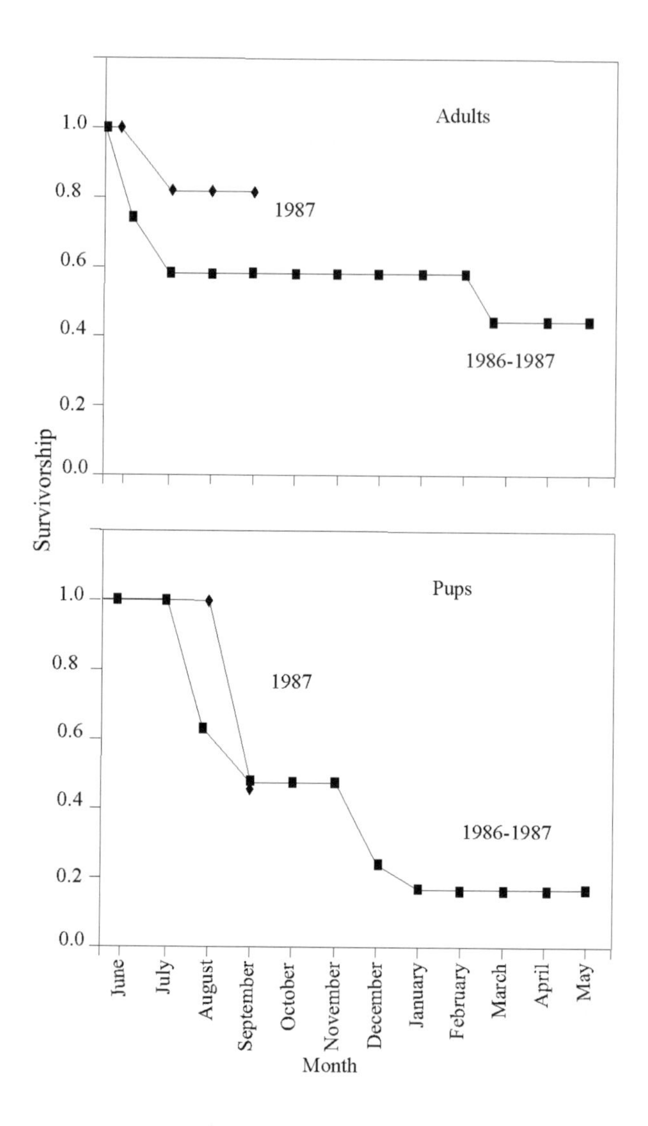

Figure 4. Staggered-entry Kaplan-Meier survival rate estimates for adult (n = 8) *and juvenile* (n = 14) *swift foxes on the intensive study area of the Piñon Canyon Maneuver Site in southeastern Colorado, 1986–1987.*

Litter size, estimated from litters that emerged at dens (n = 4 in 1986 and 1987) and from the number of fetuses in a female that was killed by a vehicle (1986), averaged 3.4 (range = 2–5). This estimate is conservative as mortalities may have occurred prior to the time that young emerged. Although mean litter sizes as large as 5 and 5.7 have been reported (Kilgore 1969), most populations average somewhat smaller litters (3.8—Covell 1992, 3.4—Hillman and Sharps 1978), and Scott-Brown et al. (1987) reported an average litter size of only 2.4 young in 37 litters at a captive breeding facility in Alberta. Based on

Table 3. Age of swift foxes harvested during the 1986–87 trapping season in southeastern Colorado, based on *cementum annuli* from canine teeth of skulls collected from trappers.

	Age (Years)				
Location	Pups	Yearling	2	3	4
Cheyenne Wells County	21	4	3	0	2
Springfield County	34	5	2	1	1
Total	55 (75%)	9 (12%)	5 (7%)	1 (1%)	3 (4%)

Table 4. Percent volume of food items found in swift fox scats ($n = 582$) by month 1986–1987 at the Piñon Canyon Maneuver Site, Colorado.

	Winter			Spring			Summer			Fall		Winter
Food Item	Jan (25)	Feb (34)	March (22)	Apr (30)	May (26)	June (88)	July (164)	August (67)	Sept (38)	Oct (31)	Nov (33)	Dec (24)
Cricetidae	42.3	0	0	35.5	9.4	0	3.2	6.1	0	2.2	11.8	17.3
Geomyidae[a]	4.1	36.7	24.3	22.5	11.3	20.3	19.2	22.3	3.0	7.6	17.1	10.1
Heteromyidae[b]	5.2	23.8	63.1	12.0	12.9	0	6.2	0	10.2	50.7	17.2	0
Sciuridae	0	0	0	0	1.3	3.3	6.3	0	0	0	0	0
Leporidae	2.4	10.5	0	0	0	1.3	10.5	7.6	4.1	2.1	9.2	12.4
Unidentified	9.7	7.6	7.1	0	6.1	22.2	2.6	2.2	15.3	31.2	24.0	39.7
Total Mammals	63.7	78.6	94.5	70.0	41.0	47.1	48.0	38.2	32.6	93.8	79.3	79.5
Birds	5.9	1.8	4.1	1.0	33.7	26.3	11.6	3.0	6.3	0	1.7	1.0
Arthropods	25.6	13.2	0.7	6.3	9.5	16.8	34.3	50.3	49.2	2.5	12.0	5.0
Vegetation	0	0.7	0	1.0	0	1.2	0.5	0	0	0	0	0.4
Soil	4.8	5.7	0.7	21.7	15.8	5.4	4.6	7.1	5.7	3.8	7.0	14.1
Reptiles	0	0	0	0	0	3.2	1.0	1.4	6.2	0	0	0

[a] Yellow-face pocket gopher *(Pappogeomys castanops)*.
[b] Silky pocket mouse *(Perognathus flavus)* and Ord's kangaroo rat *(Dipodomys ordii)*.

timing of emergence above ground (pups on our study area were born in early to late May), and assuming a gestation period of 51 days (Scott-Brown et al. 1987), breeding probably occurred between 1 and 15 March.

Although our sample sizes were small, pregnancy rates of swift fox on the PCMS were low. Whereas all 3 females captured on the area in 1986 were pregnant, only 2 of 5 females in 1987 produced young. No young were found for female number 23 (see Table 1 for animal identification numbers), a successful breeder in 1986, and females numbers 8 and 17 failed to rear young in 1987. Although these latter females were only 1-year old, both females and males are capable of breeding during their first year (Scott-Brown et al. 1986). Furthermore, female number 17 was lactating when captured on 26 June 1987, but it is unclear when her pups were lost. Female number 8 did not have a mate during 1987 when breeding normally would have occurred.

Food Habits

We examined 582 scats to identify prey items. Food items varied considerably among months, years, and family groups, but generally were comparable to previous food habits studies for both swift foxes (Cutter 1958, Zumbaugh et al. 1985, Uresk and Sharps 1986, Hines and Case 1991, also see Egoscue 1979) and kit foxes (White et al. 1996, and sources therein). Overall, mammals were the most frequent food item in scats (Table 4), making up >50% by volume of scats for 7 months. Both black-tailed jackrabbits (*Lepus californicus*) and desert cottontails (*Sylvilagus audubonii*) occur on the site, and, although we found only cottontail remains at den sites and made no attempt to identify lagomorph hairs to species in scats, swift foxes will prey on both genera of lagomorphs opportunistically (Cutter 1958). Insects were the second most frequent prey and accounted for up to 50% by volume of scats during August and September. However, the importance of insect prey probably is exaggerated in scat analysis because of the large proportion of indigestible chitin in arthropods (Scott-Brown et al. 1987). Like most previous studies, our results show that swift foxes will take birds opportunistically. Zumbaugh et al. (1985) reported that avian prey were present in almost 20% of stomachs they examined from foxes collected from trappers and fur dealers in Kansas, and Cutter (1958) found birds in 10% of scats examined from northern Texas. However, unlike the population of swift fox studied by Uresk and Sharps (1986) in South Dakota, where avian prey was an important food item throughout the year, swift fox on our study site preyed heavily on ground-nesting birds only during May and June, when avian material comprised 33% of scats. Soil appeared consistently in scats but was most common in April, May, and December, which may be a result of individuals ingesting soil while cleaning out or enlarging dens in anticipation of a litter. We observed this digging in 1987 at dens frequented by females, and noted that even unmated females sometimes dug additional entrances and enlarged dens.

In interpreting our data on food habits, one should bear in mind that volumetric analysis of scats can bias results and overestimate the importance of species with indigestible body parts such as hair, feathers, and exoskeletons, and underestimate highly digestible forms like soft-bodied invertebrates. Given that we, like previous studies based at least partially on analysis of stomach contents for this species (e.g., Cutter 1958, Kilgore 1969, Zumbaugh et al. 1985), found vertebrates to be the major prey items, we feel that our results are probably a conservative estimate of the importance of vertebrate prey because mammals and birds have a greater percentage of body mass that is easily digestible compared to arthropods, and foxes can selectively avoid indigestible components owing to the prey's larger size.

Den Activity

Swift fox on the PCMS spent most of the daylight hours in or very near a den but typically used multiple dens. Two individual foxes used >20 different dens each. Members of a pair often were found in the same den (45% of 214 locations for 4 females where both members of a pair were radiocollared) and they were more likely to be in the same den in the winter than in late summer. Male number 3 was in the den with his mate (number 2) all 12 times that she was located in January 1987. In contrast, 3 females were located in dens 22 times, but only twice with mates. Given that predation seems to be the cause of most mortality of swift fox on the PCMS, it may be that the more dens available throughout a pair's area of activity, the higher is the survival rate of that pair and their offspring (Waser 1980). Access to a den may be important in evading predation and swift fox spend most of their time in the vicinity of a den. Recent work by Wires (1995) suggests that visibility from the den location to allow foxes to detect approaching predators was a key feature in den use by kit foxes in California (see also Cypher and Spencer 1998), and this may also be important for swift fox.

Management Considerations

Although drastic range reductions have resulted in concern about the status of swift fox throughout much of their historic range, relatively little is known about their population dynamics. Our results show that coyote predation was a significant source of mortality in both adults and juveniles in a swift fox population in southeastern Colorado during a period when the coyote population on the study area was not being exploited through hunting or trapping (Gese et al. 1989). Samples of swift fox trapped in adjacent areas where coyote harvest was not restricted had higher proportions of juveniles than suggested by survival rates we found on the PCMS. Low adult and juvenile survival rates might necessitate immigration from surrounding areas to maintain the swift fox population on the PCMS. High coyote densities and predation rates might play a major role in limiting density and growth rates of swift fox populations. Availability and distribution of suitable dens and den sites may influence predation pressure, and may be important considerations in swift fox conservation.

Acknowledgments

We thank L. Carbyn, D.F. Covell, S.R. Emmons, L. Fox, L.L. Kinkel, and M. Sovada for critically reviewing previous versions of the manuscript, and W.R. Mytton, B.D. Rosenlund, T.L. Warren, S.R. Emmons, and A.R. Pfister for their help in coordinating and supporting this and other projects on the PCMS. O.J. Rongstad was instrumental in initiating this project on the PCMS, and provided guidance and direction. K.M. Firchow, and T. Kolash assisted in the field. Funding for this study was provided by the U.S. Army, Environment, Energy and Natural Resources Division, Fort Carson, Colorado, through the Colorado Fish and Wildlife Assistance Office and the Wisconsin Cooperative Wildlife Research Unit of the U.S. Fish and Wildlife Service (currently U.S. Geological Survey—Biological Resources Division). Support also was provided by the College of Agricultural and Life Sciences at the University of Wisconsin-Madison and the McGraw Wildlife Foundation.

Literature Cited

Bunker, C.D. 1940. The kit fox. Science 92:35–36.

Cameron, M.W. 1984. The swift fox (*Vulpes velox*) on the Pawnee National Grassland: its food habits, population dynamics, and ecology. Thesis, University of Northern Colorado, Greeley, Colorado.

Carbyn, L.N. 1986. Some observations on the behavior of swift foxes in reintroduction programs within the Canadian prairies. Alberta Naturalist 16:37–41.

Carbyn, L.N., H.J. Armbruster, and C. Mamo. 1994. The swift fox reintroduction program in Canada from 1983 to 1992. Pp. 247–271 *in* Restoration of Endangered Species, M.L. Bowles and C.J. Whelan, editors. Cambridge University Press, Cambridge, UK.

Chambers, G.D. 1978. Little fox on the prairie. Audubon 80:62–72.

Covell, D.F. 1992. Ecology of the swift fox (*Vulpes velox*) in southeastern Colorado, Thesis, University of Wisconsin, Madison, Wisconsin.

Cutter, W.L. 1958. Food habits of the swift fox in northern Texas. Journal of Mammalogy 39:527–532.

Cypher, B.L. and J.H. Scrivner. 1992. Coyote control to protect endangered San Joaquin kit foxes at the Naval Petroleum Reserves, California. Proceedings Vertebrate Pest Conference 15:42–47.

Cypher, B.L., and K.A. Spencer. 1998. Comparative interactions between coyotes and San Joaquin kit foxes. Journal of Mammalogy 79:204–214.

Disney, M. and L.A. Spiegel. 1992. Sources and rates of San Joaquin kit fox mortality in western Kern County, California. Transactions of the Western Section of the Wildlife Society 28:73–82.

Dixon, K.R., and J.R. Chapman. 1980. Harmonic mean measures of animal activity areas. Ecology 61:1040–1044.

Eberhardt, L.E., W.C. Hanson, J.L. Benston, R.A. Garrott, and E.E. Hanson. 1982. Arctic fox home range characteristics in an oil-development area. Journal of Wildlife Management 46:183–190.

Egoscue, H.J. 1962. Ecology and life history of the kit fox in Tooele County, Utah. Ecology 43:481–497.

——. 1975. Population dynamics of the kit fox in western Utah.

Bulletin of the Southern California Academy of Science 74:122–127.

——. 1979. *Vulpes velox*. Mammalian Species No. 122:1–5.

Firchow, K.M. 1986. Ecology of pronghorn on the Piñon Canyon Maneuver Site, Colorado. Thesis, Virginia Polytechnic Institute and State University, Blacksburg, Virginia.

Friedman, P.D. 1985. History and oral history studies of the Fort Carson-Piñon Canyon Maneuver Area, Las Animas County, Colorado. National Park Service/Rocky Mountain Region Report no. CX-1200-3-A066.

Gese, E.M. 1987. Ecology of coyotes in southeastern Colorado. Thesis, University of Wisconsin, Madison, Wisconsin.

Gese, E.M., O.J. Rongstad, and W.R. Mytton. 1989. Population dynamics of coyotes in southeastern Colorado. Journal of Wildlife Management 53:174–181.

Gittleman, J.L., and P.H. Harvey. 1982. Carnivore home-range size, metabolic needs and ecology. Behavioral Ecology and Sociobiology 10:57–63.

Glass, B.P. 1956. Status of the kit fox (*Vulpes velox*) in the high plains. Proceedings of the Oklahoma Academy of Science 37:162–163.

Hillman, C.N., and J.C. Sharps. 1978. Return of the swift fox to the northern Great Plains. Proceedings of the South Dakota Academy of Science 57:154–162.

Hines, T.D. 1980. An ecological study of *Vulpes velox* in Nebraska. Thesis, University of Nebraska, Omaha, Nebraska.

Hines, T.D., and R.M. Case. 1991. Diet, home range, movements, and activity periods of swift fox in Nebraska. Prairie Naturalist 23:131–138.

Kerwin, L. 1972. Kit fox in Saskatchewan. Blue Jay 30:200.

Kilgore, D.L. 1969. An ecological study of the swift fox (*Vulpes velox*) in the Oklahoma panhandle. American Midland Naturalist 81:512–534.

Linhardt, S.B., and W.B. Robinson. 1972. Some relative carnivore densities in an area under sustained coyote control. Journal of Mammalogy 53:880–884.

Long, C.A. 1965. The mammals of Wyoming. University of Kansas Publication, Lawrence, Kansas.

Loy, R.R. 1981. An ecological investigation of the swift fox (*Vulpes velox*) on the Pawnee National Grassland, Colorado. Thesis, University of Northern Colorado, Greeley, Colorado.

Martin, E. P., and G. F. Sternburg. 1955. A swift fox *Vulpes velox* (Say) from western Kansas. Transactions of Kansas Academy of Science 58:345–346.

McGrew, J.C. 1979. Distribution and habitat characteristics of the kit fox (*Vulpes macrotis*) in Utah. Thesis, Utah State University, Logan, Utah.

Mech, L.D. 1983. Handbook of animal radio-tracking. University of Minnesota Press, Minneapolis, Minnesota.

Miller, E.P., and C.J. McCoy. 1965. Kit fox in Colorado. Journal of Mammalogy 46:342–343.

Mohr, C.O. 1947. Table of equivalent populations of North American small mammals. American Midland Naturalist 37:233–249.

Morrel, S. 1972. Life history of the San Joaquin kit fox. California Fish and Game 58:162–174.

O' Neal, G.T., J.T. Flinders, and W.P. Clary. 1987. Behavioral ecology of the Nevada kit fox (*Vulpes macrotis nevadensis*) on a managed desert rangeland. Current Mammalogy 1:443–481.

Pfeifer, W.K., and E.A. Hibbard. 1970. A recent record of the swift fox (*Vulpes velox*) in North Dakota. Journal of Mammalogy 51:835.

Pollock, K.H., S.R. Winterstein, C.M. Bunck, and P.D. Curtis. 1989. Survival analysis in telemetry studies: the staggered entry design. Journal of Wildlife Management 53:7–15.

Ralls, K., and P.J. White. 1995. Predation on San Joaquin kit foxes by larger canids. Journal of Mammalogy 76:723–729.

Russell, R.H., and G.W. Scotter. 1984. Return of the native. Natural Canada 13:7–13.

Samuel, D.E., and B.B. Nelson. 1982. Foxes. Pp. 475-499 *in* J.A. Chapman, and G.A. Feldhamer, editors. Wild mammals of North America, Biology-Management-Economics. Johns Hopkins University Press, Baltimore, Maryland.

Scott-Brown, J.M., and S. Herrero, and C. Mamo. 1986. Monitoring released swift foxes in Alberta and Saskatchewan. Final Report 1986. Canadian Wildlife Service Report, Edmonton, Alberta.

Scott-Brown, J.M., S. Herrero, and J. Reynolds. 1987. Swift fox. Pp. 432–441 *in* M. Novak, J.A. Baker, M.E. Obbard, and B. Malloch, editors. Wild Furbearer Management and Conservation in North America. Ontario Trappers Association, North Bay, Ontario.

Seton, E.T. 1929. Lives of game animals. Volume 1, Part II. Cats, wolves, and foxes. Doubleday, Dorand, and Company, New York, New York.

Shaw, R.B., and V.E. Diersing. 1990. Tracked vehicle impacts on vegetation at the Piñon Canyon Maneuver Site, Colorado. Journal of Environmental Quality 19:234–243.

Southwood, T.R.E. 1966. Ecological methods with particular reference to the study of insect populations. Methuen, London, UK.

Spiegel, L.K. and M. Bradbury. 1992. Home range characteristics of the San Joaquin kit fox in western Kern County, California. Transactions of the Western Section of the Wildlife Society 28:83–92.

Uresk, D.W., and J.C. Sharps. 1986. Denning habitat and diet of the swift fox in western South Dakota. Great Basin Naturalist 46:249–253.

Waser, P.M. 1980. Small nocturnal carnivores: ecological studies in the Serengeti. African Journal of Ecology 18:167–185.

White, G.C., and R.A. Garrott. 1990. Analysis of wildlife radio-tracking data. Academic Press, San Diego, California.

White, P.J., K. Ralls, and R.A. Garrott. 1994. Coyote-kit fox interactions as revealed by telemetry. Canadian Journal of Zoology 72:1831–1836.

White, P.J., K. Ralls, and C.A. Vanderbilt White. 1995. Overlap between habitat and food use between coyotes and San Joaquin kit foxes. Southwestern Naturalist 40:342–349.

White, P.J., C. A. Vanderbilt White, and K. Ralls. 1996. Functional and numerical responses of kit foxes to a short-term decline in mammalian prey. Journal of Mammalogy 77:370–376.

Wires, L.R. 1995. Patterns of den site selection in the San Joaquin kit fox, *Vulpes macrotis mutica*, on the Carrizo Plain Natural Area, California. Thesis, University of Minnesota, St. Paul, Minnesota.

Worton, B.J. 1989. Kernel methods for estimating the utilization distribution in home-range studies. Ecology 70:164–168.

Young, S.P. 1944. Their history, life habits, economic status, and control. Part 1. Pp. 1–368 *in* The wolves of North America. American Wildlife Institute, Washington, D.C.

Zoellick, B.W., and N.S. Smith. 1986. Capturing desert kit foxes at dens with box traps. Wildlife Society Bulletin 14:284–286.

——. 1992. Size and spatial organization of home ranges of kit foxes in Arizona. Journal of Mammalogy 73:83–88.

Zumbaugh, D.M., J.R. Choate, and L.B. Fox. 1985. Winter food habits of the swift fox on the Central High Plains. Prairie Naturalist 17:41–47.

Home Range, Habitat Use, Litter Size, and Pup Dispersal of Swift Foxes in Two Distinct Landscapes of Western Kansas

■ Marsha A. Sovada, Christiane C. Slivinski, Robert O. Woodward and Michael L. Phillips

Abstract: We radiocollared 41 adult swift foxes (Vulpes velox) *and 25 pups to investigate home range size during breeding/pup-rearing (March–August), use of habitats, and pup dispersal for populations in 2 distinctly different landscapes in western Kansas. One study area was dominated by dryland crops (76%; hereafter Cropland Area), the other was predominantly pasture (87%; hereafter Rangeland Area). We did not detect difference in home range size of individual foxes (*P *= 0.58) or families (*P *= 0.60) between the 2 landscapes. The 95% adaptive kernal estimates (ADK) of home range size (breeding/pup-rearing period) of adult foxes averaged 15.9 km² (n = 21, SE = 1.6) overall. Core areas of use, as defined by the 50% ADK estimates ($\bar{x}$ = 2.2 km², SE = 0.3), were also similar in size between the 2 landscapes (P = 0.74). Family home range size averaged 17.8 km² (n = 8, SE = 2.1). There was minimal overlap between home ranges of adjacent fox families (16%) and core areas were nearly exclusive. Swift foxes primarily used fallow/stubble and small grain fields in the Cropland Area and grasslands in the Rangeland Area. Mean litter size of pups observed at emergence from dens was 3.1 (n =11, SE = 0.4). Five of 10 monitored pups dispersed (i.e., did not return to family home range). Average dispersal distance was 14.7 km (SE = 4.8). The average date that pups first moved from their family home range was 26 October (range = 1 Oct–3 Dec). The average dispersal date was 5 November (range = 1 Oct–27 Dec).*

Knowledge of swift fox home range and habitat use is essential to understanding their population dynamics and habitat features necessary to sustain populations. Variation in home range size across a landscape might reflect the capacity of the habitat to support a population (Macdonald 1983). Optimal habitat for swift foxes is believed to be shortgrass and mixed-grass prairie in gently rolling to level landscapes (Cutter 1958, Kilgore 1969, Hillman and Sharps 1978, Hines 1980). However, populations of swift foxes are occasionally found in what is considered atypical landscapes for swift foxes, such as mixed agricultural areas (Kilgore 1969, Hines 1980), and sagebrush steppe and shortgrass prairie transition (Olson and Lindzey 2002). Favorable areas for swift foxes provide holes for shelter and protection (Scott-Brown et al. 1987) and have low potential for contact with humans and predators, such as coyotes (*Canis latrans*, Hillman and Sharps 1978).

Many factors likely were responsible for the decline in swift fox numbers, including inadvertent poisoning (aimed at gray wolves [*Canis lupus*]), intensive trapping, and increasing numbers of predators (Sovada et al. 1998). Modification of native grassland and the associated decline in prey species are also implicated in the decline of swift foxes (Egoscue 1979). European settlers converted large expanses of prairie to cropland; less conversion occurred in drier shortgrass prairie than tall or mixed-grass prairies (Samson and Knopf 1994, Samson et al. 1998). Replacement of native grazers such as American bison (*Bison bison*) and prairie dogs (*Cynomys* spp.) by domestic cattle, which have different grazing impacts on the prairies (Schwartz and Ellis 1981), might also have affected swift foxes and their prey. Habitat alteration likely caused increased risk of predation on swift foxes where escape habitat was destroyed and could have affected spacing patterns of swift foxes, thus influencing their encounters with predators such as coyotes. Furthermore, the number of coyotes (primary predator of swift foxes) has increased since presettlement which is directly related to human activity (Johnson and Sargeant 1977, Sovada et al. 1998).

In western Kansas swift foxes occupy areas with large expanses of gently rolling grasslands, as well as extensive areas of cultivated land. This successful occupation of a mostly cultivated landscape is unusual in the current distribution of swift foxes in North America (*see* Swift Fox Conservation Team 1997). Fox and Roy (1995) suggested that dryland winter wheat/fallow rotation sustains swift foxes in this intensively cropped region of Kansas. It is not clear why mixed agricultural areas in other parts of the swift fox distribution are not supporting populations of swift foxes.

We had an opportunity to study swift fox ecology in their traditional grassland landscape and adjacent areas of highly cultivated cropland. Our objectives were to estimate home range size and evaluate habitat use by swift foxes in these 2 strikingly different landscapes in western Kansas. We wanted to determine if foxes were behaving and living differently in the 2 landscapes. That is, do foxes in cropland compensate in some way for this apparent stark environment in which they reside. Additionally, we report litter sizes, timing of dispersal, and dispersal distances for pups from both landscapes.

Study Areas

We studied populations of swift foxes on two 259-km^2 study areas in Sherman and Wallace counties of western Kansas during March 1996 through January 1997. The Cropland Area was relatively flat, approximately 76% cultivated fields (Fig. 1). Most fields were in a dryland winter wheat–fallow rotation; others were irrigated corn and sunflowers, milo, and sorghum. Ten percent of the land was enrolled in the U.S. Department of the Agriculture's Conservation Reserve Program (CRP) and seeded primarily to big bluestem (*Andropogon gerardii*), Indian grass (*Sorghastrum nutans*), and switch grass (*Panicium virgatum*). There were 125 km of roads and 20 occupied residences. The Rangeland Area was characterized by rolling hills, approximately 87% moderately to heavily grazed native pasture. Primary grasses included buffalo grass (*Buchloe dactyloides*), blue grama (*Bouteloua gracilis*), and hairy grama (*B. hirsuta*). A few cropland fields (wheat, sunflowers) were interspersed in the Rangeland Area. There were 66 km of roads and 10 occupied residences. Both study areas had few trees (<1% of landscape).

Annual precipitation was 50.5 cm (long-term average = 46.2 cm) and occurred primarily in spring and summer. Temperatures were characteristic of continental climate, with January the coldest month ($\bar{x}$ = -2.6°C, long-term average = -2.0°C) and July the warmest ($\bar{x}$ = 23.1°C; long-term average = 24.2°C; National Climate Data Center 1997).

Methods

We used 81-cm× 25-cm× 30-cm live traps to capture adult swift foxes beginning in March 1996. Traps were modified to a smaller mesh (2.5 × 2.5 cm) to reduce chance of injury to captured foxes. Traps first were placed within 3 randomly selected 10-km^2 blocks in each study area; blocks were separated from each other and study area boundaries by ≥1.6 km. Trapping effort expanded outward from the 10-km^2 blocks as capture of unmarked foxes within a block subsided. Our goal was to radiocollar 25 adult swift foxes in each study area. With the same traps, we began capturing juvenile foxes in August 1996 near dens of radiocollared adults. We attempted to capture as many pups as possible. Numbered eartags and a 39-g radiocollar detectable up to 3.2 km and containing a mortality sensor were attached to each captured swift fox. Gender, reproductive status (e.g., lactating), and general health were recorded for each fox. All foxes (adults and pups) alive in late January 1997 were recaptured and collars were removed. We noted any fur wear or abrasion caused by collars for all recaptured foxes.

We monitored radiocollared adult swift foxes from March 1996 through January 1997 from vehicles equipped with null-peak directional antennas. Foxes were systematically monitored for ≥4 nights every 2 weeks on each study area, and never on consecutive nights on the same area. We monitored foxes in two 6-hr nighttime periods, approximately 1900–0100 hr and 0100–0700 hr, with equal number of sampling periods in each 2-week period. Our goal was to locate each monitored fox 2 or 3 times in a 6-hr period, with locations for individual foxes ≥2 hours apart. We also attempted to obtain daytime (0700-1900 hr) locations of all adults approximately every 3 days, not coinciding with nighttime monitoring. When a den location was determined, we recorded the habitat the den was in. We attempted to locate each radiomarked pup 2 or 3 times per week.

Permanent tracking stations were established at every road/trail intersection, along roads at 0.4 km intervals, and at areas of higher elevation affording improved reception. Universal Transverse Mercator (UTM) coordinates were predetermined for each station with a Precision Lightweight Geographic Positioning System receiver (Rockwell International, Cedar Rapids, Iowa). Each fox location was estimated from bearings taken in rapid succession from 2–4 permanent tracking stations by one observer. To reduce error of the bearing angle, orientation of the tracking vehicle was determined at each station using a compass and the bearing was adjusted accordingly. Animal locations and 95% error ellipses were estimated using LOCATE II software (Nams 1990). Locations based on 2 bearings were estimated using a fixed standard deviation determined for each crew member based on field tests given early in the field season.

To reduce error in telemetry locations, we eliminated locations that were ≤3 km from a tracking station. We included locations 3 km from a tracking station only if the angle of intersection of bearings was >45° and <135°. The interval between first and last bearings for each triangulation was typically ≤3 minutes. Locations with areal error estimates >50 ha (LOCATE II) were eliminated from analysis during recording or processing of data.

Home Range Estimation

We estimated home range size and core area (defined below) with CALHOME software (Kie et al. 1996) based on nighttime location coordinates collected during the breeding/pup-rearing period (March–August). We considered only this period for home range estimation and analyses because sample sizes were insufficient for other periods. We calculated the adaptive kernel (ADK) estimates (Worton 1989, Gallerani Lawson and Rogers 1997) of home range. All analyses of home range were conducted with 95% ADK estimates. Core area was defined by the 50% ADK estimate. We also provide minimum convex polygon (MCP; Mohr 1947, White and Garrott 1990) estimates of home range to permit comparison with home

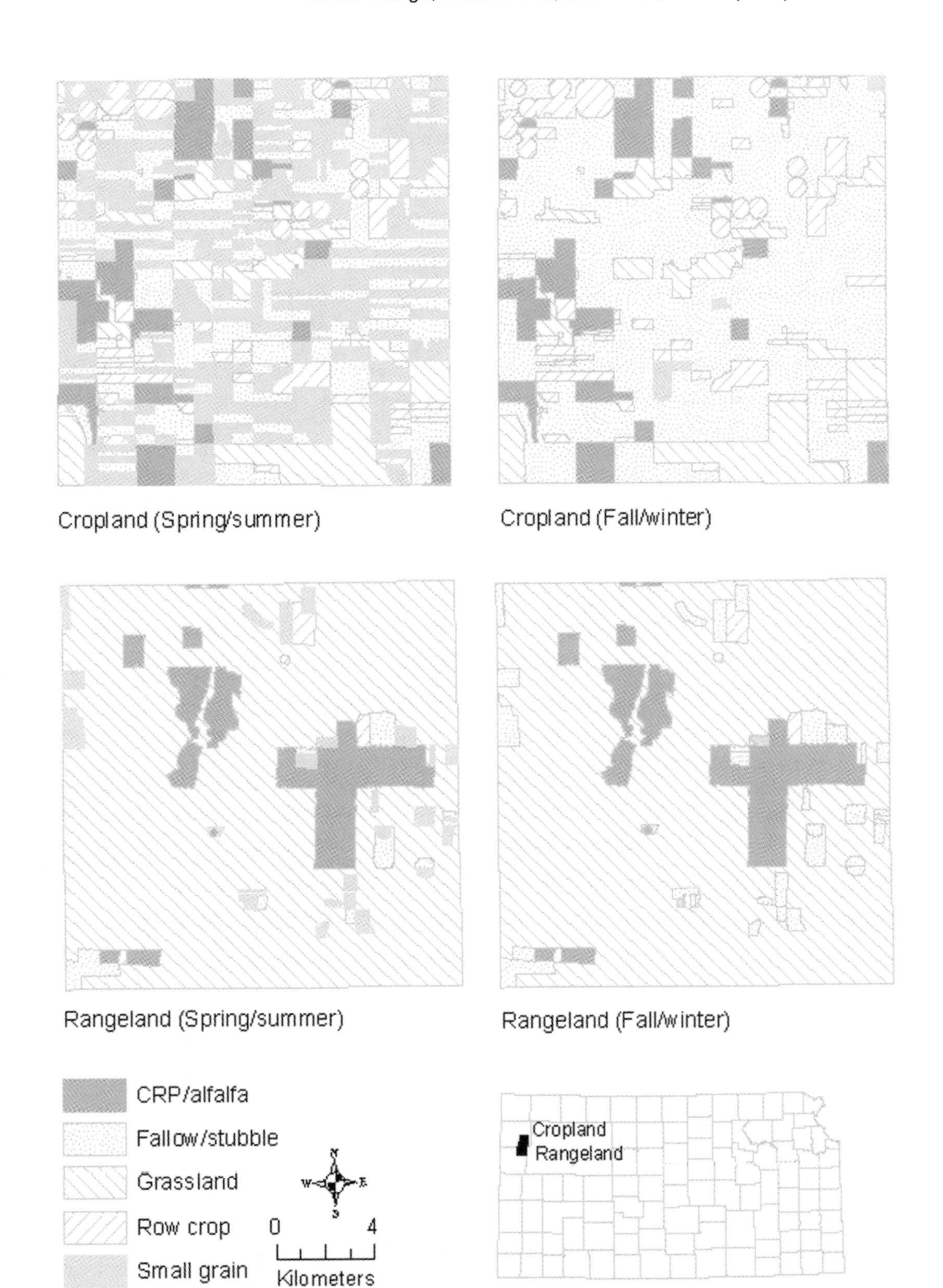

Figure 1. Cropland and Rangeland study areas (each 259 km²) in Sherman and Wallace counties of western Kansas with habitats delineated for 2 periods of the study, spring/summer (1 March–15 July) and fall/winter (16 July–31 January). UTM coordinates for the northwest corner of each study area are: Cropland—4355640 N, 273068 E; Rangeland—4336794 N, 269121 E.

range estimates among published studies of swift foxes (Andersen et al. 2002, Hines and Case 1991, Pechacek et al. 2000, Zimmerman 1998). MCP estimates are less suitable biologically as a descriptive statistic (White and Garrott 1990:148).

INDIVIDUALS: For individual foxes, we estimated home range and core area for animals with ≥60 locations and monitored throughout the breeding/pup-rearing period. Effect of landscape composition (Cropland Area vs. Rangeland Area) and gender on home range size were evaluated with analysis of variance (Proc GLM; SAS Institute 1988). For the analyses, we weighted the size estimates by the square root of the number of locations to compensate for differences in sample sizes. Within pairs, we examined variation in home-range size between male and female members of each pair with a paired t-test.

FAMILIES: We identified swift fox families based on use of shared dens and observations of radiomarked foxes. We estimated home range and core areas used by swift fox families that met the criteria of ≥60 locations collected throughout the breeding/pup-rearing period. We evaluated effect of landscape composition on home range size with analysis of variance (Proc GLM; SAS Institute 1988), weighted by the square root of the number of locations to compensate for differences in sample sizes.

We estimated overlap area between home ranges of adjacent swift fox families with a Geographic Information System (GIS; Arc/View, Environmental Systems Research Institute, Redlands, CA). Overlap was calculated as the mean of the product of the ratios of overlap size to overall home range size (Minta 1992, 1993). Overlap values potentially range from 0 to 1, with a value of 1 indicating adjacent home ranges of identical size with 100% overlap. We converted these values to percentages.

Habitat Use

We delineated habitats of study areas (and surrounding areas when needed for analysis) from aerial photography (1:40,000) entered into GIS. We generated databases for habitat composition of each study area for each of 2 periods (spring/summer: 1 March–15 July; fall/winter: 16 July–31 January). We specified these periods because wheat harvest was completed about 15 July, which marked a significant and sudden change in the landscape and habitats available to foxes. Verification of habitat determinations was conducted by ground surveillance in mid-June and again in early September. Habitat classes, based on vegetation structure were defined as: row crop (corn, sunflowers, sorghum), small grain (wheat, milo), stubble/fallow, grassland (pasture), CRP/alfalfa, and other (e.g., farmyards). Swift foxes were never found in the few farmyards on the study areas, therefore we did not include this habitat class in the analyses. We did not include roads as a habitat class and the adjacent habitat on either side of roads was assigned to mid-road.

We defined available habitats based on a series of concentric circles (i.e., buffers: 1000, 2000, and 3000-meter radius) around the mean UTM location for each animal (see Phillips et al., In press). The buffering technique avoided the assumptions about availability of habitat based on home ranges that are determined by a prior selection of habitat (White and Garrott 1990). Nonrandom use was evaluated using multivariate analysis of variance (Rencher 1995) for each of the buffer distances. The smallest buffer distance up to 3000 m that indicated nonrandom use by swift foxes was used in the compositional analysis.

We used compositional analysis (Aitchison 1982, Aebischer et al. 1993) to examine habitat selection by adult swift foxes within each study area in spring/summer and fall/winter. The statistical technique treats individual foxes as the experimental unit rather than each location. Because sample sizes of locations were not the same for all animals, we weighted the differences in proportions by the square root of the number of locations (Aebischer et al. 1993). Given nonrandom habitat use, we then ranked habitat classes by making pairwise comparisons between all habitat classes for all animals and tested for differences among habitat classes. Threshold level of selection was determined by the inverse of the number of habitat categories.

For habitats with no detected use, we replaced the habitat's proportion of use (zero) with the value 0.0001 (spring/summer) or 0.000001 (fall/winter), depending on the smallest detected proportion of use. The replacement value was an order of magnitude smaller than the smallest observed proportion and thus, preserves the meaning of no use in the analysis (Aebischer et al. 1993). Similarly, for habitats that were not available to an individual fox (occurred only 8 of 210 times), we substituted the value 0.0001 (spring/summer) or 0.000001 (fall/winter). By using this substitution, we assumed that the individual fox responded to that habitat at random.

Use of Roads

We examined distance of locations to roads as a measure of the use of roads as travel lanes or foraging sites. The distance to a road was determined for every point from the nighttime telemetry locations within the buffer distance (3000 m, see Results) resulting from our analysis of habitat selection for individual foxes. We generated files with all possible coordinates (>10,000 points) within each fox's 3000 m buffer. The output cell size was set at 7 meters, which is the same cell size as the original habitat grid developed from the GIS. We then computed the mean distance from each generated point to the nearest road within the 3000 m buffer. Comparisons between the mean distance to roads for individual fox locations and the mean distance to roads for all possible generated points were made with a paired t-test.

Litter Size and Dispersal

Litter size was determined through observations at den sites at the time of emergence. Capture and radiomarking of pups was not attempted until pups were nearly full grown. We defined the onset of dispersal as when a pup left its family home range, without subsequent return movements. Pup locations were reviewed chronologically in relation to the family's 95% ADK home range. The date of dispersal was determined as the median of the date the pup was last known to be within the family home range and the date it was first known to be continually outside the family home range (Koopman et al. 2000). We measured dispersal distances from the point of capture to the furthest location.

Results

We radiocollared 41 adult foxes, 19 in the Cropland Area and 22 in Rangeland Area. Three of the 41 animals either died (1 killed by coyote) or their radios failed (evidenced by signal abnormality just prior to signal loss and observations of collared fox with no signal) within a week of capture. One male was captured in the Cropland Area, but was most frequently found in areas adjacent to that study area. Habitat composition where it resided was more similar to the Rangeland Area than the Cropland Area, therefore we assigned that fox to Rangeland Area for analyses (Fig. 2, see F8).

We radiocollared 25 pups (12 females, 13 males) from families with a radiomarked adult. One female pup captured on 31 October was not positively associated with a marked adult, but we assigned her to the family we assumed she belonged to. We gathered an average of 2.7 locations per week for each monitored pup. Of the 25 pups, 14 died, 1 radio failed, and 10 were tracked until the end of January when collars were removed (see Sovada et al. 1998 for mortality rates and causes). At recapture, all pups and adults were in good health and there was no evidence that collars caused adverse wear on the neck. Only slight hair matting was evident.

Home Range

INDIVIDUALS: Data from 11 (5 females, 6 males) adult foxes in the Cropland Area and 10 (4 females, 6 males) in the Rangeland Area met criteria for home range estimation. We observed that foxes could move >5 km in 1.5–2 hours, the interval between our locations, thus we assumed successive locations for individual foxes were independent (Swihart and Slade 1985). MCP results are given in Table 1.

The mean home range size (95% ADK) for adult foxes was 15.9 km^2 (SE = 1.6) and did not vary between the Cropland Area ($\bar{x}$ = 16.8 km^2, SE = 2.4) and the Rangeland Area ($\bar{x}$ = 15.0 km^2, SE = 2.1; $F_{1,19}$ = 0.32 P = 0.58; Table 1). We detected no effect of gender overall ($\bar{x}$ = 14.1, SE =2.2 [females], $\bar{x}$ = 17.2, SE =2.1 [males]; $F_{1,19}$ = 1.04, P = 0.32), and there was no interaction between area and gender ($F_{1,17}$ = 1.20, P = 0.29). MCP estimates of home range are given in Table 1.

The mean core area size (50% ADK) for adult foxes was 2.2 km^2 (SE = 0.3) and did not differ between the Cropland Area ($\bar{x}$ = 2.3 km^2, SE = 0.4) and the Rangeland Area ($\bar{x}$ = 2.1 km^2, SE = 0.3; $F_{1,19}$ = 0.11, P = 0.74; Table 1). Female core areas overall tended to be smaller ($\bar{x}$ = 1.6, SE = 0.3 [females], $\bar{x}$ = 2.7, SE = 0.4 [males]; $F_{1,19}$ = 4.20, P = 0.06). There was no interaction between area and gender ($F_{1,17}$ = 0.37, P = 0.55) on core area size.

FAMILIES: We identified 10 families in the Cropland Area and 11 families in the Rangeland Area (Fig. 2). Of the 21 families, 7 lost a pair member early in the study; 3 of the 7 surviving pair members associated with a new mate. When the new pair bonds were identified, location data for these foxes were assigned to the new family and home range estimations were calculated accordingly. Radiocollars failed on females of 2 families (F4 and F8, Fig. 2), but pups were present and both females were observed

Table 1. Average 95% and 50% adaptive kernal (ADK) estimates and 100% minimum convex polygon (MCP) estimates of home range size (km ^{2}) during the breeding/pup-rearing period (March–August), and standard errors for female and male swift foxes, all foxes combined, and swift fox families residing in Cropland and Rangeland study areas and overall in western Kansas.

	n	$\bar{x}$ no. locations	95% ADK $\bar{x}$	95% ADK SE	50% ADK $\bar{x}$	50% ADK SE	100% MCP $\bar{x}$	100% MCP SE
Cropland								
Female	5	176	13.1	2.2	1.6	0.4	16.5	4.3
Male	6	176	19.7	3.7	2.9	0.6	23.6	3.7
All foxes	11	176	16.8	2.4	2.4	0.4	20.4	2.9
Families	5	279	18.7	3.1	2.5	0.3	25.3	4.1
Rangeland								
Female	4	193	15.2	4.6	1.7	0.4	19.1	6.9
Male	6	206	14.9	2.1	2.4	0.4	24.2	1.2
All foxes	10	201	15.0	2.1	2.1	0.3	22.3	2.7
Families	3	270	16.3	2.6	2.9	0.6	23.2	1.4
Overall								
Female	9	184	14.1	2.3	1.6	0.3	17.6	3.6
Male	12	191	17.2	2.1	2.7	0.4	23.9	1.9
All foxes	21	188	15.9	1.6	2.2	0.3	21.3	1.9
Families	8	276	17.8	2.1	2.7	0.3	24.5	2.5

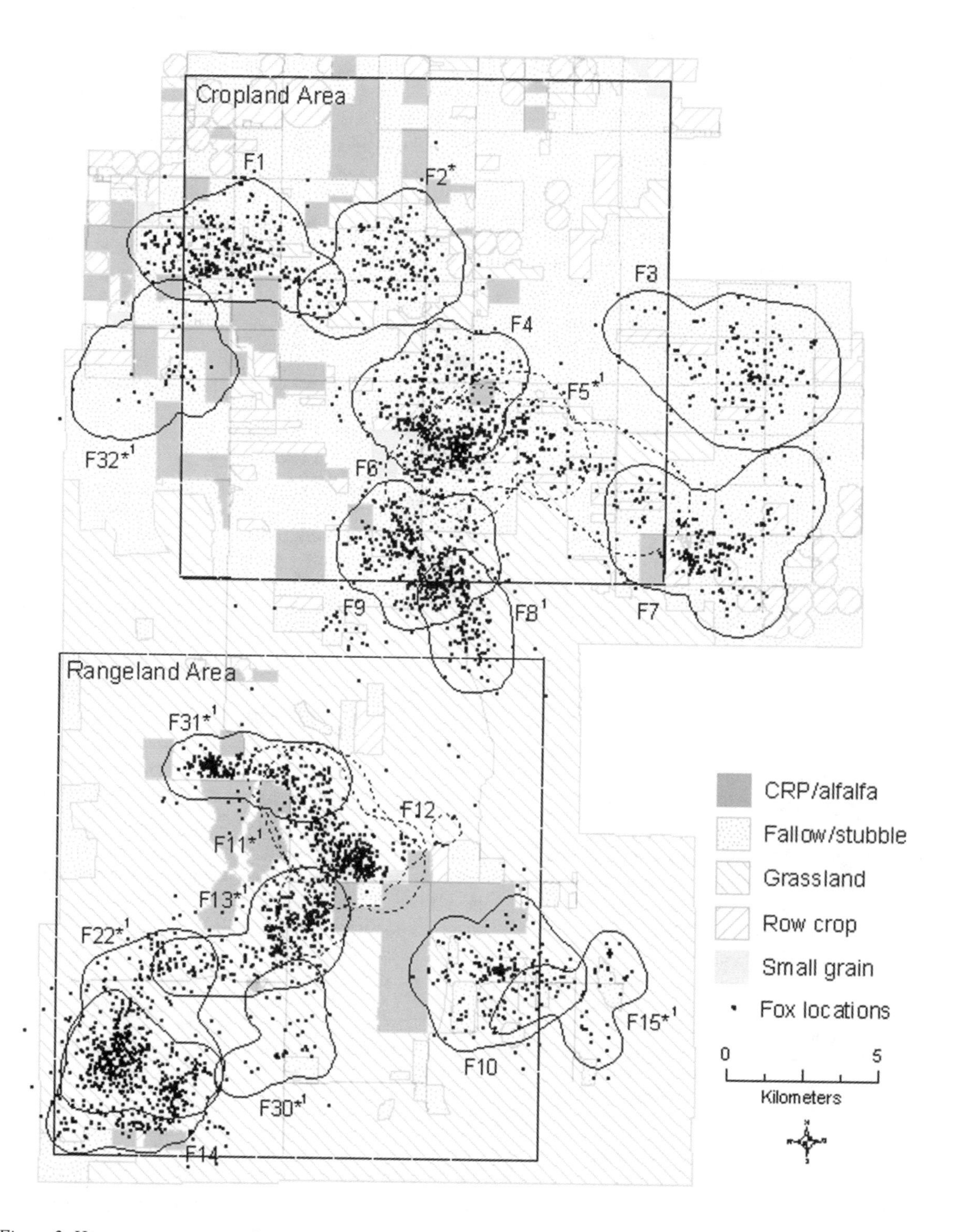

*Figure 2. Home ranges, estimated with the 95% adaptive kernal method, of swift fox families in the Cropland and Rangeland study areas of western Kansas, March 1996 through January 1997. Superscript 1 indicates only 1 member of the pair was monitored and * indicates data for the family did not meet the criteria (60 locations and monitored throughout the breeding/pup- rearing period) for analyses. Families 5 and 6 in Cropland and 11 and 12 in Rangeland (dashed home ranges) each had a shared pair member, i.e., a mate died in 1 family and the remaining pair member associated with a new fox forming a new family unit. Habitats in study areas and outside the study area were identified for the fall/winter (July–31 January) period of the study. Note: T22, this female's mate died in April and she did not attach to a new male and did not appear to be a helper to Family 14.*

throughout the study. We considered these to be stable functioning family units containing a pair and pups. We used telemetry locations from the males to estimate area used, however these estimates are not included in analyses of home range size. We did not know the status of one pair member of 3 families. One family included a second female (yearling) that seemed to be a "helper" female, because she remained in the family home range and occasionally shared a den with the adults (Kilgore 1969, Covell 1992).

For estimation of family home range size (95% ADK), criteria were met by 5 families in the Cropland Area ($\bar{x}$ = 18.7 km², SE = 3.1) and 3 in the Rangeland Area ($\bar{x}$ = 16.3 km², SE = 2.6; Table 1). Home range size of families was similar between the Cropland and Rangeland areas ($F_{1,6}$ = 0.30, P = 0.60). The mean home range size for all families was 17.8 km² (n = 8, SE = 2.1). The size of core area (50% ADK) used by families was also similar between the Cropland Area ($\bar{x}$ = 2.5 km², SE = 0.3) and the Rangeland Area ($\bar{x}$ = 2.9 km², SE = 0.6, $F_{1,6}$ = 0.26, P = 0.63). For 8 families with sufficient data, we detected no difference in home range size between females ($\bar{x}$ = 15.0 km², SE = 2.4) and males ($\bar{x}$ = 17.0 km², SE = 1.9) within families (t_7 = 0.67, P = 0.52). The females tended to have smaller core areas ($\bar{x}$ = 1.6, SE = 0.3) than males ($\bar{x}$ = 2.5 km², SE = 0.3; t_7 = -2.13, P = 0.07).

OVERLAP OF FAMILY HOME RANGES: Our data included 5 pairs of families with adjacent home ranges appropriate (fit criteria for home range estimation) for evaluating home-range overlap. The mean overlap of home ranges was 16% (range 0–32%, SE = 6) and 2 of the 5 families had core areas that overlapped (3% and 18%).

Habitat Selection

We observed consistent nonrandom habitat use by swift foxes in both landscapes in both time periods only within the 3000-m buffered area (P <0.10; Table 2). This

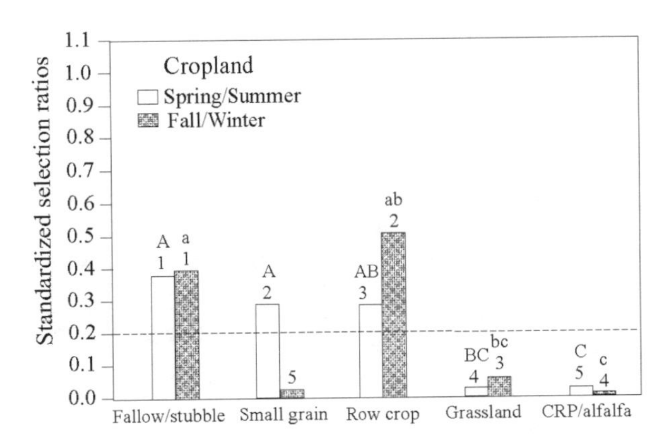

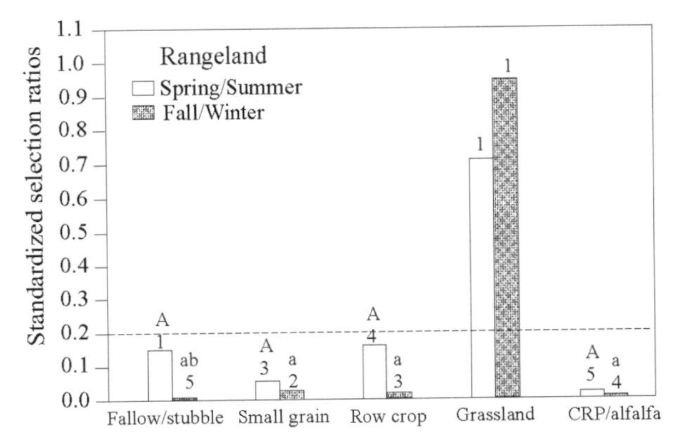

Figure 3. Standardized selection ratios for swift foxes on Cropland and Rangeland study areas in western Kansas in 2 periods (spring/summer: 1 March–15 July; fall/winter: 16 July–31 January) in 1996–1997. Numbers above the histogram bars are the rank of each habitat within the period. Habitats with the same letter above the bar are not significantly different from each other (P >0.05) within each period. The dashed line represents a threshold level of selection determined by the inverse of the number of habitat categories (i.e., 1/5 = 0.2).

Table 2. Results of MANOVA tests evaluating habitat selection by swift foxes at each buffer distance.

	Buffer Distance (m)		
	1000	2000	3000
Spring/Summer			
Cropland			
Fox (n=13)			
Wilks' Λ	0.545	0.634	0.289
P	0.199	0.340	0.016
Rangeland			
Fox (n=11)			
Wilks' Λ	0.646	0.329	0.239
P	0.484	0.068	0.025
Fall/Winter			
Cropland			
Fox (n=10)			
Wilks' Λ	0.695	0.404	0.280
P	0.643	0.184	0.069
Rangeland			
Fox (n=8)			
Wilks' Λ	0.550	0.214	0.104
P	0.355	0.118	0.030

MIKE BLAIR, KANSAS DEPT. OF WILDLIFE AND PARKS

Swift fox den in the Cropland Study Area.

MIKE BLAIR, KANSAS DEPT. OF WILDLIFE AND PARKS

Swift fox den in the Rangeland Study Area.

area included 88% of locations in spring/summer and 95% of locations in fall/winter. During spring/summer in the Cropland Area, fallow/stubble, small grain, and row crop where the highest ranked habitats in their use by foxes, but these habitats did not differ from each other (Fig. 3, $P > 0.10$). Grassland and CRP/alfalfa were the lowest ranked habitats used by foxes and fell below the selection threshold. They were selected less than fallow/stubble, small grain, and row crop. In the Rangeland Area in spring/summer, grassland had the highest ranked selection factor and was selected more often than all other habitats (Fig. 3, $P < 0.01$). Fallow/stubble, small grain, row crop, and CRP/alfalfa were not different from each other ($P > 0.33$–0.88).

During fall/winter in the Cropland Area, fallow/stubble ranked first in selection by foxes among habitats (Fig. 3), followed by row crop, grassland, CRP/alfalfa, and small grain. No difference was detected in relative use of fallow/stubble and row crop ($P = 0.76$) or row crop and grassland ($P = 0.08$), but there was a difference between fallow/stubble and grassland ($P = 0.05$). Foxes in the Rangeland Area in fall/winter selected habitats similarly to that of spring/summer; grassland had the highest rank and was different from all other habitat categories ($P < 0.02$).

Daytime telemetry revealed an average of 6.1 dens/family in the Cropland Area and 5.6 dens/family in the Rangeland Area during the spring/summer however, we mightn't have located all dens based on our sampling schedule. In the Cropland Area, dens were predominately located in fallow/stubble habitat (31 of 43 dens) and in the Rangeland dens were largely in grassland (28 of 34 dens). Additional information regarding dens on the study site can be found in Jackson and Choate (2000).

Use of Roads

The distances of fox locations from roads were not different from random locations ($P > 0.17$), although there was variability among foxes and no clear pattern was evident. Male foxes in the Rangeland Area tended to be located closer to roads than expected ($t_7 = 2.08$, $P = 0.08$).

Litter Size and Dispersal

Ten families raised ≥1 pup (post-emergence average litter size = 3.1, SE = 0.4) with an average of 3.1 pups (n = 7, SE = 0.5) in the Cropland Area and 3.0 pups (n = 3, SE = 0.7) in the Rangeland Area. Three females died prior to whelping, revealing 7, 4, and 3 embryos upon necropsy and one female that died post-partum showed 3 recent placental scars. Information on pups was unknown for 2 families, and 2 others probably did not have pups.

Average dispersal distance was 14.7 km (SE = 4.8) for 10 pups (4 Rangeland Area, 6 Cropland Area). Of these, a female and 2 males dispersed ≥20 km (20, 22, 32 km respectively) from their family home ranges. Two female pups dispersed, but stayed near (5 and 6 km) the edge of their family home range. Two male pups moved from the family home range (3 and 5 km), but continued occasional use of the periphery of the family home range. One female pup remained in the family home range, and 2 other females appeared to remain in the area used by their parents, but we had insufficient data to adequately estimate their family home range.

The average date pups first left the family home range was 26 October (n = 10, range = 1 Oct–3 Dec). The average date pups dispersed from their family home range was 5 November (n = 5, range = 1 Oct–27 Dec).

Discussion

The average home range size of 15.9 km² that we observed for swift foxes was midrange compared to other studies. However, other estimates are quite variable and difficult to compare because of differences in estimation techniques, criteria used, and periods of evaluation (Gallerani Lawson and Rodgers 1997). Using the same 95% ADK method as we used, Kitchen et al. (1999) reported an average home range size of 7.6 km² for swift foxes (>30 animals and >60 locations per breeding, rearing, and dispersal seasons) on the Piñon Canyon Maneuver Site in southeastern Colorado during 1997–98. Pechacek et al. (2000) estimated home range sizes of 11.7 km^2 (95% ADK) and 7.7 km^2 (100% MCP) for 10 swift foxes in southeastern Wyoming. Hines and Case (1991) reported an average MCP home range size of 32.3 km^2 (range 7.7–79.3 km^2) for 7 swift foxes in Nebraska, which is larger than our MCP home range estimate ($\bar{x}$ = 21.3 km^2). However, >50% of their study animals were followed ≤5 nights and only in winter or very early spring. Andersen et al. (2002) reported a similar average MCP home range size (29.0 km^2, range 12.8–34.3 km^2) on the Piñon Canyon Maneuver Site (1986–87), reporting on 5 swift foxes with 34 locations ($\bar{x}$ = 188 locations) over a minimum period of 7 months. Zimmerman et al. (2002) estimated an average MCP home range size of 10.4 km^2 (range 7.3 to 16.9 km^2) for 5 swift foxes in Montana.

Swift foxes use a broad array of plant and animal foods (Hines and Case 1991, Kitchen et al. 1999, Sovada et al. 2001), and there is no evidence that food is a limiting factor. Initially, we had expected home range size would be larger in the Cropland Area than the Rangeland Area based on different types and distribution of food resources. Spiegel (1996) suggested that spacing patterns for kit foxes (*Vulpes macrotis*) are influenced by differences in fox density, and type and abundance of prey. However, Sovada et al. (2001) reported no fundamental differences in foods used by foxes between cropland areas and rangeland areas. Although there are no data on abundance of prey, we can assume similar (or more than adequate) abundance of prey based on similar diet and similar home ranges. Therefore, there must be some other factors influencing spacing patterns. However, without long-term studies, it is difficult to assess the impact of abundance of prey on fox densities.

Similar to Spiegel (1996), Sargeant (1972) indicated that territory size for red foxes (*Vulpes vulpes*) is a reflection of population density and he further proposed that red foxes have innate minimum and maximum spatial requirement. Studies of red fox spatial patterns confirm this suggestion (Sargeant et al. 1975, Trewhella et al. 1988) and our findings suggest the same might be true for swift foxes. Because food resources are not limiting, except perhaps under extreme environmental conditions, swift fox spacing patterns might be more related to fox density on our study areas. Other accounts in the literature suggest fox densities might be influenced by factors such as disease (Lindström 1991), human inflicted mortality (Sargeant 1982), and interspecific competition (Voigt and Earle 1983, Sargeant et al. 1987, Harrison et al. 1989). The swift fox populations in this study suffered relatively high mortality due to predation by coyotes (Sovada et al. 1998), and thus may be at a lower density than the landscape can actually support. Further research is necessary to determine mechanisms regulating densities.

Earlier studies suggested that swift foxes are not territorial (Cameron 1984, Hines 1980); however more recent data, including ours, provide evidence to the contrary. Andersen et al. (2002) reported nearly total exclusion of an individual swift fox's core area (50% MCP estimates) from other same-sex individuals. Pechacek et al. (2000) found that the area used by mated pairs had minimal overlap with areas used by adjacent pairs (95% ADK).

Although considered a hallmark species of mixed-grass and shortgrass prairie, swift foxes have adapted to a variety of habitats (e.g., Kilgore 1969, Olson and Lindzey 2002). Results of this study suggest that swift foxes were well adapted to use fallow, stubble, and small grain fields in landscapes dominated by dryland cropping practices (winter wheat–fallow rotation) and minimal grassland habitat. In the cropland landscape, fallow/stubble fields likely provide excellent habitat for foraging and denning and we expected and recorded significant use of this habitat. We had expected foxes to use grasslands more than they did the cropland landscape despite the limited availability of this habitat. The grassland, not surprisingly, was the selected habitat in areas where rangeland dominated the landscape.

We expected the low selection of CRP/alfalfa. Several fox family home ranges bordered or included CRP fields, but we seldom located foxes using the CRP field (see Fig. 2). CRP lands in western Kansas were originally seeded to tall grass species that are not native to the area, primarily big blue stem, Indian grass, and switch grass. Swift foxes prefer open plains habitat that allows relatively good visibility to detect potential predators (Kilgore 1969, Hines 1980), thus CRP likely provided little security for the foxes. For conservation of the species in Kansas and similar areas, we recommend CRP lands be planted to native shortgrass and mixed-grass species rather than the currently applied tall grass species. If the landowners are given incentives under the CRP provisions, landowners might be more likely to plant native shortgrass and mixed-grass species that benefit swift foxes and other native wildlife species.

Although we do not have sufficient data to closely examine use of row crops during various stages of growth, we believe that foxes avoided these fields when their visibility was obstructed by the vegetation. Foxes were located in row crop habitat mostly (80% of row crop locations) when the crops were very short (prior to July) or after harvest (after September).

We did not observe frequent use of roads by foxes. However, other studies have indicated that swift fox movements and den locations are associated with roads (Hines and Case 1991, Pruss 1999). Road associations were a major source of vehicle-related mortality for juvenile foxes in the Cropland Area where there was more vehicle traffic (Sovada et al. 1998). Adult foxes seemed to be more cautious of roads.

We found that litter sizes ($\bar{x}$ = 3.1) were medial in relation to published average swift fox litter sizes of 2.3–5.7 (Kilgore 1969, Hillman and Sharps 1978, Covell 1992, Carbyn et al. 1994, Anderson et al. 2002, Olson and Lindzey 2002). Similar to our study, sample sizes were small in most published accounts. Regardless, our data suggest litter size was similar between landscapes.

Average dispersal distances that we found were similar to other studies ($\bar{x}$ = 12.6, Schauster 2001; $\bar{x}$ = 12.1, Moehrenschlager 2000). Timing of the onset of pup dispersal that we observed was later than reports from southern portions of the swift fox range. Kilgore (1969) reported dispersal in August/September in Oklahoma and Covell (1992) reported dispersal in September/October in southern Colorado. These differences might be related to earlier whelping in more southerly populations (Asa and Valdespino 2002). Schauster (2001) found the juvenile females dispersed later than males, often delaying dispersal until next breeding/gestation period, when they were considered adults. In our study, 6 of 7 pups that remained close to or within their family home range were females. We were unable to determine if they might have delayed dispersal until late winter/early spring or until after their first breeding/gestation.

Swift fox populations in Kansas have experienced major changes since presettlement. Despite trapping efforts, poisoning campaigns, and significant habitat modification, this small prairie fox has survived and adapted to a new and vastly different landscape within its historic range. However, the future of swift fox populations in Kansas remains uncertain despite the present, relatively healthy, populations. Agricultural cropping practices are changing toward more irrigation rather than dryland farming. Larger monotypic crop fields are becoming more common, replacing the smaller fields in dryland fallow–winter wheat rotations. Such changes likely will not benefit the swift fox that presently is able to use fallow fields for foraging and denning. However, there are suitable habitats within the region that currently are not occupied by the species (Sovada and Scheick 1999). Factors limiting or delaying expansion of swift foxes into these unoccupied portions of their historic range in Kansas continue to be unknown. Given the high mortality rates and unremarkable reproductive rates for the swift fox in Kansas, their ability to expand current distribution in the state can only be a slow unsure process, if not impossible without help from wildlife managers. We recommend continued long-term research and monitoring of swift fox populations to gain insight into population dynamics, dispersal abilities, and other aspects of ecology in landscapes with diverse habitat composition and changing agricultural practices. Ultimately, knowledge gained will be useful to guide management decisions to ensure a thriving population of swift foxes in Kansas.

Acknowledgments

Special thanks go to the landowners in western Kansas who allowed access to their property and provided valuable assistance. We thank M.D. Combs, F.E. Durbian, V.L. Jackson, and especially J.R. Gillis, J.B. Bright, J.S. Gillis and B.K. Scheick for field assistance; B.K. Telesco, D.J. Telesco, A.L. Zimmerman for data handling; L.B. Fox for consultation; and L.N. Carbyn, L.B. Fox, and R.J. Greenwood for reviews of this manuscript. This study was funded by the U.S. Geological Survey's Biological Resources Division, Northern Prairie Wildlife Research Center, Kansas Department of Wildlife and Parks, North Dakota Department of Game and Fish, and the U.S. Fish and Wildlife Service.

Literature Cited

Aebischer, N.J., P.A. Robertson, and R.E. Kenward. 1993. Compositional analysis of habitat use from animal radio-tracking data. Ecology 74:1313–1325.

Aitchison, J. 1982. The statistical analysis of compositional data (with Discussion). Journal of the Royal Statistical Society (B) 44:139–177.

Andersen, D.E., T.R. Laurion, J.R. Cary, R.S. Sikes, M.A. McLeod, and E.M. Gese. 2002. Aspects of swift fox ecology in southeastern Colorado. Pp. 139–148 *in* L.N. Carbyn and M.A. Sovada, editors. Proceedings of the Swift Fox Symposium: ecology and conservation of swift foxes in a changing world. Canadian Plains Research Center, University of Regina, Saskatchewan.

Asa, C.S., and C. Valdespino. 2002. A review of small canid reproduction. Pp. 117–123 *in* L.N. Carbyn and M.A. Sovada, editors. Proceedings of the Swift Fox Symposium: ecology and conservation of swift foxes in a changing world. Canadian Plains Research Center, University of Regina, Saskatchewan.

Cameron, M.W. 1984. The swift fox (*Vulpes velox*) on the Pawnee National Grasslands: its food habits, population dynamics, and ecology. Thesis. University of Northern Colorado, Greeley, Colorado.

Carbyn, L.N., H.J. Armbruster, and C. Mamo. 1994. The swift fox reintroduction program in Canada from 1983 to 1992. Pp. 247–271 *in* M.L. Bowles and C.J. Whelan, editors. Restoration of endangered species: conceptual issues, planning and implementation. Cambridge University Press, Cambridge, United Kingdom.

Covell, D.F. 1992. Ecology of the swift fox (*Vulpes velox*) in southeastern Colorado. Thesis, University of Wisconsin-Madison, Madison, Wisconsin.

Cutter, W.L. 1958. Denning of the swift fox in northern Texas. Journal of Mammalogy 39:70–74.

Egoscue, H.J. 1979. *Vulpes velox*. Mammalian Species 122:1–5.

Fox, L.B. and C.C. Roy. 1995. Swift fox (*Vulpes velox*) management and research in Kansas: 1995 annual report. Pp.39–51 *in* S.H. Allen, J.W. Hoagland, and E.D. Stukel, editors. Report of the Swift fox conservation team 1995. North Dakota Game Fish and Parks, Bismarck, North Dakota.

Gallerani Lawson, E.J., and A.R. Rodgers. 1997. Differences in home-range size computed in commonly used software programs. Wildlife Society Bulletin 25:721–729.

Harrison, D.J., D.A. Bissonette, and J.A. Sherburne. 1989. Spatial relationships between coyotes and red foxes in eastern Maine. Journal of Wildlife Management 53:181–185.

Hillman, C.N., and J.C. Sharps. 1978. Return of the swift fox to the Northern Plains. Proceedings of the South Dakota Academy of Science. 57:154–162.

Hines, T.D. 1980. An ecological study of *Vulpes velox* in Nebraska. Thesis, University of Nebraska, Lincoln, Nebraska.

Hines, T.D., and R.M. Case. 1991. Diet, home range, movements, and activity periods of swift fox in Nebraska. Prairie Naturalist 23:131–138.

Jackson, V.L., and J.R. Choate. 2000. Dens and den sites of the swift fox, *Vulpes velox*. Southwestern Naturalist 45:212–220.

Johnson, D.H., and A.B. Sargeant. 1977. Impact of red fox predation on the sex ratio of prairie mallards. U.S. Fish and Wildlife Service Research Report 6.

Kie, J.G., J.A. Baldwin, and C.J. Evans. 1996. CALHOME: a program for estimating animal home ranges. Wildlife Society Bulletin 24:342–344.

Kilgore, D.L., Jr. 1969. An ecological study of the swift fox (*Vulpes velox*) in the Oklahoma Panhandle. American Midland Naturalist 81:512–534.

Kitchen, A.M., E.M. Gese, and E.R. Schauster. 1999. Resource partitioning between coyotes and swift foxes: space, time, and diet. Canadian Journal of Zoology 77:1645–1656.

Koopman, M.E., B.L. Cypher, and J.H. Scrivner. 2000. Dispersal patterns of San Joaquin kit foxes. Journal of Mammalogy 81:213–222.

Lindström, E.R. 1991. Diet and demographics of the red fox (*Vulpes vulpes*) in relation to population density—the sarcoptic mange event in Scandinavia. Pp. 922–931 *in* D.R. McCullough and R.H. Barrett, editors. Wildlife 2001:populations. Elsevier Applied Science, New York, New York.

Macdonald, D.W. 1983. The ecology of carnivore social behavior. Nature 301:379–384.

Minta, S.C. 1992. Tests of spatial and temporal interaction among animals. Ecological Applications 2:178–188.

——. 1993. Sexual differences in spatio-temporal interactions among badgers. Oecologia (Berlin) 96:402–409.

Moehrenschlager, A. 2000. Effects of ecological and human factors on the behavior and population dynamics of reintroduced Canadian swift foxes (*Vulpes velox*). Dissertation. University of Oxford, Oxford, UK.

Mohr, C.O. 1947. Table of equivalent populations of North American small mammals. American Midland Naturalist 37:223–249.

Nams, V.O. 1990. LOCATE II user's guide. Pacer, Truro, Nova Scotia.

National Climatic Data Center. 1997. Climatological data annual summary: Kansas. National Climatic Data Center, Asheville, North Carolina.

Olson, T.L., and F.G. Lindzey. 2002. Swift fox survival and production in southeastern Wyoming. Journal of Mammalogy 83:199–206.

Pechacek, P., F.G. Lindzey, and S.H. Anderson. 2000. Home range size and spatial organization of swift Fox *Vulpes velox* (Say, 1823) in southeastern Wyoming. Zeitschrift Fur Saugetierkunde 65:209–215.

Phillips, M.L., W.R. Clark, M.A. Sovada, D.J. Horn, R.R. Koford, R.J. Greenwood. In press. Predator selection of prairie landscape features and its relation to duck nest success. Journal of Wildlife Management.

Pruss, S.D. 1999. Selection of natal dens by the swift fox (*Vulpes velox*) on the Canadian prairies. Canadian Journal of Zoology 77:646–652.

Samson, F.B., and F.L. Knopf. 1994. Prairie conservation in North America. BioScience 44:418–421.

Samson, F.B., F.L. Knopf, and W.R. Ostlie. 1998. Grasslands. Pp. 437–472 *in* M. Mac, P. Opler, C. Puckett Haecker, and P. Doran, editors. Status and trends of the nation's biological resources. Volume 2. U.S. Department of the Interior, U. S. Geological Survey, Reston, Virginia.

Sargeant, A.B. 1972. Red fox spatial characteristics in relation to waterfowl predation. Journal of Wildlife Management 36:225–236.

Sargeant, A.B. 1982. A case history of a dynamic resource–the red fox. Pp. 121–137 *in* G.C. Sanderson, editor. Midwest Furbearer Management. Proceedings Midwest Fish and Wildlife Conference, Wichita, Kansas. Central Mountains and Plains Section and Kansas Chapter, The Wildlife Society.

Sargeant, A.B., S.H. Allen, and J.O. Hastings. 1987. Spatial relations between sympatric coyotes and red foxes in North Dakota. Journal of Wildlife Management 51:285–293.

Sargeant, A.B., W.K. Pfeifer, and S.H. Allen. 1975. A spring aerial census of red foxes in North Dakota. Journal of Wildlife Management 39:30–39.

SAS Institute. 1988. SAS/STAT user's guide. Version 6. Volume 1. Fourth edition. Cary, North Carolina.

Schauster, E.R. 2001. Swift fox (*Vulpes velox*) on the Piñon Canyon Maneuver Site, Colorado: population ecology and evaluation of survey methods. Thesis, Utah State University, Logan, Utah.

Schwartz, C.C., and J.E. Ellis. 1981. Feeding ecology and niche separation in some native and domestic ungulates on the shortgrass prairie. Journal of Applied Ecology 18:343–353.

Scott-Brown, J.M., S. Herrero, and J. Reynolds. 1987. Swift fox. Pp. 433–441 *in* M. Novak, J.A. Baker, M.E. Obbard, and B. Malloch, editors. Wild furbearer management and conservation in North America. Ontario Trappers Association, North Bay, Ontario.

Sovada, M.A., C.C. Roy, J.B. Bright, and J.R. Gillis. 1998. Causes and rates of mortality of swift foxes in western Kansas. Journal of Wildlife Management 62:1300–1306.

Sovada, M.A., C.C. Roy, and D.J. Telesco. 2001. Seasonal food habits of swift foxes in cropland and rangeland habitats in western Kansas. American Midland Naturalist 145:101–111.

Sovada, M.A., and B.K. Scheick. 1999. Preliminary report to the swift fox conservation team: historic and recent distribution of swift foxes in North America. Pp. 80–118 + appendix *in* C.G. Schmitt, editor, 1999 Annual report of the Swift Fox Conservation Team. New Mexico Department of Game and Fish, Santa Fe, New Mexico.

Spiegel, L.K. 1996. Spatial ecology and habitat use of San Joaquin kit fox in lands of Kern County, California. Pp. 93–114 *in* L.K. Spiegel, editor. Studies of the San Joaquin kit fox in undeveloped and oil-developed areas. California Energy Commission, Sacramento, California.

Swift Fox Conservation Team. 1997. Conservation assessment and conservation strategy for swift fox in the United States. R. Kahn, L. Fox, P. Horner, B. Giddings, and C.R. Roy, editors. Montana Fish, Wildlife and Parks, Helena, Montana.

Swihart, R.K., and N.A. Slade. 1985. Influence of sampling interval on estimates of home-range size. Journal of Wildlife Management 49:1019–1025.

Telesco, R.L., and M.A. Sovada. 2002. Immobilization of swift foxes with ketamine hydrochloride-xylazine hydrochloride. Journal of Wildlife Diseases 38:764–768.

Trewhella, W.J., S. Harris, and F.E. McAllister. 1988. Dispersal distance, home-range size and population density in the red fox (*Vulpes vulpes*): a quantitative analysis. Journal of Applied Ecology 25:423–434.

Voigt, D.R., and B.D. Earle. 1983. Avoidance of coyotes by red fox families. Journal of Wildlife Manangement 47:852–857.

White, G.C., and R.A. Garrott. 1990. Analysis of wildlife radio-tracking data. Academic Press, San Diego, California.

Worton, G.J. 1989. Kernel methods for estimating the utilization distribution in home-range studies. Ecology 70:164–168.

Zimmerman, A.L. 1998. Reestablishment of swift fox (*Vulpes vulpes*) in north central Montana. M.S. Thesis. Montana State University, Bozeman, Montana.

Zimmerman, A.L., L. Irby, and B. Giddings. 2002. The status and ecology of the swift fox in northcentral Montana. Pp. 49–59 *in* L.N. Carbyn and M.A. Sovada, editors. Proceedings of the Swift Fox Symposium: ecology and conservation of swift foxes in a changing world. Canadian Plains Research Center, University of Regina, Saskatchewan.

Using Tooth Sectioning to Age Swift Fox

■ **Melissa Richholt and Ludwig Carbyn**

Abstract: Thirty-eight swift fox skulls were used in this study. Of these, 9 were of known age. The lower left canine tooth was used whenever possible. The teeth were preserved in neutral 10% formalin, decalcified by first soaking in 5% nitric acid and then buffered formic acid. The teeth were frozen at -16°C and cut into 10-micron longitudinal sections. Eight sections per tooth were affixed to albuminized slides, and stained with toluidine blue. Readings were made with a wet-mount cover slip. To test precision, 3 readers were given the same slide. One reader was a novice at canine age determination. We used this as a check for ease of use for the method. All 3 readers were in agreement on 87% of the readings. With the known aged teeth, accuracy was achieved 89% of the time.

Introduction

With the success of the reintroduction program of swift foxes (*Vulpes velox*) in Canada, it has become necessary for biologists and wildlife managers to have an accurate method to determine the age of swift foxes. With released swift foxes now breeding in the wild, maintaining records of age on each fox is no longer possible. Accurate methods of determining ages are needed to evaluate the survival potential of the population. Ageing canids by tooth annulation is not a new method (Klevezal' and Kleinburg 1967, Linhart and Knowlton 1967, Jensen and Nielson 1968, Monson et al. 1973).

Our objective was to test canine cementum annulation for accuracy in the ageing of swift foxes. Preparation of the teeth prior to sectioning, a description of the reading techniques, and the results of the study are presented.

Methods

Thirty-eight frozen swift fox heads were provided by the Canadian Wildlife Service for the study. All were either captive breeding animals from Cochrane Wildlife Reserve or foxes released to the wild in a reintroduction program. Many carcasses had been found near the release area in southern Alberta and Saskatchewan. These foxes had died from a variety of causes, such as predation, starvation, or highway casualty. The heads were in various states of decomposition.

We skinned the heads and removed the lower left canines from the jaw using side cutters to snip away the mandibular bone from around the tooth root, and easing the tooth out. If the lower left canine was not present or was damaged, the right lower canine was selected. Roberts (1978) found that the canines of red foxes (*Vulpes vulpes*) have the best definition and the thickest cemenum of all teeth, so it is anticipated that swift fox canines yield better results than other teeth.

The canines were placed in a vial containing a neutral solution of 10% formalin (Table 1), for a minimum of 72 hours, with a volume of preservative ten times the volume of the tooth. We then placed each tooth into a perforated plastic container (with identification) submerged in one liter of 5% nitric acid (Preece 1965). The acid solution was gently agitated occasionally in the first eight hours. After 26-28 hours the teeth were carefully tested for flexibility by exerting light finger pressure at each end. If sufficiently flexible to bend slightly, a sharp scalpel was used to incise the crown at the gumline, and remove from the root.

Decalcification of the root was continued by placing each plastic container in a separate beaker containing buffered formic acid (Table 1) at 100 times the volume of the tooth root. Buffered formic acid gives a slow controlled rate of decalcification reducing acid damage to the cementum. We tested the end point of decalcification daily. After the teeth had been in the solution ≥6 hours, we made a solution using 5 ml of formic acid from the bottom of the container next to the tooth, mixed with 1 ml of 5% ammonium oxalate. The tooth root was added to this solution and allowed to stand for 30 minutes to several hours. The formation of a precipitate was used as evidence that decalcification was not complete; if precipitate was present, the tooth was returned to fresh buffered formic acid. Once a clear solution was observed, decalcification was considered complete. We rinsed the decalcified teeth 12-15 hours in running water, and then stored the teeth in 70% ethanol. Before sectioning, we rinsed the teeth again for 12 hours in running water. They were then embedded in OTC compound, frozen and sectioned in a cryostat/mictrotome at about -16°C. Ten-micron longitudinal sections were taken near the root canal. The sections were floated in basic water (pH 8.5) for 20 minutes, and fixed onto albuminized glass slides, eight sections per slide. Staining was done using fresh toluidine blue stain (Table 1). When readings were done, a cover slip was wet-mounted with water over the slide.

For the first 16 slides, we used 3 readers (2 experienced and 1 inexperienced) to estimate the age of foxes. Each reader worked separately without knowledge of the

Table 1. Preparation of solutions.	
Preparation of Neutral 10% Formalin Solution [1]	
Formalin (37–40% formaldehyde)	100 ml
Sodium phosphate diabasic (anhydrous)	6.5 g
Sodium phosphate monobasic	4.0 g
Distilled water	900 ml
Preparation of Buffered Formic Acid Solutions [2]	
A. Sodium citrate	200 g
Distilled water	1000 ml
B. 90% Formic acid	500 ml
Distilled water	500 ml
Make A and B as stock solutions then mix 1:1 when needed.	
Preparation of Toluidene Blue Stain [3]	
Toluidine blue 0	80 mg
pH 8.5 water	250 ml
Mix and filter	

1. Preece 1965; 2. Ibid.; 3. Luna 1968.

other's findings. Each tooth root was examined 3 times by each reader, and readers recorded their results. After the first 16 slides, only 1 experienced and 1 inexperienced reader continued.

We used a 40x objective lens on a compound microscope to read the prepared slides. Slides contained 8 sections of the same tooth, each section giving a slightly different view. Since sectioning is never perfect, we used multiple views to make accurate determination easier. At times, the annuli appeared to run together or split. If the annuli were followed from the side, up to the tip and around, accuracy improved. The tip of the root usually had the thickest cementum, and the clearest view of annulation.

Annulation occurs in the cementum of the tooth root, visible as dark lines roughly parallel to the surface of the root. On the outer edge of the cementum, a layer of lacy-looking connective tissue (periodontal membrane) can be seen (Figs. 1–4). Annulation takes place over the course of months, and older animals generally have narrower annuli. This variance in annuli makes it difficult to determine completion. For the study, if there was a layer of cementum at the edge, the annuli were considered complete. This was recorded as a "+" in addition to the number of completed annuli (1+ meaning 1 complete annuli plus cementum). If the annuli went right to the edge, it was considered to be incomplete, and was recorded as "+1" in addition to the number of completed annuli (2+1 meaning 2 complete annuli plus 1 incomplete).

Table 2. Annuli in Tooth Cementum of Swift Fox (*Vulpes velox*) in Alberta and Saskatchewan, 1996.

Fox No.	Number of Annuli			Known age (if applicable)	Death (dd/mm/yy)	Predicted Age (yr)	Concurrence? Yes or No
	Reader 1	Reader 2	Reader 3				
L001	2+1	2+1	2+1	2yr,4.5mon	13/09/83	2–3	Y
L002	4+1	4+	4+1	4yr,5mon	26/09/84	4–5	Y
L004	1+1	1+1, 2+	1+1	2yr,5mon	25/09/84	1–2	Y
L005	6+1	6+1	7+1	n/a	21/05/88	6–7	N
L006	0+1	0+1	0+,0+1	n/a	35285	<1	Y
L007	7+1	7+1	7+1	n/a	35316	7–8	Y
L008	1+1	1+1	1+1	n/a	35285	1–2	Y
L009	0+	0+	0+	n/a	27/11/88	<1	Y
L010	5+1,6+1,7+1	7+1,8+,8+1	8+1	7yr,7mon	27/11/88	7–8	Y
L011	2+1	5+1	5+1	7yr,7mon	28/11/88	5–6	Y
L012	5+1	5+1,6+1	5+1	6yr,7mon	29/11/88	5–6	Y
L013	0+1	0+1	0+1	8.5mon	18/01/89	<1	Y
L014	5+1, 6+1	7+1,8+	5+1	≥8yr	23/08/88	6–8	N
L015	6+	8+1,9+1	6+1,7+1	≥8yr	23/08/88	7–9	N
L016	0+1	0+1,1+	0+	n/a	31625	<1	Y
L030		2+1	2+1	n/a	n/a	2–3	Y
L031		2+	2+	n/a	n/a	2–3	Y
L032		1+1,2+	1+1	n/a	n/a	1–2	Y
L033		1+1	1+1	n/a	n/a	1–2	Y
L035		1+1	1+1	n/a	n/a	1–2	Y
L036		5+1	5+1	n/a	n/a	5–6	Y
L037		0+1,1+	0+1	n/a	n/a	1	Y
L038		1+1	1+	n/a	n/a	1–2	Y
L039		2+1	2+1	n/a	n/a	2–3	Y
L040		1+1	1+1	n/a	n/a	1–2	Y
L041		1+1,2+	1+1	n/a	n/a	1–2	Y
L042		0	0	n/a	n/a	<1	Y
L044		8+1	8+1	n/a	n/a	8–9	Y
L045		9+1	7+	n/a	n/a	7–9	N
L046		0	0	n/a	n/a	<1	Y
L048		2+1	2+1	n/a	n/a	2–3	Y
L050		7+1	7+,8+	n/a	n/a	7–8	Y
L053		0+1	0+1	n/a	n/a	<1	Y
L056		1+1	1+1	n/a	n/a	1–2	Y
L059		0	0	n/a	n/a	<1	Y
L061		3+1	3+	n/a	n/a	2–3	Y
L062		0+1	0+1	n/a	n/a	<1	Y
L067		4+1	3+1	n/a	n/a	3–4	N

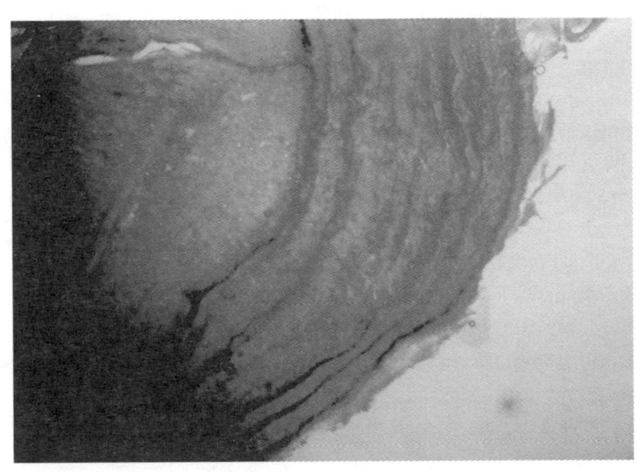

Figure 1. Good view of annulation Sectioning does not always give a perfect slice. Annuli appear to run together, split, blur or disappear entirely, but here distinct lines are seen.

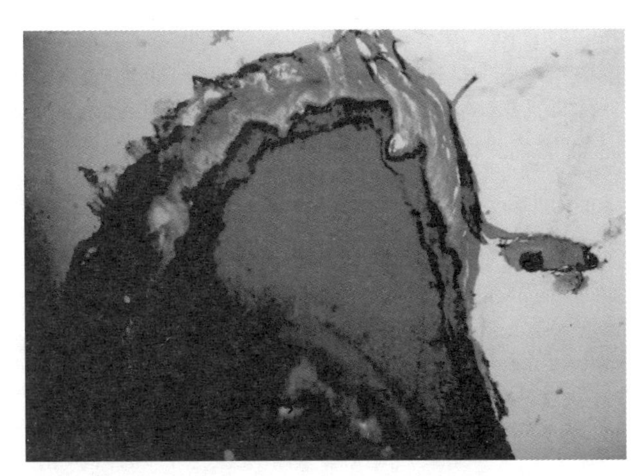

Figure 2. Root tip annulation. A tooth from a young fox with clearly distinguished annulation.

Figure 3. Side view of root. Following the annuli to the side of the root can help to distinguish the actual number of annuli.

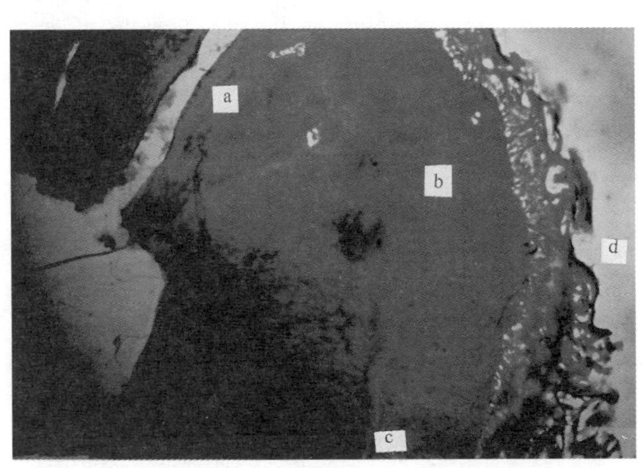

Figure 4. This tooth root is from a very young fox. Notice the open root (a) (swift fox have open roots until 8 months). Also visible are a large layer of cementum (b), the interface of the cementum and tooth root (c), the absence of annuli, and the layer of epithelial tissue (d). Annulation in swift fox forms in the first year, and cementum after four months.

Results

Of the 38 swift fox used in this study, 9 were of known age. The 3 readers were in agreement on number of annuli counted for 8 of the 9 foxes (Table 2). In one case, after sectioning, the cementum was absent making a reading impossible (Specimen L003). This tooth was removed from the study. The 2 readers who participated throughout the study were in agreement 87% of the time as to determination of age.

Discussion

The technique we used was accurate with 8 of the known age swift foxes (89%). There was agreement between the 2 readers (87%) for all teeth examined. Desiccation of some teeth may have had influence on results, since cementum is a soft tissue. Many of the fox carcasses had been frozen for up to 14 years. Fresh specimens will likely yield the best results. Care must be taken to prevent damage to the cementum during extraction. One apparent anomaly was suspected in a tooth that was quite gray when extracted. This tooth required a very short decalcification period, and had very thin dentine.

Factors that can influence the accuracy of the reading include the condition of the tooth, sectioning technique, and experience of the reader. If the tooth is in poor condition, annuli may not take the stain well. Poor sectioning will yield a more difficult reading. As readers become more experienced, they are less likely to be confused by false annuli, splitting or complex annuli, or apparent disappearance of annuli.

Age determination of mammals has been done for many species using annulation in tooth cementum. This technique has been documented in gray wolves (Canis lupus; Patriquin and Carbyn 1976, Goodwin and Ballard 1985), coyotes (*Canis latrans*; Linhardt and Knowlton 1967), red fox (Monson et al. 1973), and arctic fox (*Alopex lagopus*; Bradley et al. 1981). This study used a slightly different technique than those in these prior studies. Other methods included boiling as a means of removing the canine teeth from the mandible (Patriquin and Carbyn 1976) and use of different stains (i.e., Paragon stain which is no longer available or Tolvin blue which must be prepared daily; Patriquin and Carbyn 1976) for defining the annuli. However, the results of these studies have shown that cementum annulation is an accurate way of determining age in swift foxes.

Cementum on teeth is produced throughout the lifetime of the animal. Generally the root tip has the greatest accumulation of cementum and the best definition of annuli (Roberts 1978). This was found to be the case in swift fox as well. Cementum was thicker on the root tip, and the number of annuli was most easily determined at the root tip, or slightly to the side (Fig.1 and 2). Often it was necessary to follow the annuli from the side up to the tip to determine if complex annuli were 1 annulus or 2 separate annuli (Fig 3). Having 8 sections on each slide was valuable in seeing the annuli from more than 1 view to make a reliable estimate.

In North America, mammals produce the dark cementum band in winter (Matson 1981). The annuli of these swift foxes examined were generally produced between September and February. On only 1 occasion was annulation not complete by February (Table 2, Specimen L011). When known, the date of death can be used to clarify the age determined by annulation.

Teeth of young animals are usually quite distinct from those of adults. An open root tip can identify young coyotes until at least 8 months of age (Linhart and Knowlton 1967). Coyotes from 8 months to 1.5 years usually have a closed root, but no cementum. Swift fox in this study showed annulation forming in the first year, and cementum growth after 4 months. Allen and Kohn (1976) found that in North Dakota, coyotes formed cementum and annuli much younger than those reported in Linhart and Knowlton's study (1967). They found cementum and annulation in the first year, as we did (Fig. 4).

Tooth cementum annuli are not the only technique used to age canids. Allen (1974) used a combination of 2 techniques: a measurement from the enamel line to alveolar socket and counting cementum annuli. However, this technique only distinguished juveniles from adults, a result available using cementum alone. Determining age by tooth wear and eruption (Nellis et al. 1978) can also only reliably separate juveniles from adults. Annulation in the epiphyseal cartilage of long bones is less accurate than cementum annulation and is usually only beneficial in distinguishing juveniles and adults. Nellis et al. (1978) compared 4 techniques for ageing coyotes in central Alberta. First, they used the eruption dates of milk teeth, which could only be used in the first seven months of a coyote's life. Second, they used tooth socket tightness, which decreased with age. This method was only found to be accurate in distinguishing juveniles from adults, and was a subjective appraisal not a measurement. The third method was measuring "canine drop." As the canine tooth grows and cementum is deposited on the root tip, the tooth moves away from the gumline. By measuring from the gumline to the top of the arch of tissue around the tooth, juveniles could be distinguished from adults. The advantage of this method is that age classes can be determined without mutilation of a living animal. The fourth method was tooth cementum annulation. As in other studies, this was found to be an accurate method of determining age beyond juvenile and adult classification.

Root width is another method used for identifying young canids. The pulp cavity for arctic foxes <1 year old is 1.5 mm or more (Bradley et al. 1981). Although root width was not measured in the swift fox study, it was seen that root width was consistently larger in young animals.

Conclusion

Tooth annulation has been used in other mammals for some time. It has been assumed to be an accurate method for swift foxes as well. Our documentation of the accuracy of this technique for this species verifies for field biologists and researchers that they can determine the age of swift fox with confidence.

The long-term survival of swift foxes as a species in northern ranges is still uncertain. Reintroduction programs in North America start with well-documented foxes. Released foxes have reproduced with some success, and the age of their offspring is not necessarily documented. With a reliable method of age determination, researchers will be able to make judgments on the status of the swift fox, individually and as a population.

Acknowledgments

We wish to thank Axel Moerenschlager and Cleo Smeeton for providing skulls complete with teeth for us to use. Also thanks to Elaine Street for much help on this project, Harry Armbruster (Canadian Wildlife Service) for being a reader in the early portion of the study, and to Wendy Calvert (Canadian Wildlife Service) for assistance in many ways throughout the study.

Literature Cited

Allen, S.H. 1974. Modified techniques for ageing red fox using canine teeth. Journal of Wildlife Management 38:152–154.

Allen, S.H. and S.C. Kohn. 1976. Assignment of age-classes in coyotes from canine cementum annuli. Journal of Wildlife Management 40:796–797.

Ballard, W.B., G.M. Matson, P.R. Krausman. 1995. Comparison of two methods to age gray wolf teeth. Pp. 455–459 *in* L.N. Carbyn, S.H. Fritts, and D.R. Seip, editors. Ecology and conservation of wolves in a changing world. Canadian Circumpolar Institute, Occasional Publication No. 35.

Bradley, J.A., D. Secord, and L. Prins. 1981. Age determination in the arctic fox (*Alopex lagopus*). Canadian Journal of Zoology 59:1976–1979.

Goodwin, E.A., and W.B. Ballard. 1985. Use of tooth cementum for age determination of gray wolves. Journal of Wildlife Management 49:313–316.

Jensen, B., and L.B. Nielsen. 1968. Age determination in the red fox (*Vulpes vulpes* L.) from canine tooth sections. Danish Reserve Game Biology 5(6):1–15.

Klevezal', G.A., and S.E. Kleinenberg. 1967. Age determination of mammals from annual layers in teeth and bones. Academy of Sciences of the USSR, Severtson Institute of Animal Morphology, Moscow, Russia.

Linhardt, S.B., and F.F. Knowlton. 1967. Determining age of coyotes by tooth cementum layers. Journal of Wildlife Management 31:362–365.

Luna, L.G., editor. 1968. Manual for histologic staining methods of the Armed Forces Institute of Pathology. Third edition. McGraw-Hill, New York, New York.

Matson, G.M. 1981. Workbook for cementum analysis. Matson's. Milltown, Montana.

Monson, R.A., W.B. Stone, and E. Parks. 1973. Ageing red fox (*Vulpes fulva*) by counting annular cementum rings of their teeth. New York Fish and Game Journal 20: 54–61.

Nellis, C.H., S.P. Wetmore, and L.B. Keith. 1978. Age-related characteristics of coyote canines. Journal of Wildlife Management 42:680–683.

Patriquin, D., and L.N. Carbyn. 1976. Age determination of wolves by tooth cementum layers. Unpublished report. Canadian Wildlife Service, Edmonton, Alberta.

Preece, A. 1965. A Manual for histologic technicians. Second edition. Little, Brown & Co., Boston, Massachusetts.

Roberts, J.D. 1978. Variation in coyote age determination from annuli in different teeth. Journal of Wildlife Management 42:454–456.

Den Ecology of Swift, Kit and Arctic Foxes: A Review

■ **Magnus Tannerfeldt, Axel Moehrenschlager and Anders Angerbjörn**

Abstract: The availability and use of denning sites are important aspects of the ecology of most canids. Swift, kit and arctic foxes are closely related genetically, are similar in size, and share a number of behavioral and ecological traits. Yet, there are many differences between the species, which can be used in comparative studies. In this review, we examine differences and similarities in the den ecology of these species, in order to analyze the relationship between den use and other ecological parameters. We also discuss implications for the management of these foxes. We have found 2 different den ecology strategies, where swift and kit foxes have small litters and regularly change dens during the breeding season, while arctic foxes have large litters in large dens. The primary function of a breeding den is most likely to provide protection against predators. Sufficient escape routes can be achieved either by having several small satellite dens within each home range or by having large dens with many openings. These different den ecology strategies also involve territoriality, and are related to differences in a number of ecological parameters, such as predation rates, availability of dens, food resources and litter sizes. Identification and classification of den sites is a means of making surveys and population estimates more effective, especially for the arctic fox. An analysis of den sites is important for habitat protection and as a preparatory task for re-introduction programs for all 3 species.

Species which are closely related and share many characteristics while differing in others, can provide useful comparisons for ecological or evolutionary studies and for management purposes. Three of the smaller foxes make strong candidates for such a comparison, viz. the swift fox (*Vulpes velox*) of the shortgrass prairies, the kit fox (*V. macrotis*) of the desert, and the arctic fox (*Alopex lagopus*) of the tundra. The former 2 are restricted to North America, while the arctic fox has a circumpolar distribution. Despite differences in geographic range, habitat and physiology, there are striking similarities between the 3 species in genetics, behavior and ecology.

For example, swift, kit and arctic foxes share a strong dependence on good denning sites for breeding. They all live in open habitats and are subject to harsh climatic conditions. Furthermore, they live sympatrically with larger mammalian carnivores that often act as predators or competitors. Underground dens provide shelter for these small foxes, especially during breeding. The young are born blind and are dependent on their parents for approximately 2 months. Their den dependence has become a useful tool for ecologists, as den surveys can give good estimates of reproductive success in a population. For arctic foxes, it is a widely used method for population estimates (e.g., Elton 1924, Macpherson 1969, Angerbjörn et al. 1995, Tannerfeldt 1997). This den dependence has also been used in extermination campaigns (Hersteinsson 1984, Bailey 1992). For swift and kit foxes, den surveys lead to assessments of breeding frequency and litter sizes, which serve as indicators of reproductive success between years and regions (Egoscue 1975, Covell 1992, White and Ralls 1993, White and Garrott 1997).

This paper reviews and compares the den ecology of swift, kit and arctic foxes. It is essential to have an understanding of how availability and distribution of dens affect fox life histories and population dynamics. For example, increasing agricultural and industrial land use in sensitive areas might interfere with fox management and conservation goals. We will thus examine the den as a resource for these small foxes. Our interest was in factors that determine the distribution of dens, their structure and how they are used. How important are den sites to these foxes and what are the relationships between dens, reproductive output and other ecological parameters? Finally, we will try to determine the implications of our results for the management of these species.

The Species

Recent research suggests that the arctic fox should be included in the genus *Vulpes* to form a monophyletic group, and that arctic, kit and swift foxes are very closely related (Martin 1989, Geffen et al. 1992, Mercure et al. 1994). This genetic similarity is the closest that exists between any of the *Vulpes*-like species (Wayne and O'Brien 1987). They share an ancestor possibly adapted to an open desert and prairie habitat. From this, the arctic fox evolved into an exclusively tundra-dwelling species. Recent studies recognize only 3 subspecies of the arctic fox, 2 of which are indigenous to the isolated Commander Islands (Ginsberg and Macdonald 1990). The taxonomy of swift and kit foxes has been intensively debated. Based on protein-electrophoretic methods, Dragoo et al. (1990) concluded that swift and kit foxes were the same species but they argued that morphological differences warranted classification into separate subspecies. More recent mitochondrial DNA analyses, however, suggest that swift and kit foxes should be considered as separate species, namely *Vulpes velox* and *V. macrotis* (Mercure et al. 1994).

Swift and Kit Foxes

Swift and kit foxes are located in topographically flat, arid regions of North America. Historically, swift foxes occupied the Great Plains ranging north-south from central Alberta to central Texas and east-west from eastern North Dakota to central Colorado (Allardyce and Sovada 2003). Swift foxes are separated from kit foxes by the Rocky Mountains but interbreeding does occur within a limited hybridization zone in New Mexico (Rohwer and Kilgore 1973, Mercure et al. 1994). Kit foxes range from southern Idaho and Oregon in the United States to Durango, Zacatecas, and Nuevo in northern Mexico (O'Farrell et al. 1986, List 1998). The San Joaquin kit fox (*Vulpes macrotis mutica*) in southern California is topographically isolated from the main kit fox continuum.

Swift and kit foxes share common ecological requirements and similar threats to their persistence. Comparisons of the northern and southern peripheries of the swift/kit fox complex illustrate that habitat requirements are broadly similar throughout the range (O'Farrell 1987; Scott-Brown et al. 1987). Swift and kit foxes are morphologically similar although kit foxes have slightly longer ears (Dragoo et al. 1990) and smaller body weights than swift foxes (Moehrenschlager 2000). Along with the Fennec fox (*Fennecus zerda*) and Blanford's fox (*Vulpes cana*), these species are amongst the smallest of canids (Ginsberg and Macdonald 1990). Although the individual prey types differ between areas, both fox species primarily consume rodents, lagomorphs, birds, and insects (O'Farrell 1987, Scott-Brown et al. 1987). The species live in arid conditions which typically receive less than 400 mm of precipitation (Moehrenschlager and List 1996), although the San Joaquin kit fox obtains up to 1500 mm of precipitation which primarily falls between November and April (O'Farrell et al. 1986). While home range sizes can be highly variable within species, swift fox home ranges are generally larger than those of kit foxes (Hines and Case 1991, Zoellick and Smith 1992, List 1998, Kitchen et al. 1999, Moehrenschlager 2000).

The current distribution of swift foxes may represent only 38–41 % of its historical range (Sovada and Scheick 2000). The remaining core areas exist in Wyoming, Colorado, and Kansas where foxes appear to be abundant and populations are contiguous (Kahn and Beck 1996). A petition to list swift foxes as endangered was deemed as "warranted but precluded" in the United States, but has recently been overturned (Federal Register 1995). In Canada, this species was extirpated in the 1930's (Herrero et al. 1986), until reintroduction releases began in 1983. A small and apparently stable population has now been established (Cotterill 1997, Moehrenschlager and Moehrenschlager 1999, Moehrenschlager and Moehrenschlager 2001). Kit foxes are declared as threatened in Mexico (List 1998) and the San Joaquin kit fox is endangered (Cypher and Spencer 1998).

The reasons suggested for the decline in swift and kit fox numbers has been habitat loss and fragmentation (O'Farrell et al. 1986; Scott-Brown et al. 1987). Moreover, both species were susceptible to intensive poisoning programs which were primarily aimed at larger predators (Scott-Brown et al. 1987, Ginsberg and Macdonald 1990). The ecosystem perturbation that resulted from such predator control programs caused a significant shift in the canid community composition. The extirpation of the wolf (*Canis lupus*) allowed for the expansion of coyotes (*C. latrans*) (Schmidt 1991) which are now the primary mortality factor of swift and kit foxes in all telemetry studies (Ralls and White 1995, White and Garrott 1997, Sovada et al. 1998, Kitchen et al. 1999, Moehrenschlager 2000).

The Arctic Fox

The arctic fox inhabits most Arctic land areas above the timber line, including polar deserts and islands far away from the mainland (Preble and McAtee 1923, Lavrov 1932, Chesemore 1975, Hersteinsson 1984, Prestrud 1992a). The arctic fox has many physical adaptations to the Arctic environment, including the best insulative fur of all mammals (Prestrud 1991, Klir and Heath 1992). In the Holarctic range of the arctic fox, productivity is generally low, but food resources can be extremely abundant in small patches and during short time periods. The dominant pattern in these resource fluctuations is determined by rodent population fluctuations. In continental areas, the main prey species in summer are lemmings, (*Lemmus* and *Dicrostonyx* spp.), but also voles, (*Microtus* and *Clethrionomys* spp.) and carcasses of reindeer (*Rangifer tarandus*). In winter, the most important food resources are reindeer carcasses and ptarmigans (*Lagopus mutus* and *L. lagopus*; Macpherson 1969, Kaikusalo and Angerbjörn 1995, Elmhagen et al. 2000). In other areas, arctic fox populations are sustained on more stable summer food resources, mostly at bird cliffs and along shore lines, where food is washed up by the sea at regular intervals (Hersteinsson and Macdonald 1982, Prestrud 1992c, Hersteinsson and Macdonald 1996).

For the arctic fox, the world population is in the order of several hundred thousand individuals (Tannerfeldt 1997), but the species is endangered in some areas. On the Commander Islands (Russia), the threat is a result of an introduced disease (Kruchenkova and Formozov 1995). On the Fennoscandian peninsula (Norway, Sweden, Finland), arctic fox numbers have not recovered from a drastic population decline caused by over-hunting more than 80 years ago (Angerbjörn et al. 1995, Frafjord 1998, Löfgren and Angerbjörn 1998). This population is partly isolated from the Siberian mainland, and the lack of recovery can be explained by the combination of very low fox numbers (below 100 individuals), a lack of large predators leaving carcasses, and an absence of lemming population peaks.

Methods

We have compiled data from a large number of studies on the den ecology of these 3 fox species. Of these, more than 50 studies contained quantitative data which we have used in our analyses (Table 1). We focused our comparisons on parameters which were studied by many authors and for which data were collected in a comparable manner. Estimates of litter size have been accepted only for populations with a sample size larger than 5 and when collected around the time of weaning. Estimates of home range is the mean range given for resident (denning) animals during the breeding season. Juvenile mortality is the percentage of young which were reported to have died before the age of 2 months. We have also included unpublished or re-analyzed data from our own research in the Swedish Arctic Fox Project (see Angerbjörn et al. 1991, Tanner-feldt 1997) and the Canadian Swift Fox Reintroduction Program (see Moehrenschlager 2000).

Results and Discussion

Structure and Location of Dens

SWIFT FOX DENS. Swift fox dens are generally found in elevated areas with well-drained soils, but den sites will differ depending on their function. Extensive natal dens will have numerous entrances whereas dens that are utilized for escape from predators may frequently only have 1 opening (Kilgore 1969; A. Moehrenschlager, personal observation). Swift foxes frequently use or expand the burrows of ground squirrels (*Spermophilus* spp.), prairie dogs (*Cynomys* spp.), or badgers (*Taxidea taxus*), but they can readily dig their own dens (Cutter 1958, Kilgore 1969). Den entrances have a diameter of approximately 20 cm (Cutter 1958, Pruss 1999). The dens are normally situated near the tops of gently sloping hills (Cutter 1958, Uresk and Sharps 1986, Moehrenschlager 2000). Swift foxes primarily den in shortgrass prairie habitat, but also in midgrass prairie (Uresk and Sharps 1986) and cultivated areas (Sovada et al. 1998). In Kansas, the number of den entrances did not differ between cultivated and rangeland sites (Jackson and Choate 2000), while Kilgore (1969) determined that agricultural sites harbored relatively few den entrances in Oklahoma. Rangeland dens in Kansas were characterized by higher and denser vegetation, more rolling topography, smoother den surfaces, and more extensive burrow tailings than dens in cultivated habitats (Jackson and Choate 2000). Excavated dens in shortgrass areas had more and longer branches and extended to a greater depth than those in cultivated sites (Cutter 1958, Kilgore 1969). Swift fox dens can have as many as 17 branches and up to 2 chambers at depths extending to 1 meter (Kilgore 1969). In Nebraska, den entrances primarily had eastern or western exposures (Hines and Case 1991), while in Colorado and Alberta they were randomly oriented (Rongstad et al. 1989, Pruss 1999). Swift fox dens are frequently located in anthropogenic areas such as near roads, in culverts, pipes and buildings (Kilgore 1969, Hines and Case 1991, Zimmerman and Giddings 1997, Pruss 1999, Moehrenschlager 2000). Since coyotes avoid human habitation, such den sites can offer additional protection for swift foxes. However, they may induce additional costs as adults, and pups in particular, are frequently killed by cars and occasionally taken by domestic dogs (*Canis familiaris*; Sovada et al. 1998, Moehrenschlager 2000).

KIT FOX DENS. Kit fox dens are strikingly similar to those of swift foxes. Kit fox dens can also be variable in size and dens utilized for pup-rearing have more entrances than those used for other purposes (Morrell 1972). Like swift foxes, kit fox dens are normally only occupied by the natal pair or family groups (Morrell 1972, Egoscue 1975). Kit foxes frequently utilize or expand the burrows of other animals such as prairie dogs, kangaroo rats (*Dipodomys* spp.), and badgers (O'Neal et al. 1987, List 1998). While some authors consider the digging ability of kit foxes to be poor, others believe it is excellent (O'Neal et al. 1987). Most likely, digging frequency depends on the need and possibility to do so. In areas where numerous holes have been created by other species or where soils are not well drained, foxes may dig only rarely. Kit fox dens in Utah are situated in well-drained areas (O'Neal et al. 1987). Two excavated kit fox dens had grass-lined food and sleeping chambers which were as deep as 2.5 m below the surface (O'Neal et al. 1987). Dens of San Joaquin kit foxes at Camp Roberts had openings with an average height of 20 cm and an average width of 21 cm. Most dens had 2 to 5 entrances and 36% showed signs of fox activity (Reese et al.1992). More frequently than expected, this fox population had dens in grassland and low to medium density woodland, while medium to high density oak woodlands were avoided. Kit foxes were primarily found in well-drained soils and the average slope of occupied hillsides was 19 degrees (Reese et al. 1992). Kit foxes in Mexico's Chihuahuan desert preferred creosote habitat and Mimbres-Tome soil, which is well drained. Steep slopes of 5% to 20% were avoided. Dens were primarily found on slopes oriented to the northwest, and den openings were primarily oriented towards the southeast or northwest (Rodrick and Mathews 1999).

ARCTIC FOX DENS. Arctic fox dens can be impressive geographical features, with a hundred entrances in a mound or ridge and lush green vegetation contrasting the barren tundra. Arctic fox dens have therefore been monitored through aerial surveys, in some areas with good results (e.g., Macpherson 1969, Garrott et al. 1983, Ericson 1984,

Table 1. Data from 43 different populations of small foxes: kit fox *Vulpes macrotis*, swift fox *V. velox* and arctic fox *Alopex lagopus*. Given are the means for each parameter ± SE, with sample sizes in italics and range (in parentheses). Maximum and mean li tter sizes are at the time of weaning. Home ranges are given in km ^{2} to nearest integer by Minimum Convex Polygon method, unless otherwise stated.

Species	Main Food Resources	Site	Litter size max / n	Litter size mean / n	Max no. dens in breeding season	Reference	Home range (km^2) in breeding season / n	Annual adult mortality	Annual juvenile mortality
V macrotis		Naval Petr. Res., California	6/101	3.8/101		(Cypher et al. 2000)		0.56	0.86 first 9.5 months
V macrotis	Fluctuating rodents	Carrizo Plain, California	3/4	2/4	>50 [1] 4	(Ralls et al. 1990; White and Ralls 1993)	pairs: 11 ± 1.2	(0.36-0.45)	(0.46-0.60)
V macrotis	Rodents, rabbits	Sonoran Desert, Arizona			16	(Zoellick and Smith 1992; Zoellick et al. 1989)	grid-cell method: 11 ± 1/7		
V macrotis	Rodents	Camp Roberts, California			50 [2]	(Reese et al. 1992)			
V macrotis	Jackrabbits and cottontails	Naval Petr. Res., California		4.3/84		(O'Farrell et al. 1986)			
V macrotis	Fluctuating jackrabbits	Western Utah	6/17	3.8/17		(Egoscue 1975)			
V macrotis	Rodents, rabbits, birds	Kern County, California	5/5	4/5	5	(Morrell 1972)			
V macrotis	Rodents, rabbits, birds	Desert Exp. Range, Utah			17	(O'Neal et al. 1987)			
V velox	Rodents, insects	Northern Montana	7/3			(Zimmerman and Giddings 1997)			
V velox		SE Colorado		pairs: 2.4/13 trios: 4.2/5		(Covell 1992)		0.47 +/- 0.05	0.87 +/- 0.13
V velox	Rodents, birds, rabbits	Beaver County, Oklahoma	6/4	4.3/4		(Kilgore 1969)			
V velox	Mammals, birds, insects	Las Animas County, Colorado	5/5	3.4/5	20+	(Rongstad et al. 1989)		048	0.95
V velox		Alberta/Saskatchewan	7/12	4.2/12		(Brechtel et al. 1993; Carbyn et al. 1994)		(0.33-0.50)	
V velox	Rodents, birds, rabbits	Alberta/Saskatchewan	8/29	3.9/29	22	(Moehrenschlager 2000)	fixed kernel estimate: 24/18	(0.46-0.64)	
A lagopus	Fluctuating rodents	Hudson Bay, NWT	12/9	7.6/9		(Hall 1989)			
A lagopus	Fluctuating rodents	Keewatin, Aberdeen Lake, NWT	14/38	6.5/38		(Macpherson 1969; Speller 1972)			
A lagopus	Fluctuating rodents	Kildin Island, Russia	13/48	6.5/48		(Lavrov 1932)			
A lagopus	Fluctuating rodents	Prudhoe Bay and Colville River Delta, USA		4.8/11		(Burgess 1980; Eberhardt et al. 1983; Fine 1980; Garrott and Eberhardt 1982)	21/2 and 21 ± 6/4		dead pups in 19% of 79 litters
A lagopus	Fluctuating rodents	Snøhetta, Norway				(Landa et al. 1998)	21/21 (6-60) males: 27/7 females 18/14		
A lagopus	Fluctuating rodents	Norrbotten county, Lapland, Sweden	14/60	6.4/60		(Angerbjörn et al. 1995)			
A lagopus	Fluctuating rodents	Västerbotten county, Lapland, Sweden	15/88	6.1/88	2	(Angerbjörn et al. 1997; Angerbjörn et al. 1995; Tannerfeldt et al. 1994)	21/5 (15 - 36)	(33.3 - 60.0)	(0.08-1.00)
A lagopus	Fluctuating rodents	Jämtland county, Lapland, Sweden	16/67	6.4/67		(Angerbjörn et al. 1995)			
A lagopus	Fluctuating rodents	Wrangel Island, Russia	18/53	6.5/53		(Chernyevski and Dorogoi 1981; Dorogoi 1987)			(0.23 - 0.62)
A lagopus	Fluctuating rodents	Yamal, Russia				(Smirnov 1968)			(0.32 - 0.99)
A lagopus	Fluctuating rodents	Yugov Peninsula, Russia	16/117	7.8/117		(Nasimovich and Isakov 1985)			
A lagopus	Birds and fluctuating rodents	Kokechik Bay, Yukon-Kuskokwim delta, USA				(Anthony 1996; Anthony 1997)	7 ± 2/26 males: 10 ± 6, females: 5 ± 2		
A lagopus	Sea birds, ptarmigan, littoral	Iceland	10/309	4.2/309		(Hersteinsson 1984; Hersteinsson and Macdonald 1982)	16/3 (9 - 19)		
A lagopus	Sea birds, littoral	Bering Island, Russia	9/12	5.5/12		(Frafjord and Kruchenkova 1995)			
A lagopus	Sea birds, littoral	Mednyi Island, Russia	10/17 8/9	6.4/17 4.5/9		(Barabash-Nikiforov 1938; Frafjord and Kruchenkova 1995)			
A lagopus	Sea birds, littoral	Pribilof Islands, USA	11/22			(Preble and McAtee 1923)			
A lagopus	Sea birds, littoral	Rat Islands, USA	5/16	2.8/16		(Berns 1969)			
A lagopus	Sea birds, littoral	Svalbard (Spitzbergen), Norway	8/35 8/5	5.3/35 5.8/5		(Frafjord and Prestrud 1992; Prestrud 1992b; Prestrud 1992c; Prestrud 1992d)	54/7 (10 - 125) females: 48 ± 9/3 (36 - 50) non-breeding females: 12/2 (18 - 24)		
A lagopus	Sea birds, littoral	W Greenland				(Birks and Penford 1990)			

[1] 3–6 unique dens every month over 500 day study period.
[2] In 365–400 day study period.

Table 2. Distribution and physical characteristics of arctic fox dens in North America, Siberia, Svalbard, Greenland and Scandinavia. Based on Dalerum et al. (2001). Values in parentheses indicate total range.

Area	Site	Habitat	Geographic Region	Den Type	Latitude	No. of Dens	Density (dens/km^2)	Den openings[a]	Den area[a] (m^2)	Source
Northern Alaska	Prudhoe Bay and Colville River Delta	Coastal tundra	Arctic	Burrows	70°N	38–50	1/12 and 1/34	33 (1-85)	256 (1-625)	Garrott et al. 1983, Eberhardt et al. 1983
Northern Alaska	Teshekpuk lake area	Coastal tundra	Arctic	Burrows	70°N	50		4 (1-26)	30 (1-100)	Chesemore 1969
Northern Canada	Herschel Island	Coastal tundra	Arctic	Burrows	69°N	17	1/3	20 ± 14	123 ± 122	Smits et al. 1988
Northern Canada	Yukon coastal plain	Coastal tundra	Arctic	Burrows	69°N	25	1/102	19 ± 9	130 ± 116	Smits et al. 1988
Western Alaska	Yukon-Kuskokwim delta	Coastal tundra	Sub-arctic	Burrows	61°N	11	1/5	5 ± 3 (2-10)		Anthony 1996
Northern Siberia	Wrangel Island	Coastal tundra	Arctic	Burrows	70°N	41		31 (5-67)	70 (15-220)	Dorogoi 1987
Svalbard	Nordenskiöldland	Coastal area	Arctic	Combined rock/burrows	77°N	59	1/13	10 ± 9 (1-35)	52 ± 76 (2-630)	Prestrud 1992d
West Greenland	Disko Island	Coastal area	Arctic	Combined rock/burrows	69°N	17		18 ± 18 (1-63)	196 (3–1134)	Nielsen 1994
Norway	Hardangervidda	Mountain tundra	Sub-arctic	Burrows	60°N	31		(1-40)	(10-50000)[b]	Østbye et al. 1978
Sweden	Vindelfjällen	Mountain tundra	Sub-arctic	Burrows	66°N	69	1/21	27 ± 22 (2-89)	277 ± 237 (20-1085)	Dalerum et al. 2001

[a] Mean ± sd.
[b] Includes "den complexes," presumably with several dens.

Anthony 1996). The most important limitation to denning in the Arctic is permafrost, which often lies less than 50 cm below ground. Therefore, productive arctic fox dens are usually situated on elevated mounds, ridges, eskers, pingos, or river banks. The common characteristic of good denning sites is that they lie above the permafrost layer, accumulate comparatively little winter snow and are sun-exposed, often facing south (Macpherson 1969, Bannikov 1970, Østbye et al. 1978, Underwood and Mosher 1982, Anthony et al. 1985; Prestrud 1992d, Dalerum et al. 2001). Dens have also been found to face away from dominant summer winds (Nielsen et al. 1994). Preferred soil materials are glacifluvial sand and silt (Chesemore 1969, Østbye et al. 1978, Nielsen et al. 1994, Dalerum et al. 2001). Dens are gradually excavated deeper and deeper, as the permafrost table gradually drops due to increased air ventilation and water drainage. The thawing depth beneath a den can be twice the normal depth (Skrobov 1960, Chesemore 1969). The construction process thus takes many years, and large arctic fox dens have been described as active for hundreds of years (Lönnberg 1927, Zetterberg 1945, Macpherson 1969). Macpherson (1969) suggested an average life-span of 330 years for each den. Good den sites are limited (Smits and Slough 1993) and new arctic fox dens are constructed mainly in peak population years (Dorogoi 1987). In permafrost areas, dens are often quickly eroded and thus remain small. If necessary, dens can also be placed under large rocks and boulders (Prestrud 1992d). In some areas, artificial structures are used for denning (Eberhardt et al. 1983). In 9 studies covering a total of 539 arctic fox dens, the mean number of openings per den varied from 4 to 44 and mean den area varied from 30 to 277 m^2 (summarized in Table 2, see Dalerum et al. 2001). In Sweden, where little permafrost exists, there are dens with 147 openings, and den areas covering up to 1085 m^2 (mean=44 openings and 277 m^2, N=77; Dalerum et al. 2001). In such large dens, there is often a succession of burrows from freshly dug to completely collapsed openings (Macpherson 1969).

The openings of arctic fox dens in soft ground are round or slightly oval, 15–20 cm in diameter (Smits et al. 1988). Usually, there is little bare ground outside them, as the dug-out material is spread rather thinly, allowing the vegetation to sprout through (Chesemore 1969). The arctic fox is not as "messy" as, for instance, the red fox (*Vulpes vulpes*), and normally leaves few food remains outside the den. Even old bones and dry skin are cached in or near the den, except when food is extremely abundant (M. Tannerfeldt and A. Angerbjörn, personal observation). In rocky areas, dens are therefore difficult to find. The best signs of a whelping den are strong fox odor in the openings, extensive trampling in the vegetation between openings, and small scats. In addition, arctic fox pups often bark from below ground when approached by a human.

The excavation of arctic fox dens have consequences on plants. Poaceae and *Dryas* thrive on many sites, which are nutrient rich from fox droppings and food remains, have a warmer soil, better drainage, lower permafrost and better airing than the surrounding ground (Chesemore 1969, Macpherson 1969, Garrott et al. 1983, Smits et al. 1988, Anthony 1996). Due to the comparatively high productivity, arctic fox dens can also locally be important to grazers such as reindeer/caribou and small rodents. On the Yamal Peninsula in Siberia, there are about 24,000 dens covering on average of 250 m^2, totaling 6 km^2 of lush "arctic fox meadow" (Skrobov 1960). In other habitats, arctic fox dens may be small, ephemeral and difficult to find. This mainly occurs in areas of extensive permafrost, flooded areas (Skrobov 1960, Anthony 1996) and in rocky arctic deserts (Prestrud 1992d).

Despite their prominence in the landscape, there is little information on what arctic fox dens look like below ground. According to Skrobov (1960), dens occupy larger areas in northern Yamal (around 300 m^2) than in the south (around 100m^2). Den area is directly proportional to the severity of the conditions and especially to the proximity of the upper permafrost horizon (Skrobov 1960). This suggests that arctic fox dens extend horizontally where their downward extension is limited. On Wrangel Island near Bering Strait, a complete den was excavated, revealing a 10 x 8-m tunnel grid connecting 17 openings (Dorogoi 1987). The deepest tunnels were only 64 cm below ground, indicating a 2-dimensional structure. There was 1 widened whelping chamber in the den. Also from the Russian northeast, Nasimovich & Isakov (1985) described a cross-section of a typical den, which followed the ground contour of a slope. The tunnel systems in more complex dens have an intricate, 3-dimensional structure (Høst 1935 *in* Østbye et al. 1978), and foxes can have underground connection between most burrows also in dens with more than 50 active openings (M. Tannerfeldt, personal observation). Boitzov (1937) also excavated dens: "In the nests were found dry grass, moss, feathers and all kinds of bones. However remains of food were also found throughout the whole labyrinth." Non-natal dens, often called satellite dens, are usually smaller than natal dens for arctic foxes (Smits and Slough 1993). Arctic foxes usually have only a few satellite dens within their territory. Satellite dens can be small escape holes or day resting sites, but the category also includes alternative rearing dens.

Distribution and Use of Dens

SWIFT FOXES. Swift fox dens are usually clustered (Cutter 1958, Hines and Case 1991). Two successively used swift fox dens in Oklahoma were only 200 m apart (Kilgore 1969). Pruss (1994) found that den shifts maximally spanned 500 m. Moehrenschlager (2000) found that most movements reflected this pattern, but 1 fox pair moved 7 pups across a highway for a total distance of 1.9 km. Swift foxes use larger dens during the pup-rearing season than at other times of the year (Kilgore 1969). In southeastern Wyoming, 75.1% of dens were located within the core area of individual swift fox home ranges and common core areas of 4 fox pairs contained 84.6% of shared dens (Pechacek et al. 2000).

KIT FOXES. Kit fox dens are also clustered. O'Neal et al. (1987) found that 7–17 dens which were no more than 100 m apart formed clusters that were smaller than 2 km^2. Egoscue (1975) noted that natal dens of neighboring kit foxes were at least 3.2 km apart. Kit foxes in Arizona utilized 3–16 dens per individual and den sites were further from riparian habitat than expected (Zoellick et al. 1989). The mean movement distance between successive San Joaquin kit fox dens in California was 711 m and some foxes utilized over 50 different dens (Ralls et al. 1990).

Kit foxes use numerous dens and change between them frequently. In Utah, kit foxes changed dens as often as 33 times in 1 breeding season (O'Neal et al. 1987). Morrell (1972) found that San Joaquin kit foxes used up to 4–5 dens per month, that dens were switched most commonly during the dispersal period, and that larger dens were used during the breeding season than at other times of the year. Ralls (1990) found that San Joaquin kit foxes switched den sites after a mean period of 3.1 days and that approximately half of the dens were only utilized for 1 day at a time. Foxes infrequently reused the same sites and individuals used 3–6 unique dens per month. The rate of den switching can differ between seasons and age classes. Adults and juveniles tended to remain in the same den longest during April and September. Within each month, adults used approximately 1 more den than juveniles (Ralls et al. 1990). San Joaquin kit foxes used an average of 11.8 dens/year; the largest number of dens was used during the dispersal season whereas few were utilized during breeding and pup-rearing. Individual dens were only used for an average of 10.0% of the year, and an average of 46.6% of dens used annually had not been used by the same fox in the previous year (Koopman et al. 1998). In contrast, kit foxes in the Chihuahuan desert of Mexico used more dens during breeding and pup-rearing seasons than at other times of the year. Natal dens and satellite dens were similar, although natal dens had taller den entrances and fewer cactus species surrounding the den than non-natal sites (Rodrick and Mathews 1999).

ARCTIC FOXES. Arctic foxes maintain territories during the breeding season, sometimes all year round. Territory size and shape are determined by food availability (Hersteinsson 1984, Angerbjörn et al. 1997). There is little overlap between territories and the borders are strongly defended, although trespassing occurs when the territory owners are out of sight (Eberhardt et al. 1983, Hersteinsson 1984, Prestrud 1992b). Within each territory, there are usually 2–3 potential natal dens and several small non-natal (satellite) dens. Outside the breeding season, arctic foxes keep only a few holes open in the snow, also in large dens (M. Tannerfeldt, personal observation). Most authors agree that landscape features, substrate properties and food dispersion govern the distribution of arctic fox denning sites. Den distribution is therefore sometimes random (Fine 1980), sometimes more widely spaced than random (Macpherson 1969, Prestrud 1992b, Dalerum et al. 2001) and sometimes clumped (Prestrud 1992b, Anthony 1996). Successful denning sites on Svalbard were clustered along valley sides and the coast. Within each year, however, breeding dens were more widely spaced than random, as a result of territoriality (Prestrud 1992b). Density of arctic fox dens may vary from 1 den per 3 km^2 to 1 per 102 km^2 (Smits et al. 1988).

The large litters of arctic foxes may be split up between several dens or moved to alternative rearing dens, usually as a result of disturbance. The average distance pups are moved is 1.5–2.2 km (Boitzov 1937, Eberhardt et al. 1983, Prestrud 1992b). However, a 6-week-old arctic fox litter of 9 young moved 6 km, following red fox predation at the natal den (Elmhagen 2001). Pups may also be moved back and forth several times (Prestrud 1992b). Anthony (1996) described 1 litter of 8 which was split into 2 dens, a litter of 9 into 5 dens, and a litter of 10 into 7 dens. On Svalbard, litters were split between 2 dens in 8 cases, and entire litters moved from the primary den in 5 cases (Prestrud 1992b). Eberhardt et al. (1983) suggested that splitting of broods could reduce the risk of losing an entire litter to predation or reduce disease transmission.

Dens for Protection

The smaller canids and the pups of the larger species often run the risk of being killed by raptors and larger mammalian predators, especially when living in open habitats (e.g., Carbyn 1986, Thurber et al. 1992, Lindström et al. 1995, Ralls and White 1995, Palomares and Caro 1999). Swift and kit foxes live in moderately open habitats and are at peril from coyote, red fox, domestic dog, and golden eagle (*Aquila chrysaëtos*) (Covell 1992, Disney and Spiegel 1992, Carbyn et al. 1994, Ralls and White 1995, Moehrenschlager and Moehrenschlager 1999). To our knowledge, every telemetry study on swift or kit foxes has documented coyote-induced mortalities, and coyotes are responsible for up to 87% of fox mortalities (White and Garrott 1997). Survival rates of the latter fox species may depend on a combination of interspecific, resource-dependent home range webs and the availability of dens as escape routes (Moehrenschlager 2000). Arctic foxes live in an extremely open habitat and are killed by red fox, wolf, domestic dog, brown/grizzly bear (*Ursus arctos*), polar bear (*U. maritimus*), wolverine (*Gulo gulo*), snowy owl (*Nyctea scandiaca*) and golden eagle (Lavrov 1932; Bannikov 1970; Garrott and Eberhardt 1982; Frafjord et al. 1989; Menyushina 1994a,b).

There can also be interspecific competition for dens between canids, sometimes accompanied by food competition (e.g., Smits and Slough 1993). If the mere presence of a larger canid can reduce an animals willingness to use an area, it can be excluded from the dens in that area. This might seem to contradict other hypotheses, which state that size differences can permit otherwise similar carnivore species to coexist (e.g., Rosenzweig 1966). But a theoretical model showed that these are not mutually exclusive, but can be 2 cases in a continuous spectrum of situations with varying prey and predator sizes (Wilson 1975). Coyotes have shown interspecific territoriality towards red foxes (Dekker 1983, Voigt and Earle 1983, Sargeant et al. 1987). Similarly, coyotes avoided wolf pack territories and, like white-tailed deer (*Odocoileus virginianus*), lived mainly on wolf territory boundaries (Fuller and Keith 1981). In these situations, it is difficult to distinguish between competition and predation, but the resulting distribution is the same.

Swift foxes in Colorado (Kitchen et al. 1999) and in Canada, and kit foxes in Mexico (Moehrenschlager 2000), cannot escape from coyotes through habitat partitioning. Coyote home ranges in Mexico and Canada completely enveloped fox home ranges. However, coyotes moved randomly relative to simultaneously tracked swift foxes (Kitchen et al. 1999), and relative to swift fox or kit fox den sites (Moehrenschlager 2000). Swift and kit fox mortalities in Canada and Mexico, respectively, appear to be a function of interspecific encounter rates. Although coyotes are the main cause of swift and kit fox mortalities, they may be essential to the long-term persistence of these species because they exclude red foxes (Ralls and White 1995).

In Alberta and Saskatchewan, coyotes and swift foxes are sympatric while red foxes are normally peripheral in agricultural habitat. Nevertheless, red foxes can invade swift fox areas and take over swift fox dens (Moehrenschlager 2000). Den characteristics of swift foxes, red foxes and coyotes differed significantly. Red fox dens had significantly greater slopes than swift foxes, and red fox dens were closer to human habitation than those of the other canids. Although swift fox dens were found at all distances, red foxes always denned within 3 km of ranches, whereas all coyote dens were at least 3 km from these sites (Moehrenschlager 2000).

For arctic foxes, the main competitor for dens is the red fox. Being larger, red foxes can chase away or kill arctic foxes (Rudzinski et al. 1982, Frafjord et al. 1989, Hersteinsson et al. 1989, Hersteinsson and Macdonald 1992), thereby excluding them from important parts of their fundamental niche (Elmhagen 2001). High quality arctic fox dens were inhabited less often when red foxes were breeding within an 8-kilometer radius from the dens, than when red foxes were not present (Elmhagen 2001). Also wolves, wolverines and bears can take over arctic fox dens for breeding (Macpherson 1969, Angerbjörn and Isaksson 1995).

The examples of red foxes invading arctic and swift fox den sites illustrate the importance of the den as a crucial resource, especially during the breeding season. If dens are critical, then the intensity of competition should depend upon the availability of these sites. Since dens are more available to swift and kit foxes than to arctic foxes, we hypothesize that arctic foxes are more susceptible to competitive exclusion by red foxes than the smaller *Vulpes* species. Based on the interspecific relationships of swift and red foxes with coyotes in Canada, such pressure on arctic foxes might only be alleviated through the presence of the historic apex canid in the arctic: the wolf.

Litter Sizes

On the population level, reproductive rates are

determined by the proportion of breeding females and their litter sizes, both of which are influenced by food availability in Canidae (Macpherson 1969, Bannikov 1970, Englund 1970, Lindström 1989, Hersteinsson and Macdonald 1992, Tannerfeldt et al. 1994, Geffen et al. 1996). Litter sizes of Canadian swift foxes are correlated to breeding season body weights; consequently, winter severity can affect reproductive output in the subsequent year (Moehrenschlager 2000). Among kit foxes, breeding probabilities are closely linked to resource availability (Egoscue 1975, White and Ralls 1993). In Utah, the proportion of breeding individuals and litter size changed in response to the abundance of black-tailed jackrabbits (*Lepus californicus*) (Egoscue 1975). The prey base of kit foxes can be severely impacted by drought, which is directly manifested on the level of fox reproduction (White and Ralls 1993, Warrick and Cypher 1998).

In swift and kit foxes, litter sizes are disproportionately small compared to other canids, and so is total litter weight relative to body size (Geffen et al. 1996). The mean litter size for kit foxes and swift foxes is 3.5 and maximum litter sizes are 6 and 8 foxes, respectively (Table 1). This may be related to the food resources of swift and kit foxes, which are quite stable in comparison with the lemming fluctuations that govern most arctic fox populations, with up to 1000-fold increases in prey density (Krebs 1993). Although prey populations also for kit and swift foxes may undergo temporary crashes (White and Ralls 1993), the comparative diversity of swift and kit fox diets may allow these species to sustain smaller declines than arctic foxes, whose populations are closely tied to lemming population fluctuations.

But litter sizes are not only determined by food availability. In arctic foxes, it has been noted that coastal populations generally have smaller litters than inland foxes. Arctic foxes can have up to 19 young, which is among the largest known litter size in the order Carnivora (Ewer 1973, Ovsyanikov 1993). Also, when total litter weight, controlled for gestation time, is plotted against female weight, the arctic fox has the highest values among the Canidae (Geffen et al. 1996). A review of 16 arctic fox studies showed that unpredictable food resources (i.e., fluctuating lemming populations) were associated with larger litter sizes than more stable food resources were (unpredictable: mean litter sizes 2.8–6.4, maxima 5–11, N=5 studies; stable: mean litter sizes 5.0–11.2, maxima 7–18, N=11 studies; Tannerfeldt and Angerbjörn 1998). There were significant differences also in variances, as well as in placental scar count means. It was therefore suggested that litter size in the arctic fox is determined by adaptive plasticity. In short, according to the jackpot hypothesis, foxes with unpredictable food resources generally will have larger litter sizes at a given food resource level (Tannerfeldt and Angerbjörn 1998).

Dalerum et al. (2001) analyzed reproductive data from 16 years in 31 arctic fox breeding dens. There were 43 (58.1%) unoccupied dens which were not included in the analysis. Within breeding dens, standardized number of arctic fox litters was positively related to den area. Also, arctic fox litter size was positively related to den area. Further, for all 74 arctic fox dens in the area, the number of openings and the area covered by an den were positively related (Fig.1; Dalerum et al. 2001). Also Anthony (1996) found natal dens to be larger than non-natal dens. This confirms the conventional wisdom that large dens are good dens for arctic foxes. Arctic foxes thus chose large dens for breeding, which significantly improves their reproductive output.

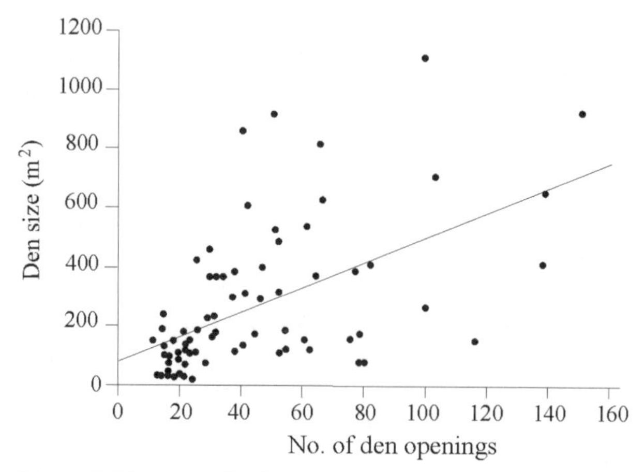

Figure 1. For arctic fox dens, the number of den openings and the area covered by a den are positively related (F (1, 72) = 37.1, $R^2 = 0.33$, $p < 0.001$, $y = 87.8 + 4.30x$, n = 74). From Dalerum et al. (2001).

Den Quality

As shown in the previous section, there are clearly good and bad dens for arctic foxes. For swift and kit foxes, distinctions into good and bad den sites have not been made. However, den sites will be used several years (Pruss 1994) and specific dens have been used at least 5 consecutive years (A. Moehrenschlager, unpublished data). Yet, Canadian swift foxes during 1995–1998 never used the same den for the birth of different litters (A. Moehrenschlager, unpublished data).

The arctic foxes' preference for specific dens could be used for management purposes. It would be advantageous to be able to identify "hot spots" for breeding, upon which surveys and management actions could focus. Angerbjörn et al. (1995) investigated arctic fox den data collected during 2 decades from an area of approximately 32,500 km^2 in Sweden. Each den was assigned a quality index (QI), based on its rate of occupancy (years producing litters divided by years monitored). As the basic population cycle had a 4-year period, dens which had been monitored less than 4 years were excluded, and the remaining 154 dens were assigned to 1 of the following 4 categories (with percentage of dens in parentheses): Cat1, QI = 0 (38.3%);

Cat2, QI ≤0.25 (38.3%); Cat3, 0.25 <QI ≤0.5 (20.8%); Cat4, QI >0.5 (2.6%). This index was used to check a monitoring program for biases (Angerbjörn et al. 1995). It would also be interesting to test whether this quality index is related to size or other den characteristics.

Macpherson (1969) categorized arctic fox dens from an airplane as either "youthful" (no characteristic vegetation and few burrows), "mature" (well-developed with good mat of vegetation), "old" (large den with many burrows and rich vegetation), or "senile" (not active with collapsed burrows). Strangely enough, however, he reported an occupancy rate of 21.4% for the latter category. The word "senile" suggests that there is a definite end to the use of a particular den. However, if good den sites are valuable to arctic foxes, why abandon them? They may be abandoned when they have been excavated by other animals (Macpherson 1969). Destroyed dens could be brought back into use after the tunnels have fallen in completely and the material on the den site has stabilized. In some large arctic fox dens, 1 side of a hill or ridge has an active den, while the other side contains old and collapsed burrows. This implies that the same den site is used continuously, although specific burrows may be left to collapse. It is also possible that dens are abandoned for a number of years to avoid a heavy parasite load (Butler and Roper 1996). Parasite infestation is an interesting aspect of den ecology, which has received little attention so far.

Social Organization

Within the canid family, helpers and/or communal feeding are reported for at least 10 species (see Kleiman [1977] for a review; also for kit fox see Cutter [1958]; dhole [*Cuon alpinus*] see Johnsingh [1982]; arctic fox see Hersteinsson [1984]). This apparently alloparental behavior is a phenomenon which has generated many theories, models and hypotheses (e.g., Brown 1983; Ferguson et al. 1983; Pyearah 1984; Schantz 1984a,b; Kruuk and Macdonald 1985). But for the smaller Vulpes, there is little evidence of extra adults providing valuable care for non-offspring (e.g., Strand et al. 2000). For several species, the number of helpers is larger when food is more abundant (Harrington et al. 1983). But, assuming that help is given by feeding the pups or the mother, the benefit to the parents should be largest at intermediate levels of food abundance (Hersteinsson 1984). Thus, there can be a conflict between parents and offspring. Taking into consideration the different options open to parents, helpers and offspring in the flexible social systems of canids, the interactions between individual and kin selection become very complex (Emlen 1978). A general model has shown that group sizes are not always of the size that would maximize individual fitness, but larger (Rodman 1981). This, Rodman concluded, is an effect of kin selection.

Swift and kit foxes are mainly monogamous (O'Farrell 1987, Scott-Brown et al. 1987), but more than 2 foxes are frequently seen at den sites. In Colorado, swift fox trios and a quad consisting of a male and accompanying females were observed in a coyote-control area. Litter sizes of groups with multiple females were significantly larger than those of pairs (Covell 1992). In Utah, a yearling female kit fox was a helper (O'Neal et al. 1987). Egoscue (1975) observed 1 polygamous trio which produced pups. Among San Joaquin kit foxes, trios contained either 2 males or 2 females (Ralls et al. 1990). Since these observations were made during a period of resource scarcity which virtually eliminated reproductive output, trio formation is not necessarily linked to abundant resources.

In Wyoming, 70% of dens belonging to mated, male swift foxes and 82% of dens belonging to females were shared with their mates (Pechacek et al. 2000). Females with dens were located with their mates approximately 60% of the time. Koopman et al. (1998) found that San Joaquin kit fox mates denned together for 45% of the year and mated adults denned together less often during dispersal than during the breeding season. Adult males and females denned with their offspring for up to 17 and 18 months, respectively. Den sharing with the yearlings decreased as the next litter was born but returned to previous levels after 2 months. Den sharing by siblings was as common in the second year as in the first, but ceased after 21 months of age.

Similar to red foxes, arctic foxes can increase group size at high population densities, usually by allowing additional adults at breeding dens (Zetterberg 1953; Eberhardt et al. 1983; Macdonald 1983; Schantz 1984a,b; Lindström 1986; Hersteinsson and Macdonald 1992; Strand et al. 2000). The Mednyi Island arctic fox population, however, is an interesting extreme in this respect. The island earlier had a very dense population of several thousand animals, with 5 individuals per km^2 (Boitzov 1937). In this population, arctic fox adults lived in large groups that shared dens. In the 1980's, numbers dropped to a few dozen due to an introduced ear tick (Kruchenkova and Formozov 1995). However, the complex social system remains and in 33 examined dens, the number of adults were as follows: 2 adults 39%, 3 adults 36%, 4 adults 15%, 5 adults 6% and 6 adults 3% (Frafjord and Kruchenkova 1995). In the Mednyi Island population, it is apparently common that related females join their litters (Kruchenkova and Formozov 1995). In Norway, it was confirmed that pups in 1 den suckled from 2 lactating females (Strand et al. 2000). The authors assumed this to be 2 different litters raised together. Two litters in 1 den have also been recorded in red foxes (Macdonald 1980). In Sweden, arctic fox females were observed to join litters with neighbors on 2 occasions. In 1 case, it was a 2-year-old daughter who moved her 2 pups to join her mother and her litter after predation attempts from red foxes at the daughter's den. The next day, 1 of the daughter's pups died as a result of bite wounds from a red fox (Elmhagen 2001).

Concluding Discussion

SIMILARITIES BETWEEN THE SPECIES. Swift, kit and arctic foxes share a number of behavioral and ecological traits, with similarities in their den ecology. The primary function of breeding dens is most likely to provide protection against predators. In addition, dens provide protection for juveniles against harsh weather in open landscapes. For all 3 species, natal dens are larger than satellite dens. Furthermore, arctic foxes choose the largest dens for breeding and reproductive output is positively correlated with den size. This relationship holds true also for dens with more than 100 openings, which indicates that large burrow systems with many escape routes are important. In swift and kit foxes, the importance of predator avoidance is instead manifested through a large number of escape routes in a 'survival sieve' of satellite dens within each home range (Moehrenschlager 2000).

Another similarity between these foxes is that natal dens are exclusively used by resident family groups. There are some exceptions to this, where breeding pairs may share a den. However, this phenomenon seems to be restricted to close relatives, and foxes are normally strongly territorial when breeding. Swift and arctic foxes, and possibly also kit foxes, tend to use den sites with lush vegetation near hilltops. As discussed for the arctic fox, lush vegetation around the den may result from the foxes' own activity. Hilltops are well drained, which reduces the risk of flooding. High sites also allow for better predator detection.

DIFFERENCES BETWEEN THE SPECIES—TWO DEN ECOLOGY STRATEGIES. We have also discussed differences in the den ecology of arctic, kit and swift foxes. Swift and kit foxes depend on dens throughout the year. Apparently, they need dens for protection of both adults and pups. Also arctic foxes may stay in their territory throughout the winter, but often leave their breeding grounds in winter. They may even spend several months as scavengers far out on the pack ice, following polar bears (Chesemore 1968; Pulliainen 1965).

Swift and kit foxes utilize more dens than arctic foxes do. This is probably the result of several factors. In the permafrost areas of arctic foxes, there is often a very limited numbers of potential den sites. Further, it is easier to enlarge small dens than to dig new dens in the Arctic, due to increasing thawing depth. On the lower latitudes of swift and kit foxes, there are many other species that dig dens which can be taken over by foxes (Cutter 1958, Kilgore 1969). When larger denning predators such as red foxes or badgers are common, dens may be taken over and it might be better to have many small dens than to invest time in digging large dens. This might be the reason why the arctic fox has difficulties in coping with an expanding red fox population (Hersteinsson et al. 1989, Hersteinsson and Macdonald 1992), while swift and kit fox populations can persist or even increase despite large numbers of competitors and predators. In undisturbed tundra systems, the only terrestrial predator to seriously compete with the arctic fox for dens is the wolf. The weight ratio between arctic foxes and wolves is around 1:10 (Ginsberg and Macdonald 1990). Their differences in home range sizes and den use are probably so large that competition for dens has been of little importance in shaping arctic fox den ecology. Swift and kit foxes, on the other hand, have evolved in areas with several den dependent competitors, but also with den providers, such as badgers and prairie dogs. Swift and kit foxes should therefore be less vulnerable than arctic foxes to increasing den competition.

Den switching is common among canids, and numerous related factors have been considered as explanations. Ryon (1986) reviewed 5 potential causes: 1) disturbance; 2) flea infestation; 3) leaking dens; 4) shifting towards food sources; and 5) predator avoidance. Although foxes will move in response to some disturbances, swift foxes in Canada utilized the same dens immediately before and after intensive pipeline construction (Moehrenschlager 2000). Swift and kit foxes have heavy flea and tick parasite loads (A. Moehrenschlager, personal observation) compared to arctic foxes, where fleas and ticks are normally not found during handling (Aguirre et al. 2000; M. Tannerfeldt and A. Angerbjörn, personal observation). The frequent, short-distance den changes of swift and kit foxes may thus be the result of parasite avoidance. Changes in food resources and leaking dens can cause den switching (Pruss 1994), but cannot account for the majority of movements. While den shifting in response to food shortages can explain long-distance movements, most consecutively used swift fox dens are close together (Pruss 1994, Moehrenschlager 2000) and short transfers would not provide substantial hunting advantages. In Canada, coyotes or signs from them were seen at 75 inspections of swift fox dens, so den switching may be a response to high predation pressure. However, predator presence does not necessitate den switching and most Canadian swift foxes did not abandon dens after coyote visits. For arctic foxes, den switching and splitting of litters is often related to predation events or human disturbance (Eberhardt et al. 1983, Prestrud 1992b).

We have thus found what could be seen as 2 different den ecology strategies in these fox species. In most areas, the arctic fox has large dens, few satellite dens and seldom move between dens during the breeding season. Swift and kit foxes, on the other hand, have small dens, use many satellite dens and readily move between them. This difference between den ecology strategies is connected to differences in a number of life history traits. The strategies also involve territoriality, and are related to differences in a number of ecological parameters such as predation rates, availability of dens, food resources and litter sizes (Table 3). To cope with fluctuating food resources, arctic foxes

Table 3. Life history traits and ecological parameters related to denning for 3 fox species. The table depicts the most common situation for each species; for the arctic fox, populations feeding in coastal areas are excluded. Data from Table 1 and references therein.

Trait or parameter	Arctic fox	Kit fox	Swift fox
Den size	Large	Small	Small
Satellite dens	Few	Many	Many
Den changes in breeding season	Rare	Very common	Very common
Litter size	Very large	Small	Small
Territory stability (between year)	Strong	Weak	Weak
Territory size	Large	Small	Small
Availability of dens	Low	High	High
Food resources	Fluctuating	Relatively stable	Relatively stable
Predation rate on young	Low-medium	High	High
Main mortality cause for adults	Starvation	Predation	Predation

have developed the ability to produce very large litters (Tannerfeldt and Angerbjörn 1998). This is facilitated by the access to large and relatively safe dens. In contrast, swift and kit foxes have smaller litters than expected by allometric relationships. This is perhaps a necessity when changing dens as often as every 3 days (Ralls et al. 1990).

In conclusion, denning swift and kit foxes often move between a number of small dens, as a result of higher predation risk and stronger competition for dens. Parasite avoidance might also play a role in short-range moves. These foxes have a good availability of already-dug dens. Arctic foxes, in the other hand, live in a less diverse ecosystem with fewer dens, predators and competitors. They have large litters in large dens and mostly stay in 1 den during the breeding season, unless disturbed by humans or predators.

Management Implications

Den ecology is an important aspect of the management of all 3 fox species. Dens are crucial as shelter against harsh weather for the young and for protection against predators for both young and adult foxes. Habitat protection for these foxes should therefore focus on important den sites. For arctic foxes, we have shown that important dens can be identified quite readily. Although there are indications of preferred den sites also for swift foxes (Pruss 1999), both kit and swift foxes move frequently between dens, so their entire home range areas should ideally be protected.

Identification and classification of den sites is a means of making surveys and population estimates more effective, especially for the arctic fox. An analysis of den sites is also an important preparatory task for re-introduction programs, as den availability is a crucial aspect of the suitability of an area for foxes. If necessary, it may also be possible to construct or improve den sites. As pointed out by Pruss (1999), information on preferred den sites could further be used to improve conditions for captive foxes.

There is a large number of studies of fox dens in the literature, but many aspects of foxes' use of dens warrant further investigation. For example, how do foxes utilize their home range and its dens, and how does this change with predation pressure and food availability? Does the minimum density of dens necessary for breeding change with predation pressure, and is this related to the quality of available dens? Swift foxes den successfully in some agricultural areas but not in others; does this depend on the availability of dens? Finally, it is worth considering whether human-made dens in some areas could be a tool to increase survival of endangered populations of arctic, swift or kit foxes.

Acknowledgments

Magnus Tannerfeldt and Anders Angerbjörn wish to thank the EU LIFE-Nature fund, WWF Sweden and the foundations of Carl Trygger, Magnus Bergvall, Ebba och Sven Schwartz, Oscar och Lili Lamms Minne and Hierta Retzius for financial support to the Swedish Arctic Fox Project SEFALO. We also express our gratitude towards A.B. Dogman, Fjällräven AB, Bestfood Nordic, Tågkompaniet and Cloetta Fazer for valuable support. Axel Moehrenschlager wishes to acknowledge the support of Alberta Environmental Protection, the Alberta Sport Recreation Parks and Wildlife Foundation, the Canadian Wildlife Service, Express Pipelines, Green Plan International, the People's Trust for Endangered Species, the Rocky Mountain Elk Foundation, Saskatchewan Environment and Resource Management, the Swift Fox Conservation Society, the University of Alberta Biodiversity Fund, Wildlife Preservation Trust Canada, and WWF Canada. Axel Moehrenschlager also thanks J. Johnson, J. Michie, C. Moehrenschlager, J. Scharlemann, C. Stokke, and I. Welsh for dedicated support. We further thank an anonymous referee for valuable comments on the manuscript.

Literature Cited

Aguirre, A.A., A. Angerbjörn, M. Tannerfeldt, and T. Mörner. 2000. Health evaluation of endangered arctic fox (*Alopex lagopus*) cubs in Sweden. Journal of Zoo and Wildlife Medicine 31(1):36–40.

Angerbjörn, A., B. Arvidson, E. Norén, and L. Strömgren. 1991. The effect of winter food on the reproduction in the arctic fox, *Alopex lagopus*: a field experiment. Journal of Animal Ecology 60:705–714.

Angerbjörn, A., and E. Isaksson. 1995. The abundance of wolves in northern Siberia. Pp. 122–127 *in* E. Grönlund and O. Melander, editors. Swedish-Russian Tundra Ecology-Expedition-94. A Cruise Report, Stockholm. Swedish Polar Research Secretariat, Stockholm, Sweden.

Angerbjörn, A., J. Stroman, and D. Becker. 1997. Home range pattern in arctic foxes. Journal of Wildlife Research 2:9–14.

Angerbjörn, A., M. Tannerfeldt, A. Bjärvall, M. Ericson, J. From, and E. Norén. 1995. Dynamics of the arctic fox population in Sweden. Annales Zoologici Fennici 32:55–68.

Anthony, R.M. 1996. Den use by arctic foxes (*Alopex lagopus*) in a subarctic region of western Alaska. Canadian Journal of Zoology 74(4):627–631.

——. 1997. Home range and movements of arctic foxes (*Alopex lagopus*) in western Alaska. Arctic 50:147–157.

Anthony, R.M., A.A. Stickney, and K. Kertell. 1985. Den ecology, distribution, and productivity of foxes at Kokechik Bay, Alaska. U.S. Fish and Wildlife Service, Alaska Office of Fish and Wildlife Research, Anchorage, Alaska.

Bailey, E.P. 1992. Red foxes, *Vulpes vulpes*, as biological control agents for introduced arctic foxes, *Alopex lagopus*, on Alaskan Islands. Canadian Field Naturalist 106(2):200–205.

Bannikov, A.G. 1970. Arctic fox in the U.S.S.R.: biological premises of productivity. *In* W. A. Fuller and P. G. Kevan, editors. Productivity and conservation in northern circumpolar lands. International Union for Conservation of Nature and Natural Resources, Morges, Switzerland.

Barabash-Nikiforov, I. 1938. Mammals of the Commander Islands and the surrounding sea. Journal of Mammalogy 19: 423–429.

Berns, V.D. 1969. Notes on the blue fox of Rat Island, Alaska. Canadian Field-Naturalist 83:404–405.

Birks, J.D.S., and N. Penford. 1990. Observations on the ecology of arctic foxes *Alopex lagopus* in Eqalummiut Nunaat, West Greenland. Medd. Grönland. Bioscience 32:1–28.

Boitzov, L.V. 1937. The arctic fox: biology, food habits, breeding. Trudy Arkticheskogo Nauchno-issledovatelskogo Instituta 65:7–144.

Brechtel, S.H., L.N. Carbyn, D. Hjertaas, and C. Mamo. 1993. Canadian swift fox reintroduction feasibility study: 1989–1992. Scientific and Technical Documents Division, Canadian Wildlife Service, Ottawa, Ontario.

Brown, J.L. 1983. Cooperation—a biologists dilemma. Advances in the Study of Behavior 13:1–37.

Burgess, R.M. 1980. Ecology of arctic foxes (*Alopex lagopus*) at Demarcation Bay. Alaskan Wildlife Project Reports.

Butler, J.M., and T.J. Roper. 1996. Ectoparasites and sett use in European badgers. Animal Behaviour 52(3):621–629.

Carbyn, L.N. 1986. Some observations on the behaviour of swift foxes in reintroduction programs within the Canadian prairies. Alberta Naturalist 16:37–41.

Carbyn, L.N., H.J. Armbruster, and C. Mamo. 1994. The swift fox reintroduction program in Canada from 1983 to 1992. Pp. 247–271 *in* M.L. Bowles and C.J. Whelan, editors. Restoration of endangered species: conceptual issues, planning and implementation. University of Cambridge Press, Cambridge, UK.

Chernyevski, F.B., and I.V. Dorogoi. 1981. Ecology of the arctic fox. *In* V.G. Krivoskeev, editor. Ecology of mammals and birds in the Wrangel Island. Vladivostok, Russia.

Chesemore, D.L. 1968. Distribution and movement of white foxes in northern and western Alaska. Canadian Journal of Zoology 46:849–853.

——. 1969. Den ecology of the arctic fox in northern Alaska. Canadian Journal of Zoology 47:121–129.

——. 1975. Ecology of the arctic fox (*Alopex lagopus*) in North America—a review. Pp. 143–163 *in* M.W. Fox, editor. The Wild Canids. their systematics, behavioral ecology and evolution. Van Nostrand Rheinhold, New York, New York.

Cotterill, S.E. 1997. Population census of swift fox (*Vulpes velox*) in Canada: winter 1996–1997. Unpublished report for the Swift Fox National Recovery Team. Alberta Environmental Protection (Natural Resources Service), Edmonton, Alberta.

Covell, D.F. 1992. Ecology of the swift fox (*Vulpes velox*) in southeastern Colorado. Thesis, University of Wisconsin, Madison, Wisconsin.

Cutter, W.L. 1958. Denning of the swift fox in northern Texas. Journal of Mammalogy 39:768–774.

Cypher, B.L, and K.A. Spencer. 1998. Competitive interactions between coyotes and San Joaquin kit foxes. Journal of Mammalogy 79:204–214.

Cypher, B.L, G.D. Warrick, M.R.M. Otten, T.P. O'Farrell, W.H. Berry, C.E. Harris, T.T. Kato, P.M. McCue, J.H. Scrivner, and B.W. Zoellick. 2000. Population dynamics of San Joaquin kit foxes at the Naval Petroleum Reserves in California. Wildlife Monographs 145:1–43.

Dalerum, F., M. Tannerfeldt, B. Elmhagen, D. Becker, and A. Angerbjörn. 2001. Distribution, morphology and use of arctic fox dens in Sweden. Wildlife Biology, in press.

Dekker, D. 1983. Denning and foraging habits of red foxes, *Vulpes vulpes*, and their interaction with coyotes, *Canis latrans*, in central Alberta, 1972–1981. Canadian Field-Naturalist 97:303–306.

Disney, M., and L.K. Spiegel. 1992. Sources and rates of San Joaquin kit fox mortality in western Kern County, California. Transactions of the Western Section Wildlife Society 28:73–82.

Dorogoi, I.V. 1987. Ecology of rodent eating predators of the Wrangel Island and their role in lemming number's dynamics. DVO AN SSSR, Vladivostok, Russia.

Dragoo, J. W., J.R. Choate, T.L.Yates, and T.P. O'Farrell. 1990. Evolutionary and taxonomic relationships among North American arid land foxes. Journal of Mammalogy 71:318–332.

Eberhardt, L.E., R.A. Garrott, and W C. Hanson. 1983. Den use by arctic foxes in northern Alaska. Journal of Mammalogy 64:97–107.

Egoscue, H.J. 1975. Population dynamics of the kit fox in western Utah. Bulletin of the Southern California Academy of Science 74:122–127.

Elmhagen, B. 2001. Competition between arctic and red foxes. Thesis, Stockholm University, Stockholm, Sweden.

Elmhagen, B., M. Tannerfeldt, P. Verucci, and A. Angerbjörn. 2000. The arctic fox (*Alopex lagopus*)—an opportunistic specialist. Journal of Zoology 251:139–149.

Elton, C.S. 1924. Period fluctuations in the numbers of animals: their causes and effects. British Journal of Experimental Biology 2(1):119–163.

Emlen, S.T. 1978. Cooperative breeding in birds and mammals. Pp. 305–339 *in* J.R. Krebs and N.B. Davies, editors. Behavioural ecology—an evolutionary approach. Blackwell Scientific Publications, Oxford, United Kingdom.

Englund, J. 1970. Some aspects of reproduction and mortality rates in Swedish foxes (*Vulpes vulpes*), 1961–63 and 1966–69. Swedish Wildlife (Viltrevy) 8(1):1–82.

Ericson, M. 1984. Fjällräv. Viltnytt 19:13-20.

Ewer, R.F. 1973. The carnivores. Cornell University Press, New York, New York.

Federal Register. 1995. Endangered and threatened wildlife and plants: 12-month finding for a petition to list the swift fox as endangered. Federal Register 60:31663–31666.

Ferguson, J.W.H, J.A.J. Nel, and M.J. De Wet. 1983. Social organization and movement patterns of black-backed jackals *Canis mesomelas* in South Africa. Journal of Zoology 199:487–502.

Fine, H. 1980. Ecology of arctic foxes at Prudhoe Bay, Alaska. Thesis, University of Alaska, Fairbanks, Alaska.

Frafjord, K., and G. Rofstad. 1998. Fjellrev på Nordkalotten. Nordkalottrådets rapportserie 47:1–39.

Frafjord, K., D. Becker, and A. Angerbjörn. 1989. Interactions between arctic and red foxes in Scandinavia—predation and aggression. Arctic 42(4):354–356.

Frafjord, K., and E. Kruchenkova. 1995. The Commander Islands: the tragedy of Bering and Steller's arctic foxes. Fauna 48(4):190–203.

Frafjord, K., and P. Prestrud. 1992. Home Range and movements of arctic Foxes Alopex-Lagopus in Svalbard. Polar Biology 12(5):519–526.

Fuller, T.K., and L.B. Keith. 1981. Non-overlapping ranges of coyotes and wolves in northeastern Alberta. Journal of Mammalogy 62(2):403–405.

Garrott, R.A., and L.Eberhardt. 1982. Mortality of arctic fox pups in northern Alaska. Journal of Mammalogy 63(1): 173–174.

Garrott, R.A., L.Eberhardt, and W. Hanson. 1983. Arctic fox den identification and characteristics in northern Alaska. Canadian Journal of Zoology 61:423–426.

Geffen, E., M.E. Gompper, J.L. Gittleman, L. Hang-Kwang, D.W. Macdonald, and R.K. Wayne. 1996. Size, life-history traits, and social organization in the Canidae: a reevaluation. American Naturalist 147(1):140–160.

Geffen, E., A. Mercure, D.J. Girman, D.W. Macdonald, and R.K. Wayne. 1992. Phylogenetic relationships of the fox-like canids: mitochondrial DNA restriction fragment, site and cytochrome b sequence analyses. Journal of Zoology 228:27–39.

Ginsberg, J.R, and D.W. Macdonald. 1990. Foxes, wolves, jackals, and dogs: an action plan for the conservation of canids. International Union for Conservation of Nature and Natural Resources, Gland, Switzerland.

Hall, M.N. 1989. Parameters associated with cyclic populations of arctic fox (*Alopex lagopus*) near Eskimo Point, Northwest Territories: morphometry, age, condition, seasonal and multiannual influences. Thesis, Laurentian University, Sudbury, Ontario.

Harrington, F.H., L.D. Mech, and S.H. Fritts. 1983. Pack size and wolf pup survival: their relationship under varying ecological conditions. Behavioral Ecology and Sociobiology 13:19–26.

Herrero, S., C. Schroeder, and M. Scott-Brown. 1986. Are Canadian foxes swift enough? Biological Conservation 36:159–167.

Hersteinsson, P. 1984. The behavioural ecology of the arctic fox (*Alopex lagopus*) in Iceland. Dissertation, University of Oxford, Oxford, UK.

Hersteinsson, P., A. Angerbjörn, K. Frafjord, and A. Kaikusalo. 1989. The arctic fox in Fennoscandia and Iceland: management problems. Biological Conservation 49:67–81.

Hersteinsson, P., and D.W. Macdonald. 1982. Some comparisons between red and arctic foxes, *Vulpes vulpes* and *Alopex lagopus*, as revealed by radio tracking. Symposia of the Zoological Society of London 49:259–289.

——. 1992. Interspecific competition and the geographical distribution of red and arctic foxes *Vulpes vulpes* and *Alopex lagopus*. Oikos 64:505-515.

——. 1996. Diet of arctic foxes (*Alopex lagopus*) in Iceland. Journal of Zoology 240(3):457–474.

Hines, T.D., and R.M. Case. 1991. Diet, home range, movements, and activity periods of swift fox in Nebraska. Prairie Naturalist 23(3):131–138.

Høst, P. 1935. Trekk av dyearelivet på Hardangervidda. Norsk Jaeger o Fisker Forenings Tidsskrift 64:76–84, 137–143, 201–211, 296–317, 410–415, 515–527.

Jackson, V.L., and J.R. Choate. 2000. Dens and den sites of the swift fox, *Vulpes velox*. Southwestern Naturalist 45(2): 212–220.

Johnsingh, A.J.T. 1982. Reproductive and social behaviour of the dhole, *Cuon alpinus* (Canidae). Journal of Zoology 198: 443–463.

Kahn, R., and T. Beck. 1996. Swift fox investigations in Colorado. Pp.10–15 *in* B. Luce and F. Lindzey, editors. 1996 Annual report of the swift fox conservation team. Wyoming Game and Fish Department, Lander, Wyoming.

Kaikusalo, A., and A. Angerbjörn. 1995. The arctic fox in Finnish Lapland, 1964–93. Annales Zoologici Fennici 32:69–77.

Kilgore, D.L., Jr. 1969. An ecological study of swift fox (*Vulpes velox*) in the Oklahoma Panhandle. American Midland Naturalist 81:512–534.

Kitchen, A.M., E.M. Gese, and E.R. Schauster. 1999. Resource partitioning between coyotes and swift foxes: space, time, and diet. Canadian Journal of Zoology 77:1645–1656.

Kleiman, D.G. 1977. Monogamy in mammals. Quarterly Review of Biology 52:39–69.

Klir, J.J., and J.E. Heath. 1992. An infrared thermographic study of surface temperature in relation to external thermal stress in three species of foxes: the red fox (*Vulpes vulpes*), arctic fox (*Alopex lagopus*), and kit fox (*Vulpes macrotis*). Physiological Zoology 65(5):1011–1021.

Koopman, M.E., J.H. Scrivner, and T.T. Kato. 1998. Patterns of den use by San Joaquin kit foxes. Journal of Wildlife Management 62:373–379.

Krebs, C.J. 1993. Are lemmings large Microtus or small reindeer? A review of lemming cycles after 25 years and recommendations for future work. Pp. 247–260 *in* N.C. Stenseth and R.A. Ims, editors. The biology of lemmings. Academic Press, London, UK.

Kruchenkova, E., and N. Formozov. 1995. The arctic foxes of Mednyi (Copper) Island. Russian Conservation News 2:19–20.

Kruuk, H., and D.W. Macdonald. 1985. Group territories of carnivores: empires and enclaves. *In* R.M. Sibly and R.H. Smith, editors. Behavioural ecology—ecological consequences of adaptive behaviour. Blackwell Scientific Publications, Oxford, UK.

Landa, A., O. Strand, J.D.C. Linnell, and T. Skogland. 1998. Home-range sizes and altitude selection for arctic foxes and wolverines in an alpine environment. Canadian Journal of Zoology 76:448–457.

Lavrov, N.P. 1932. Pesets (The arctic Fox) [In Russian]. J.D. Jackson, translator. G.M. Zhitkov, editor. Translation in Elton Library, Oxford University State Foreign-Trade Publishing House "Vneshtorgisdat," Oxford, UK.

Lindström, E. 1986. Territory inheritance and the evolution of group-living in carnivores. Animal Behavior 34:1825–1835.

——. 1989. Food limitation and social regulation in a red fox population. Holarctic Ecology 12:70–79.

Lindström, E., S. M. Brainerd, J. O. Helldin, and K. Overskaug. 1995. Pine marten—red fox interactions: a case of intraguild predation? Annales Zoologici Fennici 32(1):123–130.

List, R. 1998. Ecology of kit fox (*Vulpes macrotis*) and coyote (*Canis latrans*) and the conservation of the prairie dog ecosystem in northern Mexico. Thesis, University of Oxford, Oxford, UK.

Löfgren, S., and A. Angerbjörn. 1998. Åtgärdsprogram för fjällräv. Stockholm: Naturvårdsverket.

Lönnberg, E. 1927. Fjällrävsstammen i Sverige 1926. Royal Swedish Academy of Sciences, Uppsala, Sweden.

Macdonald, D.W. 1980. Social factors affecting reproduction amongst red foxes. *In* E. Zimen, editor. Biogeographica,

Volume 18.

Macdonald, D.W. 1983. The ecology of carnivore social behaviour. Nature 301:380–384.

Macpherson, A.H. 1969. The dynamics of Canadian arctic fox populations. Canadian Wildlife Service Report 8:1–49.

Martin, L.D. 1989. Fossil history of the terrestrial Carnivora. Pp. 536–568 *in* J.L. Gottleman, editor. Carnivore behavior, ecology, and evolution. Comstock Publishing Associates, Ithaca, New York.

Menyushina, I.E. 1994a. Interspecies relation of the polar fox (*Alopex lagopus* L.) and the snowy owl (*Nyctea scandiaca* L.) during the breeding season in the Vrangel Island, I. Lutreola 3:15–21.

——. 1994b. Interspecies relation of the polar fox (*Alopex lagopus* L.) and the snowy owl (*Nyctea scandiaca* L.) during the breeding season in the Vrangel Island, II. Lutreola 4:8–14.

Mercure, A., K. Ralls, K.P. Koepfli, and R.K. Wayne. 1994. Genetic subdivisions among small canids: mitochondrial DNA variation of swift, kit, and arctic foxes. Evolution 47:1313–1328.

Moehrenschlager, A. 2000. Effects of ecological and human factors on the behaviour and population dynamics of reintroduced Canadian swift foxes (*Vulpes velox*). Dissertation, University of Oxford, Oxford, UK.

Moehrenschlager, A., and R. List. 1996. Comparative ecology of North American prairie foxes—conservation through collaboration. Pp. 22–27 *in* D.W. Macdonald and F.H. Tattersall, editors. The WildCRU Review: The tenth anniversary report of the Wildlife Conservation Research Unit at Oxford University. George Stress Press, Stafford, UK.

Moehrenschlager, A., and C. Moehrenschlager. 2001. Demographic changes of reintroduced swift foxes in Canada and northern Montana. Alberta Environmental Protection Report, Edmonton, Alberta, in press.

Moehrenschlager, C., and A. Moehrenschlager. 1999. Canadian swift fox (*Vulpes velox*) population assessment: winter, 1999. Alberta Environmental Protection, Edmonton, Alberta.

Morrell, S. 1972. Life history of the San Joaquin kit fox. California Fish and Game 58:162–174.

Nasimovich, A., and Y. Isakov, editors. 1985. Arctic fox, red fox and racoon dog: distribution of resources, ecology, use and conservation (in Russian). Janka, Moscow, Russia.

Nielsen, S.M., V. Pedersen, and B.B. Klitgaard. 1994. Arctic fox (*Alopex lagopus*) dens in the Disko Bay area, West Greenland. Arctic 47(4):327–333.

O'Farrell, T.P. 1987. Kit fox. Pp. 422–431 *in* M. Novak, J.A. Baker, M.E. Obbard, and B. Malloch, editors. Wild furbearer management and conservation in North America. Ontario Trappers Association, North Bay, Ontario.

O'Farrell, T.P., C. E. Harris, T.T. Kato, and P.M. McCue. 1986. Biological assessment of the effects of petroleum production at maximum efficient rate, Naval Petroleum Reserve 1 (Elk Hills), Kern County, California, on the endangered San Joaquin kit fox (*Vulpes macrotis mutica*). Report nr EGG10282–2107.

O'Neal, G.T., J.T. Flinders, and W.P. Clary. 1987. Behavioral ecology of the Nevada kit fox (*Vulpes macrotis nevadensis*) on a managed desert rangeland. Pp. 443–481 *in* H.H. Genoways, editor. Current Mammalogy, Volume 1. Plenum Press, New York, New York.

Østbye, E., H.J. Skar, D. Svalastog, and K. Westby. 1978. Fjellrev og Rödrev på Hardangervidda; hiökologi, utbredelse og bestandsstatus. Meddelelser fra norsk viltforskning 3(4):1–66.

Ovsyanikov, N.G. 1993. Behaviour and social organization of the arctic fox. Dissertation, Institute of Animal Evolution, Morphology, and Ecology, Academy of Science, Moscow, Russia.

Palomares, E., and T.M. Caro. 1999. Interspecific killing among mammalian carnivores. American Naturalist 153:492–508.

Pechacek, P., F.G. Lindzey, and S.H. Anderson. 2000. Home range size and spatial organization of Swift fox *Vulpes velox* (Say, 1823) in southeastern Wyoming. Zeitschrift feur Säugetierkunde 65:209–215.

Preble, E.A., and W.L. McAtee. 1923. Birds and mammals of the Pribilof Islands. North American Fauna 46:1–128.

Prestrud, P. 1991. Adaptations by the arctic fox (*Alopex lagopus*) to the polar winter. Arctic 44(2):132–138.

——. 1992a. Arctic foxes in Svalbard: population ecology and rabies. Dissertation, Norsk Polarinstitutt, Oslo, Norway.

——. 1992b. Denning and home-range characteristics of breeding arctic foxes in Svalbard. Canadian Journal of Zoology 70(7):1276–1283.

——. 1992c. Food habits and observations of the hunting behaviour of arctic foxes, *Alopex lagopus*, in Svalbard. Canadian Field-Naturalist 106(2):225–236.

——. 1992d. Physical characteristics of arctic fox (*Alopex lagopus*) dens in Svalbard. Arctic 45(2):154–158.

Pruss, S.D. 1994. An observational natal den study of wild swift fox (*Vulpes velox*) on the Canadian Prairie. Thesis, University of Calgary, Alberta.

——. 1999. Selection of natal dens by the swift fox (*Vulpes velox*) on the Canadian prairies. Canadian Journal of Zoology 77:646–652.

Pulliainen, E. 1965. On the distribution and migrations of the arctic fox (*Alopex lagopus* L.) in Finland. Aquilo. Serie Zoologica 2:25–40.

Pyearah, D. 1984. Social distribution and population estimates of coyotes in north-central Montana. Journal of Wildlife Management 48(3):679–690.

Ralls, K., and P.J. White. 1995. Predation on San Joaquin kit foxes by larger canids. Journal of Mammalogy 76:723–729.

Ralls, K., P.J. White, J. Cochran, and D.B. Siniff. 1990. Kit fox—coyote relationships in the Carrizo Plain Natural Area. Report nr Permit PRT 702631, Subpermit Rallk-4.

Reese, E.A., W.G. Standley, and W.H. Berry. 1992. Habitat, soils and den use of San Joaquin kit fox (*Vulpes velox macrotis*) at Camp Roberts Army National Guard Training Site, California. US Department of Energy Topical Report, EG&G/EM Santa Barbara Operations. Report nr EGG10617-2156.

Rodman, P.S. 1981. Inclusive fitness and group size with a reconsideration of group sizes in lions and wolves. American Naturalist 118:275–283.

Rodrick, P.J., and N.E. Mathews. 1999. Characteristics of natal and non-natal kit fox dens in the northern Chihuahuan desert. Great Basin Naturalist 59(3):253–258.

Rohwer, S.A., and D.L.J. Kilgore. 1973. Interbreeding in the arid-land foxes, *Vulpes velox* and *Vulpes macrotis*. Systematic Zoology 22:157–165.

Rongstad, O.J., T. R. Laurion, and D.E. Andersen. 1989. Ecology of swift fox on the Piñón Canyon Maneuver Site, Colorado. Final report to the U.S. Army, Directorate of Engineering and Housing, Fort Carson, Colorado.

Rosenzweig, M.L. 1966. Community structure in sympatric carnivora. Journal of Mammalogy 47(4):602–612.

Rudzinski, D.R., H.B. Graves, A.B. Sargeant, and G.L. Storm. 1982. Behavioral interactions of penned red and arctic foxes.

Journal of Wildlife Management 46(4):877–884.

Ryon, J. 1986. Den digging and pup care in captive coyotes (*Canis latrans*). Canadian Journal of Zoology 64(7):1582–1585.

Sargeant, A.B., S.H. Allen, and J.O. Hastings. 1987. Spatial relations between sympatric coyotes and red foxes in North Dakota. Journal of Wildlife Management 51:285–293.

Schantz, T. von. 1984a. 'Non-breeders' in the red fox *Vulpes vulpes*: a case of resource surplus. Oikos 42:59–65.

——. 1984b. Spacing strategies, kin selection, and population regulation in altricial vertebrates. Oikos 42:48–58.

Schmidt, R.H. 1991. Gray wolves in California: their presence and absence. California Fish and Game 77:79–85.

Scott-Brown, J.M., T.P. O'Farrell, and K.L. Hammer. 1987. Swift fox. Pp. 433–441 *in* M. Novak, J.A. Baker, M.E. Obbard, and B. Malloch, editors. Wild furbearer management and conservation in North America. Ontario Trappers Association, North Bay, Ontario.

Skrobov, V.D. 1960. On interrelations of the polar fox and fox in the tundra of the Nenetsnational region. Zoologicheskii Zurnal 39:469–471.

Smirnov, V.S. 1968. Analysis of arctic fox population dynamics and methods of increasing the arctic fox harvest. Problems of the North 11:70–90.

Smits, C.M.M., and B.G. Slough. 1993. Abundance and summer occupancy of arctic fox, *Alopex lagopus,* and red fox, *Vulpes vulpes*, dens in the northern Yukon Territory 1984–1990. Canadian Field-Naturalist 107:13–18.

Smits, C.M.M., C.A.S. Smith, and B.G. Slough. 1988. Physical characteristics of arctic fox (*Alopex thern lagopus*) dens in north Yukon Territory, Canada. Arctic 41(1):12–16.

Sovada, M.A., C.C. Roy, J.B. Bright, and J.R. Gillis. 1998. Causes and rates of mortality of swift foxes in western Kansas. Journal of Wildlife Management 62(4):1300–1306.

Sovada, M.A., and B.K. Scheick. 2000. Preliminary report to the swift fox conservation team: historic and recent distribution of swift foxes in North America. Pp. 80–118 *in* G. Schmitt, editor. Swift Fox Conservation Team 1999 Annual Report. New Mexico Department of Game and Fish, Sante Fe, New Mexico.

Speller, S.W. 1972. Food ecology and feeding behavior of denning arctic foxes at Aberdeen Lake, Northwest Territories. Dissertation, University of Saskatchewan, Saskatoon, Saskatchewan.

Strand, O., A. Landa, J.D.C. Linnell, B. Zimmermann, and T. Skogland. 2000. Social organization and parental behavior in the arctic fox. Journal of Mammalogy 81(1):223–233.

Tannerfeldt, M. 1997. Population fluctuations and life history consequences in the arctic fox. Dissertation, Stockholm University, Stockholm, Sweden.

Tannerfeldt, M., and A. Angerbjörn. 1998. Fluctuating resources and the evolution of litter size in the arctic fox. Oikos 83:545–559.

Tannerfeldt, M., A. Angerbjörn, and B. ArvidSon. 1994. The effect of summer feeding on juvenile arctic fox survival—a field experiment. Ecography 17:88–96.

Thurber, J.M., R.O. Peterson, J.D. Woolington, and J.A. Vucetich. 1992. Coyote coexistence with wolves on the Kenai Peninsula, Alaska. Canadian Journal of Zoology 70(12):2494–2498.

Underwood, L., and J.A. Mosher, editors. 1982. Arctic fox. Pp. 491–503 in Wild animals of North America: biology, management, and economics. John Hopkins Press, Baltimore, Maryland.

Uresk, D.W., and J.C. Sharps. 1986. Denning habitat and diet of the swift fox in western South Dakota. The Great Basin Naturalist 46:249–253.

Warrick, G.D., and B. Cypher. 1998. Factors affecting the spatial distribution of San Joaquin kit foxes. Journal of Wildlife Management 62(2):707–717.

Wayne, R.K., and S.J. O'Brien. 1987. Allozyme divergence within the Canidae. Systematic Zoology 36:339–355.

White, P.J., and R.A. Garrott. 1997. Factors regulating kit fox populations. Canadian Journal of Zoology 75:1982–1988.

White, P.J., and K. Ralls. 1993. Reproduction and spacing patterns of kit foxes relative to changing prey availability. Journal of Wildlife Management 57(4):861–867.

Wilson, D.S. 1975. The adequacy of body size as a niche difference. American Naturalist 109:769–784.

Voigt, D.R., and B.D. Earle. 1983. Avoidance of coyotes by red fox families. Journal of Wildlife Management 47:852–857.

Zetterberg, H. 1945. Två Fredlösa. Almquist & Wiksell, Uppsala, Sweden.

——. 1953. Fjällräven. Pp. 204–215 *in* G. Notini and B. Haglund, editors. Svenska Djur. Däggdjuren. Uddevalla: Bohuslänningen AB.

Zimmerman, A.L., and B. Giddings. 1997. Preliminary findings of swift fox studies in Montana. P. 125 *in* B. Giddings, editor. Swift Fox Conservation Team Annual Report. Montana Department of Fish, Wildlife & Parks, Helena, Montana.

Zoellick, B.W., and N.S. Smith. 1992. Size and spatial organization of home ranges of kit foxes in Arizona. Journal of Mammalogy 73:83–88.

Zoellick, B.W, N.S. Smith, and R.S. Henry. 1989. Habitat use and movements of desert kit foxes in western Arizona. Journal of Wildlife Management 53:955–961.

Coyote and Kit Fox Diets in Prairie Dog Towns and Adjacent Grasslands in Mexico

■ **Rurik List, Patricia Manzano-Fischer and David W. Macdonald**

Abstract: Kit fox and coyote diets were determined from the percent occurrence of food items in scats collected at a prairie dog complex and surrounding grasslands in Chihuahua, Mexico. Black-tailed prairie dogs were the most frequently occurring species in the scats of both canids. Insects, kangaroo rats, small mammals, lagomorphs and ground squirrels also were important food items for the kit fox, whereas kangaroo rats, cattle and lagomorphs comprised a large proportion of the food items in coyote scats. There were seasonal differences in the occurrence of food items, and prairie dog remains occurred most frequently in coyote scats within prairie dog towns. Destruction of prairie dog towns could increase cattle predation by coyotes and reduce kit fox numbers.

Many canids have a varied, omnivorous diet (Sheldon 1992). Kit foxes (*Vulpes macrotis*) and coyotes (*Canis latrans*) are no exception, with prey items including mammals, birds, reptiles, amphibians, arthropods and plants (Egoscue 1962, Harrison and Harrison 1984, Zumbaugh et al. 1985, MacCracken and Hansen 1987, Windberg and Mitchell 1990, Cypher et al. 1994, White et al. 1996). Coyotes are opportunistic feeders (MacCracken and Hansen 1987), and their diet varies considerably with seasonal and geographical changes in prey availability (Todd et al. 1981, Windberg and Mitchell 1990, Reichel 1991, Hernández and Delibes 1994). For example, in urban Washington they eat fruits and domestic animals or rodents associated with human habitation (Quinn 1997), while in farmland, carrion dominates (Pérez-Gutierrez et al. 1982, Lafón 1983, Roy and Dorrance 1985, Vela 1985). Kit foxes seem to specialize largely on small and medium-sized mammals (Davis 1960, Egoscue 1962, Jiménez-Guzmán and López-Soto 1992), and consequently suffer reduced reproductive rates during times of prey scarcity (White and Ralls 1993, White et al. 1996, White and Garrott 1998).

The North American kit fox and swift fox (*V. velox*) complex is increasingly threatened: the San Joaquin kit fox is endangered, while swift foxes have been reintroduced into the wild in Canada and have little protection in the United States. Kit/swift fox declines are associated with the loss of prairie habitat to farmland and possibly to increases in coyote numbers, which are responsible for 65–85 % of swift fox mortalities (O'Neal et al. 1987, Ralls and White 1995, Moehrenschlager and List 1996). In Mexico, the status of kit foxes is not well known, but may be subject to similar pressures. In particular, prairie dog (*Cynomys ludovicianus*) towns may be important, but declining, habitat for kit foxes (Moehrenschlager and List 1996, List and Macdonald 1998). In this paper, we assess the diet of kit foxes and coyotes between prairie dog towns and adjacent grassland without prairie dogs in the largest extant prairie dog complex in North America (Ceballos et al. 1993).

Study Area

Our study area encompassed the largest remaining complex of prairie dog towns in North America: Janos-Casas Grandes, comprising over 50,000 ha of prairie dog towns (Ceballos et al. 1993) on privately and communally owned land. Janos-Casas Grandes lies within the Chihuahuan Desert in the northwest of Chihuahua, Mexico (30°57.8'N, 30° 37.5'N, 108°12.5'W, 108°40.3'W) (Fig. 1). The area is a plain, limited to the north and west by the mountains of the Sierra Madre Occidental, and to the south and east by the arid lands of northern Mexico.

The climate is arid and temperate with hot summers and winter rains (García 1973). Mean annual temperature is 15.7°C (6.0°C in January, 26.1°C in June). Mean annual

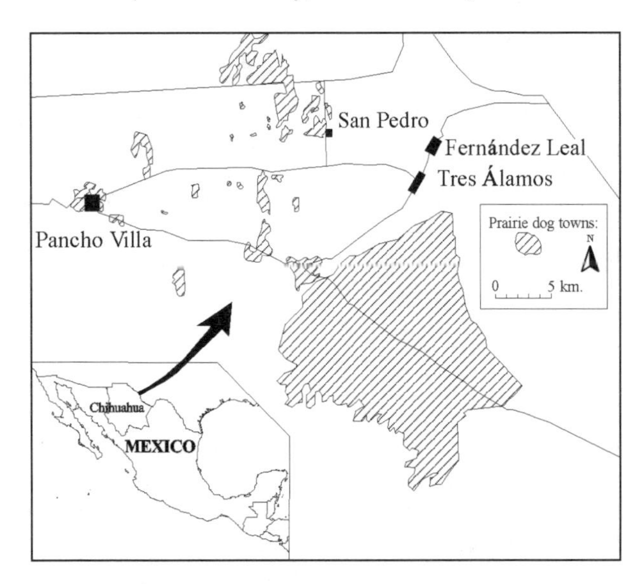

Figure 1. Location of prairie dog towns in the study area at the Janos-Casas Grandes Complex, Chihuahua, Mexico.

Table 1. Frequency (Fq.) and proportion of occurrence (%) of food items found in 303 kit fox scats from northwestern Chihuahua, Mexico, during February 1994 –June 1996.

	Total		Dogtowns		Grassland		Spring		Summer		Autumn		Winter	
Food Item	Fq.	%	Fq.	%	Fq.	%	Fq.	%	Fq.	%	Fq.	%	Fq.	%
Prairie dog	63	17.8	11	20.0	51	17.2	13	14.0	7	14.6	19	22.6	23	18.3
Insect	65	18.4	12	21.8	53	17.9	7	7.5	17	35.4	16	19.0	25	19.8
Kangaroo rat	50	14.2	5	9.1	45	15.2	14	15.1	7	14.6	13	15.5	16	12.7
Ground squirrel	38	10.8	7	12.7	31	10.5	15	16.1	2	4.2	16	19.0	5	4.0
Lagomorph	43	12.2	11	20.0	32	10.8	16	17.2	2	4.2	9	10.7	16	12.7
Small mammal	51	14.4	6	10.9	44	14.9	17	18.3	5	10.4	6	7.1	22	17.5
Cattle	18	5.1	2	3.6	16	5.4	4	4.3	4	8.3	1	1.2	9	7.1
Bird	21	5.9	1	1.8	20	6.8	7	7.5	3	6.3	3	3.6	8	6.3
Plant	4	1.1	0	0.0	4	1.4	0	0.0	1	2.1	1	1.2	2	1.6
Total	353	99.9	55	99.9	296	100.1	93	100.0	48	100.1	84	99.9	126	100.0
Mammal	245	69.4	40	72.7	203	68.6	75	80.7	23	48.0	63	74.9	82	65.2
Cattle, bird & plant	43	12.1	3	5.4	40	13.6	11	11.8	8	16.7	5	6.0	19	15.0

Table 2. Frequency (Fq.) and proportion of occurrence (%) of food items found in 76 coyote scats from northwestern Chihuahua, Mexico, during February 1994–June 1996.

	Total		Dogtowns		Grassland		Spring		Summer		Autumn		Winter	
Food Item	Fq.	%	Fq.	%	Fq.	%	Fq.	%	Fq.	%	Fq.	%	Fq.	%
Prairie dog	26	27.7	20	48.8	4	8.9	3	11.5	3	42.9	12	33.3	6	35.3
Insect	6	6.4	3	7.3	3	6.7	3	11.5	0	0.0	3	8.3	0	0.0
Kangaroo rat	12	12.8	3	7.3	8	17.8	5	19.2	0	0.0	3	8.3	3	17.6
Ground squirrel	4	4.3	3	7.3	0	0.0	0	0.0	2	28.6	1	2.8	0	0.0
Lagomorph	11	11.7	3	7.3	8	17.8	6	23.0	1	14.3	2	5.6	2	11.8
Small mammal	7	7.4	2	4.9	5	11.1	3	11.5	0	0.0	3	8.3	1	5.9
Cattle	11	11.7	1	2.4	8	17.8	3	11.5	0	0.0	4	11.1	2	11.8
Bird	7	7.4	2	4.9	5	11.1	3	11.5	0	0.0	2	5.6	2	11.8
Plant	10	10.6	4	9.8	4	8.9	0	0.0	1	14.3	6	16.7	1	5.9
Total	94	100.0	41	100.0	45	100.1	26	100.0	7	100.1	36	100.0	17	100.1
Mammal	60	63.9	31	75.6	25	55.6	17	65.3	6	85.8	21	58.3	12	70.6
Cattle, bird & plant	28	29.7	7	17.1	17	37.8	6	23.0	1	14.3	12	33.4	5	29.5
Prairie dog							3	11.5	3	60.0	12	34.3	6	35.3
Other wild mammal							14	53.8	1	20.0	8	22.8	6	35.3
Other							9	34.6	1	20.0	15	42.8	5	29.4
Total							26	100.0	5	100.0	35	100.0	17	100.0

rainfall is 381 mm, but during our study period (1994–1996) rainfall was unusually low, with drought affecting the entire region. The main habitats within the Janos-Casas Grandes Complex were native shortgrass prairie dominated by grasses and forbs, mesquite (*Prosopis* spp.) scrub, and riparian vegetation along seasonal streams (List et al. 1999). Prairie dogs occupy areas within the shortgrass prairie and we therefore considered prairie dog towns a different habitat than the shortgrass prairie without prairie dogs, which we refer to as grasslands. We defined a prairie dog town as a group of prairie dog burrows no further than 150 m from another.

Methods

Kit fox and coyote scats were collected between February 1994 and June 1996, within grasslands and prairie dog towns at the Janos-Casas Grandes Complex. Only recent scats, with no signs of color loss or perceptible erosion, were collected. Standardized transect methods of scat collection (Andelt et al. 1987, Gese et al. 1988, Windberg and Mitchell 1990) proved highly inefficient as the animals rarely used roads. Instead, kit fox scats were collected twice a month in the vicinity of dens occupied by radiocollared foxes (List and Macdonald 2003), or opportunistically from dens where other kit foxes were seen. Coyote scats were collected opportunistically from areas more than 3 km from human habitation.

Scats were air-dried for >1 month, then torn apart and separated into component food types identified through comparison with reference material. Hair, feathers and invertebrate remains were identified following methods by Teerink (1991). All hairs were identified by one person whose consistency of identification was assessed by a blind test on which 50 slides of known species were identified to genus with an accuracy of 94%. Specimens of Muridae, Geomydae and Heteromydae (except for *Dipodomys*) were pooled, and analyzed as "small mammals," to reduce the error and to increase sample size (Harrison and Harrison 1984, MacCracken and Hansen 1987, Windberg and Mitchell 1990). Identification of feathers or eggshells was not possible below the Class level. Insects were identified to order and plants to genus or species.

We expressed the contribution of each food item to the diet as a percentage of the total number of occurrences of all food items (i.e., Witmer and DeCalesta 1986, Dibello et al. 1990, Hernández et al. 1994, Johnson and Franklin 1994, Ciucci et al. 1996, White et al. 1996). Differences in the digestibility of food items mean that their frequency of occurrence in scats is not necessarily a good indicator of their importance in the diet (Reynolds and Aebischer 1991). We used Chi-squared tests to compare the frequency of occurrence of food items in scats collected in different seasons (spring: March–May; summer: June–August; fall: September–November; winter: December–February), and in scats collected within prairie dog towns or grassland. For statistical testing, we used the categories shown in Table 1 for kit foxes (except for cattle, plants and birds which were grouped into a single category) and Table 2 for coyotes (except for testing seasonal effects, where we used 3 categories: prairie dogs, other wild mammals, and other items).

Results

Kit Foxes

In 303 kit fox scats we identified 13 species of small- and medium-sized mammals (black-tailed jackrabbit [*Lepus californicus*], desert cottontail [*Sylvilagus audubonii*], black-tailed prairie dog, spotted ground squirrel [*Spermophilus spilosoma*], pocket gopher [*Thomomys bottae*], pocket mouse [*Perognathus flavus*], kangaroo rats [*Dipodomys* spp.], mice [*Peromyscus* spp.], southern grasshopper mouse [*Onychomys torridus*], harvest mice [*Reithrodontomys* spp.], cotton rat [*Sigmodon fulviventer*], whitethroat woodrat [*Neotoma albigula*], and skunks [*Mephitis* spp.]), domestic cattle, birds, 3 invertebrate groups (Orthoptera, Hymenoptera, Coleoptera), and plant material. Overall, mammals (excluding scavenged cattle) accounted for 69.4% of 263 prey occurrences in scats (Table 1). The most frequently occurring mammal species were prairie dogs (17.8% of occurrences), kangaroo rats (14.2%) and ground squirrels (10.8%). Insects were also an important prey group, comprising 18% of occurrences.

No statistically significant differences were found in the overall frequency of occurrence of food items in kit fox scats collected within prairie dog towns versus grasslands with no prairie dogs (X^2 = 8.2, df = 6, P >0.05). There were seasonal changes in prey item occurrence, with statistically significant differences between spring and summer (X^2 = 24.3, df = 6, P <0.001), spring and fall (X^2 = 13.8, df = 6, P <0.05), spring and winter (X^2 = 16.2, df = 6, P <0.05), summer and fall (X^2 = 14.8, df = 6, P <0.05) and fall and winter (X^2 = 20.1, df = 6, P < 0.01), but not between summer and winter (X^2 = 7.5, df = 6, P > 0.05). There was particularly marked seasonal variation in the occurrence of insects, which constituted only 7.5% of 93 prey occurrences in spring, but 35% of 48 occurrences in summer. In summer, when insects were common, small and medium-sized mammals occurred least frequently in scats (48% of occurrences), primarily because of a reduction in the proportion of ground squirrel and lagomorph remains. Prairie dogs occurred most frequently in scats collected in fall (23% of 84 occurrences), but were also common in other seasons (14–18%).

Coyotes

The remains of 10 mammal species (black-tailed jackrabbit, desert cottontail, black-tailed prairie dog, ground squirrels, kangaroo rats, *Peromyscus* spp., cotton rat, whitethroat woodrat and striped skunk), domestic cattle, birds, 2 invertebrate groups (Orthoptera and Coleoptera) and 4 plant groups (Gramineae, watermelon [*Citrullus vulgaris*], mesquite [*Prosopis* spp.] and apple [*Malus malus*]) were identified in the 76 coyote scats collected. Cattle remains were the hairs of adults only, and were therefore believed to be scavenged.

Excluding cattle, mammals accounted for 63.9% of the 91 food occurrences in scats (Table 2). At 27.7% of occurrences, prairie dogs were the single most common prey item, followed by kangaroo rats (12.8%). In contrast to kit foxes, insect remains were rare (6.4%), but cattle occurred frequently (11.7%).

There was a significant difference in the occurrence of prey items in scats collected within prairie dog towns versus grasslands (X^2 = 26.1, df = 8, P <0.01). In scats collected in prairie dog towns almost half (48.8%) of the 41 prey occurrences were prairie dogs, while in grassland this species accounted for only 8.9% of 45 occurrences. Ground squirrels were only eaten in prairie dog towns (7.8% of occurrences). In grassland, kangaroo rats (17.8%), lagomorphs (17.8%) and cattle (17.8 %) were important, as were small mammals (11.1%) and birds (11.1%).

There was little seasonal variation in occurrence of food items in scats, although our sample sizes were small and in summer insufficient scats were collected for statistical analysis. There was a significant difference in the occurrence of prairie dogs, other wild mammals and other foods in scats collected in spring and fall (X^2 = 7.4, df = 2, P <0.05), but not in spring and winter (X^2 = 3.6, df = 2, P >0.05) or fall and winter (X^2 = 1.2, df = 2, P >0.05). Prairie dogs occurred more frequently in scats collected in fall (34.3% of 35 occurrences) than in spring (11.5%), when other wild mammals accounted for over half (53.8%) of the 26 prey item occurrences.

Discussion

Our data indicate that the diet of the kit fox in Mexico's Janos-Casas Grandes prairie dog complex is primarily based on mammalian prey, similar to other studies (Zumbaugh et al. 1985, Uresk and Sharps 1986, Hines and Case 1991, Jiménez-Guzmán and López-Soto 1992). However, the frequency of occurrence of prairie dog remains in scats was higher than has been reported at other prairie dog sites (e.g., 1.5–5% occurrence, Jiménez-Guzmán and López-Soto 1992), possibly because of the large population and greater size and extent of towns at the Janos-Casas Grandes Complex. The most abundant prey species were prairie dogs, with densities of 15.6/ha, compared to the rest of rodent species that had densities of 0.7–14.1/ha in prairie dog towns and 4.3/ha in grassland (Pacheco et al. 2000). We never saw a kit fox carrying a prairie dog or any other species, although we often saw them going in and coming out from prairie dog burrows. Kit foxes probably hunt prairie dogs in their burrows, because in our area the foxes are nocturnal while the prairie dogs are exclusively diurnal (Koford 1958, List 1997); however another possibility is that kit foxes capture prairie dogs during the crepuscular period (B. Cypher, Endangered Species Recovery Program, personal communication). Contrary to expectation, the proportion of prairie dog remains in kit fox scats was similar in prairie dog towns and grasslands. This could be related to the close proximity of kit fox burrows to prairie dog towns (<3 km), allowing kit foxes to forage in prairie dog towns. Prairie dogs are an important source of food for kit foxes; however they are less important when other prey are available in large numbers, like insects during the summer. Cattle (from carrion) were rarely eaten, despite the frequent presence of carcasses close (500–1500 m) to the kit fox burrows.

Coyotes are widely reported to be opportunistic predators (MacCracken and Hansen 1987), and this was reflected in our study where, as expected, prairie dogs remains were more frequent in coyote scats collected in prairie dog towns than in grasslands. Prairie dogs were the main food item detected in the scats, and coyotes were frequently observed hunting prairie dogs (List 1997). Coyotes and kit foxes used most of the same prey species in comparable proportions; this situation could lead to competitive interactions in which kit foxes could be negatively affected (Cypher and Spencer 1998), unless the extent of exploitative competition between the 2 species is minimal in the sense that none appreciably reduces the uncaptured food supply of the other (Bertram 1979). This was probably the case here, given the abundance of the main prey species. Additionally, kit fox and coyote strategies to exploit prairie dogs differed: coyotes hunted prairie dogs by day, when they were active and above ground (List 1997), whereas kit foxes apparently hunted prairie dogs underground at night.

Coyotes in our area are shot because they are believed to be an important source of mortality for cattle calves. Our findings do not substantiate that claim, because the hairs found in the scats belonged only to adult cattle. Our results contrast with the diet of coyotes in other cattle ranching areas of Chihuahua, where carrion is the main food item for the coyote (Pérez-Gutierrez et al. 1982, Vela 1985). The limited use of cattle carcasses by kit foxes, despite their availability, may have been related to the heavy use of these carcasses by coyotes. Because coyotes are a main predator of the kit fox (O'Neal et al. 1987, Ralls and White 1995, Moehrenschlager and List 1996, Cypher and Spencer 1998) and coyotes readily consumed cattle carrion, the odds of an encounter with a coyote would be higher in the vicinity of a carcass.

Our findings are consistent with the notion that prairie dogs are keystone species (Ceballos et al. 1999, Kotlier et al. 1999, Miller et al. 2000). In summary, prairie dogs are a major item in the diet of both coyotes and kit foxes, 2 of the top carnivores in the Janos-Casas Grandes prairie dog complex. The eradication of the prairie dogs could have a major impact on these canids with consequences that are hard to predict, with possibilities including an increase in coyote predation on calves, and a reduction of kit fox numbers resulting in reduced burrow availability and increased competitive pressure and predation from coyotes. The latter is particularly worrying because low kit fox predation rates from coyotes in our study area appear to be associated with the presence of prairie dogs (Moehrenschlager and List 1996).

Acknowledgments

We thank volunteers M. Doughty, M. Eaton, E. Jiménez, G. Johnson and C. Philcox. G. Ceballos, B. Miller, J. Pacheco and M. Royo assisted throughout the development of the work. R. Atkinson, L. Carbyn, B. Cypher, E. Gese, K. Hambler, E. Raganella Pellicioni, M. Sovada and F. Tattersall helped to improve the manuscript. F. Cervantes allowed the collection of hair samples from the Mammal Collection of the Instituto de Biología, UNAM. The Canadian Wildlife Service, through L. Carbyn and A. Moehrenschlager, donated radio-collars. The British Council and Consejo Nacional de Ciencia y Tecnología gave scholarships to Rurik List. The study was conducted with funds from Consejo Nacional para el Conocimiento y Uso de la Biodiversidad, Dirección General de Apoyo al Personal Académico de la Universidad Nacional Autónoma de México, The People's Trust for Endangered Species, and the United States Agency for International Development.

Literature Cited

Andelt, W.F., J.G. Kie, F.F. Knowlton, and K. Cardwell. 1987. Variation in coyote diets associated with season and success ional changes in vegetation. Journal of Wildlife Management 51:273–277.

Bertram, B.C.R. 1979. Serengeti predators and their social systems. Pp. 221–248 *in* A.R.E. Sinclair and M. Norton-Grifiths, editors. Serengeti: dynamics of an ecosystem. Chicago University Press, Chicago, Illinois.

Ceballos, G., E. Mellink, and L. Hanebury. 1993. Distribution and conservation status of prairie dogs (*Cynomys mexicanus* and *C. ludovicianus*) in Mexico. Biological Conservation 63:115–112.

Ceballos, G., J. Pacheco, and R. List. 1999. Influence of prairie dogs (*Cynomys ludovicianus*) on habitat heterogeneity and mammalian diversity in Mexico. Journal of Arid Lands 41:161–172.

Ciucci, P., L. Boitani, E. Raganella Pelliccioni, M. Rocco, and I. Guy. 1996. A comparison of scat-analysis methods to assess the diet of the wolf *Canis lupus*. Wildlife Biology 2:37–47.

Cypher, B.L., and K.A. Spencer. 1998. Competitive interactions between coyotes and San Joaquin kit foxes. Journal of Mammalogy 79:204–214.

Cypher, B.L., K.A. Spencer, and J.H. Scrivner. 1994. Food-item use by coyotes at the Naval Petroleum Reserves in California. Southwestern Naturalist 39:91–95.

Davis, W.B. 1960. The mammals of Texas. Game and Fish Commission, Austin, Texas.

Dibello, F.J., S.M. Arthur, and W.B. Krohn. 1990. Food habits of sympatric coyotes, red foxes, *Vulpes vulpes*, and bobcats, *Lynx rufus*, in Maine. The Canadian Field-Naturalist 104:403–408.

Egoscue, H.J. 1962. Ecology and life history of the kit fox in Tooele County, Utah. Ecology 43:481–497.

García, E. 1973. Modificaciones al sistema de clasificación climática de Köepen. Instituto de Geografía, Universidad Nacional Autónoma de México, Mexico, Distrito Federal. Mexico.

Gese, E.M., O.J. Rongstad, and W.R. Mytton. 1988. Home range and habitat use of coyotes in southeastern Colorado. Journal of Wildlife Management 52:640–646.

Harrison, D.J., and J.A. Harrison. 1984. Foods of adult Maine coyotes and their known-aged pups. Journal of Wildlife Management 48:322–325.

Hernandez, L., and M. Delibes. 1994. Seasonal food habits of coyotes, *Canis latrans*, in the Bolson de Mapimi, Southern Chihuahuan Desert, Mexico. Zeitschrift fuer Saeugetierkunde 59:82–86.

Hernández, L., M. Delibes, and F. Hiraldo. 1994. Role of reptiles and arthropods in the diets of coyotes in extreme desert areas of northern Mexico. Journal of Arid Environments 26:165–170.

Hines, T.D., and R.M. Case. 1991. Diet, home range, movements, and activity periods of swift fox in Nebraska. Prairie Naturalist 23:131–138.

Jiménez-Guzmán, A., and J.H. López-Soto. 1992. Estado actual de la zorra del desierto, *Vulpes velox zinseri*, en el Ejido EL Tokio, Galeana, Nuevo León, México. Publicaciones Biológicas, FCB/UANL 6:53–60.

Johnson, W.E., and W.L. Franklin. 1994. Role of body size in the diets of sympatric gray and culpeo foxes. Journal of Mammalogy 75:163–174.

Koford, C.B. 1958. Prairie dogs, whitefaces, and blue grama. Wildlife Monographs 3.

Kotlier, N.B., B.W. Baker, A.D. Whicker and G. Plumb. 1999. A critical review of assumptions about the prairie dogs as keystone species. Environmental Management 24:177–192.

Lafón, A. 1983. Composición de la alimentación del coyote. Centro de Investigaciones Forestales del Norte. Instituto Nacional de Investigaciones Forestales-SARH, Mexico.

List, R. 1997. Ecology of kit fox (*Vulpes macrotis*) and coyote (*Canis latrans*) and the conservation of the prairie dog ecosystem in northern Mexico. Dissertation, University of Oxford, Oxford, UK.

List, R., and D.W. Macdonald. 1998. Carnivora and their larger mammalian prey: species inventory and abundance in the Janos-Nuevo Casas Grandes prairie dog complex. Revista Mexicana de Mastozoología 3:95–112.

——. 2003. Homerange and habitat use of the Kit Fox (*Vulpes macrotis*) in a prairie dog town (*Cynonys ludocianus*) complex. Journal of Zoology 259:1–5.

List, R., J. Pacheco, and G. Ceballos. 1999. Status of North American porcupine (*Erethizon dorsatum*) in Mexico. Southwestern Naturalist 44:400–404.

MacCracken, J.G., and R.M. Hansen. 1987. Coyote feeding strategies in southeastern Idaho: optimal foraging by an opportunistic predator? Journal of Wildlife Management 51:278–285.

Miller, B., R. Reading, J. Hoogland, T. Clark, G. Ceballos, R. List, S. Forrest, L. Hanebury, P. Manzano, J. Pacheco and D. Uresk. 2000. The role of prairie dogs as keystone species: a response to Stapp. Conservation Biology 14:318–321.

Moehrenschlager, A., and R. List. 1996. Comparative ecology of North American prairie foxes—conservation through collaboration. Pp. 22–28 *in* D.W. Macdonald, and F.H. Tattersall, editors. The WildCRU Review. Wildlife Conservation and Research Unit, Oxford, UK.

O'Neal, G.T., J.T. Flinders, and W.P. Clary. 1987. Behavioral ecology of the Nevada kit fox (*Vulpes macrotis nevadensis*) on a managed desert rangeland. Current Mammalogy 1:443–481.

Pacheco, J., G. Coallos, and R. List. 2000. Los mamiferos de la región de Janos Casas Grandes, Chihuaua, México. Revista Mexicana de Mastozoología 4:71–85.

Pérez-Gutierrez, C., L.C. Fierro, and J.C. Treviño. 1982. Determinación de la composición de la dieta del coyote (*Canis latrans* Say) a través del año en la región central de Chihuahua por medio del análisis de contenido estomacal. Pastizales 13:2–15.

Quinn, T. 1997. Coyote (*Canis latrans*) food habits in three urban habitat types of western Washington. Northwest Science 71:1–5.

Ralls, K., and P.J. White. 1995. Predation of San Joaquin kit foxes by larger canids. Journal of Mammalogy 76:723–729.

Reichel, J.D. 1991. Relationships among coyote food habits, prey populations, and habitat use. Northwest Science 65:133–137.

Reynolds, J.C., and N.J. Aebischer. 1991. Comparison and quantification of carnivore diet by faecal analysis: a critique, with recommendations, based on study of the Fox *Vulpes vulpes*. Mammal Review 21:97–122.

Roy, L.D., and M.J. Dorrance. 1985. Coyote (*Canis latrans*) movements, habitat use and vulnerability in central Alberta. Journal of Wildlife Management 49:307–313.

Sheldon, J.W. 1992. Wild dogs: the natural history of the non-domestic canidae. Academic Press, London, UK.

Teerink, B.J. 1991. Hairs of west European mammals. Cambridge University Press, Cambridge, UK.

Todd, A.W., L.B. Keith, and C.A. Fischer. 1981. Population ecology of coyotes during a fluctuation of snowshoe hares. Journal of Wildlife Management 45:629–640.

Uresk, D.W., and J.C. Sharps. 1986. Denning habitat and diet of the swift fox (*Vulpes velox*) in western South Dakota (USA). Great Basin Naturalist 46:249–253.

Vela, C.E.L. 1985. Determinación de la composición de la dieta del coyote *Canis latrans* Say, por medio del análisis de heces en tres localidades del Estado de Chihuahua. Thesis, Universidad Autónoma de Nuevo León, Mexico.

White, P.J., and R.A. Garrott. 1998. Factors regulating kit fox populations. Canadian Journal of Zoology 75:1982–1988.

White, P.J., and K. Ralls. 1993. Reproduction and spacing patterns of kit foxes relative to changing prey availability. Journal of Wildlife Management 57:861–867.

White, P.J., C.A.V. White, and K. Ralls. 1996. Functional and numerical responses of kit foxes to a short-term decline in mammalian prey. Journal of Mammalogy 77:370–376.

Windberg, L.A., and C.D. Mitchell. 1990. Winter diets of coyotes in relation to prey abundance in southern Texas. Journal of Mammalogy 71:439–447.

Witmer, G.W., and D.S. DeCalesta. 1986. Resource use by unexploited sympatric bobcats and coyotes in Oregon. Canadian Journal of Zoology 64:2333–2338.

Zumbaugh, D.M., J.R. Choate, and L.B. Fox. 1985. Winter food habits of the swift fox (*Vulpes velox*) on the central high plains. Prairie Naturalist 17:41–47.

Appendix 1
Items in kit fox and coyote scats collected in prairie dog towns and grassland in northwestern Chihuahua, Mexico. The lowest taxa identified are mentioned. (The number in parentheses indicates the number of possible species.)

	Class	Order	Family	Species	Common Name
Animals					
	Mamallia				
		Lagomorpha			
			Leporidae		
				Lepus californicus	Black-tailed jackrabbit
				Sylvilagus audubonii	Desert cottontail
		Rodentia			
			Sciuridae		
				Cynomys ludovicianus	Black-tailed prairie dog
				Spermophilus spilosoma	Spotted ground squirrel
			Geomyidae		
				Thomomys bottae	Pocket gopher
			Heteromidae		
				Perognathus flavus	Pocket mouse
				Dipodomys spp (2)	Kangaroo rat
			Muridae		
				Peromyscus spp (2)	Mouse
				Onychomys torridus	Southern grasshopper mouse
				Reithrodontomys spp(4)	Harvest mouse
				Sigmodon fulviventer	Cotton rat
				Neotoma albigula	Whitethroat woodrat
		Carnivora			
			Canidae		
				Canis latrans	Coyote
				Vulpes macrotis	Kit fox
			Mustelidae		
				Mephitis spp (2)	Striped/hooded skunks
		Artiodactyla			
			Bovidae		
				Bos taurus	Domestic cattle
	Aves				
					Unidentified bird(s)
	Insecta				
		Orthoptera			Grasshopper
		Hymenoptera			Ants
		Coleoptera			Beetle
Plants					
		Gramineae			Grass
		Cucurbitaceae			
			Citrullus vulgaris		Watermelon
		Leguminosaceae			
			Prosopis spp		Mesquite
		Rosaceae			
			Malus malus		Apple

Assessing Restoration of Swift Fox in the Northern Great Plains

■ **Kyran Kunkel, Kevin Honness, Mike Phillips and Lu Carbyn**

Abstract: Swift foxes (Vulpes velox) *are listed as threatened in South Dakota, and thus the state is mandated to "manage, protect, and restore" the species. We assessed potential for restoration of swift fox in South Dakota following the International Union for the Conservation of Nature guidelines. We reviewed the taxonomic status of swift foxes in the northern Great Plains; reviewed previous reintroductions; assessed features of reintroduction sites; examined social, economic, and legal considerations; and developed methods for translocation. In cooperation with the U.S. Fish and Wildlife Service; U.S. Forest Service; South Dakota Department of Game, Fish and Parks; Wyoming Department of Game and Fish; Colorado Division of Wildlife; and the Lower Brule Sioux Tribe, we will translocate approximately 30 foxes a year for 6 years from Wyoming and Colorado. Coyote* (Canis latrans)*-caused mortality of swift foxes is a primary limiting factor in fox population expansion, thus we will reduce coyote populations in the area during releases to maximize fox survival. We will work with local people to ensure optimization of fox management and restoration. We will monitor foxes during reintroduction to measure the project's success and to identify influencing factors.*

Since the settlement of the Great Plains of North America, swift fox (*Vulpes velox*) have disappeared from 60–90% of their historic range (Swift Fox Conservation Team [SFCT] 1997). The viability of fox populations especially in the northern half of their range (north of the North Platte River in Wyoming and Nebraska; Hall and Kelson 1959) where they exist in small, scattered, isolated patches remains in question. South Dakota lists the species as threatened and is thus mandated to "manage, protect, and restore" the species. Swift fox status in states bordering South Dakota is similarly tenuous (SFCT 1997).

Over 75% of swift fox habitat is on private property (SFCT 1997) and, as such, innovative plans will be needed to restore populations. As private land managers in South Dakota (Turner Endangered Species Fund and Turner Enterprises), we are developing a cooperative project with state, federal, and other private entities to restore swift foxes in South Dakota. We hope to demonstrate by example that stewardship of biodiversity is economically sustainable and enhances the long-term value and conservation of private "working" lands. We hope to show that private ranchers and farmers can produce public "environmental goods" or "conservation commodities" in conjunction with food and fiber (National Governors Association 2001). These and other efforts on private and federal lands are critical steps toward removing swift foxes from the state's threatened list and assuring the long-term viability of fox populations.

We developed a study plan (Kunkel et al. 1999) based on criteria set out by International Union for the Conservation of Nature Species Survival Commission (IUCN/SSC Re-introduction Specialist Group 1998) and the SFCT to assess the feasibility of reintroducing foxes. Results from that study indicate that ecologically, the Turner-owned Bad River Ranch (BRR) in western South Dakota was suitable for a swift fox reintroduction effort (Kunkel et al. 2001). Reintroduction addresses the first of 6 primary considerations identified by the SFCT (1997; p. vi) to develop a successful conservation strategy for foxes: "expanding the distribution of swift fox where ecologically and politically feasible." Additionally, the reintroduction project addresses critical research, management, and education needs identified in that document.

We believe that the most direct and immediate way to achieve swift fox recovery is to actively expand the distribution of swift foxes in the Great Plains. We believe that fox range will remain restricted without reintroductions because swift foxes are similar to kit foxes (*V. macrotis*), a species that has poor survival during dispersal along the edges of occupied habitat (Koopman et al. 2000). Coyote-caused mortality in areas of low fox-coyote ratios is especially significant for foxes dispersing into unfamiliar areas (no experience with escape terrain; edge effect of predation; Wilcove et al. 1986, Paton 1994, White and Garrott 1999). Reintroducing foxes greatly advances swift fox recovery by directly enhancing population abundance and distribution and by providing insights into swift fox ecology through experimentation.

Objectives

1. Establish a self-sustaining population of swift foxes on and around the BRR.
2. Establish a population that serves as a source for swift fox recovery and expansion in South Dakota and neighboring states and assists in removing foxes from threatened status in South Dakota.
3. Establish a population that enhances the long-term

survival of the species, restores natural biodiversity to the area (as part of the restoration of a full array of native species), and promotes prairie conservation awareness.

4. Collect and disseminate information on reintroduction techniques and the ecological requirements for successful swift fox restoration.

5. Collect and disseminate information on the ecology of swift foxes.

Reintroductions are relatively lengthy, complex, risky, and expensive conservation endeavors (IUCN/SSC Re-introduction Specialist Group 1998). We are prepared to expend the resources and to work with stakeholders and cooperators to achieve success in this project. This assessment, along with our feasibility study, ensures that we have taken the necessary steps to achieve our objectives. The topics, timeline, and responsible parties for this project follow the recommendations in the IUCN Guidelines for Re-introductions (1998) and those specifically for swift foxes by Mamo (1987) for Canada, by Sharps and Whitcher (1984) for South Dakota, and by FaunaWest (1991) for Montana.

Study Area

The reintroduction area (10,160 km^2; Fig. 1) is situated within the Pierre Hills physiographic region (Johnson et al. 1995). Soils are primarily clays derived from Cretaceous Pierre Shale. Topography consists of flats cut by intermittent drainages, including the Bad River, and gently rolling hills. Elevation ranges from approximately 590 m to 727 m above sea level. The climate is temperate with average temperatures ranging from -4°C in winter to 23°C in summer. Mean annual precipitation is 46.0 cm/year with most occurring in June and March. Kuchler (1975) characterized the area as a western wheatgrass (*Agropyron smithii*)-needlegrass (*Stipa viridula*; dominant cool season grasses) community within the typical mixed-grass prairie community region. Buffalo grass (*Buchloe dactyloides*) is the dominant warm season grass. Deciduous woodlands dominated by plains cottonwood (*Populus deltoides*) follow the Bad River valley floodplain and tributaries in the northern floodplain forest region.

Fort Pierre National Grasslands (FPNG) is a 470-km^2 grassland in Stanley, Jones and Lyman counties in west-central South Dakota. It is administered by the Nebraska National Forest and managed as a wildlife emphasis area. FPNG consists of 59 allotments and 210 pastures managed as one unit for multiple use. Deferred and rest rotation grazing is dominant between May and October.

BRR is approximately 570 km^2 and managed by Turner Enterprises for bison (*Bison bison*) production and

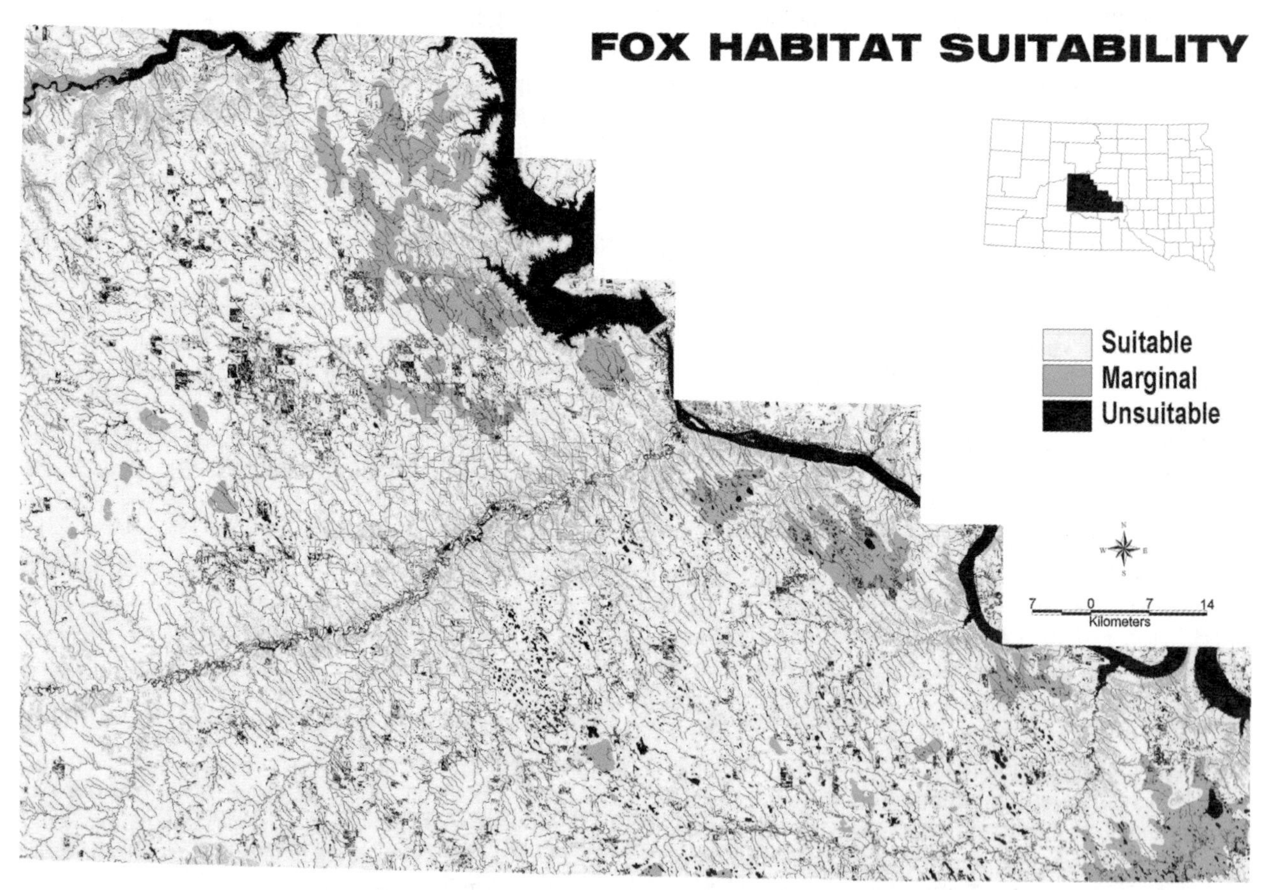

Figure 1. Relative suitability of landscape of west-central South Dakota for swift foxes based on a model of physiography and land cover.

conservation of biodiversity. The Lower Brule Reservation in Lyman and Stanley counties is 890 km^2 of primarily pasture land of which 1/4 is deeded property. Natural resources on the reservation are managed by the Lower Brule Sioux Tribe (LBST) Department of Wildlife, Fish, and Recreation to conserve and enhance the wildlife, fish and recreational resources of the reservation for the cultural, social, political and economic well being of the Lower Brule Sioux Tribe.

Biological Feasibility Study and Background Research

Taxonomic Status

Stromberg and Boyce (1986) found no justification for subspecific classification of northern (as described by Merriam 1902 and Hall and Kelson 1959) and southern swift fox but indicated that significant geographic variation among specimens may reflect genetic differences. They suggested for maintenance of genetic variability (i.e. ensure maintenance of genetics of rarer northern populations of swift fox), reintroduction in northern areas use foxes from northern areas. They cautioned that southern foxes or offspring from southern and northern crosses might not be able to endure the rigors of northern climates. Dragoo and Wayne (2002) supported this recommendation (but expanded northern areas to include Colorado) based on work by Crandall et al. (2000). There is no evidence that "southern" foxes (from southern Colorado) reintroduced into Canada have had lower survival than northern foxes (Carbyn 1998). We will reintroduce swift foxes from Wyoming and northern Colorado (latitude and climate similar to South Dakota).

We have obtained information on the history and ecology of potential source populations (Leberg 1990, Olson 2000, B. Luce, Wyoming Game and Fish [WDGF], unpublished report). We modeled various reintroduction strategies to maximize population performance (Haig et al. 1990, Kunkel et al. 2001). We will collect blood and hair samples for DNA analysis from all foxes to be released.

In order to retain >95% of the heterozygosity and allelic diversity found in the source population, we will reintroduce >25 individuals (½ males and ½ females) per year for at least 5 years (Leberg 1990). This should ensure that we have >25 founders in the population. We will maximize spacing of our traps in the source population to limit captures of related individuals to be released on BRR.

Effect of Reintroducing Foxes on the Ecosystem

Our swift fox reintroduction feasibility assessment (Kunkel et al. 2001) indicated that reduction of coyote densities on and around BRR would likely be necessary to assist in swift fox establishment. Initially (first 5 years), one primary impact on the ecosystem subsequent to the swift fox reintroduction will be a reduction in coyote density resulting from our control efforts (Henke and Bryant 1999). Coyote control will be terminated if it is not effective in assisting the establishment of swift foxes or when it is deemed no longer necessary for fox restoration (see below).

A restored population of foxes will have some level of impact on their prey (potentially reduced densities or behavioral changes in small mammals, birds, and insects). There are no reports from other occupied habitats assessing these impacts. Foxes prey to varying degrees on upland gamebirds. However, most studies indicate that birds make up a relatively small part of fox diets; in South Dakota birds comprised <6% of fox diets, thus their impact on mortality is likely relatively small and compensatory (Uresk and Sharps 1986).

Previous Reintroductions

South Dakota conducted the first recorded reintroduction of swift foxes. Eight captive-reared foxes were released in Haakon County (70 km west of BRR) in 1980 (Sharps 1984). Of these, 3 foxes died (1 shot, 1 trapped, 1 killed by car), radio contact with 3 foxes was lost within about 40 days of release, and one pair remained in the release area and raised pups. Sharps and Whitcher (1984) listed the following as criteria for site selection for fox release: open, gently rolling terrain; short/mid-grass prairie; black-tailed prairie dog (*Cynomys ludovicianus*) towns; permanent water; absence of poisoning; absence of trapping; release site >24 km from a road; and low densities of red fox (*V. vulpes*) and coyotes.

The Canadian reintroduction program released 942 swift foxes from 1983–1997 (Carbyn et al. 1994). There are now more than 300 foxes in about 58 townships in southern Alberta and Saskatchewan (Carbyn 1998). The majority of the current population are wild-born offspring of released animals. One year after release, soft-released foxes survived better than hard-released foxes (31% alive vs. 17% alive; Carbyn et al. 1994). However, 2 years post release, survival rates were similar. Coyote predation was the most significant cause of mortality (32%) for released foxes from 1987–1991 (Carbyn et al. 1994). One year post release, wild-born foxes had higher survival rates than captive-born foxes (47% alive vs. 14% alive). Captive-born foxes released in fall survived better after 1 year than foxes released in spring (14% alive vs. 4% alive). Wild-born foxes dispersed farther ($\bar{x}$ = 19.2 km) than captive-born foxes ($\bar{x}$ = 12.6 km) and all foxes dispersed farther in spring than in fall (Carbyn et al. 1994).

In fall 1998, Defenders of Wildlife and the Blackfeet Tribe hard-released 13 juvenile female and 17 juvenile male captive-reared foxes onto the Blackfeet Reservation in northwestern Montana (M. Johnson, Defenders of Wildlife, Missoula, Montana, personal communication). Protective shelters were placed and left at fox release sites for 4 days to enhance fox security (Smeeton and Weagle

2001). None of the foxes were radiocollared but surveys in 1999 indicated survival was relatively high and >1 litter had been produced. Twelve adult pairs and 3 captive-reared pups were released in August 1999; 8 of these foxes were radiocollared. Five of the radiocollared foxes remained alive in May 2000. Three litters were found in 2000. Thirty-one captive-reared foxes (16 radiocollared) were released in August 2000. At least 3 litters were produced in 2001.

Choice of Release Site and Type

The SFCT (1997) recommended expanding swift fox populations to occupy >50% of their historic range (suitable habitat). They recommended promoting dispersal or reintroductions in states that have no or severely limited swift fox population distribution. South Dakota, North Dakota, Montana, and Nebraska are the 4 states that meet this criterion. BRR is entirely within the historic range of the swift fox (SFCT 1997). Kunkel, Honness, Phillips, and Carbyn (2001) and a review of the fox sighting database for South Dakota indicate no foxes are present in the area.

Long-term Protection

All of the BRR can best be classified as "status 3 lands" (South Dakota Cooperative Wildlife and Fisheries Research Unit GAP analysis, unpublished data): "Areas having permanent protection from conversion of natural land cover for the majority of the area, but subject to extractive uses of either a broad, low-intensity type or localized intense type. It also confers protection to federally listed endangered and threatened species throughout the area." The Turner Foundation holds the BRR in trust in perpetuity. While the legal language of this classification has not been formalized, the practical application is underway: the conversion of nearly 40 km^2 of cropland back to native vegetation, the replacement of all livestock with bison, and the restoration of >10% of BRR to prairie dog colonies. Ft. Pierre National Grasslands are also classified as "status 3 lands." Together BRR and FPNG form a nearly continuous protected block of approximately 1,000 km^2. We will work to secure conservation agreements with the state and private landowners in the restoration area.

Evaluation of Reintroduction Site

Number of Foxes Reintroduction Area Can Support

Based on his observation and research in western South Dakota, Sharps (USFWS 1994) suggested that to maintain a viable fox population in South Dakota, a minimum of 1,500 individuals (250 family groups; 2 adults, 4 pups) in 10 different populations would be needed. He suggested such a population would require 1,295 km^2 of which 20–25% should contain prairie dogs. While we believe prairie dogs will be an important food base for foxes and will be actively managing for them (present prairie dog population on BRR occupies 3 km^2; Kunkel, Honness, Phillips, and Carbyn 2001), we think the evidence for the necessity of prairie dogs to foxes is less than compelling. Foxes forage on a wide variety of prey (Kilgore 1969, Sharps 1984, Kitchen et al. 1999, Sovada et al. 2001) and there is no evidence that prairie dogs are preferred. Successful reintroductions into Canada have been into regions with insignificant populations of prairie dogs (Carbyn 1998). We do, however, agree that higher densities of prey, including prairie dogs, would support higher densities of foxes.

Approximately 7,848 km^2 (77%) of the project area is suitable and 1,162 km^2 (11%) is marginal for swift foxes based on our habitat suitability model (Kunkel, Honness, Phillips, and Carbyn 2001; Fig. 1). Density of holes that can be used to escape coyotes is well above thresholds in other swift fox study areas (Kunkel, Honness, Phillips, and Carbyn 2001). Extrapolations of fox density based either on leporid availability, small mammal availability, or mean home range size (from fox populations in areas nearest BRR) in the BRR area including FPNG and the Lower Brule Reservation yielded a minimum expected density of approximately 0.10–0.25 foxes/km^2 (200–1,000 foxes) in the proposed reintroduction area with a relatively low reproductive rate (Kunkel, Honness, Phillips, and Carbyn 2001). When prey availabilities are combined, and other prey added (insects and birds), we would expect somewhat higher minimum densities and reproductive rates. Such a population, while not exceptionally large, would likely be self-sustaining especially if mortality by coyotes is not excessive.

While there is no fox recovery plan or objectives for the state of South Dakota, a population of >200 foxes in west-central South Dakota would increase the likelihood of recovery and removal from threatened status in the state. Based on population modeling, Ginsberg (1994) estimated that a population of roughly 200–600 jackals (*Canis* spp.) or foxes (i.e., asocial canids) would be required to maintain 80–90% of starting heterozygosity over 100–200 years. Restoration of foxes in west-central South Dakota can lead to a larger metapopulation in the west-central and southwestern portion of the state that connects to the contiguous population of foxes in southeastern Wyoming and northwestern Nebraska (Kunkel et al. 1999) and could expand into Montana and North Dakota, increasing the prospects for long-term viability of foxes in the northern Great Plains.

Elimination of Threats

Swift foxes and coyotes have persisted sympatrically likely since the formation of the Great Plains biome. Foxes have evolved strategies to persist with coyotes including heavy reliance on dens for escape and the ability to exploit a wide array of prey items (Kitchen et al. 1999). However, coyote predation appears to be a primary factor limiting fox population growth (Kunkel, Honness, Phillips, and Carbyn 2001), especially populations of low density that may initially require a reduction of coyote density to improve their chances of increasing to a sustainable population (Kitchen et al. 1999, White and Garrott 1999, Kunkel, Honness, and Phillips 2001). We will attempt to reduce coyote densities by >50% during reintroductions and continue control until the fox population reaches a density that allow persistence despite coyote predation.

We initiated a coyote population reduction program in late winter 2001 on the BRR and surrounding buffer areas (other private land) that will likely continue through the reintroductions following our approved animal care and use plan (Kunkel, Honness, and Phillips 2001). After killing approximately 280 coyotes on 260 km^2 in northern Texas, fox survival rates and juvenile density increased compared to pre-control and non-control areas (Kamler 2002). It appeared these control efforts turned a small, marginal sink population into a population with a surplus of dispersers. Evidence from southeastern Colorado indicates similar results after coyote control (E. Gese, Utah State University, personal communication). Predator control programs even for enhancement of endangered species are controversial both ethically and scientifically. We believe any control work must be thoroughly justified and meet stringent ethical and scientific standards (Hecht and Nickerson 1999).

Control work will consist primarily of aerial gunning during late winter. We will contract with the South Dakota Game Fish and Parks (SDGFP) Division of Animal Damage Control for this effort. Trapping and shooting from the ground will be done by local trappers and project personnel following approved protocols under the direction of a Turner Endangered Species Fund (TESF) biologist. We will attempt to target resident breeding pairs. We will intensify coyote control efforts if coyote predation remains high, and then discontinue it if this greater effort appears ineffective. Age, sex, and condition of all coyotes killed will be determined. Blood samples will be collected for disease profile analysis related to impacts on fox. We will also collect stomachs from coyotes for diet analysis.

We will estimate the percent of foxes that die due to coyote predation annually to determine the effectiveness of the predator management plan. We will measure coyote population trends via scat transects, scent station surveys, and spot light surveys (Kunkel, Honness, Phillips, and Carbyn 2001). Our goal will be to reduce and maintain coyote densities at <50% of pre-control abundance (approximately 0.20–0.40 coyotes/km^2; Kunkel, Honness, Phillips, and Carbyn 2001) or 0.10 coyotes/km^2 and to maintain coyote-caused mortality rates on foxes at <25%. This would likely mean killing approximately 50–100 coyotes/year. Coyote predation rates on foxes should decline annually as coyote densities decline due to control efforts and the swift foxes gain experience eluding the remaining coyotes. We will stop coyote control when it appears that fox density has reached a level to maintain a viable fox population capable of withstanding coyote predation (fox density >0.10/km^2), or if we have not maintained a viable fox population in 10 years. We concur with Hecht and Nickerson (1999) that "the best overall predator management strategy is an adaptive approach that monitors many factors, considers a full array of management techniques, continually appraises their potential and actual effectiveness, and makes appropriate adjustments."

Public trapping of furbearers is prohibited on BRR. We will work with local trappers to prevent incidental take of swift foxes in surrounding areas and to retrieve and release any foxes caught in traps. Habitat protection is assured by management paradigms used at BRR and on the grasslands (see above).

Surrounding landowners may use rodenticides to control prairie dogs. While zinc phosphide has been widely used to control prairie dogs, there is little indication this poses a hazard to foxes (Bell and Dimmick 1975). Schitoskey (1975) indicated kit fox survived after feeding on kangaroo rats (*Dipodomys* spp.) killed with zinc phosphide.

Diseases reported in swift foxes include plague, distemper, and mange; however, there are no confirmed cases of these diseases impacting population levels significantly (Miller et al. 2000, Pybus and Williams 2002). Plague has never been reported in western South Dakota and our sampling indicates mange levels appear low. We will continue to monitor these diseases in carnivores in the BRR area and manage appropriately (see below).

Availability of Suitable Release Stock and Assurance of No Significant Impact to Donor Populations

Wyoming Game and Fish and Colorado Division of Wildlife have conducted population surveys and identified areas with populations of foxes than can sustain the removals we propose (B. Oakleaf, WDGF, unpublished data). Monitoring of these populations will continue as long as translocations are occurring.

Socio-economic and Legal Considerations

Turner Endangered Species Fund has committed the financial resources estimated necessary for re-introduction success as defined above. The states of South Dakota, Wyoming, and Colorado and the U.S. Fish and Wildlife Service, LBST, and FPNG are committed to provide the administrative support (including permits and Memorandums of Understanding) to ensure the success of the project. Handling and collecting permits will be required from the 3 states along with an importation permit and health certificate from South Dakota.

Local support for conservation of native wildlife species is high in South Dakota. About 89% of South

Dakota residents feel that it is very (56.7%) or moderately (32.5%) important that South Dakota preserves as much wildlife as possible (Gigliotti 1998). Most (84.6%) South Dakota residents strongly (51.3%) or slightly (33.3%) agree that, "the diversity of wildlife in an area is a sign of the quality of the natural environment" (Gigliotti 1998). Most (90.4%) South Dakota residents strongly (51.6%) or slightly (38.8%) agree that, "grasslands like native prairie are a sign of the quality of the natural environment" (Gigliotti 1998). A significant majority (87%) of farmers/ ranchers agreed with the statement, "the presence of wildlife on my farm is important to me."

Public Planning and Participation

Local support is crucial to conservation efforts. We conducted 4 public meetings, 2 public hearings, and 1 field trip with local residents to provide information on fox ecology and the reintroduction proposal. We stressed the program's responsiveness to the needs, desires, and opinions of the local public and incorporated these into the program. To that end, we will incorporate the following strategy:

1) regularly update local residents by newsletter on the progress of the reintroduction and request their involvement and input.

2) produce news releases and update our website with progress.

3) develop cautionary highway signs indicating presence of swift foxes in area.

4) work with the South Dakota Natural Heritage Program to develop a voluntary swift fox–sighting network.

5) work with Watertown Zoo to disseminate prairie education packets to present to local schools.

6) develop a local organization dedicated to prairie and swift fox conservation and provide information to local ranchers and farm and ranch organizations on techniques to advance prairie conservation and swift fox restoration.

7) develop the South Dakota Swift Fox Conservation Team as a subcommittee of SFCT.

8) disseminate information on the ranges and characteristics of swift foxes to reduce the likelihood of human-caused mortality.

9) work with South Dakota Game, Fish and Parks (SDGFP) to develop swift fox management plans and prairie conservation promotional activities to update local and state political bodies.

10) work with the State, FPNG, tribes, local trappers, and South Dakota Trapper Association to reach agreements to purchase pan tension devices and develop other techniques to reduce likelihood of fox capture in coyote traps, and work with trappers to develop agreements to temporarily reduce trapping if trapping mortality become significant (not expected based on other states).

11) work with local landowners to limit potential impact of M44s ("coyote getters") on foxes.

Translocation

Acquisition Methods

We will work with Wyoming Game and Fish (priority) and Colorado Division of Wildlife to locate the best sites within each state for trapping and removing foxes. We will work with these states to monitor fox populations to ensure no significant impact to donor populations. During late summer 2002, we will set approximately 40 traps (small mesh single- or double-door tomahawk traps and wood liners [Sovada et al. 1998]) in the chosen trapping area. To maximize genetic diversity and reduce local impacts, no more than 2 foxes will be removed from 1 location (area the size of approximately 1 fox home range). We will attempt to capture 15 males and 15 females for each of 6 years from Wyoming and northern Colorado. Traps will be set in the evening and checked at dawn and then closed during the day. We will use a handling bag or blanket to remove foxes from traps. Foxes will be manually restrained and handled by 2 technicians. We will weigh foxes, assess body condition, count and collect parasites; collect blood; and measure the neck, canines, and body length. We will mark foxes with an ear tag and pit tag. If trapping success is high, we will select primarily adults for translocations.

Dr. Dave Hunter (Turner Enterprises, Bozeman, Montana) will serve as project veterinarian, providing all oversight and protocols. Parasites and diseases of wild swift foxes have not been well documented. There are no cases of confirmed overt disease in wild populations (Miller et al. 2000, Pybus and Williams 2002). We will follow the recommendations of Miller et al. (2000) and Pybus and Williams (2002) to ensure disease risks during translocation are minimized. Canine parvovirus, canine distemper virus, sylvatic plague, and rabies have been detected in swift fox from southeast Wyoming and we expect some of the foxes we capture will test positive for these diseases (Miller et al. 2000). Foxes that test positive for plague, have very high titers for distemper, or show outward signs of rabies will be returned to the capture site as per WDGF. Any fox that appear in poor condition will be released from traps at capture site. All foxes will be vaccinated for rabies, distemper, infectious hepatitis, adenovirus type 2, parainfluenza and parvovirus (J. Johnson and L. Carbyn, Canadian Wildlife Service, unpublished report). Foxes will receive Duramune Max 5 killed virus with modified live parvovirus, Rabvac 3 killed virus, and be sprayed with Frontline (fipronil) spray. Foxes will then be placed and remain in kennels for <96 hrs and then driven directly to holding pens (3.7 x 7.3 m) on BRR. Pairs of foxes will be placed into each pen (see below), but separated through the quarantine period. Pens will be approximately 4 km apart. Foxes will remain in quarantine in the holding pens for a 14-day health check period.

Release Methods

We will release animals at a number of different locations on and around BRR, including FPNG, Lower Brule and possibly Cheyenne River Reservations, clustered by year. Different locations will be used because foxes may be excluded from some areas due to interspecific competition, but persist in others because of specific behavioral adaptations that have survival value to local conditions ("fugitive" Hutchinson 1951).

As indicated in Kunkel, Honness, Phillips, and Carbyn (2001), adult survival rate is the most important factor affecting the outcome of fox reintroduction. We will strive to maximize survival by focusing on releasing foxes that have the highest probabilities of survival (wild-born adult foxes; Covell 1992, Carbyn et al. 1994, Sovada et al. 1998, Cypher et al. 2000) and through intensive management of released animals. Carbyn (1995) reported that 6 of 108 (6%) captive-raised foxes produced pups while 6 of 19 (32%) of translocated wild foxes produced pups. Additionally, the resident breeding population (as defined by den establishment) in spring 1991 in Canada indicated a roughly 1:1 ratio of captive released to translocated wild foxes despite a 5:1 release ratio (Carbyn 1995). Such patterns have been reported for other species. Stochastic models showed that extinction probabilities for Leadbeater's possum (*Gymnobelideus leadbeateri*; Burgman et al. 1995) or helmeted honeyeater (*Lichenostomus melanops cassidix*; McCarthy 1995) were reduced when adults rather than immature animals were released. Sarrazin and Legendre (2000) developed a demographic model for griffon vulture (*Gyps fulvus*) reintroductions and found that it was more efficient to release adults than juveniles, despite the overall reduction of demographic parameters following release. Caswell (1989) emphasized that the eventual population was larger at any time if the initial population was concentrated in age-classes with high reproductive rates. Adult foxes are more likely to successfully rear pups than are juveniles (Cypher et al. 2000).

Evidence from the Blackfeet and Canadian reintroductions indicates release success of captive foxes can be relatively high (see above; Smeeton and Weagle 2001). More work is needed to assess the relative value of releasing captive-reared animals. Therefore, we will attempt to attain captive animals from the Cochrane Ecological Institute in Alberta or zoos for releases. We will compare success of these releases to success of wild foxes released.

Little information is known on the relative success of soft versus hard releases of foxes. Holding foxes in pens for soft releases will have great information and educational value for local residents (L. Carbyn, personal observation). Observations of foxes will only be allowed from a distance so that stress to foxes is minimized. Additionally, holding foxes through the breeding season will allow pups to be produced in captivity and may thereby increase the survival rate for pups. Therefore, during 2002, approximately 10 wild foxes will remain in the holding pens on and around BRR for release in late May or early June 2003 (soft releases). A pair of foxes will be held in each of 10 pens. Pens will be equipped with den boxes. We will monitor tolerances among foxes and separate any foxes that have conflicts. Foxes will be allowed to breed and rear pups during spring in the holding pens (Carbyn et al. 1994). Pen doors will be locked open for releases so that foxes may continue to use den boxes. Foxes will be gradually weaned off food in the pens. The success of the soft release strategy will be assessed to determine if soft releases will continue to be used in the following years. All soft releases will occur on BRR.

Twenty (approximately 10 pairs) wild foxes will be used for hard releases. These foxes will be placed in quarantine/holding pens with panels to separate each fox. Foxes selected for hard releases will be released immediately from the holding pens after the quarantine period has ended. Some foxes selected for hard releases may be transported from holding pens in kennels to various release sites on BRR, FPNG, and LBR. Kunkel, Honness, Phillips, and Carbyn (2001) indicates escape terrain should not be limiting in the project area. Should we find this not true, we will dig escape holes for foxes in the immediate vicinity of release sites.

Criteria for Measuring Success

Initial success (1–3 years) will be reached when we achieve breeding of the first wild-born generation of foxes in the release area (Kleinman et al. 1991). Short-term criteria (3–5 years) for success will include survival and recruitment rates similar to other wild self-sustaining populations (e.g., Carbyn et al. 1994, Sovada 1998, Schauster 2001) and population growth or $r > 0$.

According to Minimum Viable Population studies, Beck et al. (1994) considered 500 free-living individuals as representing criteria for good success. This threshold is considered somewhat arbitrary without taking into account life history traits, habitat quality, or eventual metapopulation structure (Sarrazin and Babault 1996). They recommended using extinction probability estimates that combined population size, growth rate, and growth rate variance as the main criteria. We will estimate these parameters throughout the reintroduction and consider mid-term success (5–10 years) when demographic rates approach self-sustaining levels yielding extinction probabilities over 100 years of <0.20. We follow recommendations of Breitenmoser et al. (2002) and apply the IUCN Red List Categories to assess success and failure at about 10 years after completion of release. Long-term success (>10 years) will be reached when fox populations expand and connect with other populations in the region.

We will assess factors affecting survival and recruitment rates (see below) and thereby determine reasons for not meeting criteria for success. To deal with problems, we will use adaptive management to modify release and management strategies; options include modifying classes of foxes used for releases and the types of releases, intensifying and extending coyote control and supplemental feeding of foxes (e.g., providing bison carcasses; Warrick et al. 1999).

Monitoring and Research Program (Experimental Design)

We agree that reintroduction programs can provide important opportunities for real-scale hypothetico-deductive experiments in ecology (e.g., Stanley Price 1989, Kleinman et al. 1991, Sarrazin and Barbault 1996). Accordingly, we intend to implement this restoration project to maximize gain of knowledge concerning swift fox recovery throughout North America. This reintroduction will be conducted as an experiment so that we gain the most knowledge possible in the most rigorous fashion. For example, this project should provide an opportunity to better understand mechanisms of population extinction and growth. The use of reintroductions as basic experiments for building theory will eventually create better knowledge of extinction mechanisms and thereby limit the need for future reintroductions (Sarrazin and Barbault 1996).

Objectives

1. Estimate fox density annually for 10 years following first release.
2. Estimate reproductive parameters annually for 10 years following first release.
3. Estimate fox survival rates annually for 10 years following first release.
4. Determine fox diet annually for 10 years following first release.
5. Determine fox home range size and resource selection annually for 10 years following first release.

Methods

All foxes released will be fitted with standard VHF radiocollars (Cypher 1997) and ear-tagged. In order to maintain telemetry contact with all foxes, we will capture and radio-collar juveniles in the fall (fit with expandable collars) and again the following year to fit with adult-sized collars (see methods above). For the first 2 months following releases we will attempt to locate all foxes daily; thereafter, we will locate foxes 1–2 times weekly. Tracking will occur primarily by triangulation from the ground but aerial telemetry will be used to locate far-ranging foxes. Dispersing foxes will be located bi-weekly. Foxes will be located at all times of the day.

Radios will be equipped with a mortality sensor. When radios emit mortality signals, we will investigate the fox and the site to determine cause of death following standard protocols (Kunkel 1997). Landscape features at relocations, kill sites, and random sites throughout the release area will be described including slope, aspect, habitat type, vegetation height, viewshed, and percent stalking cover (Kunkel and Pletscher 2001). We will use a Geographic Positioning System to determine latitude and longitude coordinates of the site and upload this data into a Geographic Information System (GIS). GIS will be used to classify landcover (South Dakota Geographic Analysis Project [GAP analysis]), topographic ruggedness (Nicholson et al. 1997), range site (soil type), grazing intensity, distance to roads, water, settlements, and dens. We will determine location of fox den sites and escape holes and measure the same landscape attributes as measured at fox relocation sites (Pruss 1999). We will observe foxes at den sites to measure pup production and survival and attendance by adults.

Survival rates will be determined using MICROMORT (Heisey and Fuller 1985). Cause-specific mortality rates and factors affecting these will be estimated using z-tests of survival estimates and Cox proportional hazards models (Sievert and Keith 1985). Because at least initially this will be the only location in North America where we know the precise number of foxes (due to reintroduction), it will present a great opportunity to test population estimation techniques. We will test all the currently employed techniques, including catch per unit effort (Seber 1982), mark-recapture (Rexstad and Burnham 1991), scent stations, scat deposition transects, track plates, and spot light surveys.

We will continue to follow methodology outlined in the feasibility plan to monitor fox, fox prey, and fox predator trends (Kunkel et al. 1999). Swift fox and coyote scats will be collected (at fox source sites [Colorado and Wyoming] and BRR) and analyzed for disease and diet following standard techniques (Kelly and Garton 1997). Fox prey selection will be determined following Manly (1974). We will attempt to observe fox predation sequences to determine success rates and the influence of landscape and prey factors (Gese et al. 1996).

Literature Cited

Beck, B.B., L.G. Rapaport, M.R. Stanley Price, and A.C. Wilson. 1994. Reintroduction of captive born animals. Pp. 265–286 *in* P.J.S. Olney, G.M. Mace, and A.C.T. Feistner, editors. Creative conservation: interactive management of wild and captive animals. Chapman and Hall, London, UK.

Bell, H.B., and R.W. Dimmick. 1975. Hazards to predators feeding on prairie voles killed with zinc phosphide. Journal of Wildlife Management 39:816–819.

Breitenmoser, U., C. Breitenmoser-Wursten, L.N. Carbyn, and S.M. Funk. 2002. Assessment of carnivore reintroductions. Pp. 241–271 *in* J.L. Gittleman, S.M. Funk, D.M. MacDonald, and R.K. Wayne, editors. Carnivore Conservation. Cambridge University Press, Cambridge, UK.

Burgman, M., S. Ferson, and D. Lindemayer. 1995. The effect of the initial age-class distribution on extinction risks: implication for the reintroduction of Leadbeater's possum. Pp. 15–19 *in* M. Serena, editor. Reintroduction biology of Australian and New Zealand fauna. Surrey Beatty and Sons, Chipping Norton, New South Wales, Australia.

Carbyn, L.N. 1995. Swift foxes on the northern plains. Canid News 3:41–45.

——. 1998. COSEWIC status report on the swift fox. Committee on the status of endangered wildlife in Canada. Canadian Wildlife Service, Ottawa, Ontario.

Carbyn, L.N., H.J. Armbruster, and C. Mamo. 1994. The swift fox reintroduction program in Canada from 1983 to 1992. Pp. 247–271 *in* M.L. Bowles and C.J. Whelan, editors. Restoration of endangered species. Cambridge University Press, Cambridge UK.

Caswell, H. 1989. Matrix population models. Sinauer Associates, Sunderland, Massachusetts.

Covell, D.F. 1992. Ecology of the swift fox in southeastern Colorado. Thesis. University of Wisconsin, Madison, Wisconsin.

Crandall, K.A., O.R.P. Bininda-Emonds, G.M. Mace, and R.K. Wayne. 2000. Considering evolutionary processes in conservation biology. Trends in Ecology and Evolution 15:290–295.

Cypher, B.L. 1997. Effects of radiocollars on San Joaquin kit foxes. Journal of Wildlife Management 61:1412–1423.

Cypher, B.L., G.D. Warrick, M.R.M. Otten, T.P. O'Farrell, W.H. Berry, C.E. Harris, T.T. Kato, P.M. McCue, J.H. Scrivener, B.W. Zoellick. 2000. Population dynamics of San Joaquin kit foxes at the naval petroleum reserves in California. Wildlife Monographs 145.

Dragoo, J.W. and R.K. Wayne. 2003. Systematics and population genetics of kit foxes. Pp. 207–221 *in* L.N. Carbyn and M.A. Sovada, editors. Proceedings of the Swift Fox Symposium: ecology and conservation of swift foxes in a changing world. Canadian Plains Research Center, University of Regina, Saskatchewan.

FaunaWest. 1991. An ecological and taxonomic review of the swift fox (*Vulpes velox*) with special reference to Montana. Boulder, Montana.

Gese, E.M.R. Ruff, and R.L. Crabtree. 1996. Intrinsic and extrinsic factors influencing coyote predation of small mammals in Yellowstone National Park. Canadian Journal of Zoology 74:784–797.

Gigliotti, L.M. 1998. Environmental and wildlife attitudes of South Dakota residents. South Dakota Department of Game Fish and Parks Report HD-6-98. SAM.

Ginsberg, J.R. 1994. Captive breeding, reintroduction and the conservation of canids. Pp. 365–383 *in* P.J.S. Olney, G.M. Mace, and A.T.C. Feistner, editors. Creative conservation: interactive management of wild and captive animals. Chapman and Hall, London, UK.

Haig, S.M., J.O. Ballou, and S.R. Derrickson. 1990. Management options for preserving genetic diversity: reintroduction of Guam rails into the wild. Conservation Biology 4:290–300.

Hall, E.R., and K.R. Kelson. 1959. The mammals of North America. Ronald Press, New York, New York.

Henke, S.E., and F.C. Bryant. 1999. Effects of coyote removal on the faunal community in western Texas. Journal of Wildlife Management 63:1066–1081.

Hecht, A., and P.R. Nickerson. 1999. The need for predator management in conservation of some vulnerable species. Endangered Species Update 16:114–118.

Heisey, D.M., and T.K. Fuller. 1985. Evaluation of survival and cause-specific mortality rates using telemetry data. Journal of Wildlife Management 49:668–674.

Hutchinson, G.E. 1951. Copepodology for the ornithologist. Ecology 32:571–577.

IUCN/SSC Re-introduction Specialist Group. 1998. IUCN guidelines for re-introductions. IUCN, Gland, Switzerland and Cambridge, UK.

Johnson, J.R., K.F. Higgins, and D.E. Hubbard. 1995. Using soils to delineate South Dakota physiographic regions. Great Plains Research 5:309–322.

Kamler, J.F. 2002. Relationships of swift fox and coyotes in northwest Texas. Dissertation, Texas Tech University, Lubbock, Texas.

Kelly, B.T., and E.O. Garton. 1997. Effects of prey size, meal size, meal composition, and daily frequency of feeding on the recovery of rodent remains from carnivore scats. Canadian Journal of Zoology 75:1811–1817.

Kilgore, D.L., Jr. 1969. An ecological study of swift fox (*Vulpes velox*) in the Oklahoma panhandle. American Midland Naturalist 81:512–534.

Kitchen, A.M., E.M. Gese, and E.R. Schauster. 1999. Resource partitioning between coyotes and swift foxes: space, time, and diet. Canadian Journal of Zoology 77:1645–1656.

Kleinman, D.G., B.B. Beck, J.M. Deitz, and L.A. Deitz. 1991. Costs of reintroduction and criteria for success: accounting and accountability in the golden lion tamarin conservation program. Pp. 125–142 *in* J. H. W. Gipps, editor. Beyond Captive Breeding : Reintroducing Endangered Species to the Wild. Oxford University Press, Oxford, UK.

Koopman, M.E., B.L. Cypher, and J.H. Scrivner. 2000. Dispersal patterns of San Joaquin kit foxes (*Vulpes macrotis mutica*). Journal of Mammalogy 81:213–222.

Kuchler, A.W. 1975. Potential natural vegetation of the United States. American Geographic Society, Special Publication 36, New York, New York.

Kunkel, K.E. 1997. Predation by wolves and other large carnivores in northwestern Montana and southeastern British Columbia. Dissertation, University of Montana, Missoula.

Kunkel, K.E., L.N. Carbyn, K.M. Honness, and M.K. Phillips. 1999. Feasibility study plan for the re-introduction of swift foxes to Turner Properties in the Great Plains. Turner Endangered Species Fund Conservation Report 99-01, Bozeman, Montana.

Kunkel, K.E., K.M. Honness, and M.K. Phillips. 2001. Plan for restoring swift fox to west-central South Dakota. Turner Endangered Species Fund Conservation Report 01-02, Bozeman, Montana.

Kunkel, K.E., K.M. Honness, M.L. Phillips, and L.N. Carbyn. 2001. Feasibility of restoring swift fox to west-central South Dakota. Turner Endangered Species Conservation Report 01-01, Bozeman, Montana.

Kunkel, K.E. and D.H. Pletscher. 2001. Winter hunting patterns and success of wolves in Glacier National Park, Montana. Journal of Wildlife Management 65:520–530.

Leberg, P.L. 1990. Genetic considerations in the design of introduction programs. Transactions of the North American Wildlife and Natural Resources Conference 55:609–619.

Mamo, C. 1987. Swift fox habitat assessment. Unpublished report, University of Calgary, Faculty of Environmental Design, Swift Fox Research Group, Calgary, Alberta.

Manly, B.F.J. 1974. A model for certain types of selection experiments. Biometrics 30:281–294.

McCarthy, M.A. 1995. Population viability of the helmeted honeyeater: risk assessment of captive management and reintroduction Pp. 21–25 *in* M. Serena, editor. Reintroduction biology of Australian and New Zealand fauna. Surrey Beatty and Sons, Chipping Norton, New South Wales, Australia.

Merriam, C.H. 1902. Three new foxes of the kit and desert fox groups. Proceedings of the Biological Society of Washington 15:73–74.

Miller, D.S., D.F. Covell, R.G. McLean, W.J. Adrian, M. Niezgoda, J.M. Gustafson, O.J. Rongstad, R.D. Schultz, L.J. Kirk, and T.J. Quan. 2000. Serologic survey for selected infectious disease agents in swift and kit foxes from the Western United States. Journal of Wildlife Diseases 36:798–805.

National Governors Association. 2001. Private lands, public benefits principles for advancing working lands conservation. National Governors Association Center for Best Practices, National Governors Association, Washington D.C.

Nicholson, M.C., R.T. Bowyer, and J.G. Kie. 1997. Habitat selection and survival of mule deer: tradeoffs associated with migration. Journal of Mammalogy 78:483–504.

Olson, T.L. 2000. Population characteristics, habitat selection patterns, and diet of swift foxes in southeastern Wyoming. Thesis, University of Wyoming, Laramie, Wyoming.

Paton, P.W.C. 1994. The effect of edge on avian nest success: how strong is the evidence? Conservation Biology 8:17–26.

Pruss, S. 1999. Selection of natal dens by swift foxes on the Canadian prairies. Canadian Journal of Zoology 77:646–652.

Pybus, M.J., and E.S. Williams. 2002. A review of parasites and diseases of wild swift fox. Pp. 231–236 *in* L.N. Carbyn and M.A. Sovada, editors. Proceedings of the Swift Fox Symposium: ecology and conservation of swift foxes in a changing world. Canadian Plains Research Center, University of Regina, Saskatchewan.

Rexstad E., and K.P. Burnham. 1991. User's guide for interactive program CAPTURE, Colorado Cooperative Fish and Wildlife Research Unit, Ft. Collins, Colorado.

Sarrazin, F., and R. Barbault. 1996. Reintroduction: challenges and lessons for basic ecology. Trends in Ecology and Evolution 11:474–478.

Sarrazin, F., and S. Legendre. 2000. Demographic approach to releasing adults versus young in reintroductions. Conservation Biology 14:488–500.

Schauster, E.R. 2001. Swift fox (*Vulpes velox*) on the Piñon Canyon Maneuver Site, Colorado: population ecology and evaluation of survey methods. Thesis, Utah State University, Logan, Utah.

Schitoskey, F., Jr. 1975. Primary and secondary hazards of 3 rodenticides to kit fox. Journal of Wildlife Management 39:416–418.

Seber, G.A.F. 1982. The estimation of animal abundance and related parameters. Second edition, Macmillian Publishing, New York, New York.

Sharps, J.C. 1984. Northern swift fox investigations 1977–1981. South Dakota Department of Game Fish and Parks Completion Report 85–11.

Sharps, J.C. and M.F. Whitcher. 1984. Swift fox reintroduction techniques. South Dakota Department of Game Fish and Parks, unpublished report, Pierre, South Dakota.

Sievert, P.R., and L.B. Keith. 1985. Survival of snowshoe hares at a geographical range boundary. Journal of Wildlife Management 49:854–866.

Smeeton, C., and K. Weagle. 2001. The reintroduction of swift fox *Vulpes velox* to south central Saskatchewan, Canada. Oryx 34:171–179.

Sovada, M.A., C.C. Roy, J.B. Bright, and J.R. Gillis. 1998. Causes and rates of mortality of swift foxes in western Kansas. Journal of Wildlife Management 62:1300–1306.

Sovada, M.A., C.C. Roy, and D.J. Telesco. 2001. Seasonal food habits of swift fox (*Vulpes velox*) in cropland and rangeland landscapes in western Kansas. American Midland Naturalist 145:01–111.

Swift Fox Conservation Team. 1997. Conservation assessment and conservation strategy for swift fox in the United States. R. Kahn, L. Fox, P. Horner, B. Giddings, and C. Roy, editors, Montana Fish, Wildlife and Parks, Helena, Montana.

Stanley Price M.R. 1989. Animal reintroductions: the Arabian oryx in Oman. Cambridge University Press, Cambridge, UK.

Stromberg, M.R., and M.S. Boyce. 1986. Systematics and conservation of the swift fox in North America. Biological Conservation 35:97–110.

Uresk, D.W., and J.C. Sharps. 1986. Denning habitat and diet of the swift fox in western South Dakota. Great Basin Naturalist. 46:249–253.

U.S. Fish and Wildlife Service. 1994. Endangered and threatened wildlife and plants: 90 day finding on a petition to list the swift fox as endangered. Federal Register 59:28328–28329.

Warrick, G.D., J.H. Scrivner, and T.P. O'Farrell. 1999. Demographic responses of kit foxes to supplemental feeding. Southwestern Naturalist 44:367–374.

White, P.J., and R.A. Garrott. 1999. Population dynamics of kit foxes. Canadian Journal of Zoology 77:486–493.

Wilcove, D.S., C.H. McClellan, and A.P. Dobson. 1986. Habitat fragmentation in the temperate zone. Pp. 237–256 *in* M.E. Soulé, editor. Conservation biology. Sinauer, Sunderland, Massachusetts.

Captive Breeding of the Swift Fox at the Cochrane Ecological Institute, Alberta

■ **Clio Smeeton, Ken Weagle and Siân S. Waters**

Abstract: An analysis of the swift fox (Vulpes velox) *captive-breeding program at Cochrane Ecological Institute, Cochrane, Alberta from 1972 to 1997 is presented. During that period, captive breeding of swift fox provided 841 animals to the Canadian swift fox reintroduction program in southern Alberta and Saskatchewan (Smeeton and Weagle 2000). The founder population consisted of 34 individuals, the majority of which came from wild populations in the United States. The 272 litters produced averaged 3.6 ± 0.45 live kits per litter with a range of 1–7 kits. Captive-born yearling females had a lower fecundity than females >1 year old (P <0.05), with the highest fecundity in their fifth year. Captive-born yearling males also had a lower fecundity than other ages (P <0.95) with highest fecundity being found in years 9 and 10 (P <0.05). In wild-born males, fecundity was lower than all age groups except yearlings (P <0.05). The maximum number of live kits produced by a female was 33 over 8 years and 46 by a male over 14 years. Most litters (71%) were born between 16 April and 10 May with some whelping as late as 26 June. Nineteen percent of yearling females and 18% of yearling males produced litters when paired with an older animal, whereas 11% of pairs with both male and female yearlings produced litters. Females did not produce litters after 8 years of age, but males continued to produce litters to age 14.*

Captive breeding of the swift fox began in Canada in 1972 at the Wildlife Reserve of Western Canada (now the Cochrane Ecological Institute [CEI]), 6 years before the species was declared extirpated in Canada (Committee on the Status of Endangered Wildlife in Canada [COSEWIC] 1978). The specific goal of CEI is to produce swift foxes of maximum genetic heterozygosity for reintroduction into protected areas of their historic range.

Although CEI has bred swift foxes in captivity for over 25 years, little has been published on the methods or results of the program. Captive swift foxes at CEI and the Calgary Zoo had an estimated gestation period of 51–52 days (C. Schroeder, University of Calgary, Calgary, Alberta, unpublished data). Weagle and Smeeton (1995) summarized behavioral characteristics of the foxes at CEI and Teeling (1996) reported detailed observations of captive breeding behavior at CEI. Bremner (1997) developed environmental enhancement procedures for captive breeding enclosures at CEI.

Our objective was to describe the breeding biology of swift foxes in a captive breeding facility. Data were taken from the CEI studbook for swift foxes and from daily observations of the foxes. The protocols for the feeding, housing, and health care that were developed over the 25-year captive breeding program may be obtained from the senior author.

Methods

Facilities

CEI was situated in the foothills of the Rocky Mountains and within the historic range of the swift fox. The 65-hectare facility was enclosed by a 2.5-m high chain link fence with a 0.6 m wire mesh overhang and a 0.6 m chain link strip along its base. The vegetative community within the facility was 50% native mixed-grass prairie, 10% wetland, and 40% mixed woodland. The site was also used by large ungulates, including moose (*Alces alces*), white-tailed deer (*Odocoileus virginianus*), and mule deer (*Odocoileus hemionus*). There were fluctuating populations of coyote (*Canis latrans*), indigenous waterfowl, passerines, and raptors. The CEI facilities included 3 types of enclosures (single-pair, 1-ha, and 9-ha) to house foxes, an animal health building with kennels for the treatment of ill or injured foxes, and separate food preparation and storage facilities. There were also extensive quarantine pens for incoming and outgoing animals.

Breeding age swift foxes were paired and housed in 23 single-pair enclosures. These animals varied in age from juveniles to 14 years. These enclosures averaged 18 m x 12 m in size, were made of 2.5-m x 3-m chain-link panels with a 0.6 m wire mesh overhang along the top and a 0.6-m wide chain-link fencing laid on the ground and filled with rocks. Vegetation in the enclosures was native prairie, but 7 also included aspen trees (*Populus tremuloides*). Each enclosure contained 2 artificial den boxes, which consisted of 3 connected chambers covered with an insulated A-frame which is described in more detail in Smeeton and Weagle (2000).

The 1-ha enclosure was 100 m square in size and constructed as above. The vegetation consisted of aspen groves. The enclosure contained 6 artificial dens, and was used to house a fluctuating number of aged, non-breeding swift fox, mainly females.

The 9-ha enclosure was 400 m x 225 m and constructed identically to the other enclosures. Vegetation was 40% prairie, 10% bog, and 50% mixed aspen and spruce groves. The 9-ha enclosure contained 7 widely spaced artificial den boxes and 1 artificial "mound" (expanded polystyrene on wire over a 3-chambered box). Two artificial dens were in the woodland, with the remaining 6 in the prairie. Six juvenile foxes (3 males and 3 females) were housed here although only 1 litter was produced during the 2 years that they remained together in the enclosure.

Breeding Records

The data on breeding biology was obtained from the swift fox studbook maintained by CEI since 1972. Three zoos contributed data to the studbook for varying periods until they discontinued breeding the species. They were the Calgary Zoo (between 1980 and 1993), the Moose Jaw Wild Animal Park (between 1984 and 1994) and Edmonton Valley Zoo (between 1987 and 1995). The swift fox studbook was maintained using SPARKS software (ISIS 1986).

Animals were paired to guarantee maximum genetic heterozygosity in the captive population, resulting in an inbreeding coefficient of <0.05 calculated by SPARKS. Prior to that the calculations were done without the aid of SPARKS. We ensured that animals carrying the same bloodlines were not released into the same reintroduction area.

Data Collection

Daily observations of swift fox behavior were recorded during feeding (at dusk, 1700 to 2000 hr depending on season). We recorded the activity of each animal, and noted if an individual was not seen. We used these observations to estimate dates of birth. Without entering the enclosures, we noted any sign of hair removed from the females' abdomens exposing the teats, or the flush of pink in the teats, which indicates the presence of milk. We also noted changes in behavior patterns as an indication that birth had occurred. We assumed the females had whelped if competitive feeding behavior between adults stopped, if females ceased to appear for feeding, and if males began to collect food and carry it into the den. We recorded observations of female aggression toward kits when attempting to suckle as an estimate of the age of weaning.

Whelping generally occurred in the artificial den boxes or, in some cases, in underground dens dug by the pair. The male fox provided the female with food for approximately 2 weeks after whelping, when the female rarely left the den. The daily records indicated when females were no longer observed at feeding times. Once whelping occurred the males tended to spend more time out of the den and were observed collecting and carrying food into the den for the female. When this series of behaviors was observed, the date of whelping was estimated. These dates were checked by back calculations of when the kits were first observed outside of the dens. At CEI, the adults and kits were never handled during the first 6 weeks unless an injury required immediate attention.

Data Analysis

We entered swift fox studbook data into an EXCEL spreadsheet for analysis. These data included the age and number of individuals (males and females) used in the breeding program, the number of litters, average litter size, and the birth date of each litter. We conducted student's t-tests to examine differences in litter size among parents' age-categories using the statistical software package SPSS v7.1 (Morusis 1993). ANOVA was not used because it would not separate differences between years but only between all years. The term fecundity was used in this paper to denote the productivity of individual foxes. Due to the importance of limiting disturbance to lactating females, actual litter sizes at birth were unknown. Therefore, all calculations regarding litter size were based on kits that actually survived and emerged from the den.

Results and Discussion

Breeding Population

From 1972 to 1997, CEI used 97 female and 92 male foxes in their breeding program. The initial swift fox came from an animal rehabilitation center in Golden, Colorado; subsequently an additional 28 swift foxes, captured in Colorado, Wyoming, and South Dakota were added to the breeding population (Table 1). Between 1974 and 1997, 329 male and 326 female kits were produced at CEI in 272 litters. The average litter size was 3.6 ± 0.45 kits, and ranged from 1 to 7 kits (n = 272). The CEI breeding population provided 841 swift foxes to the Canadian reintroduction program for release (Smeeton & Weagle 2000).

Limited observations of swift fox life cycles in wild populations in Canada indicated that both wild and captive foxes had similar behavioral characteristics (Pruss 1993). One striking difference between the captive-born and the wild-born foxes was the life span of the animals. In the wild, death was seldom the result of old age; in contrast, death resulting from old age was the norm within our captive colony. Unpublished data (CEI) on the age of captive-bred released swift fox in the wild indicated that the oldest known captive bred male in the wild was 8 years of age;

Table 1. Sources, numbers and dates for swift foxes imported for the Cochrane Ecological Institute founder swift fox population.

Source	Males	Females	Years
Golden, Colorado (first imports)	2	4	1972
Pierre, South Dakota	1	5	1978–81
Pawnee National Grassland, Colorado	2	1	1980
Weld County, Colorado	1		1982
Lincoln County, Colorado	2	2	1982–85
Las Animas County, Colorado	1		1987
Laramie County, Wyoming	1		1990
Laramie County, Wyoming	2	2	1990–91
Laramie County, Wyoming	5	3	1994
Total	17	17	

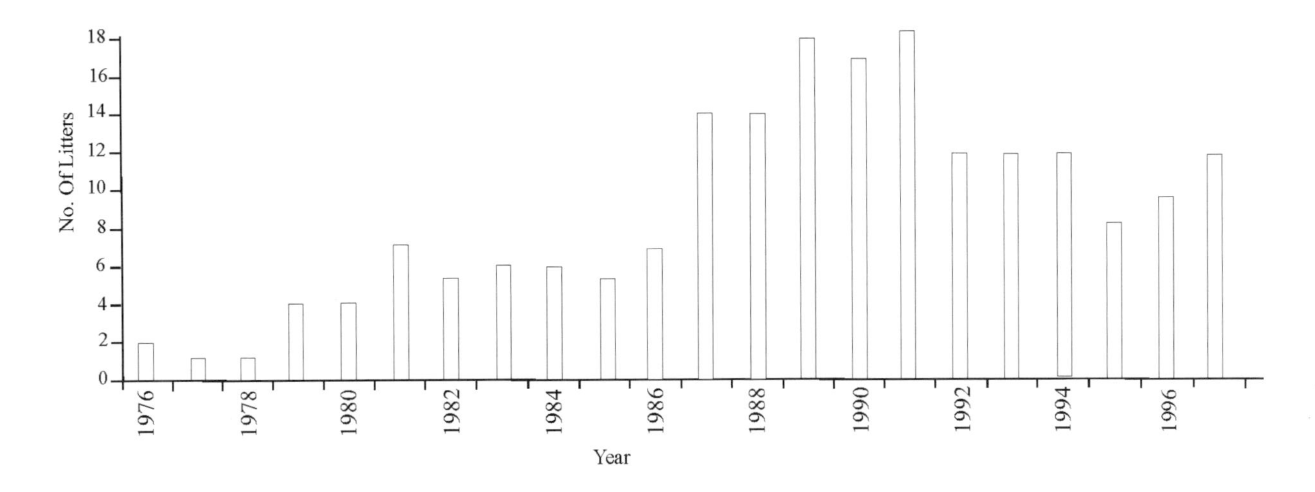

Figure 1. Numbers of litters per year born at CEI from 1976 to 1997.

the oldest known captive bred female in the wild was 6 years of age.

In 1986, the Canadian Swift Fox Recovery Team established a protocol for pairing captive foxes and instituted a minimum period for holding them. The protocol called for an inbreeding coefficient of <0.05 and that all captive-bred foxes of 4 years old should be released. Swift foxes that were >4 years old at the time of this decision remained with the colony. The goal was a breeding colony of an even age and of wide genetic heterozygosity. Swift foxes paired on a genetic basis will not necessarily breed; other factors (age, compatibility) appear to affect pair breeding success. Following compliance of the policy, the colony consisted of young, newly paired animals, older newly paired animals, and aged animals. Fewer established reproductive pairs remained, resulting in a period of reduced kit production in the captive colony after 1991 (Fig 1.).

We found that when swift foxes were newly paired there were differences in their breeding success, depending on the age of the individuals comprising the pair. In new pairs with yearling females and experienced males, 19% ($n = 37$) of the pairings produced kits in the first year. In pairings with yearling males and experienced females, 18% ($n = 27$) produced kits; and, when both in the pair were yearlings, 11% ($n = 73$) of the pairings produced kits in the first year.

Table 2 summarizes the relationship between age and fecundity for males and females. Yearling females had a lower fecundity ($P < 0.05$) than all other groups except wild-born females. Five-year-old females had a higher fecundity ($P < 0.05$) than all other groups. One-year-old males had a lower fecundity ($P < 0.05$) than most other year classes.

In both the above comparisons, the wild-born males and females in the captive breeding program were separated from the captive-born individuals because their ages could not be determined. The fecundity of the wild-born males and females was generally lower than captive-born foxes, which may be related to their ability to adjust to a captive environment. Once the program began, the wild-born animals were added to the program to maintain genetic diversity. The analysis showed that they were less able to contribute to the production of animals for the reintroduction program than the captive-born swift foxes.

Females never produced past the age of 8 years, but males produced until they died of old age (14 years; Table 2). The most productive female had an average litter size of 4.13 kits. The maximum production for a male was 46 kits with an average litter size of 4.18 kits.

Three of 14 females with an average litter size >4 were siblings but from different litters (Table 2.). A fourth sibling to these females, held at the Moose Jaw Wild Animal Park, also had an average litter size >4 kits per

Table 2. Mean litter size and number (n) of litters in sample by age class[1] and sex of captive-born and wild-born swift fox parents in the captive breeding program, Cochrane Ecological Institute, Alberta, 1972–1997.

	Females[2]		Males	
Age	$\bar{x}$	n	$\bar{x}$	n
1	3.0	35	3.0	34
2	3.4	41	3.7	38
3	3.5	43	3.9	40
4	4.0	35	3.5	28
5	4.8	23	3.8	17
6	3.8	18	3.8	16
7	3.8	18	3.1	11
8	4.0	9	3.6	9
9	0		4.1	11
10	0		4.1	10
11	0		3.8	4
12	0		4	1
13	0		4	1
14	0		5	1
Wild-born	3.1	50	2.9	51

[1] Age of wild-born adults unknown
[2] No females bred at age >8 years

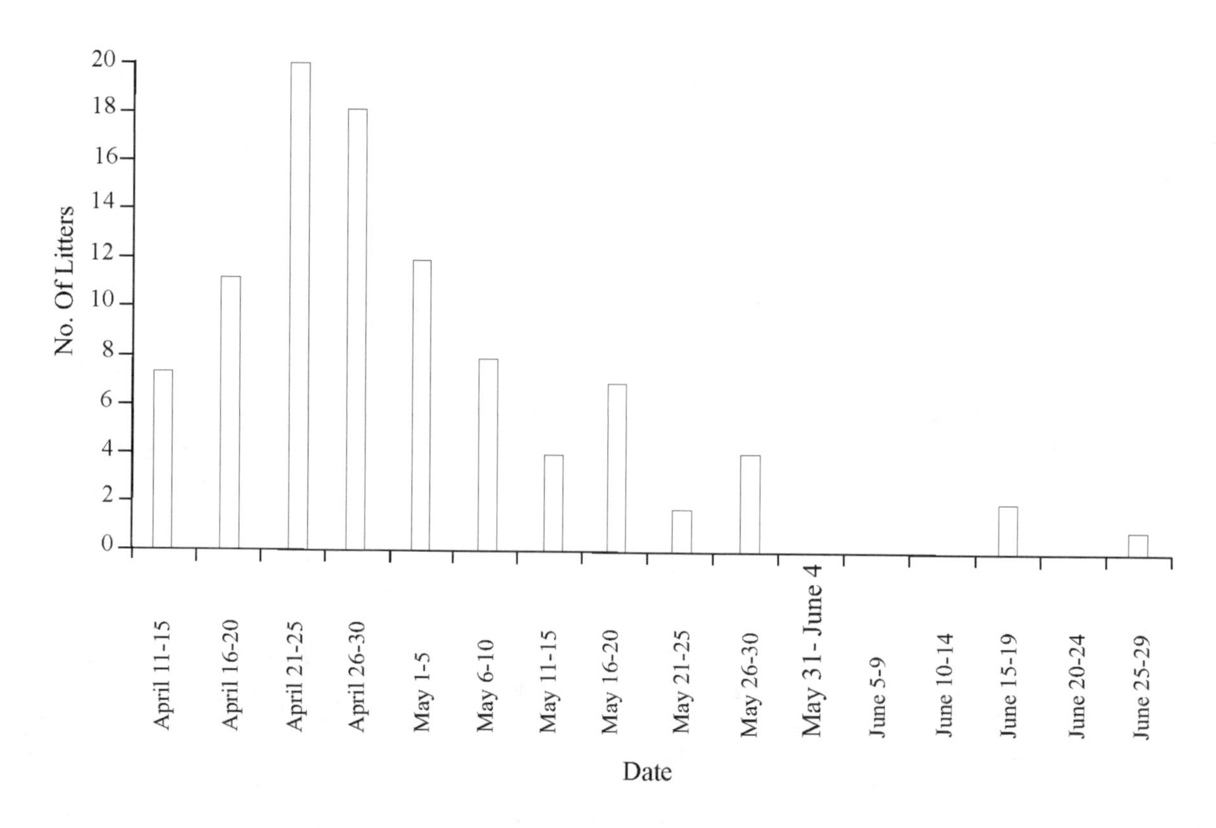

Figure 2. Distribution of birth dates for litters at CEI from 1976 to 1997.

litter ($n = 8$). It was also noted that 6 of 14 females with >4 kits per litter were from wild-born parents. Because of the lack of information on litter sizes in wild populations it is difficult to make comparisons between captive-bred and wild-born litters.

Seventy-one percent of the litters produced at CEI were born between 16 April and 10 May (Fig. 2). This indicates that the peak of breeding extended from 24 February to 20 March. Litters were born as early as 11 April and as late as 26 June. These observations coincided with the period of bay calling from the first week of February until the first weeks of April (Weagle and Smeeton 1995) within the captive breeding colony. We have observed that breeding may be affected by temperature, with more activity during warmer days, but present data are too limited to confirm this hypothesis.

Established pairs mated earlier in the season than newly paired foxes and bay calling, which was common during the breeding season, no longer occurred after successful mating. Females past breeding age did not call during the breeding season (Weagle and Smeeton 1995).

A critical period in the captive breeding program began in August. By this time, the kits had entered full adolescence and were ready to disperse. Vocal aggression, high-pitched hums, and chittering between the kits were more common. Digging in the single pair enclosures increased markedly. Injuries as a result of active digging along the wired perimeter fencing were more common. In wild populations, food (grasshoppers) was readily available and kits began to disperse from the family unit. Captive-bred kits were best released to the wild during this dispersal period (Weagle and Smeeton 1995) although this may depend on the geographical location of the captive breeding center and the release site.

Summary

1. Both males and females can reproduce in their first year. Females were able to reproduce until their eighth year and males for their entire life.

2. Swift foxes were largely monogamous and it was important that the pair bond be maintained; pairs should not be separated unnecessarily.

3. The average litter size was 3.6 ± 0.45 kits. Some females consistently produced larger litters.

4. Handling of individuals and disturbance of breeding enclosures between February and June should be kept to a minimum.

5. Extraction of nursing swift fox kits between birth and 20 days should never be undertaken.

6. Observations have shown that some swift fox individuals will undertake excessive cub carrying behavior if within sight of other pairs. Single pair breeding enclosures should be visually isolated from one other.

7. Housing should be designed to allow efficient capture to reduce handling stress and obviate the use of traps or nooses.

8. Wild animals being brought into the existing colony should be examined, immunized, and quarantined for a minimum of 60 days. Each animal should be quarantined separately. Quarantine facilities should be isolated, and each animal should be provided with an insulated den box and visually barred from seeing any other foxes.

Literature Cited

Bremner, S. 1997. Diet and hunting behaviour of captive-bred swift fox (*Vulpes velox/Vulpes velox hebes*) intended for release. Thesis, University of Edinburgh, Edinburgh, Scotland.

Committee on the Status of Endangered Wildlife in Canada. 1978. COSEWIC status reports and evaluations. Volume 1. Official classification of the swift fox as extirpated in Canada. Committee on the status of Endangered Wildlife in Canada. Canadian Wildlife Service, Ottawa, Ontario.

ISIS. 1986. Small population animal records keeping system (SPARKS). International Species Information System. 12101 Johnny Cake Ridge Road, Apple Valley, Minnesota.

Morusis, M.J. 1993. SPSS for windows. Base System Users Guide Release 6. SPSS, Chicago, Illinois.

Pruss, S.D. 1994. An observational natal den study of wild swift fox (*Vulpes velox*) on the Canadian prairie. Thesis, University of Calgary, Calgary, Alberta.

Smeeton, C., and K. Weagle. 2000. The reintroduction of the swift fox *Vulpes velox* to south-central Saskatchewan, Canada. Oryx 34:171–179.

Teeling, E. 1996. What do these swift fox really do? Captive swift fox behavior during the summer months. Thesis, University of Edinburgh, Edinburgh, Scotland.

Weagle, K., and C. Smeeton. 1995. Behavioral aspects of the swift fox (*Vulpes velox*) reintroduction program. Pp. 268–271 *in* B. Holst, editor. Proceedings of the 2nd international conference on environmental enrichment. Copenhagen Zoo, Copenhagen, Denmark.

Part V – Taxonomy/Physiology/Disease

Systematics and Population Genetics of Swift and Kit Foxes

■ **Jerry W. Dragoo and Robert K. Wayne**

Abstract: In this review, we first discuss phylogenetic relationships of kit and swift foxes to add an evolutionary framework to their taxonomy. This is followed by a detailed taxonomic history of kit and swift foxes, which adds an important perspective to interpretation of results of recent morphologic and molecular genetic studies. We then discuss phylogenetic relationships and population subdivisions between kit and swift fox populations. Because these foxes disperse over smaller distances than larger canids, such as coyotes, higher levels of population differentiation are observed. We synthesize the finding of several studies in the form of taxonomic recommendations and suggest many fewer taxonomic units than in early reports. Finally, we address conservation issues that arise from these studies, and conclude with conservation recommendations.

Relationships of Kit and Swift Foxes to Other Canids

Recent morphologic and genetic studies clearly establish foxes in the genus *Vulpes* as close relatives of the kit and swift fox (see review in Clutton-Brock et al. 1976, Wayne et al. 1997, Geffen et al. 1992). The arctic fox (*Alopex lagopus*), the fennec (*Fennecus zerda*), and the genus *Vulpes* form a monophyletic group, designated by Wayne et al. (1997) as the red fox group. This grouping is distinct from the gray fox, South American foxes, and the wolf-like canids, having diverged from them about 10 million years ago (Fig. 1). Topologic relationships within *Vulpes* are not well resolved by morphologic or genetic analyses (Clutton-Brock et al. 1977, Geffen et al. 1992, Wayne and O'Brien 1986, Wayne et al. 1987, Wayne et al. 1987, 1997). However, genetic analysis clearly establishes the arctic fox as a sister taxon to kit and swift foxes. The arctic fox has the same chromosome number, morphology, and giemsa banding patterns as kit and swift foxes (Wayne et al. 1997, Wayne and O'Brien 1987). Analysis of allozyme genetic distances and mtDNA sequences clusters the two taxa (Wayne and O'Brien 1986, 1987, Geffen et al. 1992, Wayne et al. 1997). The divergence in cytochrome *b* sequences between kit-swift and arctic fox of about 1.8% suggests a divergence time of about 900,000 years (given a divergence rate of about 2% per million years, see Mercure et al. 1993). However, the first fossil ancestors of the kit-swift foxes may be as old as 2 million years from deposits in Texas (Dalquest 1978), whereas the arctic fox first appears about 200,000 years ago in the Yukon (Kurten and Anderson 1980). This discrepancy led to the suggestion that the arctic fox is an Old World immigrant (Kurten and Anderson 1980). However, the DNA sequencing results clearly support a recent origin in the New World from a kit-swift fox-like ancestor. Both arctic and kit foxes show arid-land adaptations: kit-swift foxes are adapted to desert and prairie habitat, whereas the arctic fox is adapted to cold tundra environments (O'Farrell 1987). As a result of the overwhelming amount of genetic data cited above and the relationships determined among *Alopex*, *Fennecus*, and *Vulpes* from these data, Geffen et al. (1992) have relegated *Alopex* and *Fennecus* to synonymy with *Vulpes*.

Taxonomic History of Swift and Kit Foxes

According to Egoscue (1979), the vernacular names for kit and swift foxes have not been used consistently by various authors for these taxa of foxes. *V. velox* has been referred to as both swift and kit foxes (Say, in James 1823, Merriam 1888), whereas *V. macrotis* has been referred to as kit, long-eared, or desert foxes (Goldman 1931, Merriam 1888, 1902, Nelson and Goldman 1909, 1931). Hall et al. (1957) indicated that swift commonly is used for *V. velox* and kit is used for *V. macrotis*. We follow their usage of the common names.

Formal scientific names have been given to several species and subspecies throughout the taxonomic history of these foxes. The last taxon to be formally described was in 1938 (Benson 1938). The trend after that time has been to synonomize taxa (species relegated to subspecies and other subspecies not recognized as valid). The taxonomic history of kit and swift foxes is dynamic (Table 1). In this section, we review the taxonomic history (more or less in chronological order) of kit and swift foxes as a first step towards a coherent and consistent taxonomy for the group.

Thomas Jefferson purchased the Louisiana Territory in 1803 essentially doubling the size of the United States. Within six months after the purchase he had Meriwether Lewis and William Clark mount an expedition to explore the newly acquired territory. During the two-year expedition these captains and their corps members took extensive field notes. These notes were to be compiled by Lewis,

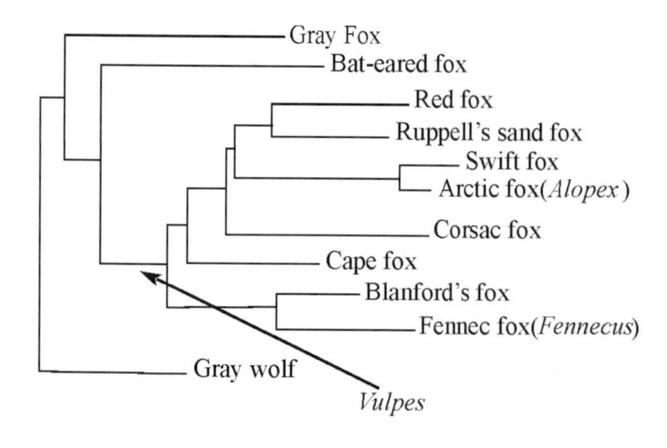

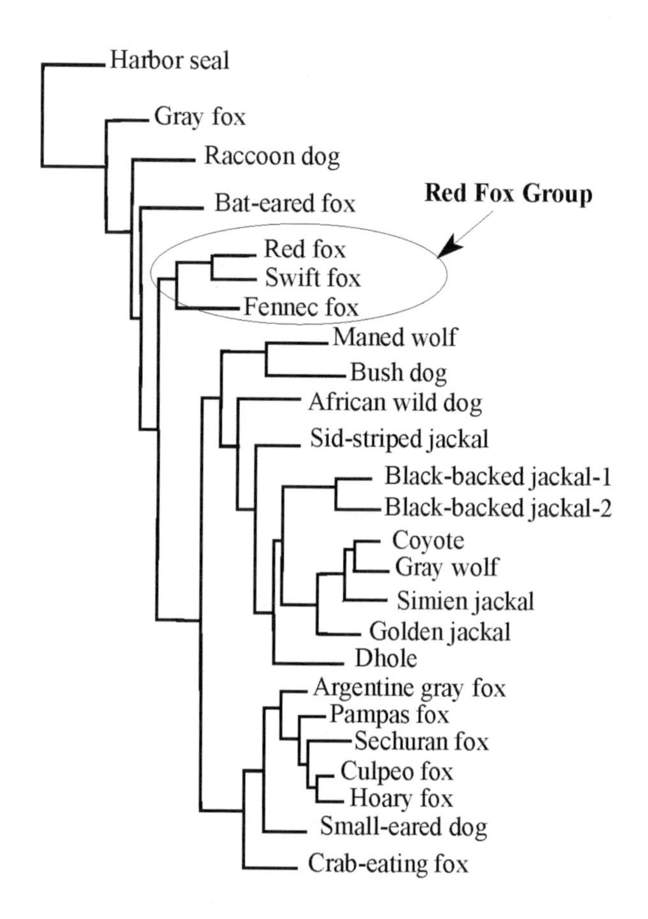

Figure 1. (top) Phylogenetic tree generated by maximum-likelihood method based on 402 bp of cytochrome b *sequence (Geffen et al. 1992). (bottom) Maximum parsimony tree of canids based on analysis of 2001 base pairs of protein coding mitochondrial DNA sequence (cytochrome* b, *cytochrome* c *oxydase I and cytochrome* c *oxydase II) from 24 canid species (Wayne et al. 1997). The harbor seal sequence is used as out-group to root the tree.*

who died before they could be published. The responsibility fell on Clark, who turned the notes over to a Philadelphia lawyer, Nicholas Biddle. From what we can ascertain from the original publication (Biddle 1814), the expedition on the return trip encountered many game species in what is today north-central Montana. Biddle (1814) noted that on Saturday, 26 July 1806, "we saw a few antelopes and wolves, and killed a buck, besides which we saw also two of the small burrowing foxes of the plains, about the size of the common domestic cat, and of a reddish brown colour, except the tail, which is black". No formal description of the fox was provided nor was a formal name given to the species.

A formal description was made by Thomas Say, who was a member of an expedition from Pittsburgh to the Rocky Mountains (James 1823). On 28 June 1820, while the members of the expedition were discussing the food rations and the possibility of going hungry, a journal footnote by Say provides the first mention of the swift fox in the literature:

> *A small fox was killed, which appears to be the animal mentioned by Lewis and Clark, in the account of their travels, under the name of the burrowing fox, (Vol, 2. p. 351 [Biddle, 1814]). It is very much to be regretted, that although two or three specimens of it were killed by our party, whilst we were within about two hundred miles of the Mountains, yet from the dominion of peculiar circumstances, we were unable to preserve a single entire skin; and as the description of the animal taken on the spot was lost, we shall endeavor to make the species known to naturalists, with the aid only of a head and a small portion of the neck of one individual, and a cranium of another, which are now before us.*

Later, skull measurements were reported and compared to the red and gray fox by Say and a name was given to the species:

> *It runs with extraordinary swiftness, so much so, that when at full speed its course has been, by the hunters, compared to the flight of a bird skimming the surface of the earth. We had opportunities of seeing it run with the antelope, and appearances sanctioned the belief, that in fleetness it even exceeded that extraordinary animal, famed for swiftness, and for the singularity of its horns. Like the corsac of Asia it burrows in the earth, in a country totally destitute of trees or bushes, and is not known to dwell in forest districts.*
>
> *If Buffon's figure of the corsac is to be implicitly relied upon, our burrowing fox must be considered as perfectly distinct, and anonymous; we would, therefore, propose for it the name of velox.*

In this discussion of the red, gray, corsac, and swift fox by Say, he referred to them as the genus *Canis* but did note that the swift fox was more similar to the red fox than the gray fox. However, naturalists prior to the acceptance of Darwin's theory of evolution by natural selection grouped species without regard to evolutionary relationships. The naming of species was done primarily to organize the diversity of organisms that was encountered. In fact, Audubon and Bachman (1851) relegated the species *velox*

Table 1. Chronology of the taxonomic history of swift and kit foxes.

Taxonomic Status	Source	Specimens Examined	Significance
Burrowing Fox	Lewis and Clark (Biddle 1814)	2 observations	First mention of swift fox
Canis velox	Say (in James 1823)	2	First official description
Vulpes velox	Audubon and Bachman (1851)		Recognized the genus *Vulpes*
V. macrotis	Merriam (1888)	1	First description of kit fox
V. velox hebes	Merriam (1902)	7	New subspecies of swift fox
V. macrotis neomexicanus	Merriam (1902)	2	New subspecies of kit fox
V. muticus	Merriam (1902)	1	New species of kit fox
V. arsipus	Elliot (1903)	1 (from a series)	New species of kit fox
V. macrotis devius	Nelson and Goldman (1909)	1	New subspecies of kit fox
V. macrotis muticus	Grinnell (1913)		Synonomized *muticus* with *macrotis*
V. macrotis arsipus	Grinnell (1913)		Synonomized *arsipus* with *macrotis*
V. macrotis mutica	Grinnell (1923)		Spelling change
V. macrotis devia	Miller (1924)		Spelling change
V. macrotis neomexicana	Miller (1924)		Spelling change
V. velox	Seton (1929)		Suggested *macrotis* as race of *velox*
V. macrotis arizonensis	Goldman (1931)	6	New subspecies of kit fox
V. macrotis nevadensis	Goldman (1931)	14	New subspecies of kit fox
V. macrotis tenuirostris	Nelson and Goldman (1931)	8	New subspecies of kit fox
California Foxes	Grinnell et al. (1937)	34	Data for 1913 conclusions
V. macrotis zinseri	Benson (1938)	3	New subspecies of kit fox
Subspecies not valid	Benson (1938)		Synonomized *arizonensis* with *arsipus*
NV subspecies not valid	Hall (1946)		Questioned recognition of 2 subspecies
Single species	Hall (1946)		Questioned recognition of 2 species
Range overlap	Packard and Bower (1970)	2	Distribution of two foxes overlap
Recognize two species	Creel and Thornton (1971)	23	Morphology support 2 species
Recognize two species	Rohwer and Kilgore (1973)	182	Intermediate character states, hybrid zone
Two species	Thornton and Creel (1975)	13 (morphology); 2 (serum proteins); 2 (chromosomes)	Morphology is main source for decision
Subspecies not valid	Waithman and Roest (1977)	127	Synonomized *arsipus*, *arizonensis*, *devia*, and *tenuirostris* under *macrotis*; recognized *mutica*
Single species	Hall (1981)		Gene flow (Rohwer and Kilgore 1973)
Subspecies not valid	Stromberg and Boyce (1986)	250	Synonomized hebes under *velox*; recognized two species
Single species Only 2 subspecies	Dragoo et al. (1990)	774 (morphology) 103 (allozymes)	*V. velox velox* and *V. velox macrotis*
Two species, 3 subspecies	Mercure et al. (1993)	256 (mtDNA)	*V. velox velox*, *V. macrotis macrotis*, and *V. macrotis mutica*
Two species Subspecies boundaries incorrect	Maldonado et al. (1997)		other subspecies need revision
Single species	Dragoo et al. (in prep.)	97 (microsatellite DNA)	*V. velox*

(and other foxes) from the genus *Canis* to the genus *Vulpes*, not for biological reasons, but rather because the genus *Canis* contained a large number of species:

> *The characters of this genus* [Vulpes] *differ so slightly from those of the genus* Canis, *that we were induced to pause before removing it from the subgenus in which it had so long remained. As a general rule, we are obliged to admit that a large fox is a wolf, and a small wolf may be termed a fox. So inconveniently large, however, is that list of species in the old genus* Canis, *that it is, we think, advisable to separate into distinct groups, such species as possess any characters different from the true Wolves.*

They also noted that the distribution of the swift fox was on the plains of the Columbia River Valley and the open country of the region and along the prairies of the eastern side of the Rocky Mountains. Their information indicated that *V. velox* was not found in New Mexico, Texas, or California.

In 1888, 65 years after Say's description, C. Hart Merriam described what he called the long-eared fox, *Vulpes macrotis*, based on a single subadult and damaged specimen from southern California, over 1000 miles from the previous type locality of *V. velox* (Merriam 1888). He suggested that the large ears were sufficient to distinguish this fox from other *Vulpes*, stating that "the animal here described differs so notably from its nearest relatives." That statement indicated some forethought about the possibility of an evolutionary relationship among other *Vulpes* species. As Audubon and Bachman (1851) believed that these foxes did not occur in California, Merriam was surprised that a mammal as large as a fox would have been missed in a relatively well-explored area such as southern

California and hypothesized that southern California might be the northernmost limits of a Mexican species. In addition to the larger ears, he described this new fox as being slightly smaller than *V. velox* but the color was almost as pale. By 1888 there were two fox species described based on a total of three incomplete specimens: *Vulpes velox* based on a head and part of a cranium (Say, in James 1823) and *V. macrotis* based on a single subadult (Merriam1888).

C. Hart Merriam had an eye for variation and described new taxa as more foxes were discovered. In 1902 he devoted two pages to describe three additional taxa of foxes. These foxes, in order of page priority, included *V. velox hebes*, *V. macrotis neomexicanus*, and *V. muticus*. By this time more information was known about the distribution of these foxes. Merriam had access to several specimens from the U.S. Biological Survey and indicated that foxes had been collected from Alberta to Colorado and from the deserts of New Mexico west to California, north of the region were the type specimen of *V. macrotis* was collected.

Merriam indicated that the "Canadian kit fox" should be recognized as a subspecies of *V. velox*. The "desert fox" of New Mexico and western Texas, he felt, was a "strongly" marked subspecies of *V. macrotis*, and the San Joaquin Valley fox was a distinct species.

The type specimen for *V. v. hebes* was collected in Calgary, Alberta. Merriam described it as being similar to *velox* but larger and grayer. It had darker patches on the side of the nose. Several of the skull characters were more pronounced. There were seven animals (including the type specimen) used in the description of this taxon.

Merriam's (1902) description of *V. m. neomexicanus* was based on the skull of a male fox collected in the San Andreas Range of New Mexico about 50 miles north of El Paso, Texas. External measurements of an adult female collected in Arizona were included in the description. This fox was similar to *V. macrotis* but was larger and heavier. However, he indicated that this fox was more similar to the new species, which he was about to describe, than it was to *V. macrotis*. He described this fox as a subspecies of *macrotis* rather than a subspecies of the San Joaquin fox because the type locality was geographically closer to the type locality of *V. macrotis*. *V. macrotis neomexicanus* was slightly smaller than the San Joaquin fox. Miller (1924) changed the spelling to *neomexicana*.

The San Joaquin fox, *Vulpes muticus*, was described from one specimen, an adult male (Merriam, 1902). This fox was similar to *macrotis* but much larger with a longer hind foot and tail. The colors of the chin and lip, top of head, and middle of back were more pronounced. The skull was larger, broader, and more massive than that of *macrotis*.

One and a half years after Merriam described three new taxa of foxes, Elliot (1903) described *V. arsipus*, which was found in the Mohave Desert. The type locality was Daggett in San Bernardino County, California, and was between the type locality of *macrotis* and *mutica*, but closer to *macrotis*. *V. arsipus* was similar to *V. macrotis* but paler and smaller and lacked the reddish summer pelage seen in *macrotis* and *hebes*. The person who collected the specimen apparently saw this fox in several localities in the Mojave Desert (Elliot 1903). However, it is not apparent that Elliot examined more than one specimen for this taxon.

During 1905 and 1906, E.W. Nelson and E.A. Goldman worked for the Biological Survey and collected several fox specimens from Baja California. They also described a new fox subspecies, which they called the "peninsula desert fox," *V. macrotis devius* (Nelson and Goldman 1909). The locality for the type specimen of this taxon was the Llano de Yrais, opposite Magdalena Island, Lower Baja California, Mexico. This subspecies was similar to *macrotis* but was a much darker specimen with a shorter pelage and smaller, more slender, tail. The rostrum also was broader and heavier than that of *macrotis*. There is no description of other specimens than the type in their description. Poole and Schantz (1942) and Miller (1924) changed the spelling to "*devia*" to match the gender of *Vulpes*.

E.T. Seton questioned the specific status of kit and swift foxes (Seton 1929). He indicated that the desert fox (*V. macrotis* and associated forms) seemed much like the "Kit-fox" [=swift fox], but that it had larger ears, and that it likely would prove to be a mere race of *velox*. He recognized only two species and a total of seven "races" within these two species. *V. velox* was comprised of *V. v. velox* and *V. v. hebes* and *V. macrotis* had five races: *macrotis*, *neomexicanus*, *mutica*, *arsipus*, and *devia*. Seton noted that:

> *The English zoologist, St. G. Mivart [no citation] considered that all of these were merely geographical races of the Northern Kit-fox* (Vulpes velox). *Its most conspicuous peculiarity is its large ears. It is interesting to note that the similar but larger environment of Africa has produced in the* Fennec *exactly the same thing on a larger scale. A parallel exaggeration of the ears is seen in the desert forms of mule deer, jackrabbit, and rock-mouse.*

On a side note, Mivart (1890) may not have been aware of the description of *Vulpes macrotis*. He mentions on *page vi*, "In the following list we have not given names to forms which we regard as being most probably mere varieties: - [list of 35 species of canids]." That list included *Canis* [=*Vulpes*] *velox* but not *V. macrotis*. However, in his description of the kit fox (*Vulpes velox*) on page 104 he does not list *V. macrotis* (Merriam) as a junior synonym. Additionally, Mivart died in 1900, prior to the description of all of the other subspecies of *V. macrotis*.

E.A. Goldman during the late 1920s and early 1930s spent some time in the collections of the biological survey (Goldman 1931):

> *The accession of specimens in recent years has materially extended the known range of the desert foxes of the*

Vulpes macrotis *group. Forms of this section of the genus occur in suitable areas from the Pacific coast east to the basin of Great Salt Lake, Utah, and the Rio Grande Valley in New Mexico and western Texas, and from the Snake River Valley, Idaho, south to southern Lower California, Sonora, and Chihuahua. Two hither-to unrecognized geographic races are described below.*

The first of these taxa was the "Arizona long-eared desert fox," *V. macrotis arizonensis*, collected near the Mexican boarder in Yuma County, Arizona, which Goldman considered was closely allied to *V. m. arsipus*. *V. m. arizonensis* and *V. m. arsipus* were separated by the Colorado River. He also suggested that *arizonensis* intergraded with *neomexicana* [spelling change] in southeastern Arizona. In his description, he included six specimens (including the type) from Arizona.

The second fox he described in the same paper (Goldman 1931) was the "Nevada long-eared desert fox," *V. m. nevadensis* from Humboldt County, Nevada. This fox also was closely allied to *arsipus*, but was darker and the skull was larger with a more fully inflated braincase. Goldman examined a total of 15 specimens from Idaho (1), Nevada (12), and Utah (2) in this diagnosis. He also noted that this taxon marks the northern limit of the *V. macrotis* group.

In 1931 Goldman published another paper with E.W. Nelson (Nelson and Goldman 1931) describing an additional taxon from Baja California—the "Trinidad Valley desert fox," *V. m. tenuirostris*—found south of the range of *V. m. devia*. This taxon was closely allied to *V. m. arsipus* and *macrotis*, but was a darker fox, presumably as a result of living in a more humid area. This fox was generally larger than the more northern *devia*. Aside from the type locality, *V. m. tenuirostris* was reported to occur in valleys along the Pacific mountain slopes of southern California. There were eight specimens examined from the type locality.

Between 1903 and 1913, three species of fox had been described in southern California: *V. macrotis*, *V. muticus*, and *V. arsipus*. Grinnell (1913) compiled a distribution list of the mammals of California and relegated *arsipus* and *muticus* as subspecies of *V. macrotis*. He provided no justification for the decision. Ten years later, he assembled another list of the mammals of California and changed the spelling of *V. m. muticus* to *V. m. mutica* (Grinnell 1923), presumably to correct agreement with gender. A list of North American recent mammals was compiled by Miller (1924) and he recognized "*mutica*" as a species, but also recognized the spelling change. Grinnell (1933) compiled another list of mammals of California, again listing the San Joaquin Valley kit fox as a subspecies of *V. macrotis* and spelled it as *mutica*. Poole and Schantz (1942) provided a list of type specimens housed at the National Museum. They listed the type specimen for *Vulpes muticus* and cited Grinnell (1933) for the name and spelling to *Vulpes macrotis mutica*.

Grinnell et al. (1937) listed the three fox taxa in California as subspecies of *V. macrotis* in their Fur-bearing Mammals of California. Two of these taxa (*V. m. macrotis* and *V. m. mutica*) were endemic to the state. Grinnell et al. (1937) examined comparative cranial and skin material and for the first time put into text their reasoning for relegating *mutica* and *arsipus* to subspecific status. They demonstrated that the type specimen for *mutica* was a "giant" and that many of the skull measurements in the series they examined overlapped with some of the other taxa (specifically *V. m. macrotis*) of kit foxes. They also showed that by 1930 the range of the San Joaquin kit fox was restricted to the driest plains of the southern and western parts of the valley, and that the type locality was no longer in the range of this subspecies. The second subspecies in California that these authors recognized was *V. m. arsipus*. This fox was the smallest of the three subspecies, but did overlap in cranial measurements with *V. m. macrotis*. It was the overlap in cranial measurements and the slight, but discernable, color variation that led the authors to conclude that there was only one species of kit fox in California. The final subspecies in California they recognized was *V. m. macrotis*, which had the smallest distribution. The last confirmed record of this taxon was from 1903, and by 1930 was designated as extinct.

Seth Benson (1938) was the last to describe a new taxon of kit/swift fox. He had collected several specimens from Sonora and Coahuila in 1937 and tried to identify them based on published records. He described *V. m. zinseri* from three specimens from the type locality in Coahuila but suggested the subspecies may occur in the desert plains of Coahuila and San Luis Potosi as well. Interestingly, Benson (1938) also compared specimens from Sonora to both *V. m. arsipus* and *V. m. arizonensis*. Using the characters that had been published to describe these two taxa, he was unable to distinguish between them. When more specimens were studied he concluded that *arizonensis* should be relegated to synonomy with *arsipus*.

Benson (1938) did not comment on the specific status of *velox* and *macrotis* because he did not have access to specimens of *V. velox*. He maintained that *V. macrotis macrotis* probably was extinct, and concluded that *V. m mutica* was distinct and that *V. m. arsipus* could be distinguished from *V. m. mutica*. He had few specimens of *V. m. nevadensis* but suggested that foxes in southern Nevada were *arsipus* and the few animals from the range of *nevadensis* appeared to be *arsipus* as well. He finally concluded that *V. m. neomexicana* was a valid form.

Benson (1938) concluded that the taxonomy and systematics of kit foxes were in need of revision. He suggested that the foxes were becoming rare and should be preserved as specimens whenever they were killed. He also suggested that the original descriptions were confusing and often the characters used to distinguish taxa were not valid. A thorough study of the variation and geographic distribution was greatly needed.

Hall in 1946 published a book on the mammals of Nevada and commented on the specific or subspecific status of the swift and kit foxes. He did not have materials at hand to address that question, but he had obviously considered the possibility that the two foxes represented the same species. Based on his description of the two subspecies (*V. m. arsipus* and *V. m. nevadensis*) that occurred in Nevada he noted a gradient in size and coloration of foxes from north to south. Northern foxes were larger and darker than southern foxes. He remarked in his description of *V. m. nevadensis* that the difference between the two "races" in Nevada was questionable and that only one subspecies should be recognized. According to Hall (1946), many of the characteristics associated with *nevadensis* that made the taxon different than *arsipus* was a result of the chemicals used to tan skins.

Modern Systematic Studies of Swift and Kit Foxes

The Rocky Mountains are a significant barrier to dispersal for swift and kit foxes. Currently, 12 states, Mexico, and Canada have populations of at least one taxon of kit or swift fox (Fig. 2). Three states—Colorado, New Mexico, and Texas—have populations of both. In Colorado, the southern Rocky Mountains separate the two forms. So, regardless of taxonomic status, swift and kit foxes are reproductively isolated by a physical barrier and have different ecological requirements. Therefore, they require different management practices (Mercure et al. 1993, Fitzgerald et al. 1994, Wayne 1996). However, in New Mexico and Texas the two foxes have the potential to be in contact and overlap (Fig. 3). Packard and Bowers (1970) reported two records of kit or swift foxes that extended the range of these taxa and recognized a zone of overlap in New Mexico. The first record extended the known range of the kit fox approximately 80 miles, suggesting that swift and kit fox populations overlapped in New Mexico. The second record was of the swift fox and extended the known range northwestward approximately 125 miles. Because these two fox taxa now are known to overlap, the taxonomy issue becomes more important, especially in New Mexico and Texas.

Creel and Thornton (1971) examined a handful of cranial material from the kit and swift foxes from New Mexico and Texas. Though the material they had did not allow them to address the sympatry of two taxa, they were able to comment on their taxonomic status. They made 10 cranial measurements and determined that the populations (16 kit and 7 swift fox skulls) were significantly distinct. They used an arbitrary cutoff ratio to distinguish between subspecies. Only three of the ten measurements were not significantly different between subspecies and were in a ratio greater than this cutoff value.

A phenetic species definition that assigns taxonomic rank based on gaps ("no matter how small") between clusters was used by Rohwer and Kilgore (1973). They were the first to analyze kit and swift foxes in the context of a species definition. The two "exceedingly similar" foxes had ranges that abutted and overlapped in New Mexico and possibly Texas (see Fig. 3). They also recognized that the ecology and life histories of these foxes were extremely similar and set out to examine the "phenetic status" of specimens collected at the contact zone in New Mexico and Texas. Most of the contact zone specimens clustered with either the *macrotis* population or the *velox* population. However, they found intermediate specimens at a high frequency and concluded that:

> *In this particular area, then, the picture seems to be one of sporadic interbreeding that probably includes some backcrossing, but the presence of specimens from this area, closely resembling the reference material, suggests that selection generally opposes hybrids and favors the maintenance of separate adaptive modes. There is no evidence that genes causing intermediate skull morphology are flowing more than 25-30 miles to either side of the presumed line of contact.*

They concluded that the phenetic difference between these two forms was being maintained and suggested that they be distinguished as separate species. However, the results of Rohwer and Kilgore's (1973) study led Hall (1981) to conclude that gene flow was occurring between the two taxa and therefore should be recognized as a single species.

Thornton and Creel (1975) desired to "present evidence in support of the premise that, while kit fox taxa are closely related, they actually represent two distinct species populations with no indication of gene exchange." They examined a grand total of 13 specimens for external characteristics such as pelage coloration, ear size and position. They also examined "electrophoregrams" of serum and hemoglobin from only 2 animals, one of each taxon. They tentatively concluded that the slight variation observed was a result of speciation rather than individual variation. Further, chromosomal analyses detected no difference in diploid number leading them to concluded that the foxes could be distinguished only on ear size and placement, head shape, and relative length of tail.

In addition to whether there was one or two species, the validity of subspecific status of several taxa was brought into question. *V. m. mutica* was listed as threatened in the early 1970s and was economically important because poisoned grain used to control ground squirrel populations could not be used in the range of this subspecies. In addition, federal and state pest control agents were unable to distinguish between *V. m. mutica* and *V. m. arsipus* and suggested that *mutica* should not be considered threatened. Waithman and Roest (1977) were interested in ascertaining the taxonomic status of the San Joaquin kit fox (*V. m. mutica*). They examined not only *mutica* and *arsipus*, but the other surrounding taxa (*macrotis*, *tenuirostris*, and *devia*) as well. They used discriminant function analyses to classify individuals into the various taxa. None of the *mutica* specimens were misclassified and none of the other

Figure 2. Representation of the geographic range (Hall 1981, Ginsberg and Macdonald 1990, Dragoo et al. 1990) of the 10 nominal subspecies of swift and kit foxes. 1 - arsipus, *2* - devia, *3* - hebes, *4* - macrotis, *5* - mutica, *6* - neomexicana, *7* - nevadensis, *8* - tenuirostris, *9* - velox, *10* - zinseri. *States and provinces with fox populations are labeled on map.*

taxa ever classified as *mutica*. This was primarily because of the larger size of the San Joaquin kit fox. However, the other four taxa misclassified on occasion and they suggested *arsipus* (including *arizonensis*), *devia*, and *tenuirostris* be synonomized with *macrotis* but that *mutica* remain as a distinct subspecies.

Stromberg and Boyce (1986) were concerned with the conservation of swift foxes in the northern part of their range. They examined the taxonomic status of the northern swift fox *V. velox hebes* and the more southern swift fox *V. v. velox*. They included an analysis of kit foxes as well and demonstrated morphological discrimination between the two forms. They also demonstrated that the swift foxes represented a gradation of morphological traits from north to south and concluded that the northern subspecies (*hebes*) was not valid. However, they did suggest that

Figure 3. Hybrid zone between kit and swift foxes. The samples were part of a microsatellite DNA study (Dragoo et al. in prep.). Triangles represent kit fox mtDNA haplotype and squares represent swift fox haplotypes. Haplotypes were determined based on cytochrome b *sequence (Dragoo et al. in prep.). Chaves County, New Mexico, represents the same hybridization zone analyzed in Mercure et al. (1993). All foxes were typed with 12 microsatellite loci. Foxes in southeastern New Mexico, regardless of mtDNA haplotype, represent a single population based on microsatellite data.*

foxes could be locally adapted genetically to their immediate environment and that moving foxes from a southern locality to a more northern one could cause a disruption of adaptive gene complexes.

In an effort to clarify the taxonomic status, both subspecific and specific, of these foxes, Dragoo et al. (1990) examined cranial measurements of 844 specimens from the ten nominal taxa of swift and kit foxes recognized at that time (Fig. 2). Their results supported the following conclusions of previous studies: 1) *V. v. hebes* is not a valid taxon (Stromberg and Boyce 1986); 2) *nevadensis* and *arsipus* represent the same taxon (Hall 1946); and 3) *macrotis*, *tenuirostis*, *arsipus*, and *devia* represent a single taxon (Waithman and Roest 1977). However, they disagreed with the latter study by suggesting that *mutica* also was a part of that same taxon (Fig. 4). Finally, the morphologic analysis supported previous studies that found morphological differences between the swift and kit foxes. Dragoo et al. (1990) also examined allozyme data from five of the nominal taxa and found that the swift and kit foxes from Kansas to Arizona were genetically identical. The most divergent populations were from the San Joaquin Valley and Nevada although these were 98.9% similar to the foxes in Kansas and Arizona. Consequently, Dragoo et al. (1990) concluded that there was only one species of arid-land fox, *Vulpes velox*, with two subspecies, *V. velox velox* and *V. velox macrotis*.

Phylogeography

In the early 1990s, the advent of the polymerase chain reaction (PCR) and methods of large-scale DNA sequencing (Sambrook et al. 1989) finally allowed populations (rather than just a few individuals from a species) to be characterized for variation in specific DNA sequences (Avise 1994). Sequencing studies, combined with earlier information on sequence variation using indirect methods such as restriction fragment length polymorphism (RFLP), led to the conclusion that species commonly are subdivided into discrete genetic units. The relationships between population genetic units often would correlate with geographic distance or the presence of topographic barriers. Moreover, DNA sequence trees commonly showed a correspondence between phylogenetic and geographic relationships among populations (e.g. Fig. 5). A new field

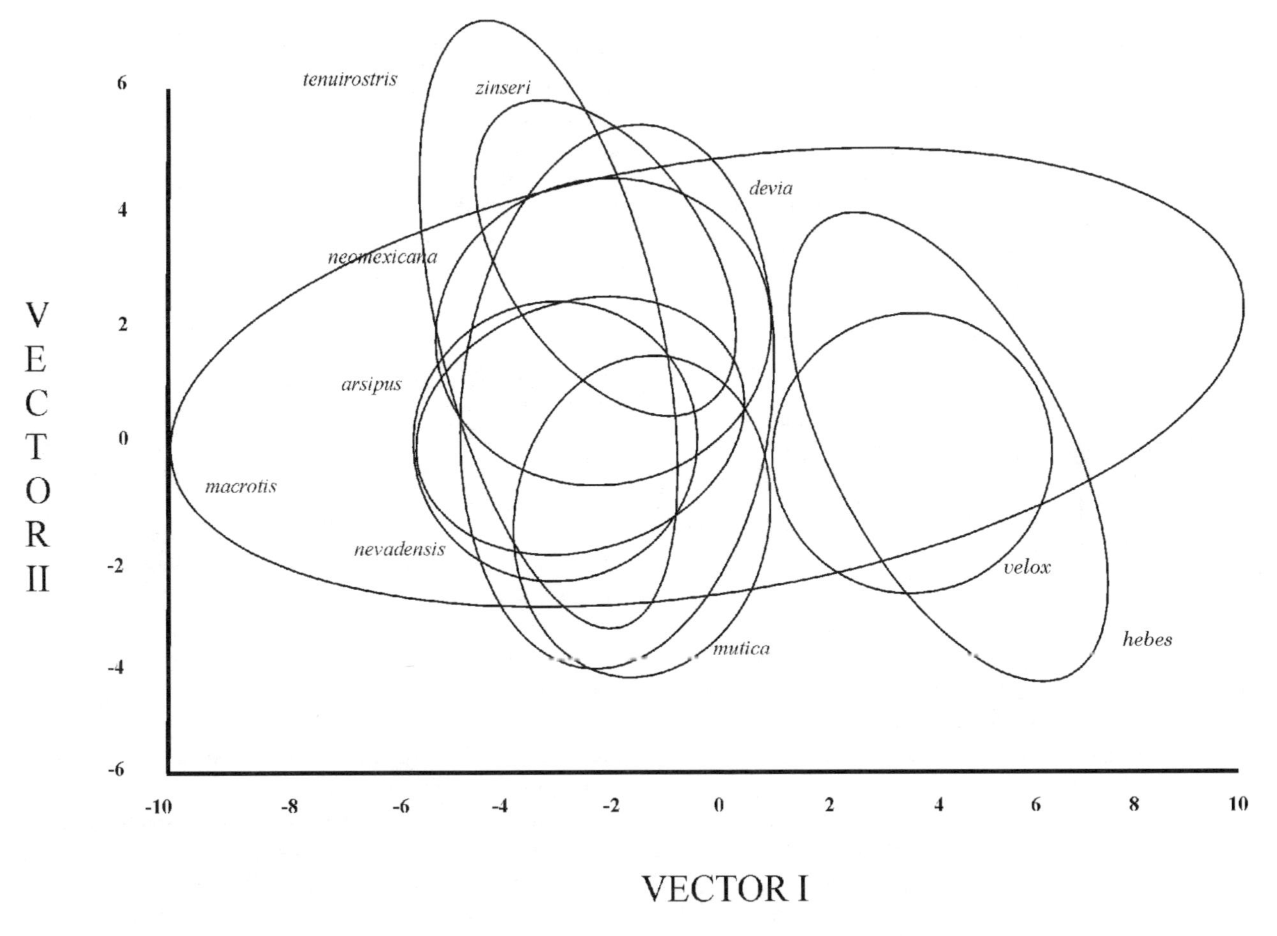

*Figure 4. Plot of the first two canonical vectors for the 10 nominal subspecies of arid-land foxes. Canonical vectors are derived from morphometric analyses (*see *Dragoo et al. 1990). The ellipses enclose 95% confidence limits around the centroid means. The large ellipse around the taxon* macrotis *is a result of low sample size (n=4). The centroid mean is within the range of the other kit fox taxa.*

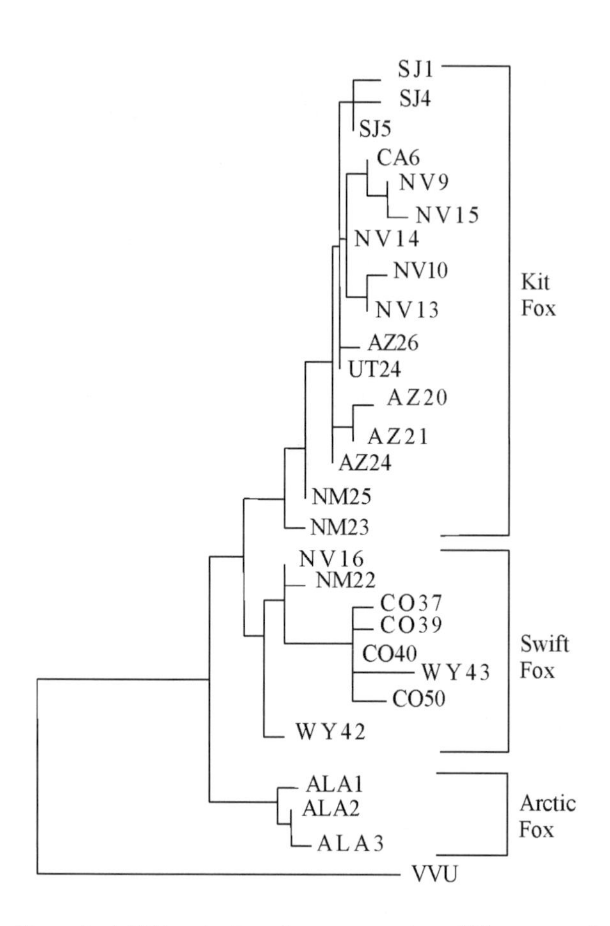

Figure 5. A 50% majority-rule consensus tree of 30 most parsimonious trees of kit and swift fox mtDNA haplotypes based on analysis of restriction site data (Mercure et al. 1993). Haplotypes were defined based on geographic locality from where they first were observed. SJ represents the San Joaquin haplotypes; AZ, CO, CA, NM, NV, UT, and WY represent the specific states. Generally, the haplotypes were found in only one locality. However, UT24 was found first in Utah and latter in Nevada. AZ21 was found in central and southeastern Arizona as well as Nevada and Utah. AZ24 was found in both southeastern Arizona and Nevada. NV16 was found in Nevada originally, but turned out to be a common haplotype throughout New Mexico, Colorado, Kansas, and Wyoming. The tree was generated using the branch-and-bound option of the PAUP program (Swofford 1989) and rooted by the red fox (VVU) restriction site data.

emerged, termed phylogeography, which studied the correspondence between geography and phylogeny within species (Avise et al. 1987). An early conclusion of phylogeographic research on carnivores was that high dispersal rates and dispersal distances in some species suppressed genetic differentiation and resulted in a lack of phylogeographic structure (*see* Wayne and Koepfli 1996 for review). Large canids, such as coyotes and gray wolves, showed weak or non-significant levels of genetic subdivision (Lehman and Wayne 1991, Wayne et al. 1992, Roy et al. 1994, Vila et al. 1999). However, because body size correlates with dispersal ability (Gittleman 1985) and kit and swift foxes are the smallest canids, they have dispersal distances nearly an order of magnitude less than large wolf-sized canids (Lehman et al. 1991, Vila et al. 1999). Consequently, genetic divisions among kit and swift populations are predicted to be more profound than in large canids (Mercure et al., 1993).

Mercure et al. (1993) studied population differentiation across much of the geographic range of kit-swift foxes including 8 of 10 subspecies. They sampled populations separated by discrete barriers such as the Rocky Mountains (kit vs. swift foxes), and the Colorado River (central and southeast Arizona vs. Nevada and California; Fig. 2). In California, Mercure et al. (1993) sampled the kit fox of the San Joaquin Valley (*V. macrotis mutica*), whose range is circumscribed by the coastal mountain range to the west and the Sierra-Nevada mountain range to the east. This subspecies is considered distinct (see above) and is protected by the U.S. Endangered Species Act (Hall 1981, O'Farrell 1987). Additionally, localities from the kit and swift fox hybrid zone in New Mexico were sampled (see Fig. 3).

Results of mtDNA analyses, including RFLP and cytochrome *b* sequencing (Mercure et al. 1993), showed that foxes west and east (i.e. kit and swift foxes, respectively) of the Rocky Mountains form distinct monophyletic groups (Fig. 5). These were reciprocally monophyletic groups differing by about 1% sequence divergence or approximately 500,000 years (Mercure et al. 1993). Significantly, the magnitude of the genetic divergence between the east (swift fox) and west (kit fox) clades is nearly as great as that between these clades and the arctic fox (*Alopex* [=*Vulpes*] *lagopus*). Within each of the kit and swift fox clades, few phylogenetic groupings (i.e. the various recognized subspecies) were apparent. The exception is the sample of 75 kit foxes from the north and south San Joaquin Valley. In this population, three genotypes were found that defined a significant monophyletic group (SJ1, SJ4, SJ5, Fig. 5). Additionally, Arizona populations east of the Colorado River appeared genetically similar to each other but distinct from populations in Nevada (Fig. 6). Moreover, most populations had unique DNA sequences, suggesting restricted gene flow in the range of 0.5 female migrants per generation. A value of less than one migrant per generation indicates that genetic drift will be a significance force in structuring genetic variation among populations (Slatkin, 1987). Genetic subdivision was more pronounced in the kit than the swift fox clade due likely to more barriers to dispersal in the western states than in the plains. However, within each of the two major kit-swift fox mtDNA clades, genetic distances among populations tended to be correlated with geographic distance (Mercure et al. 1993). Additionally, a later study of hypervariable mitochondrial control region sequences confirmed the genetic divergence between the kit and swift foxes above and suggested the Mexican kit fox (*V. macrotis zinseri*) is

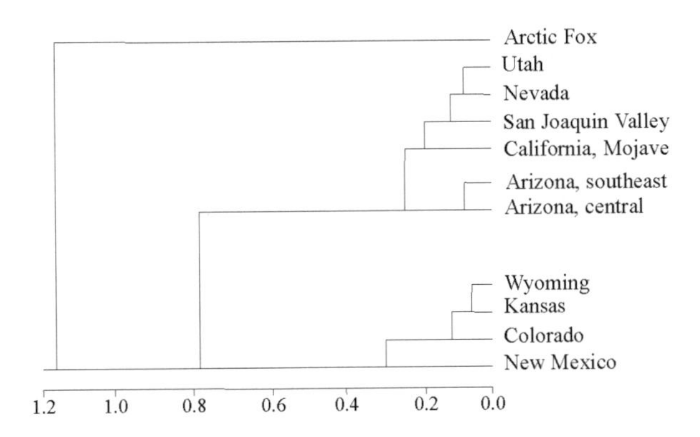

Figure 6. Genetic distance tree based on the average between-locality cytochrome b *similarity values for 10 kit-swift fox localities (*see *Mercure et al. 1993). Scale is least squares fitted sequence divergence values (Lynch and Crease 1990).*

closely related to kit foxes from Arizona and Nevada (Maldonado et al. 1997).

Finally, Mercure et al. (1993) confirmed the existence of a hybrid zone suggested by morphologic studies (Packard and Bowers 1970, Rohwer and Kilgore 1973, Thornton and Creel 1975). At the locality in southeastern New Mexico where the ranges of the two forms meet (Figs. 2 & 3), 41% (n=22) of individuals had sequences from the swift fox clades whereas the remainder had sequences from the kit fox clade. Populations from southeastern Colorado and western Kansas, approximately 400 km to the north, contained only swift fox mtDNA sequences (n=78). In contrast, the southeastern Arizona sample, approximately 440 km to the west, contained 92% kit fox sequences (n=13). These observations suppported recognition of a hybrid zone that is a few hundred kilometres in width and are consistent with descriptions by Packard and Bowers (1970), Rohwer and Kilgore (1973) and Thornton and Creel (1975) of a contact zone between kit and swift foxes in west Texas and eastern New Mexico.

Recent analysis of microsatellite loci suggests that gene flow occurs across the contact zone defined by mitochondrial DNA polymorphisms in New Mexico and that breeding is random with regard to mtDNA haplotype (Dragoo et al. in prep.). Fox samples from southeastern New Mexico in the contact zone were typed as having kit or swift fox mtDNA haplotypes based on cytochrome *b* sequence variation and were also typed for variation at 12 microsatellite loci. No difference in microsatellite allele frequencies were found between foxes grouped according to mtDNA haplotype, suggesting random breeding within the hybrid zone. Dragoo et al. (in prep.) suggested that the foxes in New Mexico represented a single species, *V. velox*. They also suggested that major barriers to dispersal, such as rivers in New Mexico, to dispersal might limit the diffusion of the mtDNA haplotypes from one form to the other. According to Ralls et al. (2001), female foxes on adjacent home ranges tend to be closely related which suggests that dispersing foxes tend to settle in a home range adjacent to their natal range when possible. This bias would restrict the movement of mtDNA haplotypes over great distances. However, the potential for female foxes to disperse over distance is evident in the finding of swift fox mtDNA haplotypes in kit fox populations well beyond the Pecos River drainage in New Mexico (Mercure et al. 1993, Rodrick 1999).

The smaller body size correlated with lesser dispersal ability (Gittleman 1985) of swift and kit foxes may explain the considerable genetic divisions in mtDNA (Mercure et al. 1993) and microsatellite loci among kit and swift populations (Dragoo et al. submitted). The phylogeographic pattern in the kit-swift fox complex contrasts with that observed in coyotes and gray wolves, where conspecific populations on opposite sides of the Sierra Nevada and Rocky Mountain ranges showed little or no detectable mtDNA differentiation (Lehman and Wayne 1991), suggesting that genetic divergence does not increase appreciably with distance (Lehman et al. 1991, Wayne et al. 1992, Vila et al. 1999, but see Forbes and Boyd 1997).

Taxonomic Considerations

Recent reports conflict concerning species status of swift and kit foxes and designate them as separate or single species (e.g. Mercure et al. 1993, separate species; Dragoo et al. in prep., single species). Regardless of the number of species recognized, many authorities recognize the 10 subspecies (see Fig. 2) presented in Miller and Kellogg (1955), Hall and Kelson (1959), Egoscue (1979), and McGrew (1979). Hall and Kelson (1959) suggested however, that the two foxes might be distinct only at the subspecific level. Hall (1981) recognized the same 10 subspecies, but considered them all to be *V. velox*. Wozencraft (1993) recognized only one species and cited Dragoo et al. (1990). The United States Fish and Wildlife Service generally follows the taxonomy consistent with Miller and Kellogg (1955), recognizing both *V. macrotis mutica* and *V. velox hebes* as endangered taxa. Several of the state wildlife agencies appear to follow Miller and Kellogg's (1955) taxonomy as well, including the state of New Mexico, which recognizes *V. velox*, *V. macrotis neomexicana*, and *V. m. arsipus*. The Texas Parks and Wildlife Department however, recognizes the taxonomy presented by Dragoo et al. (1990). Colorado recognizes two species, which are separated by the Rocky Mountains (Fitzgerald et al. 1994). Much of the published literature on kit and swift foxes since 1990 concerns the San Joaquin kit fox, referred to as *V. macrotis mutica*, which is consistent with Mercure et al. (1993) as well as Miller and Kellogg (1955). However, other taxa consistent with Miller and Kellogg (1955) have been discussed and include *V. m.*

neomexicana (Rodrick and Mathews 1999) and *V. m. zenseri* (Maldonado et al. 1997).

The taxonomic designations assigned to the genetic subdivisions observed in these foxes would differ according to which modern species/subspecies definitions are used (*see discussion in* Mercure et al. 1993). A definition that borrows from both the biological (reproduction isolation) and phylogenetic (diagnosable units) species concepts defines species as monophyletic clades that do not interbreed, or have limited interbreeding, when barriers to dispersal are removed. In contrast, subspecies will interbreed freely when barriers are removed, although they are otherwise phylogenetically distinct (Avise and Ball 1990). Mercure et al. (1993) suggested kit and swift foxes should be considered as distinct species given that they define distinct monophyletic groups and that hybridization was geographically limited, suggesting some measure of reproductive isolation. However, recent microsatellite and mtDNA analysis of the contact zone between the two forms suggested unconstrained breeding and led Dragoo et al. (in prep.) to suggest that the two foxes should be considered the same species.

In kit and swift foxes, the few apparent genetic subdivisions that are supported in mtDNA analysis can be arranged in a hierarchy reflecting sequence divergence and the strength of phylogenetic groups. The most significant division occurs between populations of swift foxes from east and populations of kit foxes from west of the Rocky Mountains. Following this basic division is the one between San Joaquin kit foxes and other kit fox populations. The former is a monophyletic grouping and could be regarded as a subspecies (Avise and Ball 1990), but here divergence levels between the San Joaquin kit fox and other kit fox populations are as little as one-tenth of one percent. Lastly, populations that are genetically divergent and have mtDNA sequence unique to them, or not widely shared, are tentative genetic management units (*sensu* Mortiz 1994). These include populations in Arizona that are separated by the Colorado River from those populations farther west which are isolated by habitat boundaries in Mexico from populations in the US.

Few of the recognized subspecific taxa of kit fox are supported by morphologic or genetic data (Dragoo et al. 1990, Mercure et al. 1993). The San Joaquin kit fox appears as the most distinctive population unit in past studies, defining a monophyletic group in mtDNA sequence trees and possessing unique morphologic characteristics. Other minimally distinct taxa include the Mexican kit fox, defined both on morphologic traits and primarily on mtDNA sequence data. However, Maldonado et al. (1997) caution that a more extensive study of intervening populations might reveal a pattern of differentiation with distance, rather than a geographically coherent subspecies, such as is observed in kit foxes of the U.S. Other conceivable taxonomic units include the southeast and central Arizona populations of kit foxes that are isolated from western populations by the Colorado River (Figs. 2 and 6).

However, the primary pattern that dominates the genetics of kit and swift fox populations is one of differentiation by distance (Mercure et al. 1993). Consequently, typological subspecies concepts may not be appropriate because with the exception of a few populations isolated by discrete and substantial geographic barriers, genetic variation among populations is continuously distributed. This pattern is reinforced by quantitative study of morphologic variation, which shows no discernable discontinuities among populations (Dragoo et al. 1990). In conclusion, the genetic data suggest many fewer subspecies of kit and swift foxes should be recognized. The most recent analysis of interbreeding in the hybrid zone formed by the taxa suggests that they should be combined into a single species contrary to previous reports.

Conservation Implications

A hierarchy has been suggested for designating genetic units for conservation (Moritz 1994, 1999). Populations that are defined by reciprocally monophyletic groups have been isolated for a long time period (about 4 times the effective female population size, Avise et al. 1987) and are considered to have evolutionary potential as new species. Such populations are designated evolutionary significant units (ESU; Moritz 1994) and the potentially unique characteristics of an ESU should be preserved at high priority. Populations of kit and swift foxes separated by geographic barriers, such as the Rocky Mountains, may represent separate ESUs. An intermediate level of distinction might be applied to the San Joaquin Valley fox because they represent a monophyletic group that has not been isolated for a sufficient time period such that they are reciprocally monophyletic with respect to other foxes (see Fig. 5).

The populations that differ significantly in mtDNA and/or nuclear gene frequencies but are not reciprocally monophyletic groups are considered genetic management units and should be managed independently if possible. The population (composed of swift and kit fox mtDNA haplotypes) of foxes in southeastern New Mexico should be considered a single management unit distinct from populations either in the northeastern or southwestern parts of New Mexico. The populations in Mexico and in Arizona east of the Colorado River also may represent distinct management units. However, more extensive sampling of populations and analysis of nuclear gene information are needed to confirm these designations.

Recently, this concept of ESU has been challenged by the suggestion that rather than evolutionary potential and isolation, adaptive divergence should be emphasized and that adaptive divergence might occur despite gene flow (Crandall et al. 2000). Therefore, a history of isolation is less important than evidence of unique adaptations. The focus of conservation planning should be on the preservation of adaptations unique to individual populations, if they are known, and on the network of genetic migration that connects them.

Swift and kit foxes have a long history of geographic isolation such that they define reciprocally monophyletic groups; however, the few characteristics in which they differ do not suggest substantial ecological differentiation. Rather, a pattern of genetic differentiation by distance dominates, which suggests the action of genetic drift and limited dispersal (Slatkin 1993). This implies that in large part, captive breeding plans may mix populations so long as individuals are obtained from closely related populations in similar ecological situations (Crandall et al. 2000). Therefore, the recent use of foxes from Colorado and South Dakota (rather than New Mexico or Texas) as a source for a reintroduction into the Canadian province of Saskatchewan appears to have been appropriate (O'Farrell 1987).

Conclusions and Future Research

Taxonomy provides a formal representation of how variation is partitioned and it has influenced conservation biology by defining potentially important units of conservation. In addition, taxonomic designations can influence management policy with respect to providing formal protection (O'Brien and Mayr 1991). Kit and swift foxes are discrete genetic entities that are considered distinct taxa at either the subspecific or specific level. However, recent morphologic and genetic studies of kit and swift foxes suggest that early typological designations of subspecies do not adequately summarize the population subdivisions existing within either taxon. Only a few genetic population units are apparent, the most distinct being the San Joaquin kit fox. The dominant pattern within kit and swift foxes is one of genetic differentiation with distance that reflects limited gene flow between populations. Consequently, conservation efforts should be aimed at maintaining historic levels of gene flow between populations unless evidence exists for long-term isolation. Future research should be aimed at characterizing variation in highly variable biparentally nuclear DNA markers, such as microsatellite loci, to contrast previous studies on maternally inherited mitochondrial DNA. Also, as discussed in Van Valkenburgh (1989) and Van Valkenburgh and Wayne (1994), functional traits, such as body size, tooth and body dimensions, and pelage, should be characterized throughout the geographic range of the kit fox to identify potentially adaptive differences among populations.

Literature Cited

Audubon, J.J. and J. Bachman. 1851. The viviparous quadrupeds of North America. Volume 2. V.G. Audubon, New York, New York.

Avise, J.C. 1994. Molecular markers, natural history and evolution. Chapman and Hall, New York, New York.

Avise, J.C., J. Arnold, R.M. Ball, E. Bermingham, T. Lamb, J.E. Neigel, C.A. Reeb, and N.C. Saunders. 1987. Intraspecific phylogeography: the mitochondrial DNA bridge between population genetics and systematics. Annual Review of Ecology and Systematics 18:489–522.

Avise, J.C., and R.M. Ball. 1990. Principles of genealogical concordance in species concepts and biological taxonomy. Pp. 45–67 *in* D. Futuyma and J. Antonovics, editors. Oxford Surveys in Evolutionary Biology. Volume 7. Oxford Press, Oxford, UK.

Benson, S.B. 1938. Notes on kit foxes (*Vulpes macrotis*) from Mexico. Proceedings of the Biological Society of Washington 51:17–24.

Biddle, N. 1814. History of the expedition under the command of Captains Lewis and Clark, to the sources of the Missouri, thence across the Rocky Mountains and down the river Columbia to the Pacific Ocean: performed during the years 1804–5–6 by order of the government of the United States/prepared for the press by Paul Allen; in two volumes. Philadelphia: Bradford and Inskeep; New York: A.H. Inskeep, Vol. 1: xxviii, 470 p.; v. 2: ix, 522 p., [6] leaves of plates, 1 folded.

Clutton-Brock, J., G.B. Corbett, and M. Hills. 1976. A review of the family Canidae with a classification by numerical methods. Bulletin of the British Museum (Natural History) (Zoology) 9:119–199.

Crandall, K.A., O.R.P. Bininda-Emonds, G. Mace, and R.K. Wayne. 2000. Considering evolutionary processes in conservation biology. Trends in Ecology and Evolution 15:290–295.

Creel, G.C., and W.A. Thornton. 1971. A note on the distribution and specific status of the fox genus *Vulpes* in west Texas. Southwestern Naturalist 15:402–404.

Dalquest, W.W. 1978. Early Blancan mammals of the Beck Ranch local fauna of Texas. Journal of Mammalogy 59:269–298.

Dragoo, J.W., J.R. Choate, T.L. Yates, and T.P. O'Farrell. 1990. Evolutionary and taxonomic relationships among North American arid-land foxes. Journal of Mammalogy 71:318–332.

Dragoo, J.W., K.D. Willoughby, K.A. Moore, C.G. Schmitt, and T.L. Yates. (In prep.) Microsatellite DNA Variation and Taxonomic Implications of Swift and Kit Foxes from New Mexico. Molecular Ecology.

Egoscue, H.J. 1979. *Vulpes velox*. Mammalian Species 122:1–5.

Elliot, D.G. 1903. Descriptions of twenty-seven apparently new species and subspecies of mammals. Field Columbian Museum, Publication 87, Zoology Series 3:239–261.

Fitzgerald, J.P., C.A. Meaney, and D.M. Armstrong. 1994. Mammals of Colorado. University Press of Colorado, Denver, xiii+467 pp.

Forbes, S.H. and D.K. Boyd. 1997. Genetic structure and migration in native and reintroduced Rocky Mountain wolf populations. Conserv. Biol. 11:1226–1234.

Geffen, E., A. Mercure, D.J. Girman, D. Macdonald, and R.K. Wayne. 1992. Phylogenetic analysis of the fox-like canids: mitochondrial DNA restriction fragment, site and cytochrome b sequence analysis. Journal of Zoology 228:27–39.

Ginsberg, J.R., and D.W. Macdonald. 1990. Foxes, wolves, jackals and dogs. An Action Plan for the Conservation of Canids. International Union for Conservation of Nature and Natural Resources, Gland, Switzerland.

Gittleman, J.L. 1985. Carnivore body size: ecological and taxonomic correlates. Oecologia 67:540–554.

Goldman, E.A. 1931. Two new desert foxes. Journal of the Washington Academy of Sciences 21:249–251.

Grinnell, J. 1913. A distributional list of the mammals of California. Proceedings of the California Academy of Sciences 3:265–390.

——. 1923. A systematic list of the mammals of California. University of California Publications in Zoology, 21:313–324.

——. 1933. Review of the recent mammal fauna of California. University of California Publications in Zoology 40:71–234.

Grinnell, J., J.S. Dixon, and J.M. Linsdale. 1937. Fur-bearing mammals of California: Their natural history, systematic status, and relations to man. 2 volumes. University of California Press, Berkeley, 777 pp.

Hall, E.R. 1946. The mammals of Nevada. University of California Press, Berkeley, xi+710 pp.

——. 1981. The Mammals of North America. Volume 2:601–1181 +90. John Wiley and Sons, New York, New York.

Hall, E.R., S. Anderson, J.K. Jones, Jr., and R.L. Packard. 1957. Vernacular names for North American mammals north of Mexico. Miscellaneous Publications, Museum of Natural History, University of Kansas 14:1–16.

Hall, E.R., and K.R. Kelson. 1959. The mammals of North America. The Ronald Press Company, New York, New York.

James, E. 1823. Account of an expedition from Pittsburgh to the Rocky Mountains: performed in the years 1819 and '20, by order of Honorable J.C. Calhoun, Secretary of War; under the command of Major Stephen H. Long. Volume I, pp. 486–487. H.C. Carey and I. Lea, Philadelphia, Pennsylvania.

Kurten, B., and E. Anderson. 1980. Pleistocene Mammals of North America. Columbia University Press, New York, New York.

Lehman, N., A. Eisenhawer, K. Hansen, L.D. Mech, R.O. Peterson, P.J.P. Gogan, and R.K. Wayne. 1991. Introgression of coyote mitochondrial DNA into sympatric North American gray wolf populations. Evolution 45:104–119.

Lehman, N., and R.K. Wayne. 1991. Analysis of coyote mitochondrial DNA genotype frequencies: Estimation of the effective number of alleles. Genetics 128:405–416.

Lynch, M., and T.J. Crease. 1990. The analysis of population survey data based on DNA sequence variation. Molecular Biology and Evolution 7:377–394.

Maldonado, J.E., M. Cotera, E. Geffen, and R.K. Wayne. 1997. Relationships of the endangered Mexican kit fox (*Vulpes macrotis zinseri*) to North American arid-land foxes based on mitochondrial DNA sequence data. Southwestern Naturalist 42:460–470.

McGrew, J.C. 1979. *Vulpes macrotis*. Mammalian Species 123:1–6.

Mercure, A., K. Ralls, K.P. Koepfli, and R.K. Wayne. 1993. Genetic subdivisions among small canids: mitochondrial DNA differentiation of swift, kit, and arctic foxes. Evolution 47:1313–1328.

Merriam, C.H. 1888. Description of a new fox from southern California. Proceedings of the Biological Society of Washington 4:5–8.

——. 1902. Three new foxes of the kit and desert fox groups. Proceedings of the Biological Society of Washington 15:73–74.

Miller, G.S., Jr. 1924. List of North American Recent Mammals 1923. Smithsonian Institution United States National Museum Bulletin 128:1–674.

Miller, G.S., Jr., and R. Kellogg. 1955. List of North American recent mammals. United States National Museum Bulletin 205:1–954.

Mivart, G.J. 1890. Dogs, Jackals, Wolves, and Foxes: a Monograph of the Canidae. R.H. Porter, London, UK.

Moritz, C. 1994. Defining evolutionarily-significant-units for conservation. Trends in Ecology and Evolution 9:373–375.

——. 1999. Conservation units and translocations: strategies for conserving evolutionary processes. Hereditas 130: 217–228.

Nelson, E.W., and E.A. Goldman. 1909. Eleven new mammals from lower California. Proceedings of the Biological Society of Washington 22:23–28.

——. 1931. New carnivores and rodents from Mexico. Journal of Mammalogy 12:302–306.

O'Brien, S.J., and E. Mayr. 1991. Bureaucratic mischief: recognizing endangered species and subspecies. Science 251:1187–1188.

O'Farrell, T.P. 1987. Kit Fox. Pp. 423–431 *in* M. Novak, J.A. Baker, M.E. Obbard, and B. Malloch (editors). Wild Furbearer Management and Conservation in North America. Ontario Ministry of Natural Resources, Toronto, Ontario.

Packard, R.L., and J.H. Bowers. 1970. Distributional notes on some foxes from western Texas and eastern New Mexico. Southwestern Naturalist 14:450–451.

Paterson, H.E.H. 1985. The recognition concept of species. Pp. 21–29 *in* E.S. Vrba (ed.), Species and Speciation. Transvaal Museum Monograph No. 4, Pretoria, Republic of South Africa.

Poole, A.J., and V.S. Schantz. 1942. Catalog of the type speciemens of mammals in the United States National Museum, including the Biological Surveys Collection. Smithsonian Institution United States National Museum, 178:xiii+1–705.

Ralls, K., K.L. Pilgrim, P.J. White, E.E. Paxinos, M.K. Schwartz, and R.C. Fleischer. 2001. Kinship, social relationships, and den sharing in kit foxes. Journal of Mammalogy 82:858–866.

Rodrick, P.J. 1999. Morphological, genetic and den site characteristics of the kit fox (*Vulpes macrotis*) on McGregor Range in south central New Mexico. Thesis. University of Wisconsin, Madison, Wisconsin.

Rodrick, P.J., and N.E. Mathews. 1999. Characteristics of natal and non-natal kit fox dens in the northern Chihuahuan Desert. Great Basin Naturalist 59:253–258.

Rohwer, S.A., and D.L. Kilgore, Jr. 1973. Interbreeding in the aridland foxes, *Vulpes velox* and *V. macrotis*. Systematic Zoology 22:157–165.

Roy, M.S., E. Geffen, D. Smith, E. Ostrander, and R.K. Wayne. 1994. Patterns of differentiation and hybridization in North American wolf-like canids revealed by analysis of microsatellite loci. Molecular Biology and Evolution 11:553–570.

Sambrook, J., E.F. Fritsch, and T. Maniatis. 1989. Molecular cloning: A laboratory manual, 2nd edition. Cold Spring Harbor Laboratory Press, Cold Spring Harbor, New York.

Seton, E.T. 1929. Lives of game animals. Volume 1 (part 2):339–640. Doubleday, Doran, & Company, Inc., Garden City, New York

Slatkin, M. 1987. Gene flow and the geographic structure of natural populations. Science 236:787–792.

——. 1993. Isolation by distance in equilibrium and non-equilibrium populations. Evolution 47:264–279.

Stromberg, M.R., and M.S. Boyce. 1986. Systematics and conservation of the swift fox, *Vulpes velox*, in North America. Biological Conservation 35:97–110.

Swofford, D.L. 1989. PAUP: Phylogenetic Analysis Using Parsimony (version 3.0). Illinois Natural History Society, Champaign, Illinois.

Thornton, W.A., and G.C. Creel. 1975. The taxonomic status of kit foxes. Texas Journal of Science 26:127–136.

Van Valkenburgh, B. 1989. Carnivore dental adaptations and diet: a study of trophic diversity within guilds. Pp. 410–436 *in* Carnivore Behavior, Ecology and Evolution, J.L. Gittleman, editor. Cornell University Press, Ithaca, New York.

Van Valkenburgh, B., and R.K. Wayne. 1994. Shape divergence associated with size convergence in sympatric East African jackals. Ecology 75:1567–1581.

Vila C., I.R. Amorim, J.A. Leonard, D. Posada, J. Castroviejo, F. Petrucci-Fonseca, K.A. Crandall, H. Ellegren, and R.K. Wayne. 1999. Mitochondrial DNA phylogeography and population history of the grey wolf *Canis lupus*. Molecular Ecology 8: 2089–2103.

Waithman, J.D., and A. Roest. 1977. A taxonomic study of the kit fox, *Vulpes macrotis*. Journal of Mammalogy 58:157–164.

Wayne, R.K. 1996. Conservation genetics in the Canidae. J.C. Avise, and J.L. Hamrick, editors. Conservation genetics: Case histories from nature. Pp. 75–118. Chapman and Hall, Inc., New York, New York; London, UK.

—— 1992. Mitochondrial DNA analysis of the eastern coyote: origins and hybridization. Pp. 9–22 in A. H. Boer (ed.), Ecology and Management of the Eastern Coyote. University of New Brunswick, Frederickton, New Brunswick.

Wayne, R.K., E. Geffen, D.J. Girman, K.P. Koepfli, L.M. Lau, and C.R. Marshall. 1997. Molecular systematics of the Canidae. Systematic Biology 46:622–653.

Wayne, R.K., and K.P. Koepfli. 1996. Demographic and historical effects on genetic variation of carnivores. Pp. 453–484 *in* J.L Gittleman, editor. Carnivore behavior, ecology, and evolution, Vol. 2. Cornell University Press: Ithaca, New York; London, UK. 1991. Mitochondrial DNA analysis supports extensive hybridization of the endangered red wolf (*Canis rufus*). Nature 351:565–568.

Wayne, R.K., N. Lehman, M.W. Allard, and R.L. Honeycutt. 1992. Mitochondrial DNA variability of the gray wolf: genetic consequences of population decline and habitat fragmentation. Conservation Biology 6:559–569.

Wayne, R.K., W.G. Nash, and S.J. O'Brien. 1987. Chromosomal evolution of the Canidae. II. Divergence from the primitive carnivore karyotype. Cytogenetics and Cell Genetetics 44:134–141.

Wayne, R.K., and S.J. O'Brien. 1987. Allozyme divergence within the Canidae. Systematic Zoology 36:339–355.

Wozencraft, W.C. 1993. Order Carnivora. Pp. 279–348 *in* Mammal species of the world: a taxonomic and geographic reference, second edition D.E. Wilson and D.M. Reeder, editors. Smithsonian Institution Press, Washington, DC.

Behavioral and Physiological Adaptations of Foxes to Hot Arid Environments: Comparing Saharo-Arabian and North American Species

■ Eli Geffen and Isabelle Girard

Abstract: The principal adaptations to the hot desert environment demonstrated by foxes are behavioral and morphological. Desert foxes are not physiologically well adapted to tolerate heat load and so avoid heat stress behaviorally. All species spend the hot hours of the day in deep burrows and delay activity to the cooler hours of the night. Their small body size increases dissipation of metabolic heat by passive conductance, but limits heat storage potential as utilized by larger animals. Foxes rely on non-evaporative heat loss for dissipating heat, and can increase conductance by behavioral or morphological mechanisms. When ambient temperatures rise above the thermal neutral zone, small canids employ evaporative cooling by panting. Small desert canids can be independent of drinking water if evaporative water loss can be restricted. Low basal metabolic rate, a wide thermal neutral zone, seasonal change in fur density and body fat, and active heat dissipation by a change in skin vasoconstriction are additional mechanisms to avoid heat load without the need to evaporate water.

Diet selection is a critical component of the survival strategy of desert-adapted fox species. Foxes are capable of maintaining water balance for an indefinite time with water input from the diet alone. All 5 fox species in the Saharo-Arabian region include fruit and vegetative material, a water-rich food component in their diet. However, kit and swift foxes feed almost exclusively on rodents. Foxes are able to produce concentrated urine, although not at the levels known in rodents. Reduced urine volume, reduced evaporative water loss, and selection of succulent food items in the diet combine to allow small canids to survive in the desert without drinking.

Five fox species inhabit the Saharo-Arabian deserts. The red fox is the largest species (2–4 kg), the pale fox and Ruppell's sand fox are intermediate in size (1.5–3.6 kg), and the fennec and Blanford's fox are the smallest (0.8–1.5 kg). The red fox is a Palearctic species that has extended its range to the Arabian Peninsula and northern Africa (Fig 1a). This species is the most opportunistic and occupies a wide range of habitat types, including human habitation, but it avoids the extreme arid deserts. The pale fox is the only species restricted to Africa, and it inhabits sandy or stony plains in the deserts and semi-deserts of the southern Sahara (Fig. 1b). Ruppell's sand fox is found in similar habitats throughout central and northern Sahara, the Arabian Peninsula, and eastwards to Afghanistan (Fig. 1b). The 2 smallest species are the most specialized in habitat type. The fennec fox occupies the extreme sandy deserts, and the Blanford's fox is restricted to rocky mountains and canyons (Fig. 1c; Harrison and Bates 1991, Kingdon 1997).

The North American congeners are the swift and kit foxes. The kit fox (1.4–3.0 kg) inhabits exclusively arid and semiarid deserts, and shrub-steppe habitats of southwestern US and northwestern Mexico. The swift fox (1.8–3.0 kg) is found in short-, mid-, and mixed grass prairies of the Great Plains in central North America (O'Farrell 1987, Scott-Brown et al. 1987).

The fox species in the Saharo-Arabian region can be divided into 2 phylogenetic lineages (Geffen et al. 1992d). The fennec and Blanford's fox lineage is about 4 million years (my), and coincides with the appearance of the deserts in this region (Fig. 2). The second lineage (red fox and Ruppell's sand foxes) suggests that Ruppell's sand fox may have entered the deserts more recently (1–2 my; Fig. 2). Both lineages are relatively old, permitting extensive time periods for all these fox species to adapt to their arid environments. In comparison, the North American kit and swift foxes are both a much more recent divergence than the red fox lineage (0.2–0.5 my; Fig. 2; Geffen et al. 1992d). These 2 species are sister taxa (0.34% divergence; Geffen et al. 1992d, Mercure et al. 1993), and are closely related to the Arctic fox.

The Physical Environment

The Saharo-Arabian region, roughly 10,000,000 km^2, consists of a series of deserts and semi-deserts. The deserts of the Arabian Peninsula, southern Iran, and Afghanistan are a continuation of the Sahara desert. These areas were disconnected from Africa about 30 million years ago with the formation of the Red Sea (Braithwaite 1987). Terrestrial faunal interchange resumed during the Miocene (5–20 my) and the peak glaciation of the Holocene (17 ky), when the shallow straits of Bab el Mandeb (130 m) were above sea level (Braithwaite 1987, Sheppard et al. 1992). A second route for faunal interchange remained open (until the construction of the Suez Canal in 1869) through the Sinai land bridge.

The Arabian deserts encompass a variety of habitats: sand dunes, flats, pebble plains, rocky plateaus, mountainous escarpments, and deep canyons. This region is characterized by extreme aridity, high temperature, and violent winds. Rainfall is notoriously unpredictable, and tends to come as sudden storms at irregular intervals. The northern fringe of the region receives rain in winter and the southern border during summer. Annual rainfall ranges between 20–150 mm. Relative mid-day humidity during the warmest months is usually below 20%. The Saharo-Arabian zone is the most extensive desert region of the world with an average of 10 hours of sunshine per day, and an average global radiation exceeding 550 Langleys per day; these high values are a consequence of the absence of

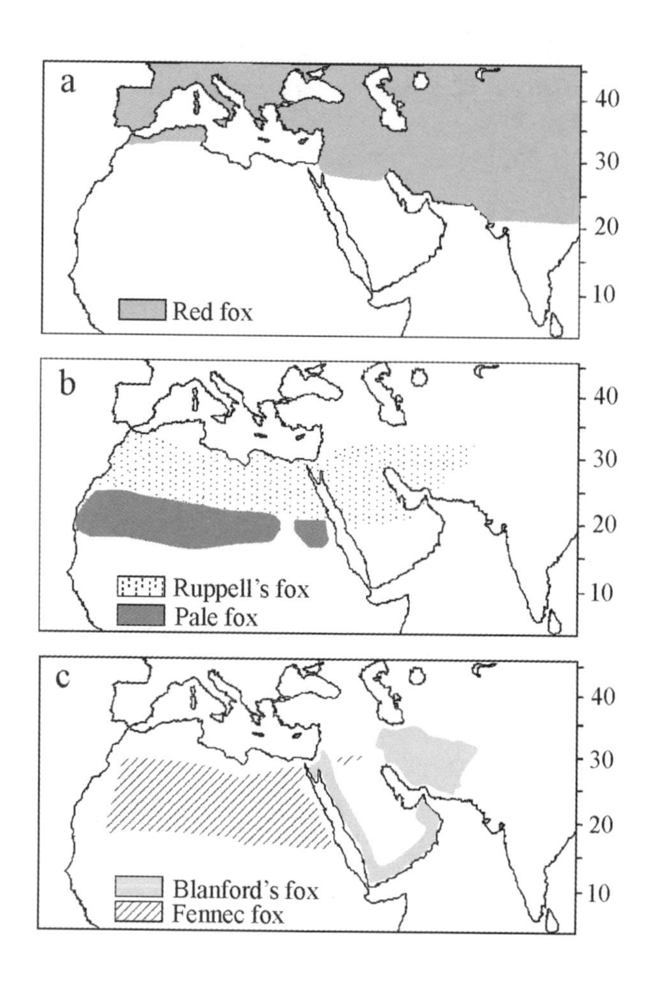

Figure 1. Distribution maps for the five Saharo-Arabian fox species. Latitude is indicated on the right side of each map.

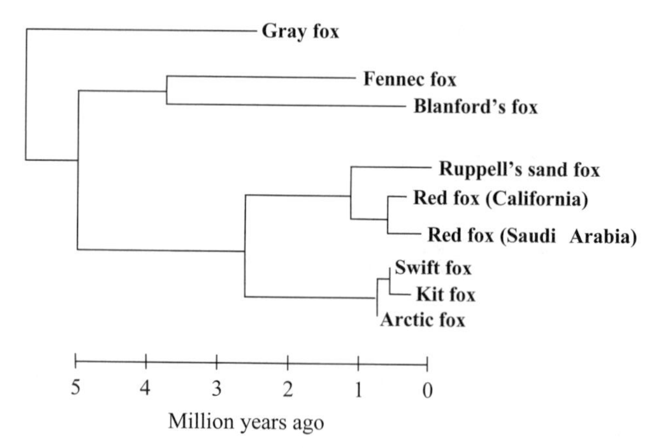

Figure 2. A phylogenetic tree (modified from Geffen et al. 1992d) for the fox species mentioned above. Phylogenetic data for the pale fox is currently not available. The gray fox (Urocyon cinereoargenteus) *was used as the outgroup.*

clouds (Smith 1984). Until 2.5–5 my ago, the Saharo-Arabian region was much wetter. The current Sahara desert was then a Savannah land with lakes and rivers. Changes in the atmospheric circulation due to several major geological events created a jet stream of dry air over the region and an increase in wind velocity. These climatic changes caused evaporation of the lakes and obliteration of the rivers by wind-blown sands (Wickens 1984, Williams 1984).

The present distribution of flora largely reflects the recolonization of the deserts following the dry late Pleistocene (20–15 ky). Mediterranean elements are still able to survive in wet and isolated pockets, but following the gradual decrease in precipitation over the past 6,000 years much of the flora has been replaced by southern species (e.g., *Acacia*). Although more than 1,000 plant species are recognized for this region, remarkably no single truly endemic family is known. At least 4 latitudinal centers of floral endemism have been defined across the Sahara and Arabia (Wickens 1984).

In North America, the decrease in rainfall had already begun during the Eocene (38–55 my), but the development of current deserts and dry prairies began in the Pliocene (5 my) with the formation of basins by a series of uplifts. These events developed isolated local rainshadows and more extended deserts in northern Mexico (Macmahon, 1979). This long period of dryness allowed the evolution of desert fauna and flora or the migration of xerophilic elements from South America. During the glaciation periods of the Pleistocene (2 my) the desert fauna and flora were driven south into desert refugia such as in central Mexico, Baja California, and the Sonoran-Sinaola coast. The Mexican highlands and peninsular Florida provided refugia for arid grassland species. Geomorphologically, the current North American deserts are very young (8,000–10,000 years old). During the last ice age maximal (around 50,000 years ago) many of the basins were filled with water and the current desert regions were much more moist. Only after the last glaciers retreated were the deserts and prairies of the Midwest were reformed (Macmahon, 1979).

The deserts in North America are divided into 2 types: northern and southern. The Great Basin desert is characterized by its northern location, high elevation, and predominantly winter moisture input in the form of snow. The southern deserts have rain in winter (Mohave desert), summer (Chihuahuan desert) or in both (Sonoran desert). Annual precipitation ranges 60–410 mm, and varies by location. Average number of days of precipitation ranges 15–91. Mean summer temperature is around 30°C, excluding the Great Basin where the temperature is 5–8°C lower. Extremely high temperatures of 57°C have been recorded only at Death Valley.

The North American deserts are characterized by the distribution of creosote bush (*Larra tridentata*). Other typical plants are the sagebrush (*Artemisia tridentata*) and shade scale (*Atriplex confertifolia*). The Sonoran desert is especially rich in cacti species, and the Mohave Desert in

Joshua trees (*Yucca brevifolia*). The North American deserts extend contiguously from southeast Oregon to central Mexico; therefore it is not surprising that the distribution of many species form a continuum. There is a great similarity in floral species composition between the southern North American desert and the "Monte" desert in Argentina (Macmahon 1979).

Interspecific Variations in Adaptation for Arid Environments

Animals living in warm and arid or semiarid environments have to cope with relatively high heat loads and accelerated water losses. Different species cope with heat and water stress by employing specialized morphological structures, modification of behavior, and a variety of physiological capabilities. Below, we review the main known adaptive mechanisms that permit these fox species to survive in the vast hot and arid lands of the Saharo-Arabian and North American deserts. The ecology and behavior of kit (*V ulpes macrotis*) and swift (*V. velox*) foxes have been intensively studied at several localities in the wild. On the other hand, pale fox (*V. pallida*) and the fennec fox (*Fennecus zerda*) have never been systematically studied in the wild, and the information about the ecology of these species is anecdotal. The ecology of Ruppell's sand fox (*V. rueppelli*) has been studied only in Oman and that of the Blanford's fox (*V. cana*) only in Israel. The ecology and behavior of red fox (*V. vulpes*) has been thoroughly investigated in Europe and North America, but within the Saharo-Arabian region it has been studied only in Israel and Saudi Arabia. The fennec and the kit foxes are the only species whose physiology have been comprehensively studied under laboratory conditions, and Blanford's, kit and swift foxes are the only species whose eco-physiology has been studied in free-ranging individuals.

BODY SIZE AND MORPHOLOGY: Fox species in the North American and the Saharo-Arabian arid regions range from 1–4 kg in weight. Small canids have substantially less heat storage potential due both to the small absolute mass and to the non-linear relationship of body volume to relative surface area (Phillips et al. 1981). In all desert fox species, heat storage by body temperature elevation is a minute fraction of the heat budget, and provides little benefit under conditions of heat stress. Consequently, in all desert fox species the small body mass hinders storage of activity heat. However, small body size facilitates a relatively high passive thermal conductance that is highly effective for dissipating exercise heat. Because heat is lost from surfaces and a small animal has a larger surface to volume ratio, conductance is relatively higher in smaller animals. For example, heat transfer across the surface of a 1 kg fox is 3 times faster than in a 5 kg fox (Bradley and Deavers 1979). In fact, arid land foxes may increase minimum summer thermal conductance from that predicted for small body mass. In the fennec, minimal thermal conductance was 77–122% of that expected by the body mass (0.023–0.0365 ml/g hr °C; Noll-Banholzer 1979a, Maloiy et al. 1982), and the kit fox (1.82 kg) had a minimum summer thermal conductance 120–145% of that expected by the body mass (Golightly and Ohmart 1983).

Body fur of Blanford's fox and swift fox during winter is much denser than in summer, thus improving body insulation (Scott-Brown et al. 1987, Geffen 1994). Seasonal changes in fur density also occur in red and Ruppell's sand foxes (personal observation). A thin coat during summer is adaptive to a sprinter, such as a fox, to enhance passive dissipation of exercise heat load. In addition, areas of the body that are permanently covered with very short hair serve as "thermal windows." These areas represent 38.16% of the total surface area in kit fox, 32.85% in red fox, and only 21.61% in the Arctic fox (*Alopex lagopus*; Klir and Heath 1992). Using infrared thermography, Klir and Heath (1992) have shown that kit fox responded to increase in ambient temperature by increasing heat dissipation, mainly through the surface regions covered by short fur. A change in circulation of blood through these areas occurred at temperatures of 20–23°C, and maximum vasodilation was reached at 33°C (Klir and Heath 1992). Vasocontrol of surface circulation and the increase in proportion of "thermal windows" in the arid land foxes appear to be adaptations to the warmer climates.

Cooling passively by conductance can also be improved by maximizing surface area. At high ambient temperature, kit fox stands in a sprawling manner with legs apart (Golightly and Ohmart 1983). This posture maximizes heat loss but elevates the metabolic expenditure. Maloiy et al. (1982) suggested that the large ears of the fennec (surface area of 228 cm^2; 16–20% of total body surface) may play a considerable role in heat dissipation. They observed that at high ambient temperature, ear temperature was lower than ambient or body temperatures, thus suggesting evaporation from the ear surface. These observations suggest that the characteristically large ears of the arid land foxes are also an adaptation to increase heat dissipation.

At low ambient temperatures, as occurs at night and in winter, the high thermal conductance of small foxes can result in substantial heat loss during cold stress. Compensatory seasonal changes in body fat content and/or in fur density can reduce conductance to conserve heat. Body weight in Blanford's fox showed significant seasonal change, with both sexes being heaviest in winter and lightest in summer; seasonal changes in activity or diet were not observed (Geffen et al. 1992a, 1992b; Geffen and Macdonald 1993). Total body water volume as a percentage of body mass in Blanford's fox was significantly higher in summer than in winter (Geffen et al. 1992c), suggesting an increase in body fat during winter. In contrast, the North American kit fox did not have extensive fat stores and their body weight did not fluctuate significantly between summer and winter (Golightly and Ohmart 1984, Girard 1998). The constant body mass and fat content of the kit fox relative to the Blanford's fox may be related to differences in body size and diet composition.

ACTIVITY: All fox species in the Saharo-Arabian and North American arid regions are nocturnal or crepuscular. Red foxes at the southern Rift Valley in Israel were active at night for 10–13 hours, and became active immediately after sunset (Assa 1990). Often, these red foxes extend their activity to the early morning hours during summer. Ruppell's sand foxes in Oman and Blanford's foxes in Israel were active only at night (Lindsay and Macdonald 1986, Geffen and Macdonald 1993). Activity duration of the Blanford's fox ranged from 8–11 hours, with onset of activity at sunset throughout the year (Geffen and Macdonald 1993). The North American kit fox is also nocturnal, with onset of activity around sunset and activity duration of 7–10 hours per night (Morrell 1972, O'Farrell 1987, Girard 1998). Girard (1998) showed that in the kit fox, emergence for activity 2 hours earlier in summer resulted in a 25% increase in daily water intake. Avoiding the heat of the day by burrowing and delaying activity to the coolest hours minimizes water intake. However, this physiological reasoning can not explain why most desert species are nocturnal year round. During winter, daytime temperatures are within the thermal neutral zone (TNZ) or lower and water intake is not expected to rise with earlier onset of activity. Geffen and Macdonald (1993) suggested that the nocturnal activity of foxes is also an adaptive behavior against raptor predation. Indeed, fox mortality can be substantial where foxes are active by day (Island gray fox; *Urocyon littoralis*) and large raptors are present (G. Roemer, University of California, Los Angeles, personal communication).

METABOLISM: Small desert-adapted fox species exhibit a relatively wide TNZ (Noll-Banholzer 1979a, Golightly and Ohmart 1983). At the TNZ range of ambient temperatures, oxygen consumption and body temperature are at the minimum. For fennec foxes, the TNZ ranged from 23–32°C. Above 32°C, body temperature rises quickly to a critical value (Noll-Banholzer 1979a). In the kit fox, the TNZ ranged from about 20–34°C (Golightly and Ohmart 1983). The temperature ranges of the TNZ indicate that both the kit fox and the fennec are intolerant of the high diurnal temperatures characteristic of the desert environment (Noll-Banholzer 1979a, Golightly and Ohmart 1983). All Saharo-Arabian and North American desert fox species make use of burrows during the day when temperatures are extreme. In the desert regions where mean ambient temperatures can reach 43°C, burrows provide an escape into an environment where metabolic rate is reduced and water loss is minimal. Excluding the Blanford's fox, which exploits the natural cavities under boulders and rock piles, all other fox species considered here actively construct their own burrows (Egoscue 1962, Lindsay and Macdonald 1986, Assa 1990, Geffen and Macdonald 1993, Kingdon 1997). The temperature range inside a fennec's burrow during summer is usually within the TNZ range, and never exceeds 34°C (Noll-Banholzer 1979a). In comparison, annual temperature in burrows of kit fox ranges 13–22°C, and mean summer burrow temperature is well within the TNZ range (Girard 1998).

The basal metabolic rate (BMR) is greatly reduced, relative to the expected, in many desert mammals (McNab 1966). A BMR lower than expected is considered an adaptation to the desert environment because it minimizes the endogenous heat load. Further, low BMR saves water needed for body cooling, and reduces the amount of energy required for body maintenance. Therefore, a reduced BMR may be advantageous during periods when food is scarce. The BMR of the fennec fox is only 61% of the level predicted on the basis of body mass (Noll-Banholzer 1979a). However, BMR in the kit fox ranges between 97–117% in summer and winter, respectively (Golightly and Ohmart 1983), whereas in the red fox BMR is unmodified by season and reaches 122% of the expected (Irving et al. 1955). Although there is some evidence that canids are capable of facultatively adjusting their BMR to different environments (Shield 1972), reducing BMR does not appear to be a widespread strategy employed by desert foxes.

Although some evidence indicates reduced field metabolic rates (FMR) among desert mammals compared with mesic mammals, the most recent review of the allometry of FMR in mammals challenges this view (Nagy 1994, Nagy et al. 1999). In a traditional analysis of FMR in 25 species of desert and 48 species of terrestrial mesic mammals, desert mammals demonstrated a steeper allometric slope; small (<1 kg) desert mammals tend to have lower FMR than similarly sized non-desert mammals, while large (>10 kg) desert mammals have relatively higher FMR. No differences in the scaling existed between desert and non-desert mammals in a comparative analysis of FMR where phylogeny was considered. Of the 5 canids for which FMR data are available (*Vulpes cana*, *V. macrotis*, *V. velox*, *Lyacon pictus*, and *Canis lupus*), 4 have FMR above that predicted for eutherian mammals (*V. cana* is the exception). This trend of elevated FMR among canids may be related to the relatively long daily movement distances typical of canids and other carnivores (Garland 1983, Goszczynski 1986), and may lead to increased energy stress in desert foxes during food shortages. Seasonal changes in FMR were observed in kit and swift foxes, but not in Blanford's foxes. FMR in Blanford's fox was similar in summer and winter (0.652–0.689 and 0.630–0.668 kJ/g day, respectively; Geffen et al. 1992c). In contrast, FMR in kit fox was significantly higher in summer than in winter (0.841–0.953 and 0.676–0.710 kJ/g day, respectively; Girard 1998). A similar trend in FMR was observed in the swift fox (0.990 in summer and 0.709 kJ/g day in winter, Covell et al. 1996). On average, FMR in the kit fox was 37% higher in summer and 6% higher in winter than FMR in Blanford's fox. The increased FMR in the kit fox is presumably related to the high summer movement distances and supported by higher prey intake rates. The reduced FMR in the Blanford's fox required a lower total energy intake and thus a reduced water input from food (Girard 1998).

WATER ECONOMY AND DIET: Water is thought to be a limiting factor and a major selective force in desert carnivores. While the high surface to volume ratio of small canids facilitates non-evaporative heat loss, the principal method for large canids to dissipate heat load is to cool evaporatively by panting. Although small canids demonstrate a reduced dependence on evaporative water loss (EWL) for temperature regulation (Golightly and Ohmart 1984), heat load and consequent EWL may be extreme during pursuit of prey or with exposure to very high ambient temperatures. The water loss in the fennec fox is rather constant under laboratory conditions (0.65–0.91 mg/g hr) below 30°C, but it increases to 3.5 times the basal rate when the ambient temperature is 38°C (Noll-Banholzer 1979a). At such high temperatures, a maximum of 75% (mean 56%) of the metabolic heat can be lost by evaporation. The fennec fox loses an increasingly greater percentage of its metabolic heat by evaporation as ambient temperature increases (Noll-Banholzer 1979a). In contrast, EWL in kit fox was shown to rise with increasing ambient temperature up to the upper limit of the TNZ, and then decreased (Golightly and Ohmart 1983). Additionally, at temperatures above the TNZ, fennec fox dissipates a greater proportion of its metabolic heat by evaporation than kit fox (at 32°C, 23% for kit fox and 36% for fennec fox; Golightly and Ohmart 1983). A unique surface area in canids is the nose, which is hairless and wet. Heat is dissipated through the nose area by evaporation, especially at high ambient temperatures (Klir and Heath 1992), and it has been suggested that the nose in foxes is part of a brain-cooling mechanism (Baker et al. 1974).

Reduction of urinary water loss is also an important adaptation to the water limitation typical of arid environments. The ability to physiologically conserve water by reducing urinary water loss is dependent on an efficient kidney. Desert rodents minimize water loss by producing highly concentrated urine (e.g., *Dipodomys* sp. 4090–5540 mOsm/l; Schmidt-Nielsen 1964), although foxes do not have the high mass-specific metabolic rates thought to be essential for producing such concentrations (MacMillen 1972). Carnivorous and insectivorous desert foxes also face substantial nitrogen loads from the high protein content of their diets, requiring urinary excretion of excess nitrogen. Fennec foxes fed on mice had a urine osmotic concentration of 1480–3828 mOsm/l (max. 4022 mOsm/l; Noll-Banholzer 1979b). Under these conditions, they were able to maintain or gain body weight without access to water or a water-rich diet (e.g., fresh fruit). Daily urine water loss in the fennec fox was reduced to only 59% of that expected based on body size (Noll-Banholzer 1979b). During dehydration, there is little change in EWL but a profound reduction in urine volume (water loss via urine is 45–48% in normal condition versus 27% in dehydrated animals; Noll-Banholzer 1979b). Fennec foxes produce a higher urine concentration than reported for other carnivores, and their kidneys have a distinct papilla (relative medullary thickness = 5.35; Noll-Banholzer 1979b). The relative medullary thickness in the red fox is shorter (4.1), and probably reflects a reduced kidney performance. In comparison, a maximum field concentration of 3,600 mOsm/l was measured in the kit fox (Girard 1998). During summer, urinary water loss is predicted to account for 39–43% of the total water loss in non-drinking kit foxes (Girard 1998). Reduction in urine volume is an adaptive mechanism when food is relatively dry and drinking water is unavailable.

Water turnover rate is largely related to availability of drinking water and type of diet. Although succulent plants and animal prey provide similar amounts of pre-formed water per gram (70–80%; Degen 1997), succulent plants contain much less energy per gram. To meet energy requirements, herbivores must consume a large mass of plant material, and consequently have high water input with food. In contrast, carnivores consume energy-dense foods, and may meet energy requirements with relatively low water input from prey. Foxes under water stress should consume large quantities of succulent but energy-poor foods, to meet water requirements without exceeding energy requirements (which would result in fat deposition, a thermally disadvantageous situation). Because passive heat dissipation is maximized in arid foxes, they may face water stress rarely or not at all. All 5 fox species in the Saharo-Arabian region are omnivorous and opportunistic feeders. The red fox is the most opportunistic, feeding on earthworms, insects, and any other live prey up to the size of a hare or turkey. It will also consume carrion, human garbage, fruit, and vegetables (Gasperetti et al. 1985, Voight 1987). In the southern Rift Valley in Israel, red foxes feed mostly on rodents, insects, and fruit (e.g., watermelons; Assa 1990). Ruppell's sand foxes in Oman feed mostly on rodents, reptiles, and grasses (Lindsay and Macdonald 1986). Diet of the pale fox consists of rodents, lizards, invertebrates, and berries (Haltenorth and Diller 1980, Kingdon 1997). The fennec fox feeds on rodents, lizards, invertebrates and plant material (Schmidt-Nielsen 1964), whereas Blanford's fox consumes mostly invertebrates and fruit (Geffen et al. 1992b). Further, Blanford's fox consumed more fruit during summer than in winter (Geffen et al. 1992c). These data show that all the fox species inhabiting the Saharo-Arabian region include fruit and succulents in their diet. In contrast, the kit fox feeds mostly on lagomorphs and rodents (O'Farrell 1987, White et al. 1996, Girard 1998). Because foxes use evaporative cooling, consuming fruit and succulents helps them maintain water balance even under extremely arid and hot conditions. However, fennec and kit foxes (and probably the other desert-adapted species as well) are capable of sustaining their water balance for an indefinite period from the body liquids of their prey, providing EWL is reduced to a fraction of total water input from food. Considering diet composition, it has been suggested that Blanford's fox forage more for water than for energy, whereas kit fox forage more for energy during winter and spring and more for water during summer (Geffen et al. 1992c, Girard 1998).

Mean water turnover rate (WTR) of free-ranging individuals of Blanford's fox was 0.11–0.13 ml/g day during summer and 0.08–0.10 ml/g day during winter (Geffen et al. 1992c). Kit foxes monitored under natural and semi-natural conditions showed similar water influxes (0.06–0.130 ml/g day; Golightly and Ohmart 1984, Girard 1998). Captive fennec foxes, kept at 25°C and fed different diets, had a WTR range of 0.05–0.06 ml/g day. Both Blanford's fox and kit fox showed a significant increase (20–27%) in WTR in summer as compared to winter (Geffen et al. 1992c, Girard 1998). This small seasonal change in WTR demonstrates a limited dependence on EWL for cooling during the hot summer.

Conclusions

1. The principal adaptations to the desert environment demonstrated by small canids are behavioral and morphological. Desert foxes are not well adapted to tolerate heat load and avoid heat stress behaviorally. All species spend the hot hours of the day in deep burrows and delay activity to the cooler hours of the night. The small body size increases dissipation of metabolic heat by passive conductance, but limits heat storage potential as utilized by larger animals. Above ambient temperatures of 32–34°C, body temperature rises quickly to a critical level. Small canids rely on non-evaporative heat loss for dissipating heat, and can increase conductance through assuming a sprawling position of the legs, by having large ears, and through use of "thermal windows" where fur is thin and short. When ambient temperatures rise above the TNZ, small canids employ evaporative cooling by panting; at ambient temperatures higher than 37°C, a maximum of 75% of the metabolic heat can be lost by evaporation, when cooling by conductance is inefficient.

2. Small desert canids can be independent of drinking water if EWL can be restricted. Evaporation is the limiting factor in the water balance of desert-adapted fox species, and the primary mechanisms for reducing EWL are the avoidance of high temperature and a large surface area to volume ratio for increased passive heat dissipation. Low BMR, a wide TNZ, seasonal change in fur density and body fat, and active heat dissipation by a change in skin vasoconstriction are additional mechanisms to avoid heat load without the need to evaporate water.

3. Diet selection is a critical component of the survival strategy of desert-adapted fox species. Only on rare occasions is drinking water available in the wild during the hot summer in the deserts, but foxes are capable of maintaining water balance for an indefinite time with water input from the diet. All 5 fox species in the Saharo-Arabian region include fruit and vegetative material, a water-rich food component in their diet. In severe water deprivation conditions (e.g., dried food in laboratory experiments), foxes would have to drink water to maintain body water balance (Noll-Banholzer 1979b, Girard 1998), but can be independent of drinking water even in the hot seasons if food contains at least 70% water, as is typical for vertebrate and invertebrate prey. Foxes are able to produce concentrated urine, although not at the levels known in rodents. In the fennec, the kidney has relatively thicker medulla and specific structure. In response to water deprivation, foxes decrease their urine output and food intake (Noll-Banholzer 1979b, Girard 1988). In normal conditions, water loss via urinary output is 30–50% of total water loss. Thus, reduced urine volume, reduced EWL, and selection of succulent food items in the diet combine to allow small canids to survive in the desert without drinking.

Literature Cited

Assa, T. 1990. The biology and biodynamics of the red fox (*Vulpes vulpes*) in the northern Arava Valley. Thesis, Tel Aviv University, Tel Aviv, Israel.

Baker, M.A., L.W. Chapman, and M. Nathanson. 1974. Control of brain temperature in dogs: effects of tracheostomy. Respiration Physiology 22:325–333.

Bradley, S.R., and D.R. Deavers. 1979. A re-examination of the relationship between thermal conductance and body weight in mammals. Comparative Biochemistry and Physiology 65A:465–476.

Braithwaite, C.J.R. 1987. Geology and palaeogeography of the Red Sea region. Pp. 22–44 *in* A.J. Edwards and S.M. Head, editors. Key environments—Red Sea. Pergamon Press, Oxford, UK.

Covell, D.F., D.S. Miller, and W.H. Karasov. 1996. Cost of locomotion and daily energy expenditure by free-living swift foxes (*Vulpes velox*): a seasonal comparison. Canadian Journal of Zoology 74:283–290.

Degen, A.A. 1997. Ecophysiology of small desert mammals. Springer, Berlin, Germany.

Egoscue, H.J. 1962. Ecology and life history of the kit fox in Tooele county, Utah. Ecology 43:481–497.

Garland, T., Jr. 1983. Scaling the ecological costs of transport to body mass in terrestrial mammals. American Naturalist 121:571–587.

Gasperetti, J., D.L. Harrison, and W. Büttiker. 1985. The carnivora of Arabia. Pp. 397–461 *in* W. Büttiker and F. Krupp, editors. Fauna of Saudi Arabia. Pro Entomologia, Natural History Museum, Basel, Switzerland.

Geffen, E. 1994. Blanford's fox, *Vulpes cana*. Mammalian Species 462:1–4.

Geffen, E., and D.W. Macdonald. 1993. Activity and time tabling in the movement patterns of Blanford's foxes, *Vulpes cana*, in Israel. Journal of Mammalogy 74:455–463.

Geffen, E., R. Hefner, D.W. Macdonald, and M. Ucko. 1992a. Morphological adaptations and seasonal weight changes in the Blanford's fox, *Vulpes cana*. Journal of Arid Environments 23:287–292.

——. 1992b. Diet and foraging behavior of Blanford's foxes, *Vulpes cana*, in Israel. Journal of Mammalogy 73:395–402.

Geffen, E., A.A. Degen, M. Kam, R. Hefner, and K.A. Nagy. 1992c. Daily energy expenditure and water flux of free-living Blanford's foxes (*Vulpes cana*), a small desert carnivore. Journal of Animal Ecology 61:611–617.

Geffen, E., A. Mercure, D.J. Girman, D.W. Macdonald, and R. K. Wayne. 1992d. Phylogenetic relationships of the fox-like canids: mitochondrial DNA restriction fragment, site and cytochrome b sequence analyses. Journal of Zoology 228:27–39.

Girard, I. 1998. The physiological ecology of a small canid, the kit fox (*Vulpes macrotis*), in the Mojave Desert. Dissertation, University of California, Los Angeles, California.

Golightly, R.T., and R.D. Ohmart. 1983. Metabolism and body temperature of two desert canids: coyotes and kit foxes. Journal of Mammalogy 64:624–635.

——. 1984. Water economy of two desert canids: coyote and kit fox. Journal of Mammalogy 65:51–58.

Goszczynski, J. 1986. Locomotor activity of terrestrial predators and its consequences. Acta Theriologica 31(6):79–95.

Haltenorth, T., and H. Diller. 1980. A field guide to the mammals of Africa, including Madagascar. Collins, London, UK.

Harrison, D.L., and P.J. Bates. 1991. The mammals of Arabia. Harrison Zoological Museum Publication, Sevenoaks, Kent, UK.

Irving, L., H. Krog, and M. Monson. 1955. The metabolism of some Alaskan animals in winter and summer. Physiological Zoology 28:173–185.

Kingdon. J. 1997. The Kingdon field guide to African mammals. Academic Press, San Diego, California.

Klir, J.J., and J.E. Heath. 1992. An infrared thermographic study of surface temperature in relation to external thermal stress in three species of foxes: the red fox (*Vulpes vulpes*), arctic fox (*Alopex lagopus*), and kit fox (*Vulpes macrotis*). Physiological Zoology 65:1011–1021.

Lindsay, I.M., and D.W. Macdonald. 1986. Behaviour and ecology of the Ruppell's fox, *Vulpes rueppelli*, in Oman. Mammalia 50:461–474.

Macmahon, J.A. 1979. North American deserts: their floral and faunal componenets. Pp. 21–82 *in* D.A. Goodall, and R.A. Perry, editors. Arid-land ecosystems: structure, functioning and management. Cambridge University Press, London, UK.

Macmillen, R.E. 1972. Water economy of nocturnal desert rodents. Pp. 147–174 *in* G. M. O. Maloiy, editor. Comparative Physiology of Desert Animals. Symposia of the Zoological Society of London (31), Academic Press, London, UK.

Maloiy, G.M.O., J.M.Z. Kamau, A. Shkolnik, M. Meir, and R. Arieli. 1982. Thermoregulation and metabolism in a small desert carnivore: the fennec fox (*Fennecus zerda*). Journal of Zoology 198:279–291.

McNab, B.K. 1966. The metabolism of fossorial rodents: a study of convergence. Ecology 47:712–733.

Mercure, A., K. Ralls, K. Koepfli, and R.K. Wayne. 1993. Genetic subdivisions among small canids: mitochondrial DNA differentiation of swift, kit and Arctic foxes. Evolution 47:1313–1328.

Morrell, S. 1972. Life history of the San Joaquin kit fox. California Fish and Game 58:162–174.

Nagy, K.A. 1994. Field bioenergetics of mammals—what determines field metabolic rates. Australian Journal of Zoology 42:43–53.

Nagy, K.A., I.A. Girard, and T.K. Brown. 1999. Energetics of free-ranging mammals, reptiles and birds. Annual Review of Nutrition 19:247–277.

Noll-Banholzer, U. 1979a. Body temperature, oxygen consumption, evaporative water loss and heart rate in the fennec. Comparative Biochemistry and Physiology 62A:585–592.

——. 1979b. Water balance and kidney structure in the fennec. Comparative Biochemistry and Physiology 62A:593–597.

O'Farrell, T.P. 1987. Kit fox. Pp. 423–431 *in* M. Novak, G. A. Baker, M. E. Obbard and B. Malloch, editors. Wild furbearer management and conservation in North America. Ministry of Natural Resources, Ontario.

Phillips, C.J., R.P. Coppinger, and D.S. Schimel. 1981. Hyperthermia in running sled dogs. Journal of Applied Physiology 51:135–142.

Scott-Brown, J.M., S. Herrero, and J. Reynolds. 1987. Swift fox. Pp. 433–441 *in* M. Novak, J.A. Baker, M.E. Obbard, and B. Malloch, editors. Wild furbearer management and conservation in North America. Ontario Trappers Association, North Bay, Ontario.

Schmidt-Nielsen, K. 1964. Desert animals. Clarendon Press, Oxford, UK.

Sheppard, C., A. Price, and C. Roberts. 1992. Marine ecology of the Arabian region. Academic Press, London, UK.

Shield. J. 1972. Acclimation and energy metabolism of the dingo *Canis dingo* and the coyote *Canis latrans*. Journal of Zoology 168:483–501.

Smith, G. 1984. Climate. Pp. 17–30 *in* J.L Cloudsley-Thompson, editor. Key environments—Sahara Desert. Pergamon Press, Oxford, UK.

Voight, D.R. 1987. Red fox. Pp. 379–392 *in* M. Novak, G.A. Baker, M.E. Obbard and B. Malloch, editors. Wild furbearer management and conservation in North America. Ministry of Natural Resources, Ontario.

White P.J., C.A.V. White, and K. Ralls. 1996. Functional and numerical responses of kit foxes to a short-term decline in mammalian prey. Journal of Mammalogy 77:370–376.

Williams, M. 1984. Geology. Pp. 31–39 *in* J.L Cloudsley-Thompson, editor. Key environments—Sahara Desert. Pergamon Press, Oxford, UK.

Wickens, G.E. 1984. Flora. Pp. 67–75 *in* J.L Cloudsley-Thompson, editor. Key environments—Sahara Desert. Pergamon Press, Oxford, UK.

A Review of Parasites and Diseases of Wild Swift Fox

■ M.J. Pybus and E.S. Williams

Abstract: Parasites and diseases of wild swift fox have not been well documented. Fleas are the most common and abundant ectoparasite. One cestode and 2 nematode species dominate the helminth fauna, and are common and numerous in most swift fox populations. The cestode is transmitted by Pulex irritans; *the ascarid nematodes have direct life cycles. Hookworms* (Ancylostoma caninum, Uncinaria *sp.*), *stomach worms* (Physaloptera *sp.*), *whipworms* (Trichuris vulpis), *and one trematode* (Alaria arisaemoides) *also have been found as well as miscellaneous protozoans and ectoparasite species. Exposure to a few infectious diseases has been documented serologically (e.g., sylvatic plague, canine distemper); however, there are only a few cases of confirmed canine distemper in wild swift fox. It is likely that swift fox share a community of parasites and diseases with sympatric canids and either have lost or have not developed a specialized suite of agents.*

It is appropriate that parasites and diseases be included in a symposium to review the body of knowledge regarding interactions of swift fox (*Vulpes velox*) with other species in the prairie community. In addition to interactions with predators, prey, and habitats, parasites simply represent another level of interspecific interactions within the community. Thus, swift fox can be viewed as mobile habitats for other species. Parasites and disease agents are inherent and integral components of natural ecosystems, neither negative nor positive, but simply part of the system. This paper will attempt to identify the variety of species that live in or on swift fox.

As with many endangered or threatened species, parasite and disease agents in wild swift fox have not been studied extensively. This may, in part, reflect the precipitous decline in fox populations in conjunction with European colonization of the central prairie regions of North America (Egoscue 1979). Thus, even early records are limited. This paper will review the documented parasites and diseases of swift fox as well as provide brief mention of others that could potentially occur. Life history strategies of parasitic and infectious species are linked critically to their occurrence in hosts and their geographic distribution. In some instances, inferences regarding the ecology of swift fox can be gained by examining the life history strategies of their parasites. In addition, a brief comparison of the parasite communities among wild canids will be presented.

Ectoparasites

Fleas (*Pulex irritans*) are by far the most common and the most often reported ectoparasite on swift fox (Rapp 1962, Kilgore 1969, Hillman and Sharps 1978, Hines 1980, Holland 1985; Table 1). In the only comprehensive study of wild foxes, Kilgore (1969) found fleas on all swift fox examined except neonatal pups. Adult foxes were heavily infested and additional fleas were collected from den entrances and debris within the den. Fleas use warm-blooded hosts as a source of blood meals for the adult stage and are common on birds and mammals that use habitual den or nest sites. Eggs occur in the environment wherever they drop. The larvae and pupae also occur off the host, wherever they can find suitable food, water, and shelter. The common occurrence of fleas reflects the extensive use of dens by swift fox, a species characterized as the most subterranean and "burrow-dependent" wild canid in North America (Seton 1929, Egoscue 1979). Hines (1980) reported an increased number of fleas on orphaned pups after the female was killed and suggested it was related to the continued use of the same den for an extended period. Swift fox burrows provide a cool, moist environment littered with prey remains and used extensively by a warm-blooded fox—near-optimum conditions from a flea's perspective.

Pulex irritans has been found on a wide range of hosts including domestic dogs and cats as well as swift fox, kit fox (*Vulpes macrotis*), red fox (*V. vulpes*), gray fox

Table 1. Ectoparasites of swift fox.

	Species	Life History	Other Hosts	Occurrence[a]
Fleas	*Pulex irritans*	Direct (in burrows)	Various	Very common
	Pulex simulans	Direct (in burrows)	Various	Common?
	Opisocrostis hirsutus	Direct	Prairie dogs, coyotes	Rare
	Foxella ignota	Direct	Pocket gophers	Rare
	Cediopsylla sp.	Direct	Rodents, lagomorphs	Rare
Ticks	*Ixodes kingi*	Direct	Rodents, mustelids, canids, felids	Uncommon
	Dermacentor andersoni	Direct	Small rodents, rabbits	Uncommon

[a]On swift fox

(*Urocyon cinereoargenteus*), coyotes (*Canis latrans*), badgers (*Taxidea taxus*), raccoons (*Procyon lotor*), and various lagomorphs (Tohm 1953, Turkowski 1974, Honess and Bergstrom 1982, Holland 1985). It occurs throughout the range of swift fox. However, fleas are relatively benign parasites and are not associated with significant damage to infested hosts (habitats). Occasionally they may serve as a vector for other parasites and disease agents.

Fleas can be spread by direct contact with infested animals, thus most fleas are undoubtedly passed from fox to fox. Fleas from prey species occasionally transfer to predators and this may explain the record on swift fox of *Opisocrostis hirsutus*, normally found on prairie dogs (*Cynomys* spp.); *Foxella ignota*, a parasite of pocket gophers (*Geomys bursarius*); and *Cediopsylla* sp., a genus of flea usually found on rodents, rabbits, and hares. An alternative explanation could be infestation of foxes during use of contaminated rodent and lagomorph burrows. In most cases, the life cycle of the flea is not perpetuated, probably due to opportunistic transfer of inadequate numbers to establish a new population on the predator species.

Some of these flea species, especially *O. hirsutus*, may serve as vectors for *Yersinia pestis*, the bacterium that causes sylvatic plague. Exchange of fleas between prairie dogs and their predators, in this case swift fox, could lead to exposure of foxes to the bacterium. In addition, it has been suggested that predators could serve to spread plague from area to area by transport of infected fleas of rodents.

There are few records of ticks on swift fox (Kilgore 1969, Hillman and Sharps 1978, Hines 1980) and it appears they do not provide particularly good habitat for this group of ectoparasites. Again, this may reflect the extensive time foxes spend underground, but with consequences quite different than with fleas. Tick life cycles include 4 stages: egg, larva, nymph, and adult. The cycles often are complicated and may involve different species of birds or mammals as hosts for each developmental stage. Generally, tick life cycles occur above ground and, as such, swift fox may be less available as hosts than other mammalian species. *Ixodes kingi*, the predominant tick reported on swift fox, has a broad distribution and occurs on various rodents, canids (including kit fox), mustelids, felids, lagomorphs, raccoons, and even on snakes (Egoscue 1962, Honess and Bergstrom 1982). Gregson (1956) characterized this species as the "common prairie *Ixodes*." Nymphs of *Dermacentor andersoni* (Rocky Mountain wood tick) were collected from a swift fox in Alberta (W. M. Samuel, University of Alberta, personal communication). This tick species is common throughout much of the province. Nymphs generally feed on small mammals, including rabbits and ground-dwelling rodents. It is unlikely that either of these tick species is maintained in swift fox populations and current records may be spillover from contact with infected prey.

Sarcoptic mange is the most conspicuous and potentially the most significant ectoparasite of wild canids. The mite, *Sarcoptes scabei*, has a cosmopolitan distribution, and transfers readily among a variety of host species that offer suitable habitat for its establishment (Sweatman 1971). Mange mites can cause extensive irritation and damage as they tunnel through the epidermal layers of skin. Additional damage is associated with scratching, rubbing, and chewing of affected areas by infested hosts, leading to hair loss and even death (Morner and Christensson 1984). In severely affected individuals, damage to the haircoat and underlying skin can be extensive. However, in most cases, infestations are self-limiting and significance is minimal. Following molt, the replacement pelage of infested individuals is not affected.

We found no report of mange on swift fox, despite occurrence of the disease in sympatric wild canids in areas such as Alberta and Wyoming (Pybus, unpublished data; Williams, unpublished data). It may be that the low population density and reduced social contact among swift fox (relative to coyotes and wolves, for example) reduces the potential for maintaining a population of mange mites or that infestations remain mild and go undocumented. Mange on red foxes is associated with lesions quite different from those in coyotes and wolves and may imply a different host response in small canids. A further alternative is that swift fox are particularly susceptible to mite infestation and mortality is undetected.

Endoparasites

The helminth community of swift fox is characterized by 1 common cestode and 1 or 2 common nematodes (Table 2). *Dipylidium caninum*, a tapeworm of domestic dogs, is frequent and abundant in swift fox populations (Cutter 1958, Kilgore 1969). Adult tapeworms occur in the gut of infected canids. Eggs released in fecal material are ingested by fleas and develop to an infective stage. Larvae are transferred to the final host when infected fleas are ingested. This most likely occurs during routine grooming by the foxes. *Pulex irritans* is a known intermediate host for *D. caninum* and, given the widespread occurrence of this flea on swift fox, it is not surprising that this cestode also is common in swift foxes.

Taenia multiceps is a cestode found in predators of lagomorphs. The larval stages occur as cysts in the connective tissues of hares and rabbits. The presence of adults of this species in swift fox reflects the use of lagomorphs as a food item. Although not primary food items, lagomorphs occur as a significant component in the diet of swift fox (Cutter 1958, Kilgore 1969, Kitchen et al. 1999). Young rabbits are particularly important in the spring diet of the foxes.

Echinococcus multilocularis is a cestode that occurs in the gut of wild canids and domestic dogs (Leiby and Dyer 1971). Primarily a subarctic species, it has been introduced

Table 2. Parasites of swift fox.

	Species	Life History	Other Hosts	Occurrence[a]
Cestodes	*Dipylidium caninum*	Indirect: fleas	Dogs	Common
	Taenia multiceps	Indirect: lagomorphs	Canids, felids	Locally common
Trematodes	*Alaria arisaemoides*	Indirect: snail, tadpole	Canids, mustelids	Unknown
Nematodes	*Toxocara canis*	Direct	Canids	Common
	Toxascaris leonina	Direct	Felids, canids	Locally common
	Ancylostoma caninum	Direct	Dogs	Uncommon
	Uncinaria sp.	Direct	Dogs	Rare?
	Physaloptera sp.	Indirect: insects	Canids, mustelids	Common
	Trichuris vulpis	Direct	Canids	Uncommon?
Protozoans	*Isospora bigemina*	Indirect: rodents	Canids	Unknown
	I. felis	Indirect: rodents	Felids	Unknown
	I. idahoensis	Indirect: rodents	Canids?	Unknown

[a]In swift fox

into the northern portions of swift fox range. Small rodents, including various species of ground squirrels (*Spermophilus* sp.) and mice, are used as intermediate hosts. Although *E. multilocularis* occurs in wild red fox within the range of swift fox (Leiby et al. 1970, Seesee and Worley 1976, Hildreth and Schneider 1989, Davidson et al. 1992a), we could find no documented case of infection in swift fox. However, given its presence in wild rodents, the potential exists for infection in swift fox.

Species of *Echinococcus* are zoonotic (infective to humans) and eggs produced by adult tapeworms in the intestine of canids can establish infections in people. Anyone handling live or dead foxes or fox feces should take precautions to prevent infection. Avoid handling fecal material or the anal region of foxes. Wear latex or rubber gloves if it is necessary to handle feces. Eggs cannot survive desiccation or high temperatures (>100°C); thus, autoclaving fecal samples to be used for food analysis studies is recommended. Detailed parasitic examination of foxes and the development of a more sensitive test to detect infected foxes (Bretagne et al. 1992) may help determine the extent of the health risk.

Mature trematodes (flukes) of the species *Alaria arisaemoides* were collected from a young (<1yr) female swift fox born in the wild in southern Alberta. *Alaria* spp. have a 3-host life cycle involving a snail, a tadpole, and a carnivore. In addition, small mammals that have eaten infected tadpoles or frogs may pass on the parasite when they are eaten by a predator. Once in the carnivore, the fluke larvae migrate to the lungs and later move to the intestine where they develop into sexually mature adults. The larval stages of *A. arisaemoides* develop in common aquatic snails, then leave the snails and penetrate into tadpoles of various species of frogs and toads (Pearson 1956). Adult flukes occur in various canids and mustelids. Thus, the infected swift fox ingested 1 or more prey items that contained larvae of *A. arisaemoides*. The species also has been recovered from red fox in Wyoming (Kingston and Honess 1982) as well as coyote and red fox in Alberta (Holmes and Podesta 1968; University of Alberta Parasite Collection, unpublished).

Ascarid nematodes (*Toxocara canis*, *Toxascaris leonina*) dominate the roundworm component of the helminth community in swift fox. Both species have a direct life cycle and eggs passed in feces of infected foxes are readily available for accidental ingestion by other foxes. In addition, infected females may pass the parasite *in utero* to the fetus. Although one or other of these species is a ubiquitous finding in most swift fox populations, there is no evidence of significant damage or impact on infected individuals. *Toxocara canis* also is zoonotic and persons handling foxes or fox feces should take precautions (as above).

The hookworms *Ancylostoma caninum* and *Uncinaria* sp. are reported infrequently from swift fox. These nematodes have a direct life cycle and transmit readily by ingestion of eggs in contaminated environments. Hookworms are more common in domestic dogs and coyotes and may occur largely as spillover into wild foxes.

Physaloptera spp. occur in the stomach of a variety of carnivores. These nematodes use insects as intermediate hosts and Cutter (1958) and Kilgore (1969) reported various insects as food items of swift fox. Infection with *Physaloptera* sp. could also occur through ingestion of small birds and mammals that contain dormant larval stages.

Whipworms, *Trichuris vulpis*, were found in swift fox collected in Wyoming (Williams, unpublished data). These worms feed on blood and heavy infections can be associated with bloody diarrhea and loss of body condition in dogs and red fox. This has not been documented in swift fox.

Canine heartworm (*Dirofilaria immitis*) is found in red and gray fox in some parts of their range (Simmons et al. 1980, Pappas and Lunzman 1985, Wixson et al. 1991) and it is probable that swift fox could become infected in these areas. Although potentially fatal in large numbers, natural infections probably would not be clinically significant.

A number of protozoans of the genus *Isospora* have been reported sporadically. There is insufficient information to determine the significance or occurrence of these infections. Clinical and pathologic changes in infected foxes are not reported.

Infectious Disease Agents

Although we could find no survey of infectious diseases of swift fox, antibodies to infectious agents are relatively common in many wild canid populations, including other fox species (Amundson and Yuill 1981, McCue and O'Farrell 1988, Davidson et al. 1992b, Garcelon et al. 1992, Cypher and Frost 1999). These bacterial and viral species probably are shared among sympatric canids or reflect exposure to infectious agents in common prey species. Antibodies often are detected in the absence of overt clinical signs or gross lesions. It is likely that exposure to infectious agents may be common but infection and illness may be rare. However, outbreaks of canine distemper virus (CDV) and infectious canine hepatitis virus (ICHV) in wild canids have been documented. These agents are transmitted by direct contact and outbreaks generally occur in high density populations or in species that are highly social. In addition, low-level enzootic situations may occur in many populations but often are undetected.

In Wyoming, live and dead swift fox were examined for evidence of infectious disease (Williams, unpublished data). Sample sizes differed for different disease agents. Fifteen of 25 serologic samples contained antibodies to sylvatic plague; 2 of 12 had titers to CDV. The following samples were seronegative: 16 sera tested for tularemia, 6 tested for toxoplasmosis, and 3 tested for leptospirosis.

Sylvatic plague occurs in various ground-dwelling rodents throughout large regions of western North America (Olsen 1981, Barnes 1982). Although well documented in Europe and Asia, there is controversy over the origin of plague in North America. Unfortunately there is insufficient evidence to determine whether it arrived relatively recently at western seaports in stowaway rats (*Rattus* spp.) and their fleas or whether it crossed the Bering land bridge and entered the continent in pre-Pleistocene times (Pollitzer 1954). Currently, the eastern edge of the distribution appears to be defined by the eastern edge of the Great Plains.

In western regions, sylvatic plague is relatively common in colonial rodents such as prairie dogs and ground squirrels. Predators that feed on infected rodents may develop circulating antibodies in response to the presence of the bacterial antigen in prey items. Most predators are resistant to sylvatic plague; however, within the range of swift fox, active cases of sylvatic plague have been identified in black-footed ferrets (*Mustela nigripes*) (Williams et al. 1994) and bobcat (*Lynx rufus*). Swift fox are likely resistant to developing clinical or infectious plague.

Plague is an acute febrile disease in which significant clinical signs quickly develop in susceptible species, including humans. The bacterium is transmitted in infected fleas, including *P. irritans* though it is considered an inefficient vector, and fleabites may be an alternative means of activating an immune response in foxes. In enzootic areas, carnivores that predate rodents (the primary hosts) may be used as sentinels to detect plague infections in prey species but are themselves resistant to the disease. There is speculation that carnivores also may play a role in disseminating fleas from infected prey; however, to date, this is not documented. Although unlikely to occur, any person exhibiting clinical signs of plague after handling swift foxes should contact a physician.

Canine distemper virus is reported from most species of wild canids. Current information suggests that different species of foxes react differently to CDV; gray fox are particularly susceptible (Hoff et al. 1974, Davidson et al. 1992b), while red fox appear relatively resistant (Davidson et al. 1992b). Swift fox are susceptible and cases of mortality due to CDV infection has been seen in some western states (Williams, unpublished data). The fatality rate of swift fox to canine distemper is unknown but at least some survive infection based on the presence of seropositive individuals (Williams, unpublished data). An outbreak of canine distemper occurred in coyotes in Wyoming during the mid 1980s at the time positive serum samples were collected from swift fox and it is likely the infection spilled over from coyotes to foxes. McCue and O'Farrell (1988) detected antibodies but no illness in kit fox in California and it may be that the virus is maintained in various fox populations without being detected. Distemper often infects young domestic dogs 3 to 9 weeks old (Gillespie and Carmichael 1968), often with fatal results. Thus, if swift fox pups become infected, mortality in the den could be easily missed.

Rabies virus in wild canids is of concern because of the fatal nature of infections and the potential to infect humans and domestic species. The virus occurs in distinct strains and geographic areas throughout much of North America (Smith and Baer 1988), but strains in terrestrial hosts tend to be relatively host specific due to inherent biological characteristics of each strain. Within the range of swift fox, rabies occurs in striped skunks throughout the northern prairie region, gray fox in eastern Texas, and more recently, coyotes in southern Texas (Clark et al. 1994). A strain adapted to swift fox has not been identified and rabies is unlikely to occur except as a sporadic event or in individual swift fox. Bat rabies occurs throughout North America and it is possible that an individual fox could become infected by eating an infected bat. However, such individuals generally die before infections are passed to other foxes or to people.

Other infectious agents with potential to infect swift fox include infectious canine hepatitis virus, canine parvovirus, tularemia, leptospirosis, and toxoplasmosis. These are known to infect other wild canids (Williams and Thorne 1996) and potentially could be transferred to swift fox.

The parasite and disease profiles of swift fox are more similar to those of coyotes and dogs than to other fox species. This may reflect the severe declines in fox

Table 3. Potential infectious diseases of swift fox.

	Life History	Other Hosts	Occurrence[a]
Sylvatic plague	Indirect: fleas or consumption	Rodents, ferrets, felids	Antibodies only
Canine distemper	Direct	Canids	Antibodies and a few documented mortalities
Infectious canine hepatitis	Direct	Canids	Not reported
Canine parvovirus	Direct	Canids	Not reported
Tularemia	Direct	Lagomorphs, rodents	Not reported
Rabies	Direct	Various (incl. canids)	Not reported
Leptospirosis	Direct	Various (incl. canids)	Not reported
Toxoplasmosis	Direct	Various (incl. canids)	Not reported

[a]In swift fox

Table 4. Generalized concerns and recommendations relative to translocation of parasites and diseases in wild swift fox.

Hazard	Risk of Translocation	Control Procedures
Sylvatic plague	Near zero	30 d quarantine, insecticide
Canine distemper	Moderate	30 d quarantine
Infectious canine hepatitis	Low	30 d quarantine
Canine parvovirus	Low	30 d quarantine
Tularemia	Low	Broad-spectrum antibiotic?
Rabies	Low	30 d quarantine
Leptospirosis	Low	30 d quarantine
Toxoplasmosis	Zero	Not necessary
Ectoparasites	High	Insecticide, acaricide
Endoparasites	High	Anthelmintic

populations at the turn of the century and the disappearance of parasites specific to swift fox. Although wolf and coyote densities were relatively high on the prairies prior to European settlement (Young and Jackson 1951), wolves were soon extirpated. The long-term extensive spatial and dietary overlap between swift fox and coyotes (Cypher and Spencer 1998, Kitchen et al. 1999) may have provided an opportunity for parasite species to transfer from coyotes to remnant swift fox populations. The introduction of domestic dogs to western North America probably offered similar opportunities for their parasites, particularly helminth parasites, to colonize swift fox, in association with sympatric use of agricultural and farmstead areas. This continued overlap ensures that the helminth species are able to maintain viable populations throughout the prairies. In many cases, intermediate hosts provide the mechanism for the sharing of helminths. In contrast, fox species appear to segregate more distinctly and the same opportunities for sharing of parasites among different foxes may not occur.

There is a lack of specific knowledge of the diseases and parasites that affect swift fox. It should not be assumed that, because no significant diseases are currently recognized in swift fox, none exist. Throughout this review, we have extrapolated what is known about some diseases in other fox species and wild canids to swift fox; however, this may be a dangerous exercise. Even species related closely enough to interbreed may respond very differently when exposed to a given pathogen (Williams and Thorne 1996). Additional study of the diseases and parasites affecting swift fox is warranted. Based on what we know to date, generalized concerns and procedures relative to translocation of parasites and diseases in wild swift fox are outlined (Table 4).

Literature Cited

Amundson, J.E., and T.M. Yuill. 1981. Prevalence of selected pathogenic microbial agents in the red fox (*Vulpes fulva*) and gray fox (*Urocyon cinereoargenteus*) of southwestern Wisconsin. Journal of Wildlife Diseases 17:17–22.

Barnes, A.M. 1982. Surveillance and control of plague in the United States. *In* M.A. Edwards and U. MacDonnell, editors. Animal disease in relation to animal conservation. Zoological Society of London 50:237–270.

Bretagne, S., J.P. Guillou, M. Morand, and R. Houn. 1992. Détection des oeufs d'*Echinococcus multilocularis* Leukart, 1863 dans les fèces de renard (*Vulpes vulpes* Linnaeus, 1758) par amplification en chaîne par polymérase. Revue Scientifique et Techinque d'Office International des Epizooties 11:1051–1056.

Clark, K.A., S.U. Neill, J.S. Smith, P.J. Wilson, V.W. Whadford, and G.W. McKirahan. 1994. Epizootic canine rabies transmitted by coyotes in south Texas. Journal of the American Veterinary Medical Association 204:536–540.

Cutter, W.L. 1958. Food habits of the swift fox in northern Texas. Journal of Mammalogy 39:527–532.

Cypher, B.L., and K.A. Spencer. 1998. Competitive interactions between coyotes and San Joaquin kit foxes. Journal of Mammalogy 79:204–214.

Cypher, B.L., and N. Frost. 1999. Condition of San Joaquin kit foxes in urban and exurban habitats. Journal of Wildlife Management 63:930–938.

Davidson, W.R., M.J. Appel, G.L. Doster, O.E. Baker, and J.F. Brown. 1992a. Diseases and parasites of red foxes, gray foxes, and coyotes from commercial sources selling to fox chasing inclosures. Journal of Wildlife Diseases 28:581–589.

Davidson, W.R., V.F. Nettles, L.E. Hayes, E.W. Howerth, and C.E. Couvillion. 1992b. Diseases diagnosed in gray foxes *Urocyon cinereoargenteus* from the southeastern United States. Journal of Wildlife Diseases 28:28–33.

Egoscue, H.J. 1962. Ecology and life history of the kit fox in Tooele County, Utah. Ecology 43:481–497.

Egoscue, H.J. 1979. *Vulpes velox*. Mammalian Species No. 122, American Society of Mammalogists, Lawrence, Kansas.

Garcelon, D.K., R.K. Wayne, and B.J. Gonzales. 1992. A serologic survey of the island fox (*Urocyon littoralis*) on the Channel Islands, California. Journal of Wildlife Diseases 28: 223–229.

Gillespie, J.H., and L.E. Carmichael. 1968. Distemper and infectious hepatitis. Pp. 11–130 *in* E.J. Catcott, editor. Canine Medicine. American Veterinary Publishers, Wheaton, Illinois.

Gregson, J.D. 1956. The Ixodoidea of Canada. Canada Department of Agriculture, Ottawa, Ontario.

Hildreth, M.B., and D. Schneider. 1989. Zoonotic helminths in red foxes from east central South Dakota. Proceedings of the South Dakota Academy of Sciences 68:109.

Hillman, C.N., and J.C. Sharps. 1978. Return of swift fox to northern Great Plains. Proceedings of the South Dakota Academy of Science 57:154–162.

Hines, T.D. 1980. An ecological study of *Vulpes velox* in Nebraska. Thesis, University of Nebraska, Lincoln, Nebraska.

Hoff, G.L., W.J. Bigler, S.J. Proctor, and L.P. Stallings. 1974. Epizootic of canine distemper virus infection among urban raccoons and gray foxes. Journal of Wildlife Diseases 10:423–428.

Holland, G.P. 1985. The fleas of Canada, Alaska and Greenland (Siphonaptera). Memoirs of the Entomological Society of Canada, No. 130. Entomological Society of Canada, Ottawa, Ontario.

Holmes, J.C., and R. Podesta. 1968. The helminths of wolves and coyotes from the forested regions of Alberta. Canadian Journal of Zoology 46:1193–1204.

Honess, R.F., and R.C. Bergstrom. 1982. Ectoparasites. Pp. 231–260 *in* E.T. Thorne, N. Kingston, W.R. Jolley, and R.C. Bergstrom, editors. Diseases of wildlife in Wyoming. Wyoming Game and Fish Department, Cheyenne, Wyoming.

Kilgore, D.L. 1969. An ecological study of the swift fox (*Vulpes velox*) in the Oklahoma panhandle. The American Midland Naturalist 81:512–534.

Kingston, N., and R.F. Honess. 1982. Platyhelminthes. Pp. 155–187 *in* E.T. Thorne, N. Kingston, W.R. Jolley, and R.C. Bergstrom, editors. Diseases of wildlife in Wyoming. Wyoming Game and Fish Department, Cheyenne, Wyoming.

Kitchen, A.M., E.M. Gese, and E.R. Schauster. 1999. Resource partitioning between coyotes and swift foxes: space, time, and diet. Canadian Journal of Zoology 77:1645–1656.

Leiby, P.D., W.P. Carney, and C.E. Woods. 1970. Studies on sylvatic echinococcosis: III. Host occurrence and geographic distribution of *Echinococcus multilocularis* in the north central United States. Journal of Parasitology 56:1141–1150.

Leiby, P.D., and W.G. Dyer. 1971. Cyclophyllidean tapeworms of wild carnivora. Pp. 174–234 *in* J.W. Davis and R.C. Anderson, editors. Parasitic diseases of wild mammals. Iowa State Press, Ames, Iowa.

McCue, P.M., and T.P. O'Farrell. 1988. Serological survey for selected diseases in the endangered San Joaquin kit fox (*Vulpes macrotis mutica*). Journal of Wildlife Diseases 24:274–281.

Morner, T., and D. Christensson. 1984. Experimental infection of red foxes (*Vulpes vulpes*) with *Sarcoptes scabiei* var *vulpes*. Veterinary Parasitology 15:159–164.

Olsen, P.F. 1981. Sylvatic plague. Pp. 232–243 *in* J.W. Davis, L.H. Karstad, and D.O. Trainer, editors. Infectious diseases of wild mammals. Second edition. Iowa State Press, Ames, Iowa.

Pappas, L.G., and A.T. Lunzman. 1985. Canine heartworm in domestic and wild canids of southeastern Nebraska. Journal of Parasitology 71:828–831.

Pearson, J.C. 1956. Studies on the life cycles and morphology of the larval stages of *Alaria arisaemoides* Augustine and Uribe, 1927 and *Alaria canis* LaRue and Fallis, 1936 (Trematoda: Diplostomidae). Canadian Journal of Zoology 34:295–387.

Pollitzer, R. 1954. Plague. World Health Organization Monograph Series #22. Geneva.

Rapp, W F. 1962. Notes on a small collection of fleas from Crowley County, Colorado. Journal of Kansas Entomological Society 35:217–218.

Seesee, F.M., and D.L. Worley. 1976. The occurrence of *Echinococcus multilocularis* Leuckart, 1863 (Cestoda: Taeniidae) in the red fox *Vulpes vulpes* L., in southwestern Montana. Proceedings of the Montana Academy of Sciences 36:145–149.

Seton, E.T. 1929. Lives of game animals. Doubleday, Doran and Company, New York, New York.

Simmons, J.M., W.S. Nicholson, E.P. Hill, and D.B. Briggs. 1980. Occurrence of *Dirofilaria immitis* in gray fox (*Urocyon cinereoargenteus*) in Alabama and Georgia. Journal of Wildlife Diseases 16:225–228.

Smith, J.S., and G.M. Baer. 1988. Epizootiology of rabies: the Americas. Pp. 267–299 *in* Rabies, J.B. Campbell and K M. Charlton, editors. Kluwer Academic Publishers, Boston, Massachusetts.

Sweatman, G.K. 1971. Mites and pentastomes. Pp. 3–64 *in* J.W. Davis and R.C. Anderson, editors. Parasitic diseases of wild mammals. Iowa State Press, Ames, Iowa.

Tohm, G.L. 1953. Some siphonaptera from Pima County, Arizona. Pan-Pacific Entomology 29:42.

Turkowski, F.J. 1974. Fleas of Arizona gray and kit foxes. Journal of the Arizona Academy of Science 9:55.

Williams, E.S., K. Mills, D.R. Kiatkowski, E.T. Thorne, and A. Boerger-Fields. 1994. Plague in a black-footed ferret (*Mustela nigripes*). Journal of Wildlife Diseases 30:581–585.

Williams, E.S., and E.T. Thorne. 1996. Infectious and parasitic diseases of captive carnivores, with special emphasis on the black-footed ferret (*Mustela nigripes*). Revue scientifique et technique d'Office International des Epizooties 15:91–114.

Wixsom, M.J., S.P. Green, R.M. Corwin, and E.K. Fritzell. 1991. *Dirofilaria immitis* in coyotes and foxes in Missouri. Journal of Wildlife Diseases 27:166–169.

Young, S.P., and H.H. Jackson. 1951. The clever coyote. Stackpole, Harrisburg, Pennsylvania.

Index

Contributors

ALLARDYCE, DAVID. Endangered Species Field Office, U.S. Fish and Wildlife Service, Pierre, SD, 57501, USA

ANDERS, ANGERBJÖRN. Department of Zoology, Stockholm University, SE-106 91 Stockholm, Sweden

ANDERSEN, DAVID E. Minnesota Cooperative Fish and Wildlife Research Unit, 200 Hodson Hall, 1980 Folwell Avenue, University of Minnesota, St. Paul, MN 55108, USA

ANDERSON, STANLEY H. Wyoming Cooperative Fish and Wildlife Research Unit, Box 3166 University Station Laramie, WY 82071, USA

ASA, CHERYL S. Saint Louis Zoo, 1 Government Drive, St. Louis, MO 63110, USA

CARBYN, LU. Canadian Wildlife Service, 4999-98 Avenue, Edmonton, Alberta T6B 2X3, Canada

CARY, JOHN R. Department of Wildlife Ecology, University of Wisconsin, Madison, WI 53706, USA

CYPHER, BRIAN L. Endangered Species Recovery Program, P.O. Box 9622, Bakersfield, CA 93389, USA

DIENI, J. SCOTT. 403 Deer Road, Evergreen, CO 80439, USA

DOOD, ARNOLD R. Montana Department of Fish, Wildlife, and Parks, 1420 E. 6th Ave., Helena, MT 59620, USA

DOWD STUKEL, EILEEN. South Dakota Department of Game, Fish and Parks, 523 East Capitol Ave., Pierre, SD 57501, USA

DRAGOO, JERRY W. Museum of Southwestern Biology, Department of Biology, University of New Mexico, Albuquerque, NM 87131-1091, USA

GAUTHIER, DAVID A. Executive Director, Canadian Plains Research Center, University of Regina, Regina, Saskatchewan S4S 0A2, Canada

GEFFEN, E. Institute for Nature Conservation Research, Faculty of Life Sciences, Tel-Aviv University, Ramat Aviv 69578, Israel

GESE, ERIC M. Department of Forest, Range, and Wildlife Sciences, Utah State University, Logan, UT 84322, USA

GIDDINGS, BRIAN. Montana Department of Fish, Wildlife, and Parks, 1420 E. 6th Ave., Helena, MT 59620, USA

GIRARD, I. Laboratory of Biomedical and Environmental Sciences, University of California, 900 Veteran Ave, Los Angeles, CA 90024, USA

HARRISON, ROBERT L. Department of Biology, University of New Mexico, Albuquerque, NM, 87131-1091, USA

HERRERO, STEPHEN. Environmental Science, Faculty of Environmental Design, University of Calgary, Alberta T2N 1N4, Canada

HONNESS, KEVIN. Turner Endangered Species Fund, 1123 Research Dr., Bozeman, MT 59718, USA

IRBY, LYNN. Department of Fish and Wildlife, Montana State University, Bozeman, MT 59717, USA

KELLY, PATRICK A. Endangered Species Recovery Program, 1900 N. Gateway Blvd., Suite 101, Fresno, CA 93727, USA

KNOWLES, CRAIG J. FaunaWest Wildlife Consultants, P.O. Box 113, Boulder, MT 59632, USA

KNOWLES, PAMELA R. FaunaWest Wildlife Consultants, P.O. Box 113, Boulder, MT 59632, USA

KUNKEL, KYRAN. 1875 Gateway S., Gallatin Gateway, MT 59718, USA

LAURION, THOMAS R. Department of Wildlife Ecology, University of Wisconsin, Madison, WI 53706, USA

LICHT, DANIEL S. Northern Great Plains Inventory Coordinator/Regional Wildlife Biologist, Mount Rushmore National Memorial, P.O. Box 268, Hwy 244, Keystone, SD 57751,USA

LINDZEY, FREDERICK G. Wyoming Cooperative Fish and Wildlife Research Unit, Box 3166 University Station Laramie, WY 82071, USA

LIST, RURIK. Instituto de Ecología, UNAM. Apartado 70-275, Ciudad Universitaria, 04510 México, D. F., México

MacDONALD, DAVID W. Wildlife Conservation Research Unit, Department of Zoology, South Parks Road, Oxford OX1 3PS, England

MANZANO-FISCHER, PATRICIA. Matamoros 14, Esquina Manuel Doblado, Colonia Pilares, 52179 Metepec, Estado de Mexico, Mexico

McLEOD, MARY A. 215 Comanche Street, Flagstaff, AZ 72204, USA

MOEHRENSCHLAGER, AXEL. Conservation Research Department, Calgary Zoological Society, 1300 Zoo Road, Calgary, Alberta T2E 7V6, Canada

MOEHRENSCHLAGER, CYNTHIA. Conservation Research Department, Calgary Zoological Society, 1300 Zoo Road, Calgary, Alberta T2E 7V6, Canada

OLSON, TRAVIS L. P.O. Box 714, Englewood, CO 80151, USA

PHILLIPS, MICHAEL K. Turner Endangered Species Fund, 1123 Research Dr., Bozeman, MT 59718, USA

PHILLIPS, MICHAEL L. University of Florida, Institute of Food and Agricultural Sciences, 3205 College Ave, Ft. Lauderdale, Florida 33314, USA

PYBUS, M.J. Alberta Natural Resources Service—Fish and Wildlife Division, 6909-116 Street, Edmonton, Alberta T6H 4P2, Canada

RICHHOLT, MELISSA. Canadian Wildlife Service, 4999-98 Avenue, Edmonton, Alberta T6B 2X3, Canada

SARGEANT, GLEN A. U.S. Geological Survey, Northern Prairie Wildlife Research Center, Jamestown, ND 58401, USA

SCHMITT, C. GREGORY. New Mexico Department of Game and Fish, P.O. Box 25112, Santa Fe, NM 87504, USA

SHAUGHNESSY, MICHAEL J., JR. Department of Biological and Environmental Sciences, Morehead State University, Morehead, KY 40351, USA

SIKES, ROBERT S. Department of Biology, University of Arkansas at Little Rock, Little Rock, AR 72204, USA

SLIVINSKI, CHRISTIANE C. 11642 Wall Rd., Caledonia, MN 55921, USA

SMEETON, CLIO. Cochrane Ecological Institute, P.O. Box 484, Cochrane, Alberta T4C 1A7, Canada

SOVADA, MARSHA A. Northern Prairie Wildlife Research Center, U.S. Geological Survey, Jamestown, North Dakota 58401, USA

TANNERFELDT, MAGNUS. Department of Zoology, Stockholm University, SE-106 91 Stockholm, Sweden

VALDESPINO, CAROLINA. Departamento Ecología y Comportamiento Animal, Instituto de Ecología, Xalapa, Veracruz, México

WATERS, SIÂN S. CEI Consulting Ltd., P.O. Box 484, Cochrane, Alberta T4C 1A7, Canada

WAYNE, ROBERT K. Department of Organismic Biology, Ecology & Evolution, 621 Charles E. Young Drive South, University of California, Los Angeles, CA 90095-1606, USA

WEAGLE, KEN. Cochrane Ecological Institute, P.O. Box 484, Cochrane, Alberta T4C 1A7, Canada

WHITAKER-HOAGLAND, JULIANNE. Oklahoma Department of Wildlife Conservation, 1801 N. Lincoln Blvd., Oklahoma City, OK 73105, USA

WHITE, P.J. National Park Service, PO Box 168, Yellowstone National Park, WY 82190

WILLIAMS, DANIEL F. Endangered Species Recovery Program, Department of Biology, California State University, Stanislaus, CA 95382, USA

WILLIAMS, E.S. Wyoming State Veterinary Laboratory, 1174 Snowy Range Road, Laramie, WY 82070, USA

WOODWARD, ROBERT O. U.S. Geological Survey, Northern Prairie Wildlife Research Center, 8711 37th Street SE, Jamestown, ND 58401, USA

ZIMMERMAN, AMY L. Northern Prairie Wildlife Research Center, U.S. Geological Survey, Jamestown, ND 58401, USA